AF413054

Arthropods form the largest and most varied assemblage of organisms on earth. They are important economically as agricultural pests, as vectors of disease and as a source of food. The study of their endocrine systems reflects their economic significance, most work having been carried out on insects, crustaceans and ticks. Knowledge in these areas has increased greatly in the last decade, with the advent of improved techniques for the isolation and study of the hormones themselves, revealing fascinating relationships between the endocrine systems of the various arthropod groups. This volume brings together contributions from many of the leading workers in the field, providing in-depth accounts of the current state of knowledge of a wide range of hormone systems. The book presents a unique summary of some of the most significant and exciting advances of the last decade, brought together in a single volume for the first time.

SOCIETY FOR EXPERIMENTAL BIOLOGY
SEMINAR SERIES: 65

RECENT ADVANCES IN ARTHROPOD ENDOCRINOLOGY

SOCIETY FOR EXPERIMENTAL BIOLOGY SEMINAR SERIES

A series of multi-author volumes developed from seminars held by the Society for Experimental Biology. Each volume serves not only as an introductory review of a specific topics, but also introduces the reader to experimental evidence to support the theories and principles discussed, and points the way to new research.

RECENT ADVANCES IN ARTHROPOD ENDOCRINOLOGY

Edited by

Geoffrey M. Coast

*Department of Biology, Birkbeck College,
University of London, London, UK*

Simon G. Webster

*School of Biological Sciences, University of
Wales Bangor, Bangor, Gwynedd, UK*

CAMBRIDGE
UNIVERSITY PRESS

PUBLISHED BY THE PRESS SYNDICATE OF THE UNIVERSITY OF CAMBRIDGE
The Pitt Building, Trumpington Street, Cambridge CB2 1RP, United Kingdom

CAMBRIDGE UNIVERSITY PRESS
The Edinburgh Building, Cambridge CB2 2RU, United Kingdom
40 West 20th Street, New York, NY 10011-4211, USA
10 Stamford Road, Oakleigh, Melbourne 3166, Australia

First published 1998

Printed in the United Kingdom at the University Press, Cambridge

Typeset in Monophoto Times NR 10/12pt [SE]

A catalogue record for this book is available from the British Library

Library of Congress Cataloguing in Publication data

Recent advances in arthropod endocrinology / edited by Geoffrey M.
   Coast, Simon G. Webster.
      p.   cm. – (Society for Experimental Biology seminar series: 65)
   Includes bibliographical references (p.   ) and index.
   ISBN 0 521 59113 9 (hb)
   1. Arthropoda – Physiology,   2, Endocrinology, Comparative.
3. Peptide hormones.   I. Coast, Geoffrey M. (Geoffrey Mason), 1941–   .
II. Webster, Simon G. (Simon George), 1956–   .   III. Series:
Seminar series (Society for Experimental Biology (Great Britain)) : 65
QL434.72.R435   1998
573.4'15–dc21   97-21516 CIP

Contents

Contributors

AUDSLEY, N.
Central Science Laboratory, Ministry of Agriculture Fisheries and Food, Sand Hutton, York Y04 1LZ, UK.
BELLÉS, X.
Centro de Investigación y Desarollo (CSIC), Jordi Girona 18, 08034 Barcelona, Spain.
BENDENA, W. G.
Department of Biology, Queens University, Kingston, Ontario, Canada K7L 3N6.
BROCK, H.
Department of Zoology, University of British Columbia, Vancouver, BC, Canada V6T 1Z4.
COAST, G. M.
Department of Biology, Birkbeck College, University of London, Malet Street, London WC1E 7HX, UK.
COATES, D.
Department of Biology, University of Leeds, Leeds LS2 9JT, UK.
DIRCKSEN, H.
Universität Bonn, Institut für Zoophysiologie, Endenicher Allee 11–13, 53115 Bonn, Germany.
DUVE, H.
School of Biological Sciences, Queen Mary and Westfield College, University of London, Mile End Road, London E1 4NS, UK.
EAST, P. D.
Division of Entomology, CSIRO, Canberra ACT 2601, Australia.
EDWARDS, J. P.
Central Science Laboratory, Ministry of Agriculture Fisheries and Food, Sand Hutton, York Y04 1LZ, UK.
GOLDSWORTHY, G. J.
Department of Biology, Birkbeck College, University of London, Malet Street, London WC1E 7HX, UK.

HOLMAN, G. M.
Veterinary Entomology Research Unit, FAPRL, Agricultural Research
Service, 2811 F&B Road, College Station, TX 778945, USA.
ISAAC, R. E.
Department of Biology, University of Leeds, Leeds LS2 9JT, UK.
JOHNSEN, A. H.
Department of Clinical Biochemistry, University State Hospital,
Copenhagen DK-2100, Denmark.
LACHAISE, F.
Laboratiore d'Embryologie Moléculaire et Cellulaire, Université Paris XI,
91405 Orsay Cedex, France.
LANGE, A. B.
Department of Zoology, Unversity of Toronto, 25 Harbord Street,
Toronto, Ontario, Canada M5S 3G5.
LEE, M. J.
Bio and Immuno Chemistry, Unilever Research Laboratorium
Vlaardingen, Olivier van Noortlaan 120, 3133 AT Vlaardingen, The
Netherlands.
LOMAS, L. O.
Department of Biochemistry, University of Liverpool, PO Box 147,
Liverpool L69 3BX, UK.
LITTLEFORD, D.†
Department of Zoology, University of British Columbia, Vancouver, BC,
Canada V6T 1Z4.
LUNDQUIST, C. T.
Department of Zoology, Stockholm University, S-10691 Stockholm, Sweden.
MACINS, A.
Department of Zoology, University of British Columbia, Vancouver, BC,
Canada V6T 1Z4.
MAESTRO, J.-L.
School of Biological Sciences, Queen Mary and Westfield College,
University of London, Mile End Road, London E1 4NS, UK.
MARSHALL, A. K.
School of Biolgical Sciences, University of Bath, Claverton Down, Bath
BA2 7AY, UK.
MEREDITH, J.
Department of Zoology, University of British Columbia, Vancouver, BC,
Canada V6T 1Z4.
MUREN, J. E.
Department of Zoology, Stockholm University, S-10691 Stockholm, Sweden.

† Deceased

NACHMAN, R. J.
Veterinary Entomology Research Unit, FAPRL, Agricultural Research Service, 2881 F&B Road College Station, TX 77845, USA.

NÄSSEL, D. R.
Department of Zoology, Stockholm University, S-10691 Stockholm, Sweden.

ORCHARD, I.
Department of Zoology, University of Toronto, 25 Harbord Street, Toronto, Ontario, Canada

PHILLIPS, J. E.
Department of Zoology, University of British Columbia, Vancouver, BC, Canada V6T 1Z4.

REES, H. H.
Department of Biochemistry, University of Liverpool, PO Box 147, Liverpool L69 3BX, UK.

REYNOLDS, S. E.
School of Biological Sciences, University of Bath, Claverton Down, Bath BA2 7AY, UK.

RING, M.
Department of Zoology, University of British Columbia, Vancouver, BC, Canada V6T 1Z4.

SCHOOFS, L.
Zoological Institute, Katholieke Universiteit Leuven, Naamsestraat 59, 3000, Leuven, Belgium.

SCOTT, A. G.
School of Biological Sciences, Queen Mary and Westfield College, University of London, Mile End Road, London E1 4NS, UK.

SEDLMEIER, D.
Universität Bonn, Institut für Zoophysiologie, Endenicher Allee 11–13, 53115 Bonn, Germany.

SEINSCHE, A.
Universität Bonn, Institut für Zoophysiologie, Endenicher Allee 11–13, 53115 Bonn, Germany.

SOMMÉ, G.
Laboratiore d'Embryologie Moléculaire et Cellulaire, Université Paris IX, 91405 Orsay Cedex, France.

THEILMANN, D.
Department of Zoology, University of British Columbia, Vancouver, BC, Canada V6T 1Z4.

THORPE, A.
School of Biological Sciences, Queen Mary and Westfield College, University of London, Mile End Road, London E1 4NS, UK.

TOBE, S. S.
Department of Zoology, University of Toronto, Toronto, Ontario, Canada M5S 3G5.
VAN DER HORST, D. J.
Department of Experimental Zoology, Utrecht University, Padualaan 8, 3584 CH Utrecht, The Netherlands.
VAN HERP, F.
Department of Experimental Zoology, Faculty of Sciences, Catholic University of Nijmegen, Toornooiveld, NL 6525 ED Nijmegen, The Netherlands.
VAN MARREWIJK, W. J. A.
Department of Experimental Zoology, Faculty of Sciences, Catholic University of Nijmegen, Toornooiveld, NL 6525 ED Nijmegen, The Netherlands.
WEAVER, R. J.
Central Science Laboratory, Ministry of Agriculture Fisheries and Food, Sand Hutton, York YO4 1LZ, UK.
WEBSTER, S. G.
School of Biological Sciences, University of Wales Bangor, Bangor, Gwynedd LL57 2UW, Wales, UK.
WILLIAMS, T. A.
INSERM Unit 36, College de France, 3 rue d'Ulm, 75005 Paris, France.
WINTHER, Å. M. E.
Department of Zoology, Stockholm University, S-10691 Stockholm, Sweden.

Preface

In the past few years, there has been a dramatic increase in our knowledge of the structure and function of arthropod neuropeptides. Just ten years ago, very few insect and crustacean neuropeptides had been identified. Rapid advances in techniques for peptide purification, sequence analysis and mass determination have greatly facilitated the characterisation of neuropeptides which are often present in small (picomole; 10^{-12} mol) amounts. Today, more than 100 arthropod neuropeptides have been fully characterised and that number is continually being added to. The availability of these peptides, together with specific antisera and cDNA probes, has expedited progress towards understanding their functions as hormones, neurotransmitters or neuromodulators at molecular, cellular and organismal levels. Although the diversity of the phylum might presage a similar diversity in neuropeptide structure and function, some neuropeptides, such as crustacean cardioactive peptide (CCAP), appear to be extraordinarily well conserved with regard to structure, function and neural network architectures. Others appear to have very different functions in divergent (or even similar) arthropod groups, although they maintain sufficient structural similarity to permit comparisons to be made. Additionally, many neuropeptides have multiple actions on a wide variety of target tissues, and new functions are regularly being reported. Indeed, it seems likely that the physiologically relevant role(s) of many neuropeptides may be quite different from those originally described. In a volume of this size, it has not been possible to cover all aspects of this rapidly expanding field, but invited contributors describe current work in selected areas of arthropod endocrinology and highlight future directions such studies might take. Endocrine mechanisms implicated in the control of intermediary metabolism, water balance, visceral and cardiac muscle activity, moulting and metamorphosis are dealt with in the first sixteen chapters, while the final two chapters are concerned with peptide processing and the development of stable lipophilic peptidomimetics. In the long term it seems likely that agonists or antagonists of arthropod neuropeptides will be of commercial value as insecticides and possibly in manipulation of growth and reproduction in crustacean aquaculture.

This volume arose from a two day symposium held as part of the 1996 annual meeting of the Society for Experimental Biology. We wish to acknowledge the financial support provided by the Company of Biologists, Cambridge University Press and the Society for Experimental Biology. We would also like to thank everyone who attended this symposium and who contributed to some very lively discussions.

Geoff Coast and Simon Webster

Table of three- and single-letter abbreviations for amino acids

Alanine	Ala	A
Arginine	Arg	R
Asparagine	Asn	N
Aspartic acid	Asp	D
Aspartic acid or asparagine (undefined)	Asx	B
Cysteine	Cys |	C
Cystine (half)	Cys or Cys |	—
Glutamine	Gln	Q
Glutamic acid	Glu	E
Glutamic acid or glutamine (undefined)	Glx	Z
Glycine	Gly	G
Histidine	His	H
Hydroxylysine	Hyl	—
Hydroxyproline	Hyp	—
Isoleucine	Ile	I
Leucine	Leu	L
Lysine	Lys	K
Methionine	Met	M
Ornithine	Orn	—
Phenylalanine	Phe	F
Proline	Pro	P
Serine	Ser	S
Threonine	Thr	T
Tryptophan	Trp	W
Tyrosine	Tyr	Y
Unknown or 'other'	Xaa	X
Valine	Val	V

Part I

Moulting, metamorphosis and reproduction

ROBERT J. WEAVER, JOHN P. EDWARDS,
WILLIAM G. BENDENA
and STEPHEN S. TOBE

Structures, functions and occurrence of insect allatostatic peptides

Introduction

The multitude of physiological processes controlled or modulated by insect juvenile hormones (JH) underlines the vital roles performed by these sesquiterpenoid hormones in insects. Predominant among the various functions of JH is the control of metamorphosis in pre-adult stages, and the regulation of reproductive physiology (especially oocyte growth) in adults. In addition, these hormones have been implicated in the determination of phase and caste polymorphisms, diapause induction, pheromone production, and several aspects of behaviour (for a recent review, see Nijhout, 1994). Furthermore, in many of those circumstances where JH-controlled systems have been investigated, there are marked changes in JH levels that can be correlated with changing physiological events. For example, in cockroaches, changes in the biosynthetic activity of the corpora allata (CA) are directly related to corresponding changes in haemolymph or whole-body JH levels (Tobe *et al.*, 1985; Edwards *et al.*, 1990) and these changes can be correlated with specific physiological events. These and other studies have also pointed to the fact that JH levels may change dramatically over relatively short periods of time. Ultimately, the finite haemolymph or tissue levels of JH is a balance between the rate at which the hormones are biosynthesised by the CA (since JH is released immediately following biosynthesis), and the rate at which these molecules are metabolised. Despite the fact that there is ample evidence that JH-degrading enzymes (predominantly, esterases and epoxide hydrolases) can play a role in determining the physiological levels of JH, it is likely that the overall control of tissue or haemolymph JH levels is more intimately linked with changes in the rate of JH biosynthesis (Tobe & Stay, 1985; Tobe *et al.*, 1985). Thus, the control of JH biosynthesis by the CA is, ultimately, the factor that modulates all JH-mediated physiological events.

The CA are retrocerebral endocrine organs that are innervated by neurones originating in the brain. Although some of these connections may be nervous, there is ample evidence that they are also neurohormonal, and that neurosecretory material passes from the neurosecretory cells of the brain directly to the CA (Scharrer, 1952; Tobe & Stay, 1985; Cassier, 1990). In many insect species, these nerves pass through the corpora cardiaca, which are themselves often in intimate association with the CA. In some species, the CA receive afferent and efferent connections from the suboesophageal ganglion (SOG). In addition, the CA are bathed in haemolymph and therefore are exposed to regulatory factors circulating in the haemolymph.

The insect brain is in direct receipt of both internal and external stimuli that provide a constant stream of information about the physiological conditions prevailing in the internal and external environment. Many of these stimuli will ultimately result in a physiological response to changes in the environment and, frequently, this response will involve modification of the endogenous JH levels. It is therefore not surprising that investigators seeking the factors controlling CA activity have concentrated their efforts on the association between the brain and retrocerebral complex and, in particular, on the neurohormones produced by the brain neurosecretory regions. Deducing from a considerable body of evidence that surgical section of the nervous connections between brain and CA could alter the rate of JH biosynthesis, and from experiments in which crude brain extracts were shown to alter the activity of the CA, researchers concentrated on the brain as the primary source of allatoregulatory factors. These early studies were greatly aided by the development of the *in vitro* JH biosynthesis assay (Pratt & Tobe, 1974; Tobe & Pratt, 1974) in which changes in the rates of JH biosynthesis are monitored by measuring changes in the rate of production of JH incorporating a radiolabelled precursor (radiolabelled [*methyl*] methionine). From such studies, it soon became apparent that brain tissue contained principles that were able either to stimulate or to inhibit JH production by the CA. These biologically active molecules were termed allatotropins and allatostatins (ASTs), respectively.

The first allatostatins to be characterised were isolated from brains of the cockroach *Diploptera punctata* (Woodhead *et al.*, 1989; Pratt *et al.*, 1991). Several closely related allatostatins have subsequently been isolated from other cockroach species including *Periplaneta americana* and *Blattella germanica* (Bellés *et al.*, 1994; Weaver *et al.*, 1994). In addition, similar peptides have been detected in a dipteran species (the blowfly *Calliphora vomitoria*; Duve *et al.*, 1993). In some species (for example, *D. punctata*), the allostatins are widely distributed in the tissues, and have been isolated or immunochemically localised in the brain, haemolymph, visceral ganglia, gut and other nervous or neuromuscular organs (Woodhead, Stoltzman & Stay, 1992; Stay *et al.*, 1994a; Stay, Tobe & Bendena, 1994b; Tobe, Yu & Bendena, 1994, Yu *et al.*, 1995). These cockroach peptides share a characteristic, highly conserved C-terminus, and show a remarkable conservation of amino acid sequence, even in different cockroach species. From the known sequences of the *D. punctata* allatostatins, researchers designed probes that eventually led to the isolation and characterisation of the gene coding for allatostatins in *D. punctata* and, subsequently, in *P. americana* (Donly *et al.*, 1993; Ding *et al.*, 1995). These genes encode 13 or 14 related peptides, respectively. All of these peptides have been synthesised and all show some degree of allatostatic activity *in vitro* against CA from *D. punctata* and *P. americana*. However, glands from different stages of *P. americana* do not show the same degree of sensitivity to synthetic *Diploptera* allatostatins irrespective of physiological status, and some of the peptides are potent inhibitors of JH biosynthesis at certain times, whereas others are not (Weaver *et al.*, 1995b). Nevertheless, probes based either on native peptides or on the encoding gene have been utilised in the search for related sequences in other organisms (Stay *et al.*, 1994b; Tobe *et al.*, 1995).

Another peptide with allatostatic activity has been identified in the moth *Manduca sexta* (Kramer *et al.*, 1991). This peptide, containing 15 amino acid residues, bears no sequence similarity to the cockroach allatostatins. Using probes based on the amino acid sequence of *M. sexta* allatostatin (Mas-AST), Tobe *et al.* (1995) and Jansons *et al.* (1996) have reported the existence of a Mas-AST-like gene in the noctuid moth *Pseudaletia unipuncta*. Similar results, using the same probe, have recently been obtained in another noctuid (*Lacanobia oleracea*), in which this probe has been shown to hybridise with mRNA isolated from larval and pupal brains (J. P. Edwards *et al.*, unpublished results). However, it is clear that the cockroach peptides are inactive as allatostatins in *M. sexta*, and likewise, the *M. sexta* allostatin shows no inhibition of JH biosynthesis by CA of the cockroach *P. americana* (Kramer *et al.*, 1991; R. J. Weaver, unpublished results). These results suggest that the Dictyoptera and Lepidoptera each utilise a separate type of allatostatic neuropeptide. However, this interpretation is complicated by the observations that the cockroach allatostatins are present in representatives from other insect taxa, including Diptera, Hymenoptera and Lepidoptera and, indeed, in other non-insect groups such as Crustacea, Arachnida and Mollusca (see Stay *et al.*, 1994b and references contained therein). Similarly, the *M. sexta* AST appears to occur in at least one other insect species – the dipteran *Drosophila melanogaster* (Zitnan, Sehnal & Bryant, 1993; Tobe *et al.*, 1995). In many of these examples, it is unlikely that the primary function of these molecules is allatostatic. Equally, the fact that the cockroach allatostatins are widely distributed in the peripheral nervous system, often in association with muscular tissues, suggests at least one other function for these molecules as modulators of neuromuscular function. Added to these complications is the discovery of yet another allatostatic substance in the moth *M. sexta* which appears to be a much larger peptide than Mas-AST, and which, unlike the cockroach allatostatins, appears to act slowly and in a non-reversible fashion prior to metamorphosis (Bhaskaran *et al.*, 1990; Unni *et al.*, 1993).

Despite the progress made in the identification of allatostatic neuropeptides from several insect species, a host of questions remain to be answered. Why do cockroaches appear to require more than a dozen different peptides (all apparently performing similar functions) whereas the Lepidoptera seem to need only one or two? Why do flies have peptides identical or similar to both cockroach and moth allatostatins, when these peptides seem to be devoid of allatostatic activity in adult dipterans? Because these peptides occur in insects in which they do not appear to be allatostatins, yet do have myomodulatory activity more or less universally, are they indeed the exclusive or only primary modulators of CA activity in cockroaches or moths? This review concentrates exclusively on the allatostatins, although at least one peptide with allatotropic activity has also been identified (Kataoka *et al.*, 1989). Studies that led to the elucidation of the structures of those molecules so far identified from insects are reviewed, and the temporal and physical distribution of these molecules and of the genes that encode the peptides are considered. The question of the role(s) of these peptides is addressed, and their mode of action and metabolism considered. Finally, an attempt is made to address some of the outstanding questions raised in this brief introduction.

Cockroach allatostatins

Early experiments, particularly those with cockroaches, provided clear indications that insect brains exerted an inhibitory effect on JH production and were most likely a major site of production of allatostatic signals acting on the CA (Scharrer, 1952; Engelmann, 1957, 1959; Stay & Tobe, 1977, 1978). Direct evidence for the occurrence of cockroach allatostatins came from *in vitro* assays of aqueous extracts of brain and retrocerebral tissue of virgin female *D. punctata* (Rankin *et al.*, 1986). Using direct radiochemical measurements of JH biosynthesis (Pratt & Tobe, 1974; Tobe & Pratt, 1974), CA were observed to be inhibited in a dose-dependent fashion by extracts of corpora cardiaca and also by 0.5 brain equivalents of a water soluble, heat-stable and trypsin-sensitive substance from protocerebral lobes (Rankin *et al.*, 1986). Subsequent studies showed that CA-inhibitory, neuropeptide-like materials were extractable from several ganglia of the central nervous system of females at different stages during the reproductive cycle (Rankin & Stay, 1987). Aqueous extracts of brains of adult males, and of final instar male and female larvae, were similarly shown to inhibit release of JH from adult female CA (Paulson & Stay, 1987; Paulson *et al.*, 1987).

Following the demonstration that simple extracts of brain tissue were able to inhibit JH biosynthesis in a reliable assay, it became possible to proceed with the isolation of functional allatostatic factors. Initially, four peptides showing strong and rapid inhibition of JH biosynthesis were isolated and characterised from high-performance liquid chromatography (HPLC) separations of extracts of brains from virgin female *D. punctata* (Woodhead *et al.*, 1989). Each of these peptides, termed allatostatins I–IV, was C-terminally amidated. They show a high degree of C-terminal identity and range in size from 8 to 18 amino acid residues (Table 1). Using similar techniques, three additional *D. punctata* allatostatins have been identified from extracts of brains of adult female *D. punctata* (Pratt *et al.*, 1991; Woodhead *et al.*, 1994) and two more from the cockroach *P. americana* (Weaver *et al.*, 1994). Between 1000 and 4500 cockroach brains were required to purify each batch of allatostatic peptides. Yields of allatostatin per cockroach brain ranged from 0.1 pmol brain^{-1} in *D. punctata* (Woodhead *et al.*, 1989) to 0.3 pmol brain^{-1} in *P. americana* (Weaver *et al.*, 1994). Four additional cockroach allatostatins, one identical to *D. punctata* allatostatin IV, were subsequently isolated and sequenced from brains of virgin adult females of a third species, *Blattella germanica* (Bellés *et al.*, 1994).

Table 1 summarises data on the cockroach allatostatins isolated thus far, their origins and their past and current designations. These peptides clearly belong to an extended intraphyletic family that shares the C-terminal sequence Tyr/Phe-Xaa-Phe-Gly-Leu/Ile-NH$_2$, where Xaa is alanine, glycine, serine or asparagine. Bioassays of preliminary reverse-phase separations indicated that the five *D. punctata* allatostatins and two *P. americana* allatostatins represented only a portion of the extractable JH-inhibitory activity in adult female cockroach brain of either species. The likelihood that additional allatostatins from the same family could be found in these extracts was confirmed by the molecular characterisation of a multiple allatostatin precursor cDNA and gene in *D. punctata* that encoded 13 very similar peptides, including the five already isolated (Donly *et al.*, 1993). Subsequently, it was shown that the homologous gene of *P. americana* encoded many of these same peptides, but had a total of 14 potential allatostatins in the primary sequence (Ding *et al.*, 1995).

The rather atypical hydrophobic nature and high tyrosine content of the octadecapeptide ASB2 (=allatostatin V=Dip-allatostatin 2=Dip-AST 2) of *D. punctata* led to the conjecture that it may have been representative of a different family of peptides (Pratt *et al.*, 1991). Molecular characterisation of the allatostatin neuropeptide precursors of *D. punctata* and *P. americana* have since indicated that this is an unlikely premise (Donly *et al.*, 1993; Ding *et al.*, 1995). Another unusual characteristic of this peptide is the presence of the dibasic cleavage site Lys^9-Arg^{10} within the internal N-terminal sequence. To date, the cleavage product Dip-AST 2 (residues 11–18) has not been identified in either *D. punctata* or in *P. americana*, although this octapeptide has been identified in the locust, *Schistocerca gregaria* (Veelaert *et al.*, 1995).

Table 1 *Cockroach allatostatins identified by isolation and purification*

Species	Designation[a]																		Sequence[b]	Ref.[c]
D. punctata	Dip-AST 2, V, ASB2	A	Y	S	Y	V	S	E	Y	K	R	L	P	V	Y	N	F	G	L-NH$_2$	[1]
	Dip-AST 4, VII									D	G	R	M	Y	S		F	G	L-NH$_2$	[2]
	Dip-AST 5, IV										D	R	L	Y	S		F	G	L-NH$_2$	[3]
	Dip-AST 7, I					A	P	S	G	A	Q	R	L	Y	G		F	G	L-NH$_2$	[3,4]
	Dip-AST 8, III									G	G	S	L	Y	S		F	G	L-NH$_2$	[3]
	Dip-AST 9, II								G	D	G	R	L	Y	A		F	G	L-NH$_2$	[3]
	Dip-AST 11, VI							Y	P	Q	E	H	R	F	S		F	G	L-NH$_2$	[2]
P. americana	Pea-AST 7, Pea-AST 1					S	P	S	G	M	Q	R	L	Y	G		F	G	L-NH$_2$	[5]
	Pea-AST 9, Pea-AST 2								A	D	G	R	L	Y	A		F	G	L-NH$_2$	[5]
B. germanica	BLAST 1												L	Y	D		F	G	L-NH$_2$	[6]
	BLAST 2										D	R	L	Y	S		F	G	L-NH$_2$	[6]
	BLAST 3					A	P	S	S	A	Q	R	L	Y	S		F	G	L-NH$_2$	[6]
	BLAST 4						A	G	S	D	G	R	L	Y	S		F	G	L-NH$_2$	[6]

Notes:

[a] Where more than one designation exists, the first cited designation is the one currently in use.

[b] Single-letter amino acid code.

[c] References: [1] Pratt *et al.* (1991); [2] Woodhead *et al.* (1994); [3] Woodhead *et al.* (1989); [4] Pratt *et al.* (1989); [5] Weaver *et al.* (1994); [6] Bellés *et al.* (1994).

Cockroach allatostatin genes

The amino acid sequence of the first allatostatin precursor was deduced from a cDNA sequence derived from mRNA isolated from the brains of virgin female *D. punctata* (Donly *et al.*, 1993; Stay *et al.*, 1994b). The specific cDNA encoding allatostatins was identified by polymerase chain reaction amplifications which used degenerate oligonucleotide primers designed by reverse-translation of purified *D. punctata* allatostatin peptide sequences (Woodhead *et al.*, 1989; Pratt *et al.*, 1991). DNA primers within the coding region of the *D. punctata* cDNA were subsequently used to amplify allatostatin-specific DNA fragments from both *P. americana* genomic DNA and brain-derived cDNA (Ding *et al.*, 1995). The coding regions for these cockroach preproallatostatin precursors are found as a single copy within the genome of each species, contained within approximately 1100 nucleotides and share 67% similarity at the nucleotide level. Genomic DNA sequences from both species are contiguous with cDNA in the region encoding the preproallatostatin precursor. However, introns have been found upstream of the initiator ATG. Conservation of DNA sequences outside of the coding region is limited but the (A+T)-rich character of the sequence 3' to the coding region is conserved in both species. Surprisingly, the only detectable brain mRNA transcripts in *D. punctata* and *P. americana* are single species of 9.2 kb and 5.0 kb, respectively. Alternative mRNA species were not detected at different developmental times or in different tissues. Despite the size differences, features of AST mRNA expression appear to be conserved. AST mRNA has been localised by *in situ* hybridisation to several cells in the medial and lateral cell regions of the protocerebrum, as well as the tritocerebrum in both species. The mRNA size difference between the two species suggests that regulatory differences occur at the level of translation or mRNA stability of the mRNA transcripts. These differences may account, in part, for the differences in JH biosynthesis associated with the reproductive mode in each species (see below). The allatostatin polypeptide precursors of the two species share 71% amino acid sequence identity and are similar in structural organisation (Fig. 1). Each precursor is organised into domains beginning with a hydrophobic signal sequence domain required for translocation into the lumen of the endoplasmic reticulum, the allatostatin peptides with associated sequence Gly-Lys-Arg required for α-amidation and endoproteolytic cleavage, and finally acidic domains which probably serve to balance the basic charge contribution of the endoproteolytic cleavage sites. The *P. americana* precursor is cleaved into 14 different allatostatin peptides compared to the 13 peptides that are derived from the *D. punctata* precursor. The difference in peptide number is the result of a carboxyl-terminal substitution/deletion in the *D. punctata* counterpart to Pea-AST 12

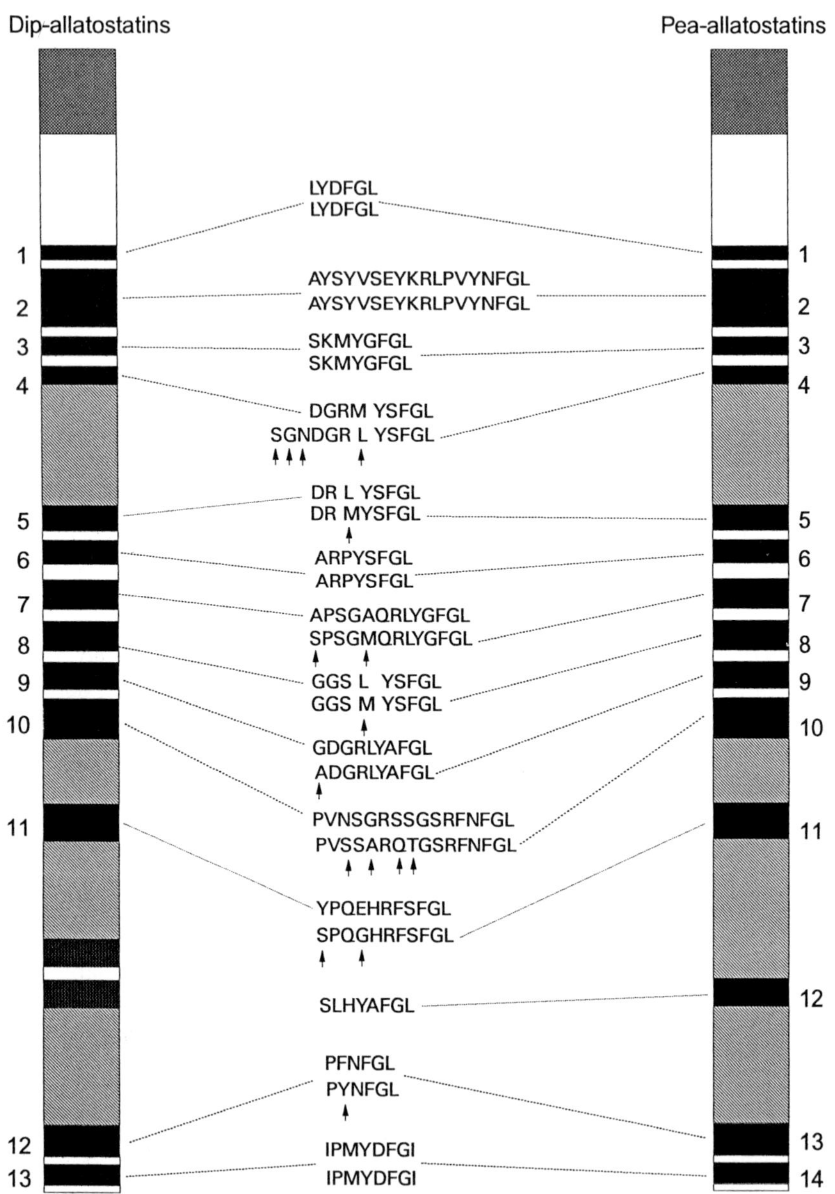

Dip-allatostatins
Pea-allatostatins
LYDFGL
LYDFGL
1
1
AYSYVSEYKRLPVYNFGL
AYSYVSEYKRLPVYNFGL
2
2
SKMYGFGL
SKMYGFGL
3
3
4
4
DGRM YSFGL
SGNDGR L YSFGL
DR L YSFGL
DR MYSFGL
5
5
ARPYSFGL
ARPYSFGL
6
6
7
7
APSGAQRLYGFGL
SPSGMQRLYGFGL
8
8
GGS L YSFGL
GGS M YSFGL
9
9
10
10
GDGRLYAFGL
ADGRLYAFGL
PVNSGRSSGSRFNFGL
11
PVSSARQTGSRFNFGL
11
YPQEHRFSFGL
SPQGHRFSFGL
SLHYAFGL
12
PFNFGL
PYNFGL
12
IPMYDFGI
13
13
IPMYDFGI
14

which has destroyed the amidation signal that is required for biological activity. Dip-AST 1, 2, 3, 6, and 13 are identical to Pea-AST 1, 2, 3, 6, and 14 in both sequence and position in the precursor. Comparison of the remainder reveals that in the amino-terminal address sequences, conservative amino acid substitutions occur that maintain either the hydrophobic or aromatic character of the position. A notable exception is Pea-AST 4, which appears to have a three amino acid residue N-terminal extension relative to Dip-AST 4. These changes may provide an initial starting point for evolutionary grouping of other insect species at the genomic level.

Structure activity studies of cockroach allatostatins

Structure activity relationships for the *Diploptera* allatostatins, in terms of their ability to inhibit JH biosynthesis, have been determined, following the isolation and identification of the cDNA and the corresponding putative peptide precursor, with its 13 peptide products (see Stay *et al.*, 1994b; Bendena *et al.*, 1996; S. S. Tobe *et al.*, unpublished results). These data indicate a wide range of biological activities for the Dip-allatostatin family, with Dip-AST 2 as the most potent, showing an ED_{50} of 10 pM, followed by Dip-AST 5 with an ED_{50} of 100–200 pmol 1^{-1} and Dip-AST 1 as the least potent, with an ED_{50} of *c.* 0.2 μmol 1^{-1} (S. S. Tobe *et al.*, unpublished results). Thus, the Dip-allatostatins are active over four orders of magnitude. Interestingly, neither Dip-AST 2 nor 5 is particularly effective in modulating myotropic activity (ranking ninth and eighth of the 13 allatostatins, respectively) (Lange, Bendena & Tobe, 1995), indicating that the rank order for different biological activities, and presumably the interaction with receptors, depends upon the particular function under study.

Fig. 1. Schematic diagram of insect allatostatin polypeptide precursors deduced from isolated DNA sequences. The polypeptide precursors of cockroaches *Diploptera punctata* (left) and *Periplaneta americana* (right) begin with an NH_2-terminal hydrophobic leader region (cross-hatched) that is presumably cleaved by signal endoproteases. Black boxes represent the individual allatostatin peptides which are numbered according to their position relative to the amino-terminus in the precursor. The sequence of each allatostatin peptide is shown between the two precursors. The differences between the *D. punctata* and *P. americana* peptide sequences are indicated by arrows. Acidic regions are indicated by diagonal lines. Incorporated within the third acidic spacer region of the *Diploptera allatostatin* precursor are two non-amidated peptides (shaded boxes). Each allatostatin, from both species, precedes a Gly-Lys-Arg sequence required for endoproteolytic cleavage and α-amidation. Single-letter code for amino acids is used.

As noted below, the core pentapeptide of the allatostatins is highly conserved, with only residue 2 showing significant variability. The pentapeptide has significant efficacy and reduced, but appreciable, potency. As such, it probably represents the minimum necessary for binding to receptor(s), with modifications to the N-terminal address sequence altering potency but not efficacy (see Stay *et al.*, 1994b). Whether this translates into differences in binding affinity, to interaction with different receptor subtypes, to synergism with other allatostatin or other peptides, or to differences in ability to effect signal transduction following binding remains to be determined. Nonetheless, conservation of both structure and function for the core sequence suggests important biological activities for all of the allatostatins, including inhibition of JH biosynthesis and modulation of myotropic activity.

Structure activity studies have also recently been completed for the *P. americana* allatostatins. In summary, all 14 of the Pea-AST are capable of inhibiting JH III biosynthesis in CA from day 5 female *P. americana* in a dose-dependent manner, but with varying degrees of efficacy (Table 2). Maximal efficacy varied from high, for example, 80–90% for Pea-AST 1, 2, 3, 4, 6, 7, 9 to 55% for Pea-AST 5, whereas ED_{50} values showed a 400-fold range, from 0.7 nmol 1^{-1} for Pea-AST 2 to greater that 0.3 μmol 1^{-1} for Pea-AST 14.

The ED_{50} values for *D. punctata* and *P. americana* allatostatins show that length of the peptide does not correlate well with biological potency. One of the smaller of the *P. americana* allatostatins, the octapeptide Pea-AST 6, shows a potency only slightly lower (less than one order of magnitude) than that of Dip-AST/Pea-AST 2, the most active endogenous peptide in both species. The observed range of ED_{50} values may indicate that allatostatins of different length address sequences interact with different types of receptor (Stay *et al.*, 1994b). However, a comparison of the dose response data for the five *P. americana* octapeptides reveals that even short address sequence variations markedly influence the potency of the peptide. The address sequences of the *P. americana* allatostatins varies from 9–13 amino-terminal residues in the case of Pea-AST/Dip-AST 2 to 2–3 amino acid residues in the case of Pea-AST 3, 5, 6, 12 and 14. The least potent peptides tested were Pea-AST 5 and 14 which, even at the highest doses tested (10 μmol 1^{-1}), were only capable of inhibiting JH biosynthesis by 50–60%. The data for Pea-AST 14 are not surprising, since this is the only allatostatin with an Ile residue at the important C-terminal position (see Stay *et al.*, 1994b). In contrast, the poor performance of Pea-AST 5 was unexpected, particularly since the equivalent Dip-AST 5 is one of the most active allatostatins when tested against *D. punctata* (Stay *et al.*, 1994b) and is also active both *in vivo* and *in vitro* in *P. americana* (Weaver *et al.*, 1995b). It would seem that the single amino

Table 2 *Relative potencies of the* P. americana *allatostatins for inhibition of JH biosynthesis by corpora allata of day 5 adult female*

Designation	Residues (*n*)	ED_{50} $(\text{mol l}^{-1})^a$	Efficacy $(\%)^b$	Rank order
Peastatin 2	18	7.0×10^{-10}	82	1
Peastatin 6	8	1.9×10^{-9}	83	2
Peastatin 7	13	3.3×10^{-9}	80	3
Peastatin 4	12	4.0×10^{-9}	84	4
Peastatin 10	16	6.0×10^{-9}	84	5
Peastatin 9	10	7.0×10^{-9}	90	6
Peastatin 11	11	5.5×10^{-9}	70	7
Peastatin 3	8	3.2×10^{-8}	82	8
Peastatin 12	8	4.0×10^{-9}	65	9
Peastatin 1	6	2.1×10^{-8}	80	10
Peastatin 8	9	1.9×10^{-8}	65	11
Peastatin 13	6	1.0×10^{-7}	70	12
Peastatin 5	8	3.0×10^{-7}	55	13
Peastatin 14	8	3.0×10^{-7}	55	14

Notes:
[a] Molar concentration of peptide required for 50% inhibition of JH release in a 3 h *in vitro* radiochemical assay with pairs of five-day virgin CA compared to groups of controls (*n*=7 to 10).
[b] Maximum recorded inhibition of JH release.

acid substitution at position 3 (Leu → Met) is highly significant in this respect.

The rank order of effectiveness of the *D. punctata* and *P. americana* allatostatins as inhibitors of JH III biosynthesis differs from that observed for inhibition of proctolin-induced contractions in the hindgut of *D. punctata* (Lange *et al.*, 1995), in which concentrations often in excess of two orders of magnitude greater were required to obtain a myoregulatory response.

The availability of synthetic Pea-ASTs allows comparisons to be made with partially purified materials isolated from cockroach brain (Weaver *et al.*, 1994, 1995a). Reversed-phase fast protein liquid chromatography (FPLC) with an acetonitrile gradient is adequate to resolve the mixture of 14 different peptides into one compound peak, three doublets and five distinct single peaks (Fig. 2). A range of different eluants, additives, pH and, when necessary, different column types can be readily employed to resolve complex

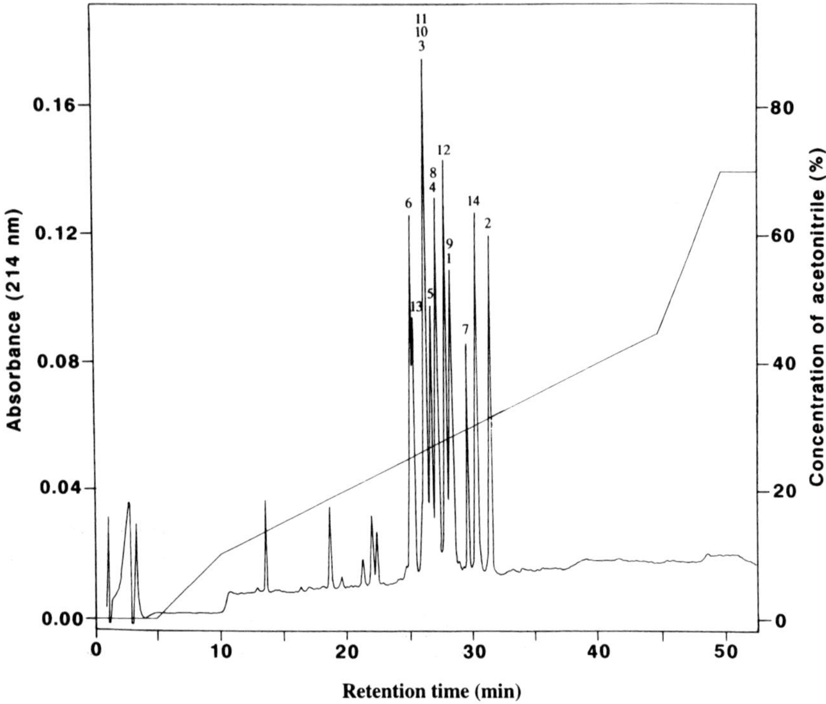

Fig. 2. Reversed-phase FPLC profile of 14 synthetic *P. americana* allatostatins. Separation was performed on a Pharmacia PepRPC HR 5/5 column using a gradient of acetonitrile (1% min^{-1}) in 0.1% trifluoroacetic acid. Flow rate, 1 ml min^{-1}, detector 0.2 AUFS at 214 nm. Numbers adjacent to each peak represent the designation of the *P. americana* allatostatins according to their relative positions within the cDNA sequence.

peaks further. Although these peptides elute over a rather broad range of acetonitrile concentrations (25–33%), this is within the same range as both identified and non-identified endogenous allatostatin bioactivity derived from *P. americana* brain (Weaver *et al.*, 1994, 1995a). Many of the same extracts are immunoreactive to Dip-AST 7 antibody, suggesting that all 14 of the Pea-ASTs may be processed and released from *P. americana* neurosecretory cells (Weaver & Freeman, 1992).

Sensitivity of cockroach CA to allatostatins

Cockroaches display several reproductive strategies that range from oviparity through ovoviviparity to viviparity (Roth & Stay, 1961; Tobe,

1980). Females of the more primitive oviparous species, *P. americana*, have continuous overlapping ovarian cycles that are accompanied by rapid on/off switching of JH biosynthesis by the CA leading to equally rapid fluctuations in JH titre (Weaver & Pratt, 1977; Edwards *et al.*, 1990). In contrast, in the viviparous species *D. punctata*, embryos are incubated in a brood sac for the duration of a prolonged pregnancy, during which time JH biosynthesis is necessarily suppressed and basal oocyte development is minimal (Tobe & Stay, 1977). There are major differences in the sensitivity of CA of both cockroach species at different stages of development and particularly during the female reproductive cycle (Pratt, Farnsworth & Feyereisen 1990; Weaver, 1991; Stay, Joshi & Woodhead 1991; Weaver *et al.*, 1995b). Although slightly different results were obtained in mated females of *D. punctata* through the use of two qualitatively different assay techniques (Pratt *et al.*, 1990; Stay *et al.*, 1991), it is clear that sensitivity to allatostatins declined significantly as JH biosynthesis increased during the early stages of the first reproductive cycle, falling to a minimum at the peak of JH production on day 4–5. The subsequent decline in JH biosynthesis observed in post-vitellogenic females is accompanied by a dramatic increase in sensitivity of the CA to allato-statins, which then slowly subsides during the ensuing ovulation and preg-nancy. Precisely equivalent data are not currently available for *P. americana*, although a similar trend of declining sensitivity of CA to an allatostatin as JH biosynthesis increases has been reported in virgin females (Weaver, 1991). In contrast, CA from mated female *P. americana* are insensitive to Dip-AST 7 during the reproductive cycle, but show good inhibitory response to Dip-AST 2, followed by a sudden loss, in the period leading up to and during the time of peak JH biosynthesis, i.e. shortly before ovulation (Weaver *et al.* 1995b). Injections of Dip-AST 5 or 7 (100 μg insect^{-1}) into day 4 virgin *P. americana* are effective in reducing JH titres at 12 h post-injection, but injec-tions of the same peptides into mid-cycle mated females were not as effective (Weaver *et al.*, 1995b). Clearly, more work is needed with adult *P. americana* CA and JH titre measurements, particularly in respect to possible changes in inhibitory response to endogenous Pea-allatostatins during the later half of the egg maturation when JH titres have been reported to decline markedly (Edwards *et al.*, 1990). In *D. punctata*, the acquisition of sensitivity of the CA to low concentrations of Dip-AST 7, observed in mated females just before choriogenesis as JH production is beginning to decline, seems to be unaffected by denervation of the CA, but may be dependent on a humoral factor (Unnithan & Feyereisen, 1995). Transplantation of CA from post-vitellogenic females into previtellogenic females prevents the acquisition of allatostatin sensitivity, whereas CA from pre-vitellogenic females become sensitive following transplantation into post-vitellogenic females. Ovariectomy in vitellogenic females also disrupts both JH biosynthesis and the acquisition of allatostatin sensitivity, whereas denervation of the CA in virgin females leads to a rapid decline in allatostatin sensitivity. These results may provide additional evidence that sensitivity to allatostatins may be humorally modulated (Unnithan & Feyereisen, 1995).

Distribution of allatostatins

Allatostatins are widely distributed in *D. punctata* (Stay, Woodhead & Chan, 1992; Woodhead *et al.*, 1992; Yu *et al.*, 1993, 1995) and in *P. americana* (Agricola *et al.*, 1992). A monoclonal antibody against Dip-AST 7 reveals tracts from the lateral neurosecretary cells in the protocerebrum leaving the brain as the nervi corporis cardiaci (NCC) II and branching within the corpora cardiaca and CA. Rather surprisingly, the NCC I from the median neurosecretary cells that innervate the corpora cardiaca are conspicuously lacking in immunoreactivity (Stay *et al.*, 1992). Antibodies against Dip-AST 7 have also been used to localise allatostatin-immunoreactive cells in the nervous system of *P. americana*. Several groups of cell bodies and associated axon tracts were immunoreactive to a polyclonal antibody in the pars lateralis of the brain, some clearly showing projections to the corpora cardiaca and CA (Schildberger & Agricola, 1992). The same monoclonal antibodies have revealed, amongst others, many cells in the suboesophageal and abdominal ganglia that are allatostatin immunoreactive. Tissues and organs that are innervated by allatostatin-immunoreactive neurones include the midgut and hindgut and midgut neurosecretary cells, the antennal heart, the prothoracic, salivary and male accessory glands and the abdominal perisympathetic organs (Agricola *et al.*, 1992; see also Yu *et al.*, 1995). In *D. punctata*, as in *P. americana*, the nerves of the antennal pulsatile organ and hindgut muscle are conspicuously immunopositive to allatostatin antibody (Woodhead *et al.*, 1992; Lange, Chan & Stay, 1993), and extracts of antennal hearts yield allatostatic factors that co-elute with synthetic allatostatins in HPLC (Woodhead *et al.*, 1992). The hindgut of *D. punctata* responds to Dip-AST 5 and 7 with a dose-dependent decrease in amplitude and frequency of spontaneous contraction. When applied simultaneously with proctolin, Dip-AST 5 or 7 antagonised the stimulatory effect of proctolin, an effect which was dose-dependent and has recently been extended to all members of the *D. punctata* allatostatin family (Lange *et al.*, 1995). In contrast to the modulation by allatostatins of spontaneous contractions of hindgut, antennal heart muscle did not show any response to application of allatostatin alone in *D. punctata* (Lange *et al.*, 1993) or in *P. americana* (Hertel & Penzlin, 1992), although allatostatins may act on the antennal heart muscle to modulate the action of a myostimulatory peptide – most likely proctolin (Lange *et al.*, 1993).

In *D. punctata*, the midgut is extensively innervated by AST-immunoreactive proctodeal neurones, which are in intimate contact with the hindgut musculature as well (Yu *et al.*, 1995). Immunocytochemistry and *in situ* hybridisation have also demonstrated the presence of numerous AST endocrine cells in the midgut (Yu *et al.*, 1995); these cells are not uniformly distributed but, rather, are concentrated in the anterior region. Release sites for the ASTs are likely to occur on the proctodeal axons although the sites of release of midgut endocrine cells require clarification.

The immunolocalisation of allatostatin-like peptides in interneurones of the central nervous system, in neurones that innervate visceral muscle, in midgut cells and in glands other than the CA, suggests that allatostatins are multi-functional neuropeptides and that inhibition of JH biosynthesis may not be their primary function in all insect species (see Duve *et al.*, this volume).

Haemolymph extracts from mated adult female *D. punctata* contain products that co-elute, using HPLC, with Dip-AST 2, 5 and 7–9 and that also inhibit CA *in vitro* (Woodhead, Asano & Stay, 1993). Yu *et al.* (1993), using an enzyme-linked immunosorbent assay, showed quantifiable changes in the content of Dip-AST 7-immunoreactive material in brains and haemolymph of mated female *D. punctata* during the reproductive cycle. The presence of allatostatin-immunoreactive material in the haemolymph suggests a humoral pathway for the action of these peptides in addition to neural pathways from the brain. *In vitro* bioassays confirm a lesser quantity of allatostatins in brain and CA during early to mid-vitellogenesis and during pregnancy. The highest brain content coincides roughly with the peak of JH biosynthesis, which occurs towards the end of vitellogenesis. The quantity of Dip-AST 7-like material in haemolymph was low during vitellogenesis (<0.2 nmol l^{-1}), increased rapidly shortly after ovulation and reached a maximum (2.25 nmol l^{-1}) early in pregnancy. In last-instar larvae, the haemolymph concentration of allatostatin was low in the first half of the stadium but increased dramatically (1.27 nmol l^{-1}) in the latter half of the stadium, an interval when JH biosynthesis becomes undetectable (Yu *et al.*, 1993). To demonstrate that the quantities of allatostatins found in haemolymph may be sufficient to inhibit the CA *in vivo*, Dip-AST 2 and 7 and the related peptide, callatostatin 5 from *Calliphora vomitoria* (Duve *et al.*, 1993) were injected into mated females of *D. punctata* at intervals after ecdysis. These treatments significantly retarded basal oocyte growth and reduced JH biosynthesis (Woodhead *et al.*, 1993).

The effect of a JH analogue (7-*S*-hydroprene) on the release of allatostatins into the haemolymph, and on JH biosynthesis by isolated CA, was determined in adult male *D. punctata* in which nerves to the CA had been severed or left intact (Stay *et al.*, 1994a). CA with intact nerves were strongly inhibited by JH analogue treatment, but those with severed nerves were less inhibited. Haemolymph allatostatins were elevated above normal only in 7-*S*-hydroprene-treated males with denervated CA. This suggests that allatostatins are normally released at nerve endings within the CA and that, following denervation of the CA, allatostatins are released directly into the haemolymph from where they inhibit JH biosynthesis less effectively.

Metabolism of allatostatins

Information on the metabolism of insect peptides is limited, in part because of their relatively recent discovery. However, in those insect systems studied, inactivation of neuropeptides by degradation no doubt plays a major role in the termination of biological effect, as a consequence of time and space constraints (that is, the time of exposure of target tissues to the peptides and distance over which they spread), as has been demonstrated in vertebrate systems. In the case of allatostatins, it is essential that degradative mechanisms exist since continued exposure of CA to allatostatins, for example, results in continued inhibition of JH biosynthesis (Stay *et al.*, 1994b); this effect is rapidly reversible upon removal of the peptide. Because allatostatins appear to act both humorally (this route is mimicked by *in vitro* experiments) and by way of direct innervation through nervous tracts (to CA and gut, for example), the presence of peptidases in both target tissue and haemolymph compartments is important for modulation of bioactivity.

Metabolism of insect neuropeptides may occur by interaction with either soluble or membrane-bound peptidases, resulting in the formation of either secondarily active or inactive products and a potential change in the functionality of the molecule. The metabolism of allatostatins by peptidases in haemolymph represents an interesting example of the generation of active products by soluble peptidases (Garside, Hayes & Tobe, 1997). Allatostatins do appear to occur in the haemolymph of *D. punctata* at biologically relevant concentrations (Yu *et al.*, 1993; Woodhead *et al.*, 1993 – see above), probably as a result of direct release from neurohaemal sites in the corpus cardiacum and gut, from endocrine cells of the gut, or leakage from target tissues.

Incubation of Dip-AST 7 and 9 at micromolar concentrations in haemolymph results in cleavage of the peptide to the C-terminal hexapeptide Leu-Tyr-Xaa-Phe-Gly-Leu-NH$_2$ and subsequently, the C-terminal pentapeptide Tyr-Xaa-Phe-Gly-Leu-NH$_2$ (Fig. 3). Interestingly, these two products contain the C-terminal pentapeptide core sequence required for inhibition of JH biosynthesis (Hayes *et al.*, 1994), and retain some potency and full efficacy in the *in vitro* assay for JH biosynthesis (Stay *et al.*, 1991). The presence of these metabolites might be expected because the C-terminus is blocked by an amide group, affording protection from enzymatic degradation by carboxypeptidases. Indeed, on the basis of peptidase inhibitor studies, it appears that there is an amastatin-sensitive aminopeptidase in the

D<u>**RL**</u>YSFGLa *(Dip-AST 5)* → **<u>LY</u>**SFGLa → YSFGLa

APSGAQ<u>**RL**</u>YGFGLa *(Dip-AST 7)* → **<u>LY</u>**GFGLa → YGFGLa

GDG<u>**RL**</u>YAFGLa *(Dip-AST 9)* → **<u>LY</u>**AFGLa → YAFGLa

Fig. 3. Metabolism of *D. punctata* allatostatins by haemolymph preparations. The allatostatins were incubated for 0–180 min with haemolymph (100-fold dilution) and products identified by co-elution with standards on HPLC or by amino acid analyses. Note the common cleavage sites Arg-Leu (RL) and Leu-Tyr (LY) (data from Garside *et al.*, 1997). Single-letter code for amino acids is used.

haemolymph which cleaves the leucine residue from the N-terminus of the hexapeptide, yielding the pentapeptide of Dip-AST 7 and 9 (Garside *et al.*, 1997; see Fig. 3). There is no evidence for the occurrence of carboxypeptidases in the haemolymph.

The data of Garside *et al.* (1997) indicate that haemolymph contains an endopeptidase, which cleaves at Arg-Leu of both Dip-AST 7 and 9 to yield the C-terminal hexapeptide. This hexapeptide is then cleaved by an amastatin-sensitive aminopeptidase to yield the C-terminal pentapeptide, which then appears to be rapidly degraded by as yet unidentified haemolymph peptidases. Many small insect peptide hormones with blocked N- or C-termini also are cleaved by endopeptidases (Fox & Reynolds, 1991; Isaac & Lamango, 1994), and the allatostatins fall into this category.

The rapid degradation of Dip-AST 7 and 9 in haemolymph suggests that these two peptides play a limited biological role in this compartment. The primary products of cleavage show reduced activity in inhibition of JH biosynthesis; alternatively they may act synergistically to modify the action of haemolymph allatostatins. The C-terminal hexapeptide and pentapeptide of Dip-AST 5 are 30× and 200× less potent than Dip-AST 5 (Stay *et al.*, 1991), and the C-terminal hexapeptide and pentapeptide of Dip-Ast 7 or 9 probably show similar reductions in bioactivity.

Dip-AST 5 shows high sequence similarity to Dip-AST 7 and 9 (Table 1), including the identical internal tripeptide Arg-Leu-Tyr, suggesting that Dip-AST 5 would be degraded by similar mechanisms. However, Dip-AST 5 at micromolar concentrations, incubated in haemolymph, showed high resistance to degradation, with a half-life 20 times that of Dip-AST 7 and 9 (Fig. 3 and Garside *et al.*, 1997). At more physiological concentrations (nanomolar), [^{3}H-Tyr]Dip-AST 5 was cleaved, with a half-life almost five times that of Dip-AST 7 or 9 (at micromolar concentrations). These data suggest the presence of an endopeptidase specific for the Arg-Leu of allatostatins, and an aminopeptidase which cleaves Leu from the hexapeptide (Fig. 3). Thus, there are enzymes in the haemolymph capable of cleaving all allatostatins tested, albeit at differing rates and at different cleavage sites. The large differences in rates of degradation suggest that the catabolic enzyme(s) in the haemolymph exhibit a high degree of substrate specificity.

The high resistance of Dip-AST 5 to cleavage by haemolymph peptidases indicates that it may play a special humoral function in *D. punctata*. For example, haemolymph Dip-AST 5 may be responsible for the maintenance of baseline or threshold levels of inhibition of JH biosynthesis. The continued presence of Dip-AST 5 in the haemolymph may also minimise carbon flow early in the biosynthetic pathway of JH (Stay *et al.*, 1994b). Alternatively, because Dip-AST 5 shows high resistance to degradation, it may function *in vivo* as an enzyme inhibitor, and may act to regulate peptidase activity and hence the half-life of the other allatostatins.

The probable occurrence of 13 allatostatins in *D. punctata* suggests that individual peptides or groups of peptides may have unique biological functions. The physiological processes that are controlled by families of peptides originating from a single precursor remain undefined in virtually all insect systems. The variable N-terminal regions or unique secondary and tertiary structures of the allatostatins may provide the structural information necessary for their specific functions at the appropriate target site(s). It will be instructive to examine the degradation of other allatostatins by haemolymph, by whole tissues and by membrane preparations, to establish whether haemolymph and membrane peptidases act in concert to degrade the family of allatostatins.

Lepidopteran allatostatin and the *Pseudaletia unipuncta* allatostatin gene

A single non-amidated peptide, pGlu-Val-Arg-Phe-Arg-Gln-Cys-Tyr-Phe-Asn-Pro-Ile-Ser-Cys-Phe-OH (Mas-AST), purified from brains of *M. sexta* strongly inhibits JH biosynthesis *in vitro* by CA of *M. sexta* fifth-stadium larvae and adult females (Kramer *et al.*, 1991). A search for a similar peptide using recombinant DNA methods was conducted in the true armyworm, *Pseudaletia unipuncta*, as in this species JH has been implicated in vitellogenesis (Cusson *et al.*, 1994b), ovarian development (Cusson & McNeil, 1989a), the biosynthesis and emission of female sex pheromone (Cusson & McNeil, 1989b), and seasonal migratory flight (Cusson, Tobe & McNeil, 1994a; McNeil *et al.*, 1995). A 1229 nucleotide *Pseudaletia* brain cDNA was isolated that specified a polypeptide of 125 amino acid residues (Fig. 4). In contrast to the cockroach allatostatin precursors, which contain a

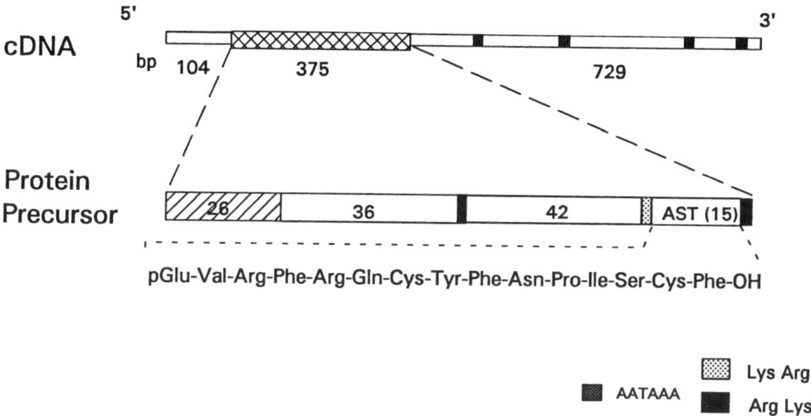

Fig. 4. The polypeptide precursor of the moth *Pseudaletia unipuncta* is derived from a 375 bp open reading frame and begins with a 26 amino acid residue hydrophobic leader region (cross-hatched). The precursor, through proteolytic cleavage at dibasic cleavage sites (black and shaded boxes as indicated), could potentially give rise to three peptides of 42, 36 and 15 amino acid residues. The 15 residue peptide corresponds to an allatostatin (AST) polypeptide previously isolated from *Manduca sexta* (Kramer *et al.*, 1991) as shown below the protein precursor.

family of related peptides, the *P. unipuncta* precursor contains a single peptide with a sequence identical to Mas-AST. This peptide would be released by endoproteolysis at dibasic cleavage sites along with two additional peptides of 36 and 42 amino acid residues (Fig. 4). These latter peptides share no sequence in common with Mas-AST or any other sequence found in the Genbank sequence database. The *P. unipuncta* gene is expressed developmentally, with steady-state mRNA levels being low in sixth instar larvae, prepupae and early pupae, then maintaining a comparatively high level in late pupae through the first five days of adult life. This pattern of expression is unexpected for an allatostatin since JH levels increase with adult age. Mas-AST inhibits JH biosynthesis *in vitro* by CA of *M. sexta* fifth-instar larvae and adult females by 100%, and by 77% for CA of *Helicoverpa zea* at concentrations of 0.1 and 0.5 μmol 1^{-1}, respectively. In contrast, 1 μmol Mas-AST 1^{-1} fails to inhibit *in vitro* JH biosynthesis by CA of *P. unipuncta* sixth-instar larvae or newly emerged adults, but does inhibit JH biosynthesis by 58% in five-day-old adult females. Unlike *M. sexta, H. zea* and *P. unipuncta* are long-distance migrants, and it has been postulated that low JH titres are essential for migratory flight (Rankin & Riddiford, 1978; Cusson, McNeil & Tobe, 1990). A complete inhibition of JH biosynthesis would not

Table 3   *Amino acid sequences of allatostatins and allatostatin-like peptides from other insect species*

	Sequence																Ref.[a]
Calliphora vomitoria																	
Leu-callatostatin 1	D	P	L	N	E	E	R	R	A	N	R	Y	G	G	G	L-NH$_2$	[1]
Leu-callatostatin 2			L	N	E	E	R	R	A	N	R	Y	G	F	G	L-NH$_2$	[1]
Leu-callatostatin 3									A	N	R	Y	G	F	G	L-NH$_2$	[1]
Leu-callatostatin 4									N	R	P	Y	S	F	G	L-NH$_2$	[1,2]
Met-callatostatin									G	P	P	Y	D	F	G	M-NH$_2$	[1]
[Hyp^3]Met-callatostatin									G	P	X	Y	D	F	G	M-NH$_2$	[2]
[Hyp^2]Met-callatostatin									G	X	P	Y	G	F	G	M-NH$_2$	[3]
des Gly-Pro Met-callatostatin											P	Y	D	F	G	M-NH$_2$	[3]
Gryllus bimaculatus																	
Grb-AST A1								A	Q	H	Q	Y	S	F	G	L-NH$_2$	[4]
Grb-AST A2							A	G	G	R	Q	Y	G	F	G	L-NH$_2$	[4]
Grb-AST B1								G	W	Q	D	L	N	G	G	W-NH$_2$	[5]
Grb-AST B2								G	W	R	D	L	N	G	G	W-NH$_2$	[5]
Grb-AST B3								A	W	R	D	L	S	G	G	W-NH$_2$	[5]
Grb-AST B4																	[5]

Notes:
[a] References: [1] Duve *et al.* (1993); [2] Duve *et al.* (1994); [3] Duve *et al.* (1995); [4] Lorenze *et al.* (1995a); [5] Lorenz *et al.* (1995b).

be expected in migratory species, and, although Mas-AST has limited allatostatic activity under the conditions tested, the possibility exists that variation in activity may occur with alterations in environmental conditions.

Allatostatins in other invertebrates

In addition to the cockroach and lepidopteran allatostatins discussed above, there are several reports describing the occurrence of other allatostatins, allatostatin-related peptides or allatostatin-like immunoreactivities in several other insects species, and in other invertebrates.

In the blowfly, *Calliphora vomitoria*, eight peptides with sequence similarity to cockroach allatostatins have been identified (Table 3). These peptides inhibit JH III biosynthesis in cockroaches, but not biosynthesis of JH III bisepoxide in the blowfly itself (Duve *et al.*, 1993, 1994, 1995 see Duve *et al.*, this volume). Immunocytochemical studies with antisera to the blowfly peptides have revealed callatostatin immunoreactivity widely distributed in a variety of neurones within the central nervous system of the blowfly and in the endocrine cells of the midgut (Duve *et al.*, 1993; Duve & Thorpe, 1994). Immunoreactive staining was particularly intense in the nerves of the hindgut. However, in contrast to the abundant evidence of cockroach allatostatins and allatostatin immunoreactivity in the brain CC–CA axis of cockroaches and crickets (Stay *et al.*, 1991, 1992; Schildberger & Agricola, 1992; Neuhäuser *et al.*, 1994; Bellés *et al.*, 1994), it appears that the Leu-callatostatins are noticeably absent from the blowfly CA. Thus, the callatostatins may not play a role in the down-regulation of JH biosynthesis in this fly.

Inhibition of JH biosynthesis by extracts of the brain of the field cricket, *Gryllus bimaculatus*, has led to the isolation and identification of a further six allatostatins (Lorenz, Kellner & Hoffmann, 1995a,b). Two of these are additional members of the cockroach allatostatin family and the others, all nonapeptides (Table 3), show high sequence similarity to the *Locusta* myoinhibiting peptide (Lom-MIP: Schoofs *et al.*, 1991). The ability of these Lom-MIP-related *G. bimaculatus* peptides to inhibit JH biosynthesis is approximately 10-fold less than the most active of the cockroach-related *G. bimaculatus* peptides (Grb-AST A1), which must call into question their possible allatoregulatory functionality.

In respect of other insect species, cockroach allatostatin immunoreactivity has been described in the nervous system of *D. melanogaster* and *M. sexta* (see Stay *et al.*, 1994b), and in the brain-retrocerebral complex, thoracic and abdominal ganglia of the locusts *Schistocerca gregaria* and *Locusta migratoria*, and of the fleshfly *Neobellieria bullata* (Veelaert *et al.*, 1995). It is also present in the brain and suboesophageal ganglia of the honeybee (B. Stay, unpublished results), in the pars intercerebralis and pars lateralis of the earwig, *Labidura riparia* (Sayah, Fayet & Laverdure, 1996), and in whole insect extracts of the aphid *Megoura viciae* (S. Tilley & R. J. Weaver, unpublished results). Recently, the identity of one of a group of peptides that inhibit CA activity in honeybees has been established (Kaatz, 1994).

Immunocytochemistry with anti-Dip-AST 7 polyclonal antibodies has demonstrated the presence of cockroach allatostatin-like peptides in the commissural, oesophageal and stomatogastric ganglia and nerves of the crab *Cancer borealis* (Skiebe & Schneider, 1994). The nature and function of the endogenous peptides has yet to be established, although four of the Dip-ASTs have actions on the rhythmic movements of the crab stomach.

Allatostatins or allatostatin-like molecules may occur also in many non-arthropods. Dip-AST immunoreactivity has been demonstrated in the ganglia of the freshwater snails *Bulinus globosus* and *Stagnicola elocles* (Rudolph & Stay, 1997), and in the nervous system of *Hydra oligactis* (Hydrozoa), *Moniezia expansa* (Cestoda), *Schistosoma mansoni* (Trematoda), *Artioposthia triangulata* (Turbellaria), *Ascaris suum* (Nematoda), *Lumbricus terrestris* (Oligochaeta), *Limax pseudoflavus* (Gastropoda), and *Eledone cirrhosa* (Cephalopoda) (Smart *et al.*, 1994). The distribution of allatostatin-like immunoreactivity within the nervous system of a range of helminths has led to speculation that they may have neurotransmitter/neuromodulatory functions with a role in locomotion, feeding, reproduction and sensory perception (Smart *et al.*, 1995), a suggestion that awaits appropriate experimental verification.

Summary, current status and future prospects

Clearly, considerable progress has been made in the identification of allatostatic peptides or their homologues in a wide range of invertebrate species. However, only in cockroaches and (arguably) in moths have these peptides been shown unequivocally to have a functional biological activity as inhibitors of JH biosynthesis by CA. At present, information about allatostatins or allatostatin-like peptides in other insect groups is sparse. Reports indicate the presence of peptides similar to the dictyopteran and lepidopteran allatostatins in species of Orthoptera, Diptera, Homoptera and Hymenoptera, but we have yet to explore other insect Orders, particularly those that contain important pest species (for example, the Coleoptera). Moreover, even in the Lepidoptera, questions remain about the universality of allatostatic function of Mas-AST throughout the Order. Despite the discovery of a gene coding for this peptide in the noctuid *P. unipuncta*, and the demonstration that this peptide has the ability (albeit weak) to inhibit JH production in adult females of this species, the apparent expression pattern of the gene does not correlate well with the changes in JH production, assuming that allatostatin is the primary regulator of JH biosynthesis. Similar apparent contradictions between brain allatostatin expression and measured fluctuations in JH titres (Edwards *et al.*, 1995) were observed in another noctuid species, *Lacanobia oleracea* (J. P. Edwards *et al.*, unpublished results). It has been suggested (see above) that the Mas-AST peptide might be involved in producing subtle changes in JH levels in adult moths that exhibit migratory behaviour. However, in non-migratory species such as *L. oleracea*, in which JH levels in adult females are high and show a distinct increase during early adult life (Edwards *et al.*, 1995), the function of an allatostatin is unclear. Mas-AST was originally isolated from trimmed heads of pharate adult *M. sexta* (Kramer *et al.*, 1991), and it is perhaps not surprising that this molecule was active against CA from 0–4 h old adults. The same compound is also active against CA from early final instar *M. sexta* larvae, but is inactive against CA from final instar *P. unipuncta* larvae (Jansons *et al.*, 1996). This reduced activity of Mas-AST in final instar larvae of a noctuid lepidopteran (in which the absence of JH seems to be critical for successful metamorphosis; Edwards *et al.*, 1995) is yet another observation which casts doubt on the function of this peptide in lepidopterans. There is clearly scope for further exploration of the identity and action of allatostatins in the Lepidoptera.

The occurrence of cockroach-like allatostatic peptides in other insect species, especially flies, in which there is no evidence for allatostatic (*sic*) activity of these peptides, in conjunction with the reports of myomodulatory activity of these compounds (see Duve *et al.*, this volume), strongly suggests that regulation of JH biosynthesis by CA may represent an altered or secondary role for these neuropeptides in the Orthoptera/Dictyoptera and, that in other orders, their role(s) may be primarily myomodulatory. This suggestion is supported by recent immunocytological studies demonstrating the presence of identical or similar peptides in tissues from a wide range of invertebrate taxa that are devoid of either JH or CA. No doubt, immunocytochemistry and molecular biology approaches will reveal the occurrence of similar peptides in a wider range of organisms. While such techniques can demonstrate the ubiquity and diversity of native peptides or their encoding genes, they cannot indicate roles or functions. The latter questions can be addressed only by the more traditional techniques of careful isolation from appropriate tissues, derived from appropriate developmental stages, and rigorously assayed with relevant bioassay techniques.

One compelling reason for the study of allatostatins is to increase the scientific understanding of the molecular, biochemical and physiological regulation of invertebrate endocrine systems. A further reason, however, is the tantalising prospect of using these peptides (and other invertebrate neurohormones) to develop a 'fourth generation' of insecticides. Pest management technologies based on neuropeptides potentailly offer a degree of biological activity, target specificity and environmental compatibility that are lacking in conventional neurotoxic insecticides (Menn, Raina & Edwards, 1989). However, peptides do not at first sight make ideal insecticides. They do not easily penetrate insect cuticle, and many are not readily soluble in traditional formulation solvents (see Nachman *et al.*, this volume). Peptides are generally unstable molecules that degrade relatively rapidly in the environment and in insect tissues. Finally, it is unlikely that such molecules would retain biological activity if administered orally. Nevertheless, if suitable vectors (for example, genetically modified micro-organisms) can be developed, and expression levels maintained, there remain intriguing prospects for using the genes coding for invertebrate neuropeptides as key contributors to a new generation of insect control techniques.

Acknowledgements

This work was supported by the Pesticides Safety Directorate of the Ministry of Agriculture, Fisheries and Food (R.J.W., J.P.E.), and by operating grants to S.S.T. and W.G.B. from the Natural Sciences and Engineering Research Council of Canada and Insect Biotechnology Canada Inc.

References

Agricola, H., Schildberger, K., Schmidt, A., Naumann, W., Reissmann, S., Huber, F. & Penzlin, H. (1992). Immunocytochemical distribution of allatostatin in the nervous system of *Periplaneata americana*. In *Rhythmogenesis in Neurones and Networks*, ed. N. Elsner & D. W. Richter, p. 949. Stuttgart, G. Thieme.

Bellés, X., Maestro, J.-L., Piulachs, M.-D., Johnsen, A. H., Duve, H. & Thorpe, A. (1994). Allatostatic neuropeptides from the cockroach *Blattella germanica* (L) (Dictyoptera, Blattellidae). Identification, immunolocalization and activity. *Regulatory Peptides* **53**, 237–247.

Bendena, W. G., Garside, C. S., Yu, C. G. & Tobe, S. S. (1997). Allatostatins: diversity in structure and function of an insect neuropeptide family. *Annals of the New York Academy of Science*, **814**, 53–66.

Bhaskaran, G., Dahm, K. H., Barrera, P., Pacheco, J. L., Peck, K. E. & Muszynska-Pytel, M. (1990). Allatinhibin, a neurohormonal inhibitor of juvenile hormone biosynthesis in *Manduca sexta*. *General and Comparative Endocrinology* **78**, 123–136.

Cassier, P. (1990). Morphology, histology and ultrastructure of JH-producing glands in insects. In *Morphogenetic Hormones of Arthropods*, Volume 1, ed. A. P. Gupta, pp. 84–194. New Brunswick: Rutgers University Press.

Cusson, M. & McNeil, J. N. (1989a). Ovarian development in female armyworm moths, *Pseudaletia unipuncta*: its relationship with pheromone release activities. *Canadian Journal of Zoology* **67**, 1380–1385.

Cusson, M. & McNeil, J. N. (1989b). Involvement of juvenile hormone in the regulation of pheromone release activities in a moth. *Science* **243**, 210–212.

Cusson, M., McNeil, J. N. & Tobe, S. S. (1990). *In vitro* biosynthesis of juvenile hormone by corpora allata of *Pseudaletia unipuncta* virgin females as a function of age, environmental conditions, calling behaviour and ovarian development. *Journal of Insect Physiology* **36**, 139–146.

Cusson, M., Tobe, S. S. & McNeil, J. N. (1994a). Juvenile hormones: their role in the regulation of the pheromonal communication system of the armyworm moth, *Pseudaletia unipuncta*. *Archives of Insect Biochemistry and Physiology* **25**, 329–345.

Cusson, M., Yu, C. G., Carruthers, K., Wyatt, G. R., Tobe, S. S. & McNeil, J. N. (1994b). Regulation of vitellogenin production in armyworm moths, *Pseudaletia unipuncta*. *Journal of Insect Physiology* **40**, 129–136.

Ding, Q., Donly, B. C., Tobe, S. S. & Bendena, W. G. (1995). Comparison of the allatostatin neuropeptide precursors in the distantly related cockroaches *Periplaneta americana* and *Diploptera punctata*. *European Journal of Biochemistry* **234**, 737–746.

Donly, B. C., Ding, Q., Tobe, S. S. & Bendena, W. G. (1993). Molecular cloning of the gene for the allatostatin family of neuropeptides from the cockroach *Diploptera punctata*. *Proceedings of the National Academy of Sciences USA* **90**, 8807–8811.

Duve, H., Johnsen, A. H., Scott, A. G., East, P. & Thorpe, A. (1994). [Hyp^3]Met-callatostatin: identification and biological properties of a novel neuropeptide from the blowfly *Calliphora vomitoria*. *Journal of Biological Chemistry* **269**, 21 059–21 066.

Duve, H., Johnsen, A. H., Scott, A. G. & Thorpe, A. (1995). Isolation, identification and functional significance of [Hyp^2]Met-callatostatin and des Gly-Pro Met-callatostatin, two further post-translational modifications of the blowfly neuropeptide Met-callatostatin. *Regulatory Peptides* **57**, 237–245.

Duve, H., Johnsen, A. H., Scott, A. G., Yu, C. G., Yagi, K. J., Tobe, S. S. & Thorpe, A. (1993). Callatostatins: neuropeptides from the blowfly *Calliphora vomitoria* with sequence homology to cockroach allatostatins. *Proceedings of the National Academy of Sciences USA* **90**, 2456–2460.

Duve, H. & Thorpe, A. (1994). Distribution and functional significance of Leu-callatostatins in the blowfly *Calliphora vomitana*. *Cell and Tissue Research* **276**, 367–379.

Edwards, J. P., Chambers, J., Short, J. E., Price, N. R., Weaver, R. J., Abraham, L. & Walter, C. M. (1990). Endogenous juvenile hormone III titres and *in vitro* rates of hormone biosynthesis by corpora allata during the reproductive cycle of adult female *Periplaneta americana*. In *Chromatography and Isolation of Insect Hormones, Pheromones and Related Substances*, ed. A. R. McCaffery & I. D. Wilson, pp. 3–8. New York: Plenum Press.

Edwards, J. P., Corbitt, T. S., McArdle, H. F., Short, J. E. & Weaver, R. J. (1995). Endogenous levels of insect juvenile hormones in larval, pupal and adult stages of the Tomato moth, *Lacanobia oleracea. Journal of Insect Physiology* **41**, 641–651.

Engleman, F. (1957). Die Steuerung der Ovarfunktion bei der Ovoviviparen Schabe *Leucophaea maderae* (Fabr.) *Journal of Insect Physiology* **1**, 257–278.

Engleman, F. (1959). The control of reproduction in *Diploptera punctata* (Blattaria). *Biological Bulletin of Woods Hole* **116**, 406–419.

Fox, A. M. & Reynolds, S. E. (1991). Degradation of adipokinetic hormone family peptides by a circulating endopeptidase in the insect *Manduca sexta. Peptides* **12**, 937–944.

Garside, C. S., Hayes, T. K. & Tobe, S. S. (1997). Degradation of Dip-allato-statins by hemolymph from the cockroach, *Diploptera punctata. Peptides* **18**, 17–25.

Hayes, T. K., Guan, X.-C., Johnson, V., Strey, A. & Tobe, S. S. (1994). Structure-activity studies of allatostatin 4 on the inhibition of juvenile hormone biosynthesis by corpora allata: the importance of individual side-chains and stereochemistry. *Peptides* **15**, 1165–1171.

Hertel, W. & Penzlin, H. (1992). Function and modulation of the antennal heart of *Periplaneta americana* (L.). *Acta Biologica Hungarica* **43**, 113–125.

Isaac, R. E. & Lamango, N. S. (1994). Peptidyl dipeptidase activity in the hemolymph of insects. *Biochemical Society Transactions* **22**, 92S.

Jansons, I., Cusson, M., McNeil, J. N., Tobe, S. S. & Bendena, W. G. (1996). Molecular characterization of a cDNA from *Pseudaletia unipuncta* encoding the *Manduca sexta* allatostatin peptide (Mas-AST). *Insect Biochemistry and Molecular Biology* **26**, 767–773.

Kaatz, H. (1994). Regulation of juvenile hormone biosynthesis in the honey bee. *Apidologie* **25**, 469–470.

Kataoka, H., Tosch8i, A., Li, J. P., Carney, R. L., Schooley, D. A. & Kramer, S. J. (1989). Identification of an allatrotropin from adult *Manduca sexta. Science* **243**, 1481–1483.

Kramer, S. J., Toschi, A., Miller, C. A., Kataoka, H., Quistad, G. B., Li, J. P., Carney, R. L. & Schooley, D. A. (1991). Identification of an allatostatin from the tobacco hornworm, *Manduca sexta. Proceedings of the National Academy of Sciences USA* **88**, 9458–9462.

Lange, A. B., Bendena, W. G. & Tobe, S. S. (1995). The effect of thirteen Dip-allatostatins on myogenic and induced contractions of the cockroach (*Diploptera punctata*) hindgut. *Journal of Insect Physiology* **41**, 581–588.

Lange, A. B., Chan, K. K. & Stay, B. (1993). Effect of allatostatin and proctolin on antennal pulsatile organ and hindgut muscle in the cockroach *Diploptera punctata. Archives of Insect Biochemistry and Physiology* **24**, 79–92.

Lorenz, M. W., Kellner, R. & Hoffmann, K. H. (1995a). Identification of two allatostatins from the cricket, *Gryllus bimaculatus* de Geer (Ensifera, Gryllidae): additional members of a family of neuropeptides inhibiting juvenile hormone biosynthesis. *Regulatory Peptides* **57**, 227–236.

Lorenz, M. W., Kellner, R. & Hoffmann, K. H. (1995b). A family of neuro-peptides that inhibit juvenile biosynthesis in the cricket, *Gryllus bimacu-latus. Journal of Biological Chemistry* **270**, 21 103–21 108.

McNeil, J. N., Cusson, M., Delisle, J., Orchard, I. & Tobe, S. S. (1995).

Physiological integration of migration in Lepidoptera. In *Insect Migration: Tracking Resources in Time and Space*, ed. V. A. Drake & A. G. Gatehouse, pp. 279–302. Cambridge: Cambridge University Press.

Menn, J. J., Raina, A. K. & Edwards, J. P. (1989). Juvenoids and neuropeptides as insect control agents: retrospect and prospects. In *Progress and Prospects in Insect Control*, ed. N. R. McFarlane, BCPC Monograph No. 43 pp. 89–106. Farnham: British Crop Protection Council.

Neuhäuser, T., Sarge, D., Stay, B. & Hoffman, K. H. (1996). Responsiveness of the adult cricket (*Gryllus bimaculatus* and *Acheta domesticus*) retrocerebral complex to allostatin-1 from a cockroach *Diploptera punctata*. *Journal of Comparative Physiology B* **164**, 23–31.

Nijhout, H. F. (1994). *Insect Hormones*. Princeton, NJ, Princeton University Press.

Paulson, C. R. & Stay, B. (1987). Inhibition of corpora allata by larval brain extract in the cockroach in *Diploptera* punctata. *Molecular and Cellular Endocrinology* **51**, 243–252.

Paulson, C. R., Stay, B., Kikukawa, S. & Tobe, S. S. (1987). *In vitro* inhibition of the corpora allata by a larval brain factor in the cockroach *Diploptera punctata*. *Insect Biochemistry* **17**, 965–969.

Pratt, G. E., Farnsworth, D. E. & Feyereisen, R. (1990). Changes in the sensitivity of adult cockroach corpora allata to brain allatostatin. *Molecular and Cellular Endocrinology* **70**, 185–196.

Pratt, G. E., Farnsworth, D. E., Fox, K. F., Siegel, N. R., McCormack, A. L., Shabanowitz, J., Hunt, D. F. & Feyereisen, R. (1991). Identity of a second type of allatostatin from cockroach brains: an octadecapeptide amide with tyrosine-rich address sequence. *Proceedings of the National Academy of Sciences USA* **88**, 2412–2416.

Pratt, G. E., Farnsworth, D. E., Siegel, N. R., Fox, K. F. & Feyercisen, R. (1989). Identification of an allostatin from adult *Diploptera punctata*. *Biochemical and Biophysical Research Communications* **163**, 1243–1247.

Pratt, G. E. & Tobe, S. S. (1974). Juvenile hormones radiobiosynthesised by corpora allata of adult female locusts *in vitro*. *Life Sciences* **14**, 575–586.

Rankin, M. A. & Riddiford, L. M. (1978). Significance of haemolymph juvenile hormone titer changes in timing of migration and reproduction in adult *Oncopeltus fasciatus*. *Journal of Insect Physiology* **24**, 31–38.

Rankin, S. M. & Stay, B. (1987). Distribution of allatostatin in the adult cockroach, *Diploptera punctata* and effects on corpora allata *in vitro*. *Journal of Insect Physiology* **33**, 551–558.

Rankin, S. M., Stay, B., Aucoin, R. R. & Tobe, S. S. (1986). *In vitro* inhibition of juvenile hormone synthesis by corpora allata of the viviparous cockroach, *Diploptera punctata*. *Journal of Insect Physiology* **32**, 151–156.

Roth, L. M. & Stay, B. (1961). Oocyte development in *Diploptera punctata* (Eschscholtz) (Blattaria). *Journal of Insect Physiology* **7**, 186–202.

Rudolph, P. H. & Stay, B. (1997). Cockroach allatostatin-like immunoreactivity in the central nervous system of the freshwater snails *Bulinus globosus* (Planorbidae) and *Stagnicola elodes* (Lymnaeidae). *General and Comparative Endocrinology* **106**, 241–250.

Sayah, F., Fayet, C. & Laverdure, A. M. (1996). Cyclic allatostatin-like peptide detection in the brain-retrocerebral complex of the earwig *Labidura riparia*, related to the reproductive cycle. *Annales d'Endocrinologie* **57**, suppl. 4, p. 64.

Scharrer, B. (1952). Neurosecretion. XI. The effects of nerve section on the intercerebralis-cardiacum-allatum system of the insect *Leucophaea maderae*. *Biological Bulletin of Woods Hole* **102**, 261–272.

Schildberger, K. & Agricola, H. (1992). Allatostatin-like immunoreactivity in the brains of crickets and cockroaches. In *Rhythmogenesis in Neurones and Networks*, ed. N. Elsner & D. W. Richter, p. 489. Stuttgart: G. Thieme.

Schoofs, L., Holman, G. M., Hayes, T. K., Nachman, R. J. & De Loof, A. (1991). Isolation, identification and synthesis of locust myoinhibiting peptide (Lom-MIP), a novel biologically active neuropeptide from *Locusta migratoria*. *Regulatory Peptides* **36**, 111–119.

Skiebe, P. & Schneider, H. (1994). Allatostatin peptides in the crab stomato-gastric nervous system: inhibition of the pyloric motor pattern and distribution of allatostatin-like immunoreactivity. *Journal of Experimental Biology* **194**, 195–208.

Smart, D., Johnston, C. F., Curry, W. J., Williamson, R., Maule, A. G., Skuce, P. J., Shaw, C., Halton, D. W. & Buchanan, K. D. (1994). Peptides related to the *Diploptera punctata* allatostatins in non arthropod invertebrates: an immunocytochemical survey. *Journal of Comparative Neurology* **347**, 426–432.

Smart, D., Johnston, C. F., Maule, A. G., Halton, D. W., Hrcková, G., Shaw, C. & Buchanan, K. D. (1995). Localization of *Diploptera punctata* allatostatin-like immunoreactivity in helminths: an immunocytochemical study. *Parasitology* **110**, 87–96.

Stay, B., Joshi, S. & Woodhead, A. P. (1991). Sensitivity to allatostatins of corpora allata from larval and adult female *Diploptera punctata*. *Journal of Insect Physiology* **37**, 63–70.

Stay, B., Sereg Bachmann, J. A., Stoltzman, C. A., Fairbairn, S. E., Yu, C. G. & Tobe, S. S. (1994a). Factors affecting allatostatin release in a cockroach (*Diploptera punctata*): nerve section, juvenile hormone analog and ovary. *Journal of Insect Physiology* **40**, 365–372.

Stay, B. & Tobe, S. S. (1977). Control of juvenile hormone biosynthesis during the reproductive cycle of a viviparous cockroach. I. Activation and inhibition of corpora allata. *General and Comparative Endocrinology* **33**, 531–540.

Stay, B. & Tobe, S. S. (1978). Control of juvenile hormone biosynthesis during the reproductive cycle of a viviparous cockroach. II. Effects of unilateral allatectomy, implementation of supernumerary corpora allata and ovariectomy. *General and Comparative Endocrinology* **34**, 276–286.

Stay, B., Tobe, S. S. & Bendena, W. G. (1994b). Allatostatins: identification, primary structure, functions and distribution. *Advances in Insect Physiology* **25**, 267–338.

Stay, B., Woodhead, A. P. & Chan, K. K. (1992). Immunocytochemical distribution of allatostatin in the brain and retrocerebral complex of *Diploptera punctata*. In *Insect Juvenile Hormone Research: Fundamental and Applied Approaches*, ed. B. Mauchamp, F. Couillard & J. C. Baehr, pp. 257–263. Paris: INRA.

Stay, B., Woodhead, A. P., Joshi, S. & Tobe, S. S. (1991). Allatostatins, neuropeptide inhibitors of juvenile hormone synthesis in brain and corpora allata of the cockroach *Diploptera punctata*. In *Insect Neuropeptides*, ed. J. J. Menn, T. J. Kelly & E. P. Masler, ACS symposium series 453, pp. 164–176. Washington, DC: American Chemical Society.

Tobe, S. S. (1980). Regulation of corpora allata in adult female insects. In *Insect Biology of the Future: VBW 80*, ed. M. Locke & D. S. Smith, pp. 355–367. New York: Academic Press.

Tobe, S. S., Garside, C. S., Jansons, I. S., Price, M. D. & Bendena, W. G. (1995). Allatostatins: metabolism, occurrence and expression. In *Molecular Mechanisms of Insect Metamorphosis and Diapause*, ed. A. Suzuki, H. Kataoka & S. Matsumoto, pp. 13–24. Tokyo: Industrial Publishing and Consulting Incorporated.

Tobe, S. S. & Pratt, G. E. (1974). The influence of substrate concentration on the rate of insect juvenile hormone biosynthesis by corpora allata of the desert locust *in vitro*. *Biochemical Journal* **144**, 107–113.

Tobe, S. S., Ruegg, R. P., Stay, B. A., Baker, F. C., Miller, C. A. & Schooley, D. A. (1985). Juvenile hormone titre and regulation in the cockroach *Diploptera punctata*. *Experientia* **41**, 1028–1034.

Tobe, S. S. & Stay, B. (1977). Corpus allatum activity in vitro during the reproductive cycle of the viviparous cockroach, *Diploptera punctata* (Eschscholtz). *General and Comparative Endocrinology* **31**, 138–147.

Tobe, S. S. & Stay, B. (1985). Structure and regulation of the corpus allatum. *Advances in Insect Physiology* **18**, 305–432.

Tobe, S. S., Yu, C. G. & Bendena, W. G. (1994). Allatostatins, peptide inhibitors of juvenile hormone production in insects. In *Perspectives in Comparative Endocrinology*, ed. K. G. Davey, R. E. Peter & S. S. Tobe, pp. 12–19. Ottawa: National Research Council of Canada.

Unni, B., Barrera, P., Muszynska-Pytel, M., Bhaskaran, G. & Dahm, K. H. (1993). Partial characterization of allatinhibin, a neurohormone of *Manduca sexta*. *Archives of Insect Biochemistry and Physiology* **24**, 171–183.

Unnithan, G. C. & Feyereisen, R. (1995). Experimental acquisition and loss of allatostatin sensitivity by corpora allata of *Diploptera punctata*. *Journal of Insect Physiology* **41**, 975–980.

Veelaert, D., Devreese, B., Vanden Broech, J., Yu, C. G., Schoofs, L., Van Becumen, J., Tobe, S. S. & De Loof, A. (1996). Isolation and characterization of schistostatin-2 (11–18) from the desert locust, *Schistocerca gregaria*. A truncated analog of schistostatin-2. *Regulatory Peptides* **67**, 195–199.

Veelaert, D., Schoofs, L., Tobe, S. S., Yu, C. G., Vullings, H. G. B., Couillard, F. & De Loof, A. (1995). Immunological evidence for an allatostatin-like neuropeptide in the central nervous system of *Schistocerca gregaria*, *Locusta migratoria* and *Neobellieria bullata*. *Cell and Tissue Research* **279**, 601–611.

Weaver, R. J. (1991). Profile of the responsiveness of corpora allata from virgin female *Periplaneta americana* to an allatostatin from *Diploptera punctata. Journal of Insect Physiology* **37**, 111–118.

Weaver, R. J., Ding, Q., Paterson, Z. A., Bendena, W. G., Tobe, S. S. & Edwards, J. P. (1995a). Further characterisation and biological activities of neuropeptide allatostatins from the cockroach *Periplaneta americana. Journal of the Netherlands Zoological Society* **45**, 50–52.

Weaver, R. J. & Freeman, Z. A. (1992). Regulation of juvenile hormone biosynthesis by brain allatostatins in adult stages of the cockroach *Periplaneta americana*. In *Insect Juvenile Hormone Research: Fundamental and Applied Aspects*, ed. B. Mauchamp, F. Couillard & J. C. Baehr, pp. 83–95. Paris: INRA.

Weaver, R. J., Freeman, Z. A., Pickering, M. G. & Edwards, J. P. (1994). Identification of two allatostatins from the CNS of the cockroach *Periplaneta americana*: novel members of a family of neuropeptide inhibitors of insect juvenile hormone biosynthesis *Comparative Biochemistry and Physiology* **107C**, 119–127.

Weaver, R. J., Paterson, Z. A., Short, J. E. & Edwards, J. P. (1995b). Effects of *Diploptera punctata* allatostatins on juvenile hormone biosynthesis and endogenous juvenile hormone III levels in virgin and mated female *Periplaneta americana. Journal of Insect Physiology* **41**, 117–125.

Weaver, R. J. & Pratt, G. E. (1977). The effects of enforced virginity and subsequent mating on the activity of the corpus allatum of *Periplaneta americana* measured in vitro, as related to changes in the rate of ovarian maturation. *Physiological Entomology* **2**, 59–76.

Woodhead, A. P., Asano, W. Y. & Stay, B. (1993). Allatostatins in the haemolymph of *Diploptera* and their effect *in vivo. Journal of Insect Physiology* **39**, 1001–1005.

Woodhead, A. P., Khan, M. A., Stay, B. & Tobe, S. S. (1994). Two new allatostatins from the brains of *Diploptera punctata. Insect Biochemistry and Molecular Biology* **24**, 257–263.

Woodhead, A. P., Stay, B., Seidel, S. L., Khan, M. A. & Tobe, S. S. (1989). Primary structure of four allatostatins: neuropeptide inhibitors of juvenile hormone synthesis. *Proceedings of the National Academy of Sciences USA* **86**, 5997–6001.

Woodhead, A. P., Stoltzman, C. A. & Stay, B. (1992). Allatostatins in nerves of the antennal pulsatile organ muscle of the cockroach *Diploptera punctata. Archives of Insect Biochemistry and Physiology* **20**, 253–263.

Yu, C. G., Stay, B., Ding, Q., Bendena, W. G. & Tobe, S. S. (1995). Immunochemical identification and expression of allatostatins in the gut of *Diploptera punctata. Journal of Insect Physiology* **41**, 1035–1043.

Yu, C. G., Stay, B., Joshi, S. & Tobe, S. S. (1993). Allatostatin content of brain, corpora allata and haemolymph at different developmental stages of the cockroach, *Diploptera punctata*: quantitation by ELISA and bioassay. *Journal of Insect Physiology* **39**, 111–122.

Zitnan, D., Sehnal, F. & Bryant, P. J. (1993). Neurones producing specific neuropeptides in the central nervous system of normal and pupariation-delayed *Drosophila. Developmental Biology* **156**, 117–135.

SIMON G. WEBSTER

Neuropeptides inhibiting growth and reproduction in crustaceans

Introduction

The eyestalk neurosecretory complex of decapod crustaceans, comprising a prominent group of neurosecretory perikarya in the medulla terminalis (the X-organ), directing axons towards a neurohaemal tissue (the sinus gland), produces and secretes a wide variety of neuropeptide hormones involved in pigment migration (red pigment-concentrating hormone, pigment-dispersing hormone), regulation of carbohydrate metabolism (crustacean hyperglycemic hormone, CHH), moulting (moult-inhibiting hormone, MIH) and gonadal growth (gonad-inhibiting hormone, GIH). In recent years, micro-analytical and molecular techniques have been very successfully used to describe the structures of all of these neurohormones in a variety of species of malacostracans (for reviews, see Keller, 1992; De Kleijn & Van Herp, 1995; also Van Herp, this volume). Perhaps the most important recent finding has been the demonstration that several of these neurohormones are highly structurally related. CHH was first fully identified using conventional microsequencing strategies by Kegel *et al.* (1989). Although expected similarities in structure with more recently sequenced CHHs from several crustaceans (see Chang, 1995, for list) were observed, a surprising finding was that neuropeptides which were involved in the control of moulting (MIH) and gonadal development (GIH), were also structurally related to CHH, together forming a group generically known as the 'CHH family'. Nevertheless, similarities in structure of the mature peptides of this group are rather less evident when (conceptual) sequences of the preprocessed hormones are compared. All functionally defined CHHs are flanked by a so-called CHH precursor (CPRP) of 33–38 residues and a signal peptide, whereas for MIH and GIH, the preprohormone consists only of a signal peptide and sequence of the peptide (for a review, see De Kleijn & Van Herp, 1995). Thus, the currently accepted view is that the MIH/GIH peptides form a related, but somewhat separate, sub-group within the CHH neuropeptide family.

From a physiological viewpoint, members of the CHH neuropeptide family are of particular importance, because MIH and GIH inhibit generally mutually exclusive events in energy partitioning, namely somatic and gonadal growth. Additionally, because CHH controls energy metabolism, there are profound implications with respect to the above mentioned processes. Thus, one would, *a priori*, expect that there are complex multi-hormonal systems which control growth and reproduction. Elucidation of these interactions is a major challenge in crustacean endocrinology.

The purpose of this review is to outline the structure and function of neuropeptides involved in the control of growth and reproduction, and to highlight areas where there is firm evidence for multi-hormonal regulation of these processes and those where evidence is at present speculative.

Moult-inhibiting hormones

It has long been known that in decapod crustaceans, eyestalk ablation often (but not always) results in acceleration of several processes associated with moulting, and ultimately in precocious ecdysis. Passano (1953) postulated that the eyestalk neurosecretory tissues were a source of a MIH, which, when released from the sinus gland (SG), negatively regulates synthesis of moulting hormones (ecdysteroids) by the Y-organs. Progression of the moult cycle from intermoult to premoult, and ultimately moulting were thus thought to occur as a result of a decreased secretion of MIH, therefore freeing the Y-organs from the inhibitory influence of MIH. This simple model of moult control has repeatedly been confirmed by experiments showing that eyestalk ablation leads to rapid increases in haemolymph ecdysteroid titre, reminiscent of those seen during premoult, and that in such animals, ecdysteroid synthesis by the Y-organs is dramatically enhanced (Chang, Sage & O'Connor, 1976; Keller & Schmid, 1979; Jegla *et al.*, 1983). However, it is unfortunate that contradictory observations on the results of eyestalk ablation have not been given due prominence in the literature. For example, eyestalk ablation does not result in accelerated moulting in female crustaceans undergoing (seasonal) vitellogenesis (Panouse, 1946; Drach, 1955; Lachaise *et al.*, 1992). Although eyestalk ablation accelerates moulting in juvenile *Homarus americanus*, such animals continue to show characteristic, albeit accelerated changes in ecdysteroid titre over several moult cycles (Chang & Bruce, 1980; Chang, 1985), and highly reproducible fluctuations of ecdysteroid titre in premoult *Uca pugilator* continue in the absence of eyestalks (Hopkins, 1983, 1986). Thus, it is readily apparent that the currently accepted model of moult control is far from complete, and that reproductive cycles are inextricably linked with moult cycles, obviously suggesting complex interendocrine mechanisms.

Notwithstanding these caveats, the simple model of moult control has been invaluable in determining the nature of MIH in brachyurans (crabs). A bioassay measuring the reduction of ecdysteroid synthesis by Y-organs *in vitro* when exposed to MIH, initially reported by Soumoff & O'Connor (1982), has allowed characterisation of MIH from the shore crab *Carcinus maenas* (Webster, 1986; Webster & Keller, 1986), leading to the elucidation of the complete sequence in this species by conventional microsequencing strategies (Webster, 1991), and by cDNA cloning (Kleijn *et al.*, 1993). Recently, the complete sequences of MIH from the blue crab, *Callinectes sapidus* (Lee *et al.*, 1995), and the edible crab, *Cancer pagurus* (Chung, Wilkinson & Webster, 1996), have been reported. The sequences of these MIHs are shown in Fig. 1a. All are highly structurally related (around 80% identity) and are 78 residue peptides with free N- and C-termini, and three intra-chain disulphide bridges. Although the structure of the preprocessed precursor for *C. pagurus* is not yet available, it is noteworthy that for *C. sapidus* and *C. maenas* these are similar, consisting of a 35 residue signal peptide (sequence identity 69%) flanked by the sequence for MIH.

Although these peptides seem to form a clearly identifiable family in brachyurans, this situation may not hold for other crustaceans. Immunochemical studies using an antiserum directed against *C. maenas* MIH have demonstrated a widespread occurrence of MIH-like immunoreactivity in brachyurans, but not in astacurans (Dircksen, Webster & Keller, 1988; S. G. Webster, unpublished observations). In the lobster *H. americanus*, a molecule which seems to be extremely similar if not identical to CHH-A (Tensen, De Kleijn & Van Herp, 1991) certainly acts as both a MIH and a CHH *in vivo* (Chang, Prestwich & Bruce, 1990). Similarly, molecules that are clearly functionally and structurally identifiable as CHHs, may act as MIHs in the crayfish *Procambarus clarkii* and *P. bouvieri* (Yasuda *et al.*, 1994; Aguilar *et al.*, 1995). However, a recent study (Nagasawa *et al.*, 1996) has identified a functional MIH from *P. clarkii* sinus glands which clearly shows some similarity to brachyuran MIHs (Fig. 1a,b).

In view of the considerable importance of penaeid prawns in aquaculture, attention is currently directed towards identification of MIH and CHH-like molecules in this group. In a recent report, an MIH-like molecule has been isolated by 3′ and 5′ rapid amplification of complementary ends (RACE) of cDNA from *Penaeus vannamei* eyestalks (Sun, 1994). This too seems to be very similar to CHHs with respect to the structure of the mature peptide and the preprocessed precursor. The complete sequence of a functionally identified CHH and partial sequences of four other CHH-related peptides have been reported from the Kuruma prawn, *Penaeus japonicus* (Yang, Aida & Nagasawa, 1995), and these workers have recently fully sequenced a functional MIH in this species (Yang *et al.*, 1996) (Fig. 1c). With regard to these studies, it appears that, although several CHH-like neuropeptides which do

(a)

```
                 1          10          20          30          40
Cam MIH:  RVINDECPNLIGNRDLYKKVEWICEDCSNIFRKTGMASLC
Cas MIH:  RVINDDCPNLIGNRDLYKKVEWICDDCANIYRSTGMASLC
Cap MIH:  RVINDDCPNLIGNRDLYKKVEWICEDCSNIFRNTGMATLC

Prc MIH:  RYVFEECPGVMGNRAVHGKVTRVCEDCYNVFRDTDVLAGC

                       50          60          70          78
          RRNCFFNEDFLWCVHATERSEELRDLEEWVGILGAGRD
          RKDCFFNEDFLWCVRATERSSDLAQLKQWVTILGAGRI
          RKNCFFNEDFLWCVYATERTEEMSQLRQWVGILGAGRE

          RKGCFSSEMFKLCLLAMERVEEFPDFKRWIGILNA-NH₂
```

(b)

```
                   1          10          20          30          40
Hoa MIH:    pEVFDQACKGVYDRNLFKKLDRVCEDCYNLYRKPFVATTCR
Hoa CHH-A:  pEVFDQACKGVYDRNLFKKLDRVCEDCYNLYRKPFAATTCR
Hoa CHH-B:  pEVFDQACKGVYDRNLFKKLNRVCEDCYNLYRKPFIVTTCR
Prc CHH/  :pEVFDQACKGIYDRAIFKKLDRVCEDCYNLYRKPYVATTCR
Prb CHH

                       50          60          70
            ENCYSNWVFRQCLDDLLLSNVIDEYVSNVQMV-NH₂
            ENCYSNWVFRQCLDDLLLSDVIDEYVSNVQMV-NH₂
            ENCYSNRVFRQCLDDLLMIDVIDEYVSNVQMV-NH₂
            QNCYANSVFRQCLDDLLLIDVVDEYISGVQTV-NH₂
```

(c)

```
               1          10          20          30          40
Pej-CHH:     SLFDPACTGI-YDRQLLRKLGRLCDDCYNVFREPKVATGCR
Pej-SGP-I:   SLFDPSCTGV-FDRQLLRRLGRVCDDCFNVFREPNVATECR
Pej-SGP-II:  SLFDPSCTGV-FDRQLLRKLGRVCDDCFNVFREPNVAMECR
Pej-SGP-V:   IVFDPSCAGV-YDRVLLGKLNRLCDDCYDVFREPDVATECR
Pej-SGP-VI:  LVFDPSCAGV-YDRVLLGKLNRLCDDCYNVFREPNVATECR
Pev-MIH:     DTFDHSCKGI-YDRELFRKLDRVCEDCYNVFREPKVATECK
Pej-MIH:     SFIDNTCRGVMGNRDYNKKVVRVCEDCTNIFRLPGLDGMCR

                       50          60          70
            SNCYHNLIFLDCLEYLIPSHLQEEHMAAMQTV-NH2
            SNCYNNPV?RQCMAY....
            SNCYNNP....
            SNCYINI....
            SNC?YN?A?VQ....
            SNCFVNKRFNVCVADLRHDVSRFLKMANSALS
            NRCFYNEWFLICLKANREDEIEKFRVWISILNAGQ
```

not display MIH activity have been identified, a late eluting peptide (Pej-SGP-VI), exhibiting both CHH and MIH activities, appears to be structurally much more similar to the CHH peptides than to MIH (Fig. 1c). The significance of these findings is discussed in more detail later when I consider the roles of CHH in the control of moulting and reproduction.

Apart from the established roles of these neurohormones on ecdysteroid synthesis, there is evidence that non-peptidic compounds may have some importance. Several years ago, Soyez & Kleinholz (1977) demonstrated that a low molecular weight compound, possibly an indole alkylamine, isolated from eyestalk extracts of the shrimp *Pandalus jordani*, had moult-inhibiting activity in the amphipod *Orchestia traskiana*. More recently, Naya *et al.* (1988, 1989, 1991) demonstrated that two tryptophan metabolites, 3-hydroxykynurenine and xanthurenic acid, isolated from eyestalks of *C. sapidus*, inhibited moulting in eyestalk-ablated crayfish (*P. clarkii*), but were ineffective in lobsters (*H. americanus*) (for structures, see Fig. 2). These

---

Fig. 1 (a) Comparison of amino acid sequences of MIHs from brachyurans and an astacuran. Cam MIH (*Carcinus maenas* MIH; Webster, 1991), Cas MIH (*Callinectes sapidus* MIH; Lee *et al.*, 1995), Cap MIH (*Cancer pagurus* MIH; Chung *et al.*, 1996), Prc MIH (*Procambarus clarkii* MIH; Nagasawa *et al.*, 1996). **Bold** lettering indicates identical residues.

(b) Comparison of amino acid sequences of CHHs from astacurans, which, in addition to hyperglycemic activity, exhibit either MIH activity (Hoa MIH, *Homarus americanus* MIH: Chang *et al.*, 1990), Prc CHH/Prb CHH *Procambarus clarkii/bouvieri* CHH; Yasuda *et al.*, 1994, Huberman *et al.*, 1993), or GSH activity (Hoa CHH-B, *H. americanus* CHH-B; Tensen *et al.*, 1989, 1991). The sequence of CHH-A from *H. americanus*, obtained by cDNA cloning (Tensen, *et al.*, 1991) is included for comparison with the almost identical MIH isolated by Chang *et al.* (1990). **Bold** lettering indicates identical residues.

(c) Comparison of amino acid sequences of CHH-like and MIH-like neuropeptides in penaeid prawns. The complete sequence of a neuropeptide identified as a CHH in *Penaeus japonicus* (Pej-CHH, Yang *et al.*, 1995) can be compared with partial sequences of other CHH-like peptides from the same species which also show hyperglycemic activity; Pej-SGP-I, II, V, VI (Yang *et al.*, 1995, 1996). The sequence of a neuropeptide which has moult-inhibiting activity *in vitro*, designated as Pej-MIH (Yang *et al.*, 1996), previously identified as Pej-SGP-IV (Yang *et al.*, 1995) is not hyperglycemic, and shows greater identity with the brachyuran MIHs than with CHH-like molecules. The MIH-like molecule identified in *Penaeus vannamei* by cDNA cloning (Sun, 1994) clearly has a greater identity with CHH-like molecules than with MIHs. **Bold** lettering indicates identical residues. Gaps (-) have been introduced to maximize sequence similarity.

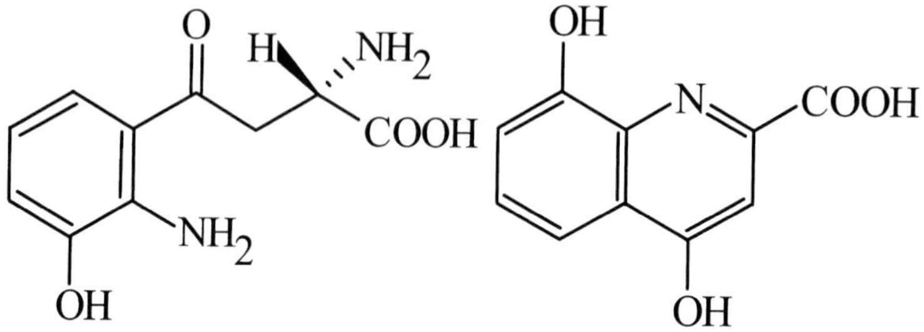

3-Hydroxy-L-kynurenine   Xanthurenic acid

Fig. 2. Structures of two tryptophan metabolites, 3-hydroxykynurenine and xanthurenic acid.

workers have also shown staggered correlations between xanthurenic acid and ecdysteroid levels in the Y-organs and haemolymph. Nevertheless, these correlations may be purely associative. It is possible that moulting and increased titres of xanthurenic acid are only linked with respect to season, because 3-hydroxykynurenine is a precursor of the ommochrome xanthommatin and seasonal photoperiodic changes might therefore be expected to lead to changes in rates of ommochrome biosynthesis.

Gonad-inhibiting hormone

The proposal that the eyestalk neurosecretory system of decapod crustaceans produces and secretes a neuropeptide which (in the female) inhibits vitellogenesis, the vitellogenesis- or gonad-inhibiting hormone (VIH/GIH) has long been accepted, primarily from the results of eyestalk removal experiments, where this operation may, according to season and physiological status of the animal, result in accelerated secondary vitellogenesis, due to increased uptake of vitellogenin (for reviews, see Adiyodi & Adiyodi, 1970; Adiyodi, 1985; Meusy & Payen, 1988). A major obstacle to the isolation and characterisation of VIH/GIH has been the lack of a suitable, rapid and sensitive bioassay. Several reports have used heterologous assays, where inhibition of ovarian growth or oocyte diameter in eyestalk-ablated animals injected with active factors, relative to destalked, saline-injected controls have been reported (Bomirski et al., 1981; Quackenbush & Herrnkind, 1983), and which could be used as a VIH/GIH bioassay. In particular, this type of assay has been used to isolate a VIH/GIH from H. americanus using the shrimp Palaemonetes varians as a test animal (Soyez, Van Deijnen & Martin, 1987).

```
                  1          10         20         30    *
Cap MOIH-1(-2)*:RRINNDCQNFIGNRAMYEKVDWICKDCANIFRKDGLLNN
Hoa VIH:        ASAWFTNDECPGVMGNRDLYEKVAWVCNDCANIFRNNDVGVM
Cap MIH:        RVINDDCPNLIGNRDLYKKVEWICEDCSNIFRNTGMATL

                 40         50         60         70
                CRSNCFYNTEFLWCIDATENTRNKEQLEQWAAILGAGWN
                CKKDCFHTMDFLWCVYATERHGEIDQFRKWVSILRA
                CRKNCFFNEDFLWCVYATERTEEMSQLRQWVGILGAGRE
```

Fig. 3. Comparison of amino acid sequences of MOIH-1 and -2 from *Cancer pagurus* (Wainright *et al.*, 1996) with MIH from the same species (Cap MIH, Chung *et al.*, 1996) and a VIH from *Homarus americanus* (Hoa VIH, Soyez *et al.*, 1991; De Kleijn *et al.*, 1994). **Bold** lettering indicates identical residues.

While heterologous *in vitro* assays which measure either uptake of colloidal gold-labelled vitellogenin by endocytosis (Jugan & Soyez, 1985) or endogenous synthesis of vitellogenins (Quackenbush & Keeley, 1988) and the possible inhibitory effect of VIH/GIH on these processes have been described, there is a real need for further development of similar types of assay in homologous systems, to facilitate the identification of further VIHs. Unfortunately, because vitellogenesis is prolonged in most large decapod crustaceans, this may be difficult. Consequently, crustaceans with rapid vitellogenic cycles, such as the caridean or penaied shrimps may be useful models for future research.

Despite the shortcomings of existing VIH/GIH assays, the characterisation of a VIH/GIH in *H. americanus* (Soyez *et al.*, 1987) has led to complete sequencing of this molecule using conventional microsequencing methods (Soyez *et al.*, 1991) and by cDNA cloning (De Kleijn *et al.*, 1994). It was found to consist of a 78 residue peptide (M_r 9135), with an amidated C-terminus. Surprisingly, this molecule showed considerable identity (53%) to *C. maenas* MIH (Webster, 1991) (Fig. 3), and because the preprocessed precursors for both molecules do not contain the CPRP present in all CHHs, it was readily apparent that these peptides belong to a separate subgroup within the CHH family. An interesting feature of *H. americanus* VIH/GIH is the existence of two isoforms of identical sequence and mass (Soyez *et al.*, 1991). This phenomenon has been observed for astacuran CHHs, where it is attributable to a stereoinversion of L-Phe[3] to the D-configuration (Soyez *et al.*, 1994; Yasuda *et al.*, 1994; Aguilar *et al.*, 1995). Thus it is entirely feasible that a similar stereoinversion may occur in VIH/GIH. The only other detailed study characterising a VIH/GIH has been reported by Aguilar *et al.* (1992), showing that in the Mexican crayfish *P. bouvieri* a SG neuropeptide

(M_r 8388±2), apparently structurally related to a CHH, inhibits endogenous vitellogenin synthesis (incorporation of ^{3}H-labelled leucine into yolk proteins) by *Penaeus setiferus* ovaries, albeit at a rather high concentration (10 SG equivalents).

For *H. americanus* VIH/GIH, immunochemical and *in situ* hybridisation studies have shown that VIH/GIH immnoreactivity and mRNA expression may sometimes be co-localised with that of CHH in both adults (Kallen & Meusy, 1989; De Kleijn *et al.*, 1992) and larvae (Rotllant *et al.*, 1993). Although the mechanisms involved in (possible) differential expression of both peptides, for example by alternative splicing, are not yet known, the finding of VIH/GIH in males, females and larvae would suggest a multiplicity of physiologically relevant roles. Thus, it now seems more appropriate to refer to this molecule as a GIH.

Crustacean hyperglycemic hormone

The CHHs are without doubt the best-known crustacean neuropeptides with regard to structure and function. Many CHHs have been structurally identified in the last few years (for reviews, see Chang, 1995; De Kleijn & Van Herp, 1995; Van Herp, this volume). Functionally, it is well known that CHHs regulate glucose levels by binding to membrane-bound receptors on the hepatopancreas and abdominal muscle (Kummer & Keller, 1993), leading to production of cAMP and cGMP, which in turn bind to protein kinases, activating phosphorylase and inhibiting glucose synthase (for reviews, see Sedlmeier, 1985; Keller & Sedlmeier, 1987). CHH may also act as a secretagogue in that amylase release from the hepatopancreas is stimulated by this hormone (Sedlmeier, 1988).

In recent years, it has become increasingly apparent that CHH has several more roles in decapods, particularly with regard to the regulation of moulting, and possibly reproduction. Indeed, because both of these processes are highly energy demanding, the suggestion of involvement of CHH might seem obvious and facile. The first indication that CHH might be of relevance in moult control was the finding that CHH inhibited ecdysteroid synthesis of *C. maenas* Y-organs, although CHH was much less effective than MIH in this respect (Webster & Keller, 1986). Subsequently, high affinity CHH receptors were found on Y-organ membrane preparations, which appeared to be similar if not identical to those on hepatopancreas membrane preparations (Webster, 1993), thus suggesting a physiologically relevant role for CHH in regulation of ecdysteroid synthesis.

Experiments have recently been performed to see whether MIH and CHH, both of which repress ecdysteroid synthesis, act synergistically. The results of a preliminary series of experiments are shown in Fig. 4. Synergism

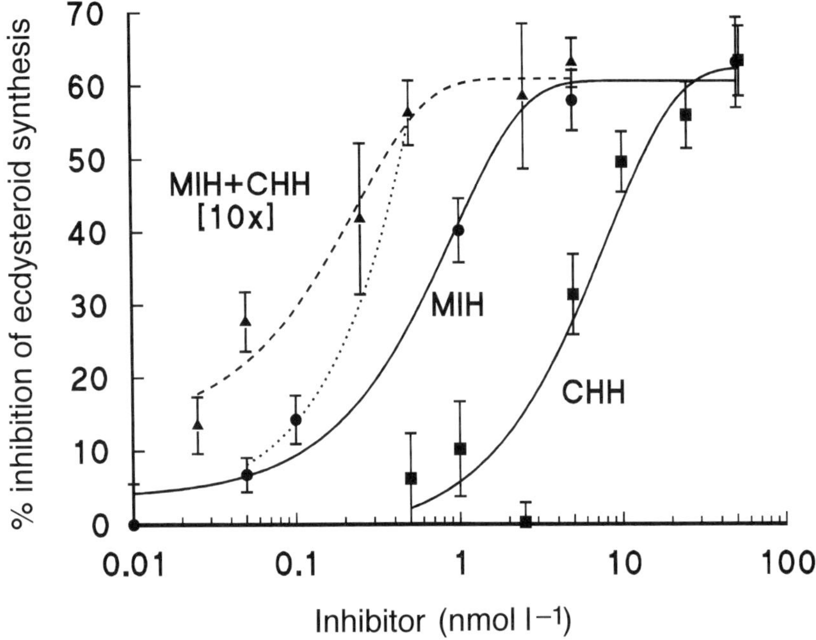

Fig. 4. Possible synergism between MIH and CHH in repressing ecdysteroid synthesis. *Carcinus maenas* Y-organs were exposed to varying concentrations of MIH, CHH, or both hormones (dashed line) at an MIH: CHH ratio of 1:10, using previously described bioassay procedures (Webster, 1986). $N=5$ to 10 Y-organ pairs for each point, bars$=\pm$SEM. The theoretical additive effect (dotted line) is to the right of the observed dose–response curve, which suggests that the hormones may act synergistically (S. G. Webster, unpublished results).

seems to occur because the inhibitory response of a mixture of MIH and CHH, mimicking the proportions of both in the SG, inhibit ecdysteroid synthesis in a greater than additive manner. However, the biological significance of this phenomenon is as yet obscure.

With regard to other crustaceans, a peptide which has been characterised as a functional MIH in *H. americanus* (Chang *et al.*, 1990) is also functionally hyperglycemic, and may be identical to CHH-A in this species (Tensen *et al.*, 1991). Similarly, a CHH which represses ecdysteroid synthesis has been identified in *P. clarkii* (Yasuda *et al.*, 1994), although a recent report has identified a functional MIH, which has some structural similarities with brachyuran MIHs in this species (Nagasawa *et al.*, 1996). In penaeids, as alluded to earlier, the situation appears somewhat similar, in the sense that both CHH and MIH-like molecules seem to have moult-inhibiting activity (Sun, 1994; Yang *et al.*, 1996).

With regard to the possible involvement of CHH in reproduction, little can as yet be said with certainty, but the recent finding of low-level transcription of mRNAs coding for CHH-A and B in ventral nerve cord of *H. americanus* (De Kleijn *et al.*, 1995) may be of significance. There are numerous reports suggesting, on the basis of simple implantation or injection experiments, that various parts of the nervous system including the brain, thoracic and abdominal ganglia are sites of production of a gonad-stimulating hormone (GSH) which accelerates ovarian maturation (Ōtsu, 1963; Gomez, 1965; Hinsch & Bennett, 1979; Eastman-Reks & Fingerman, 1984; Takayanagi, Yamamoto & Takeda, 1986; Yano *et al.*, 1988). Although the biological significance of these observations is uncertain in view of experimental procedures, which often involved heterologous bioassays, and administration of large physiologically irrelevant doses, they may have some relevance in view of the above-mentioned study, and that by Tensen *et al.* (1989), who demonstrated that a CHH from HPLC-fractionated SGs of *H. americanus* stimulated occyte growth in *P. varians*, albeit at a high dose. Recent experiments using [^{125}I]CHH in receptor assays have consistently shown that crude membrane preparations of previtellogenic oocytes of *C. pagurus* and *C. maenas* possess many high affinity binding sites for CHH which dramatically decrease in number during vitellogenesis (J. S. Chung & S. G. Webster, unpublished results). Nevertheless, more, critical work needs to be done before any gonad-stimulating effect of CHH can be firmly established.

Mandibular organ-inhibiting hormone

As befits their common ancestry, insects and crustaceans share many developmental, anatomical and physiological features, not least with regard to many common endocrine mechanisms. This common theme is well illustrated by the finding that crustaceans possess the unepoxidated precursor of insect juvenile hormone (JH III), methyl farnesoate (MF) (for structures, see Fig. 5), which has now been measured in the haemolymph of more than a dozen species (Laufer *et al.* 1987a; Laufer & Borst, 1988; Tobe, Young & Khoo, 1989; Sagi, Homola & Laufer, 1991; Tsukimura & Borst, 1992). MF is synthesised by the mandibular organ (MO) (Laufer *et al.*, 1987a,b; Borst *et al.*, 1987; Tsukimura & Borst, 1992), which although described as an endocrine tissue by Le Roux (1968) had no known function until this point. In view of the established roles of JH III in insect reproduction, it has been tacitly assumed that MF would have a similar function in crustaceans. At present, the balance of evidence would seem to favour a role in reproduction,

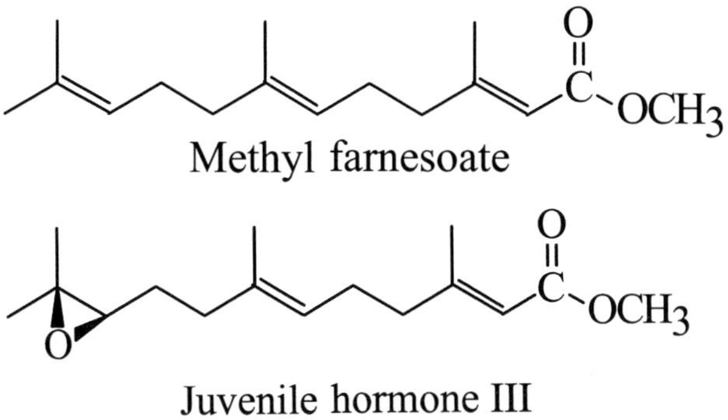

Fig. 5. Structure of methyl farnesoate (crustaceans) and juvenile hormone III (insects).

because MF titres and biosynthesis can be correlated with ovarian maturation in the spider crab, *Libinia emarginata* (Laufer *et al.*, 1987a), and morphological changes in ovarian follicle cells reminiscent of JH-stimulated patency in insects (Davey & Huebner, 1974) are observed in crabs injected with JH mimics (Hinsch, 1981). A comprehensive account of the evidence supporting a role in crustacean reproduction has been presented by Laufer, Ahl & Sagi (1993). An interesting effect of MF is its action as an ecdysiotropin. Tamone & Chang (1993) have shown that ecdysteroid synthesis by Y-organs from the Dungeness crab, *Cancer magister*, is increased during incubation with MO conditioned medium or with MF. Additionally, during premoult, the MO undergoes dramatic ultrastructural changes indicative of increased synthetic activity (Aoto, Kamiguchi & Hisano, 1974; Demeusy, 1975), perhaps suggesting that the large increases in haemolymph ecdysteroid titre observed during premoult may, at least in part, be due to an ecdysiotropic action of MF.

While this hypothesis and those concerning the role of MF in reproduction need further corroboration using other crustacean models, the inhibitory action of an eyestalk neuropeptide upon the MOs is relatively clear. It has long been known that eyestalk removal leads to hypertrophy of the MO (Byard, Shivers & Aiken, 1975; Bazin, 1976; Hinsch, 1977). More recently it has been demonstrated that eyestalk removal results in increased MF haemolymph titres in *L. emarginata* (Laufer *et al.*, 1986, 1987b), *H. americanus* and *Orconectes virilis* (Tsukimura *et al.*, 1989), and that SG extracts reduce the increase in circulating MF that result from eyestalk ablation (Tsukimura & Borst, 1992). Although initial experiments to characterise the inhibitory neuropeptide suggested that pigment-dispersing hormone (PDH) and red pigment-concentrating hormone (RPCH), respectively inhibited or promoted MF synthesis by *P. clarkii* MO *in vitro* (Landau, Laufer & Homola, 1989), recent studies have shown that a novel eyestalk neuropeptide, named mandibular organ-inhibiting hormone (MOIH) is the most likely inhibitor of MF synthesis. Using a bioassay based upon inhibition of MF synthesis by MO *in vitro*, Wainwright *et al.* (1996) have identified two MOIHs from HPLC-purified SG of *C. pagurus*, and these have been fully sequenced using a combination of automated Edman degradation of endoproteinase-derived fragments and electrospray mass spectrometry. Both peptides (M_r 9235; 78 amino acid residues) have free N- and C-termini and three intrachain disulphide bridges, and are almost identical with the exception of a Lys → Gln substitution at position 33. Although this difference could be accounted for by a single base substitution (A-C), and might suggest allelic polymorphism, HPLC analysis of SG from single crabs always shows that both peptides are present. Alignment of the MOIH sequence with other members of the CHH neuropeptide family clearly shows that MOIH is most closely related to MIHs (50–60%). The sequence alignment of *C. pagurus* MIH and MOIHs is shown in Fig. 3. Despite these similarities, MIH is inactive in the MOIH bioassay, and MOIH has a somewhat limited and variable activity in the MIH assay. Thus, these two very closely related peptides seem to be functionally distinct. With regard to the site of biosynthesis of MOIH, preliminary results (P. G. Withers, S. G. Webster & H. H. Rees., unpublished observations) using affinity-purified polyclonal antisera raised against both MIH and MOIH from *C. pagurus* clearly indicate that these neuropeptides are completely co-localised: both are expressed in neurones in the X-organ previously identified as being MIH neurones in brachyurans (Dircksen *et al.*, 1988). It now remains to be determined whether generalities regarding the structure and site of expression of MOIHs can be confirmed in a comparative context, because it is quite possible that other related molecules (CHH, MIH) may act as MOIHs in other species. Indeed, a preliminary report (Laufer, Liu & Van Herp, 1994) indicated that CHHs could act as MOIH in the crayfish *P. clarkii*, and further studies (Liu & Laufer, 1996) have identified three MOIHs in *L. emarginata*, one of which is structurally a CHH. Clearly, these findings now need confirmation in other crustacean models, to determine their generality in a comparative context.

Conclusions and perspectives

In the past ten years, it has become increasingly apparent that moulting and reproduction in crustaceans are ultimately controlled by inhibitory neuropeptides that act upon the endocrine tissues producing hormones which stimulate these events (ecdysteroids, juvenoids). While, *sensu lato*, these latter hormones control comparable processes in insects, as might be expected given the common ancestry of crustaceans and insects, the remarkable diversity of processes controlled by the CHH group of peptides is noteworthy. From a comparative viewpoint, the inhibitory activity of MIH and CHH upon ecdysteroid synthesis contrasts vividly with the stimulatory activity of prothoracicotrophic hormones in insects, and while juvenoid synthesis is inhibited by allatostatins in insects, the equivalent process appears to be inhibited by CHH group peptides (MOIH and CHH) in crustaceans. As detailed in the chapter by Phillips *et al.* (this volume), a CHH-related peptide may play an important role as an antidiuretic hormone in orthopteran insects, again highlighting the functional diversity of these peptides.

While many more CHH-like peptides will undoubtedly be identified in the next few years, the most challenging questions in crustacean endocrinology have scarcely begun to be addressed. Recent research highlighted in this review has given tantalising glimpses of complex multi-hormonal control mechanisms involved in moulting and reproduction, yet little is known about the physiological significance of any of the CHH group neuropeptides in these processes in decapods.

Acknowledgements

Work performed in my laboratory was supported by funding from the Royal Society and the Biotechnology and Biological Sciences Research Council.

References

Adiyodi, R. G. (1985). Reproduction and its control. In *The Biology of Crustacea*, ed. D. E. Bliss, pp. 147–215. New York: Academic Press.

Adiyodi, K. G. & Adiyodi, R. G. (1970). Endocrine control of reproduction in decapod Crustacea. *Biological Reviews of the Cambridge Philosophical Society* **45**, 121–165.

Aguilar, M. B., Quackenbush, L. S., Hunt, D. T., Shabanowitz, J. & Huberman, A. (1992). Identification, purification and initial characterization of the vitellogenesis-inhibiting hormone from the Mexican crayfish *Procambarus bouvieri* (Ortmann). *Comparative Biochemistry and Physiology* **102B**, 491–498.

Aguilar, M. B., Soyez, D., Falchetto, R., Arnott, D., Shabanowitz, J., Hunt, D. F. & Huberman, A. (1995). Amino acid sequence of the minor isomorph of the crustacean hyperglycemic hormone (CHH-II) of the Mexican crayfish *Procambarus bouvieri* (Ortmann): Presence of a D-amino acid. *Peptides* **16**, 1375–1383.

Aoto, T., Kamiguchi, Y. & Hisano, S. (1974). Histological and ultra-structural studies on the Y-organ and mandibular organ of the fresh-water prawn *Palaemon paucidens*, with special reference to their relation with the molting cycle. *Journal of the Faculty of Science of Hokkaido University* **19**, 295–308.

Bazin, F. (1976). Mise en évidence des charactères cytologiques des glandes stéroïdogènes dans les glandes mandibulaires et le glandes Y du crabe *Carinus maenas* (L.) normal et épédonculé. *Computes Rendus de l'Académie des Sciences Paris, Série D* **282**, 739–741.

Bomirski, A., Arendarczyk, M., Kawinska, E. & Kleinholz, L. H. (1981). Partial characterization of crustacean gonad-inhibiting hormone. *International Journal of Invertebrate Reproduction* **3**, 213–219.

Borst, D. W., Laufer, H., Landau, M., Chang, E. S., Hertz, W. A., Baker, F. C. & Schooley, D. A. (1987). Methyl farnesoate and its role in crustacean reproduction and development. *Insect Biochemistry* **17**, 1123–1127.

Byard, E. H., Shivers, R. R. & Aiken, D. E. (1975). The mandibular organ of the lobster *Homarus americanus*. *Cell and Tissue Research* **162**, 13–22.

Chang, E. S. (1985). Hormonal control of molting in decapod Crustacea. *American Zoologist* **25**, 179–185.

Chang, E. S. (1995). Physiological and biochemical changes during the moult cycle in decapod crustaceans – an overview. *Journal of Experimental Marine Biology and Ecology* **193**, 1–14.

Chang, E. S. & Bruce, M. J. (1980). Ecdysteroid titers of juvenile lobsters following molt induction. *Journal of Experimental Zoology* **214**, 157–160.

Chang, E. S., Prestwich, G. D. & Bruce, M. J. (1990). Amino acid sequence of a peptide with both molt-inhibiting and hyperglycemic activities in the lobster, *Homarus americanus*. *Biochemical and Biophysical Research Communications* **171**, 818–826.

Chang, E. S., Sage, B. A. & O'Connor, J. D. (1976). The qualitative and quantitative determination of ecdysones in tissues of the crab, *Pachygrapsus crassipes*, following molt induction. *General and Comparative Endocrinology* **30**, 21–33.

Chung, J. S., Wilkinson, M. C. & Webster, S. G. (1996). Determination of the amino acid sequence of the moult-inhibiting hormone from the edible crab, *Cancer pagurus*. *Neuropeptides* **30**, 95–101.

Davey, K. G. & Huebner, E. (1974). The response of the follicle cells of *Rhodnius prolixus* to juvenile hormone and antigonadotropin in vitro. *Canadian Journal of Zoology* **52**, 1407–1412.

De Kleijn, D. P. V., Coenen, A. J. M., Laverdure, A. M., Tensen, C. P. & Van Herp, F. (1992). Localization of mRNAs encoding crustacean hyper-glycemic hormone (CHH) and gonad-inhibiting hormone (GIH) in the X-organ sinus gland complex of the lobster *Homarus americanus*. *Neuroscience* **51**, 121–128.

De Kleijn, D. P. V., De Leeuw, E. P. H., Van Den Berg, M. C., Martens, G. J. M. & Van Herp, F. (1995). Cloning and expression of two mRNAs encoding structurally different crustacean hyperglycemic hormone precursors in the lobster *Homarus americanus. Biochemica et Biophysica Acta* **1260**, 62–66.

De Kleijn, D. P. V., Sleutels, F. J. G. T., Martens, G. J. M. & Van Herp, F. (1994). Cloning and expression of mRNA encoding prepro-gonad-inhibiting hormone (GIH) in the lobster *Homarus americanus. FEBS Letters* **353**, 255–258.

De Kleijn, D. P. V. & Van Herp, F. (1995). Molecular biology of neurohormone precursors in the eyestalk of Crustacea. *Comparative Biochemistry and Physiology* **112B**, 573–579.

Demeusy, N. (1975). Observations sur le functionnement des glandes mandibularies du Décapode brachoure *Carcinus maenas* L.: animaux témoins et animaux sans pédoncles ocularies. *Comptes Rendus de l'Académie des Sciences Paris, Serié D* **281**, 1887–1889.

Dircksen, H., Webster, S. G. & Keller, R. (1988). Immunocytochemical demonstration of the neurosecretory systems containing putative molt-inhibiting hormone and hyperglycemic hormone in the eyestalk of brachyuran crustaceans. *Cell and Tissue Research* **251**, 3–12.

Drach, P. (1955). Système endocrinien pédonculaire, durée d'intermue et vitellogénèse chez *Leander serratus* (Penn.). *Comptes Rendus des Séances de la Société de Biologie* **149**, 2079–2083.

Eastman-Reks, S. & Fingerman, M. (1984). Effects of neuroendocrine tissue and cyclic AMP on ovarian growth *in vivo* and *in vitro* in the fiddler crab, *Uca pugilator. Comparative Biochemistry and Physiology* **79A**, 679–684.

Gomez, R. (1965). Acceleration of development of gonads by implantation of brain in the crab *Paratelphusa hydromonous. Naturwissenschaften* **9**, 216.

Hinsch, G. W. (1977). Fine structural changes in the mandibular gland of the male spider crab, *Libinia emarginata* (L.) following eyestalk ablation. *Journal of Morphology* **145**, 179–187.

Hinsch, G. W. (1981). Effects of juvenile hormone mimics on the ovary in the immature spider crab, *Libinia emarginata. International Journal of Invertebrate Reproduction* **3**, 237–244.

Hinsch, G. W. & Bennett, D. C. (1979). Vitellogenesis stimulated by thoracic ganglion implants into destalked immature spider crabs, *Libinia emarginata. Tissue and Cell* **11**, 345–351.

Hopkins, P. M. (1983). Patterns of serum ecdysteroids during induced and uninduced proecdysis in the fiddler crab, *Uca pugilator. General and Comparative Endocrinology* **52**, 350–356.

Hopkins, P. M. (1986). Ecdysteroid titers and Y-organ activity during late anecdysis and proecdysis in the fiddler crab, *Uca pugilator. General and Comparative Endocrinology* **63**, 362–372.

Huberman, A., Aguilar, M. B., Brew, K., Shabanowitz, J. & Hunt, D. F. (1993). Primary structure of the major isoform of the crustacean hyperglycemic hormone (CHH-1) from the sinus gland of the Mexican crayfish *Procambenis* bouvieri (Ortmann): interspecies comparison. *Peptides* **14**, 7–16.

Jegla, T. C., Ruland, K., Kegel, G. & Keller, R. (1983). The role of the Y-organ and cephalic gland in ecdysteroid production and control of molting in the crayfish *Orconectes limosus*. *Journal of Comparative Physiology B* **152**, 91–95.

Jugan, P. & Soyez, D. (1985). Démonstration in vitro de l'inhibition de l'endocytose ovocytaire par un extrait de glandes du sinus chez la Crevette *Macrobrachium rosenbergii*. *Comptes Rendus de l'Académie des Sciences, Paris* **300**, 705–709.

Kallen, J. L. & Meusy, J. J. (1989). Do the neurohormones VIH (vitellogenesis-inhibiting hormone) and CHH (crustacean hyperglycemic hormone) of crustaceans have a common precursor? Immunolocalization of VIH and CHH in the X-organ sinus gland complex of the lobster *Homarus americanus*. *Invertebrate Reproduction and Development* **16**, 43–52.

Kegel, G., Reichwein, B., Weese, S., Gaus, G., Peter-Katalinic, J. & Keller, R. (1989). Amino acid sequence of the crustacean hyperglycemic hormone (CHH) from the shore crab, *Carcinus maenas*. *FEBS Letters* **255**, 10–14.

Keller, R. (1992). Crustacean neuropeptides: structures, functions and comparative aspects. *Experientia* **48**, 439–448.

Keller, R. & Schmid, E. (1979). In vitro secretion of ecdysteroids by Y-organs and lack of secretion by mandibular organs of the crayfish following molt induction. *Journal of Comparative Physiology* **B 130**, 347–353.

Keller, R. & Sedlmeier, D. (1987). A metabolic hormone in crustaceans: the hyperglycemic neuropeptide. In *Invertebrate Endocrinology*, Vol. 2, *Endocrinology of Selected Invertebrate Types*, ed. H. Laufer & R. G. H. Downer, pp. 315–326. New York: A. R. Liss Inc.

Kleijn, J. M., Mangerich, S., De Kleijn, D. P. V., Keller, R. & Weidemann, W. M. (1993). Molecular cloning of crustacean molt-inhibiting hormone (MIH) precursor. *FEBS Letters* **334**, 139–142.

Kummer, G. & Keller, R. (1993). High-affinity binding of crustacean hyperglycemic hormone (CHH) to hepatopancreatic plasma membranes of the crab *Carcinus maenas* and the crayfish *Orconectes limosus*. *Peptides* **14**, 103–108.

Lachaise, F., Goudeau, M., Carpentier, G., Saïdi, B. & Goudeau, H. (1992). Eyestalk albation in female crabs: effects on egg characteristics, *Journal of Experimental Zoology* **261**, 86–96.

Landau, M., Laufer, H. & Homola, E. (1989). Control of methyl farnesoate in the mandibular organ of the crayfish *Procambarus clarkii*: evidence for peptide neurohormones with dual functions. *Invertebrate Reproduction and Development* **16**, 165–168.

Laufer, H. Ahl, J. S. B. & Sagi, A. (1993). The role of juvenile hormones in crustacean reproduction. *American Zoologist* **33**, 365–374.

Laufer, H. & Borst, D. W. (1988). Juvenile hormone in Crustacea. In *Invertebrate Endocrinology*, Vol. 2, *Endocrinology of Selected*

Invertebrate Types, ed. H. Laufer & R. G. H. Downer, pp. 305–313. New York: A. R. Liss Inc.

Laufer, H., Landau, M., Borst, D. & Homola, E. (1986). The synthesis and regulation of methyl farnesoate, a new juvenile hormone for crustacean reproduction. In *Advances in Invertebrate Reproduction 4*, ed. M. Porchet, J. C. Andries & A. Dhainaut, pp. 135–143. Amsterdam: Elsevier.

Laufer, H., Borst, D., Baker, F. C., Carrasco, C., Sinkus, M., Reuter, C. C., Tsai, L. W. & Schooley, D. A. (1987a). Identification of a juvenile hormone-like compound in a crustacean. *Science* **235**, 202–205.

Laufer, H., Landau, M., Homola, E. & Borst, D. W. (1987b). Methyl farnesoate: its site of synthesis and regulation of secretion in a juvenile crustacean. *Insect Biochemistry* **17**, 1129–1131.

Laufer, H., Liu, L. & Van Herp, F. (1994). A neuropeptide family that inhibits the mandibular organ of Crustacea and may regulate reproduction. In *Insect Neurochemistry and Neurophysiology*, ed A. J. Borkovec & M. J. Loeb, pp. 203–206. Boca Raton: CRC Press.

Lee, K. J., Elton, T. S., Bej, A. K., Wattes, S. A. & Watson, R. D. (1995). Molecular cloning of a cDNA encoding putative molt-inhibiting hormone from the blue crab, *Callinectes sapidus*. *Biochemical and Biophysical Research Communications* **209**, 1126–1131.

Le Roux, A. (1968). Description d'organes mandibularies nouveaux chez les Crustacés Décapodes. *Comptes Rendus de l'Académie des Sciences, Paris, Séries D* **266**, 1317–1320.

Liu, L. & Laufer, H. (1996). Isolation and characterization of sinus gland peptides with both mandibular organ inhibiting and hyperglycemic effects from the spider crab *Libinia emarginata*. *Archives of Insect Biochemistry and Physiology* **32**, 375–385.

Meusy, J.-J. & Payen, G. G. (1988). Female reproduction in malacostracan Crustacea. *Zoological Science* **5**, 217–265.

Nagasawa, H., Yang, W.-J., Shimizu, H., Aida, K., Tsutsumi, H., Terauchi, A. & Sonobe, H. (1996). Isolation and amino acid sequence of a molt-inhibiting hormone from the American crayfish, *Procambarus clarkii*. *Bioscience, Biotechnology and Biochemistry* **60**, 554–556.

Naya, K., Kishida, K., Sugiyama, M., Murata, M., Miki, W., Ohnishi, M. & Nakanishi, K. (1988). Endogenous inhibitor of ecdysone synthesis in crabs. *Experientia* **44**, 50–52.

Naya, Y., Ohnishi, M., Ikeda, M., Miki, W. & Nakanishi, K. (1989). What is molt-inhibiting hormone? The role of an ecdysteroidogenesis inhibitor in the crustacean molting cycle. *Proceedings of the National Academy of Sciences USA* **86**, 6826–6829.

Naya, K., Ohnishi, M., Ikeda, M., Miki, W. & Nakanishi, K. (1991). Physiological role of 3-hydroxykynurenine and xanthenuric acid upon crustacean molting. In *Kynurenine and Serotonin Pathways*, ed. R. Schwarcz, pp. 309–318. New York: Plenum Press.

Ōtsu, T. (1963). Bihormonal control of the sexual cycle in the freshwater crab, *Potamon dehaani*. *Embryologia* **8**, 1–20.

Panouse, J. B. (1946). Reserches sur les phénomènes humoraux chez les crustacés. L'adaptation chromatique et la croissance ovarienne chez le crevette *Leander serratus. Annales de l'Institut Oceanographique, Monaco* **23**, 65–147.

Passano, L. M. (1953). Neurosecretory control of molting in crabs by the X-organ sinus gland complex. *Physiologia Comparata et Oecologia* **3**, 155–189.

Quackenbush, L. S. & Herrnkind, W. F. (1983). Partial characterization of eyestalk hormones controlling molt and gonadal development in the spiny lobster *Panulirus argus. Journal of Crustacean Biology* **3**, 34–44.

Quackenbush, L. S. & Keeley, L. L. (1988). Regulation of vitellogenesis in the fiddler crab, *Uca pugilator. Biological Bulletin* **175**, 321–331.

Rotllant, G., De Kleijn, D. P. V., Charmantier-Daures, M., Charmantier, G. & Van Herp, F. (1993). Localization of crustacean hyperglycemic hormone (CHH) and gonad-inhibiting hormone (GIH) in the eyestalk of *Homarus gammarus* larvae by immunocytochemistry and in situ hybridization. *Cell and Tissue Research* **271**, 507–512.

Sagi, A., Homola, E. & Laufer, H. (1991). Methyl farnesoate in the prawn *Macrobrachium rosenbergii*: synthesis by the mandibular organ in *in vitro*, and titers in the hemolymph. *Comparative Biochemistry and Physiology* **99B**, 879–882.

Sedlmeier, D. (1985). Mode of action of the crustacean hyperglycemic hormone. *American Zoologist* **25**, 223–232.

Sedlmeier, D. (1988). The crustacean hyperglycemic hormone releases amylase from the crayfish midgut gland. *Regulatory Peptides* **20**, 91–98.

Soumoff, C. & O'Connor, J. D. (1982). Repression of Y-organ secretory activity by molt-inhibiting hormone in the crab *Pachygrapsus crassipes. General and Comparative Endocrinology* **48**, 432–439.

Soyez, D. & Kleinholz, L. H. (1977). Molt-inhibiting factor from the crustacean eyestalk. *General and Comparative Endocrinology* **31**, 233–242.

Soyez, D., Van Deijnen, J. E. & Martin, M. (1987). Isolation and characterization of a vitellogenesis-inhibiting factor from sinus glands of the lobster, *Homarus americanus. Journal of Experimental Zoology* **244**, 479–484.

Soyez, D., Le Caer, J.-P., Noël, P. Y. & Rossier, J. (1991). Primary structure of two isoforms of the vitellogenesis-inhibiting hormone from the lobster *Homarus americanus. Neuropeptides* **20**, 25–32.

Soyez, D., Van Herp, F., Rossier, J., Le Caer, J.-P., Tensen, C. P. & Lafont, R. (1994). Evidence for a conformational polymorphism of invertebrate neurohormones; D-amino acid in crustacean hyperglycemic peptides. *Journal of Biological Chemistry* **269**, 18 295–18 298.

Sun, P. S. (1994). Molecular cloning and sequence analysis of a cDNA encoding a molt-inhibiting hormone-like neuropeptide from the white shrimp *Penaeus vannamei. Molecular Marine Biology and Biotechnology* **3**, 1–6.

Takayanagi, H., Yamamoto, Y. & Takeda, N. (1986). An ovary-stimulating factor in the shrimp, *Paratya compressa. Journal of Experimental Zoology* **240**, 203–209.

Tamone, S. L. & Chang, E. S. (1993). Methyl farnesoate stimulates ecdysteroid secretion from crab Y-organs *in vitro*. *General and Comparative Endocrinology* **89**, 425–432.

Tensen, C. P., Janssen, K. P. C. & Van Herp, F. (1989). Isolation, characterization and physiological specificity of the crustacean hyperglycemic factors from the sinus gland of the lobster, *Homarus americanus* (Milne-Edwards). *Invertebrate Reproduction and Development* **16**, 155–164.

Tensen, C. P., De Kleijn, D. P. V. & Van Herp, F. (1991). Cloning and sequence analysis of cDNA encoding two crustacean hyperglycemic hormones from the lobster *Homarus americanus*. *European Journal of Biochemistry* **200**, 103–106.

Tobe, S. S., Young, D. A. & Khoo, H. W. (1989). Production of methyl farnesoate by the mandibular organs of the mud crab, *Scylla serrata*: validation of the radiochemical assay. *General and Comparative Endocrinology* **73**, 342–353.

Tsukimura, B. & Borst, D. W. (1992). Regulation of methyl farnesoate in the hemolymph and mandibular organ of the lobster, *Homarus americans*. *General and Comparative Endocrinology* **86**, 297–303.

Tsukimura, B., Martin, M., Frinsko, M. & Borst, D. W. (1989). Measurement of methyl farnesoate (MF) levels in crustacean hemolymph. *American Zoologist* **29**, 49A.

Wainwright, G., Webster, S. G., Wilkinson, M. C., Chung, J. S. & Rees, H. H. (1996). Structure and significance of mandibular organ-inhibiting hormone in the crab, *Cancer pagurus*; involvement in multihormonal regulation of growth and reproduction. *Journal of Biological Chemistry* **271**, 12749–12754.

Webster, S. G. (1986). Neurohormonal control of ecdysteroid biosynthesis by *Carcinus maenas* Y-organs *in vitro*, and preliminary characterization of the putative molt-inhibiting hormone (MIH). *General and Comparative Endocrinology* **61**, 237–247.

Webster, S. G. (1991). Amino acid sequence of putative moult-inhibiting hormone from the crab *Carcinus maenas*. *Proceedings of the Royal Society. London B* **244**, 247–252.

Webster, S. G. (1993). High-affinity binding of putative moult-inhibiting hormone (MIH) and crustacean hyperglycaemic hormone (CHH) to membrane bound receptors on the Y-organ of the shore crab *Carcinus maenas*. *Proceedings of the Royal Society of London B* **251**, 53–59.

Webster, S. G. & Keller, R. (1986). Purification, characterisation and amino acid composition of the putative moult-inhibiting hormone (MIH) of *Carcinus maenas* (Crustacea, Decapoda). *Journal of Comparative Physiology B* **156**, 617–624.

Yang, W.-J., Aida, K. & Nagasawa, H. (1995). Amino acid sequences of a hyperglycemic hormone and its related peptides from the Kuruma prawn, *Penaeus japonicus*. *Aquaculture* **135**, 205–212.

Yang, W.-J., Aida, K., Terauchi, A., Sonobe, H. & Nagasawa, H. (1996). Amino acid sequence of a peptide with molt-inhibiting activity from the Kuruma prawn *Penaeus japonicus*. *Peptides* **17**, 197–202.

Yano, I., Tsukimura, B., Sweeney, J. N. & Wyban, J. A. (1988). Induced ovarian maturation of *Penaeus vannamei* by implantation of lobster ganglion. *Journal of the World Aquaculture Society* **19**, 204–209.

Yasuda, A., Yasuda, Y., Fujita, T. & Naya, Y. (1994). Characterization of crustacean hyperglycemic hormone from the crayfish (*Procambarus clarkii*): multiplicity of molecular forms by stereoinversion and diverse functions. *General and Comparative Endocrinology* **95**, 387–398.

FRANÇOIS VAN HERP

Molecular, cytological and physiological aspects of the crustacean hyperglycemic hormone family

Introduction

In malacostracan crustaceans, neuroendocrine centres are found in the eye-stalk, brain, suboesophageal ganglion and throughout the remainder of the central nervous system. The neurohaemal organs, supplied by several groups of neuroendocrine cells are: the sinus gland which stores and releases neuropeptides produced by the so-called X-organ; the post-commissural organ and pericardial organs, which are respectively innervated by neurosecretory perikarya in the brain and thoracic ganglia. The X-organ sinus gland complex produces several neuropeptides belonging to two functionally different families, firstly the chromatophorotropins, including the red pigment-concentrating hormone (RPCH) and the pigment-dispersing hormone (PDH), secondly members of the crustacean hyperglycemic hormone (CHH) family, which include not only functionally defined CHHs, but also the moult-inhibiting hormone (MIH), the gonad- or vitellogenin-inhibiting hormone (GIH/VIH) and the mandibular organ-inhibiting hormone (MOIH). The pericardial organ is reponsible for the accumulation and release of cardioactive neuropeptides (for a review on crustacean cardioactive peptide, CCAP, see Dircksen, this volume).

This review focuses on three aspects of the biochemistry and physiology of the CHH family. The first part summarises recent results on the chemical structure of the members of this neuropeptide family; a survey of the neuroendocrine cells in the X-organ of the eyestalk producing these neurohormones is given in the second part; the role of CHH and related neuropeptides in the regulation of several physiological processes is reviewed shortly in the third part. A detailed account of the roles of members of the CHH hormone family in moulting and reproduction of crustaceans is presented by Webster (this volume).

Molecular aspects of the members of the CHH neuropeptide family

Techniques in the field of microanalytical chemistry such as high-performance liquid chromatography (HPLC), gas-phase microsequencing, fast atom bombardment and electrospray mass spectrometry have been successfully used in the last few years to isolate, purify, and elucidate the primary structures of several members of the CHH neuropeptide family. However, major advances have come from application of molecular biological methods such as cDNA cloning and the polymerase chain reaction (PCR), and have resulted in the elucidation of the preprohormones of several of the respective neurohormones. Recent reviews on the structure of CHH family peptides and precursors have been given by Keller (1992) and De Kleijn & van Herp (1995).

CHH has been characterised in the shore crab, *Carcinus maenas* (Kegel *et al.*, 1989), in the American lobster, *Homarus americanus* (Tensen, De Kleijn & Van Herp, 1991b), in the isopod *Armadillidium vulgare* (Martin, Sorokine & Van Dorsselaer, 1993) and in three species of crayfish *Orconectes limosus* (Kegel *et al.*, 1991), *Procambarus bouvieri* (Huberman *et al.*, 1993) and *Procambarus clarkii* (Yasuda *et al.*, 1994)). All these CHHs have a similar number of amino acid residues (72 or 73) and comparable molecular weights, a conserved location of six cysteine residues interconnected by three disulphide bridges, several identical motifs of amino acid residues in the central part of the molecule and a blocked N-terminus. These common characteristics show that the structure of CHH is highly conserved among crustaceans. The percentage identity between functionally defined CHHs from different species is greater than 55%. Several peptides with hyperglycemic activity have been found in individuals of different species indicating the presence of polymorphic forms of CHH. For example, bioactive isoforms of both CHH A and B are produced by the X-organ sinus gland complex of the lobster *H. americanus*. Soyez *et al.* (1994) reported that these isoforms differ by a change in the configuration of a single amino acid residue. The third residue is either L- or D-phenylalanine and, in consequence, two L-Phe containing and two D-Phe containing CHHs are present in the sinus gland of the lobster. Interestingly, these authors have shown that the biological activity of the two L- and D-isoforms differ in the kinetics of their hyperglycemic activities, thus suggesting that isomerisation of Phe^3 may be of physiological significance in that this post-translational modification could generate functional diversity. A comparable polymorphic phenomenon has also been described in the crayfish *P. clarkii* (Yasuda *et al.*, 1994) and *P. bouvieri* (Aguilar *et al.*, 1995).

The complete primary amino acid sequence of GIH (VIH) was elucidated by Soyez *et al.* (1991) in the lobster *H. americanus*. Lobster GIH was isolated by reversed-phase HPLC and identified by an *in vivo* heterologous assay developed in the grass shrimp *Palaemonetes varians* (Soyez, Van Deijnen & Martin, 1987). The gonad-inhibiting effect was associated with only one peptide, consisting of a chain of 77 amino acid residues, with a free N-terminus, three disulphide bridges, a molecular mass of 9135 Da and an isoelectric point of 6.8. A second peptide with the same amino acid sequence, but without such biological activity, was also purified and shows that polymorphism also exists for lobster GIH, although this peptide has yet to be functionally identified. In the crayfish *P. bouvieri*, Aguilar *et al.* (1992) have isolated a peptide which inhibits the growth of oocytes from *Penaeus vannamei, in vitro*. This neuropeptide contains 72–74 amino acid residues and has a molecular weight of 8388 Da and a blocked N-terminus. Structurally, this crayfish GIH is more closely related to the CHHs than to lobster GIH. This peptide was active in a heterologous bioassay only at high (non-physiological) levels and its significance as a GIH remains equivocal.

The primary structure of MIH has been studied in the lobster and in the crab. In the lobster *H. americanus* (Chang, Prestwich & Bruce, 1990), the MIH-active peptide, which also displays hyperglycemic activity resembles one of the identified lobster CHH isoforms, which is known as CHH A. In the shore crab, *C. maenas* (Webster, 1991), MIH is structurally and functionally distinct from CHH, although the latter peptide also inhibits ecdysteroid synthesis by Y-organs (see Webster, this volume). It should be noted that whilst lobster MIH has been shown to exert an MIH effect *in vivo*, crab MIH has thus far been shown to act only *in vitro*.

Molecular biological studies dealing with the complete preprohormone structures of CHH have been reported for *C. maenas* (Weidemann, Gromoll & Keller, 1989), *O. limosus* (De Kleijn *et al.*, 1994a) and *H. americanus* (De Kleijn *et al.*, 1995a). Klein *et al.* (1993a) isolated a cDNA encoding the complete precursor of the putative MIH of the shore crab, while a cloning and expression study of mRNA encoding preprogonad-inhibiting hormone (GIH) has been reported in the lobster *H. americanus* by De Kleijn *et al.* (1994b). Alignment of the sequences of the preprohormones (Fig. 1) shows the relationship between the neurohormones, the signal peptides and the additional peptide sequences. Concerning the amino acid sequence of the neurohomones, a number of structural features such as an identical position of five of the six cysteine residues and some other conserved amino acid motifs indicate that all seven neuropeptides belong to the same family. The existence of two distinct subgroups within the family can be deduced from the comparison of the complete preprohormone sequences. The prepro-hormones from GIH and MIH show a 53% amino acid sequence identity, which is similar to the identity of 55% found when the different CHHs are compared. The much lower degree of identity (19%) between both groups, and especially the lack of a CHH-precursor-related peptide (CPRP) in the precursors of MIH and GIH, and the position of the first Cys residue, justify the recognition of a CHH subfamily versus a GIH/MIH subfamily. These differences may reflect an early evolutionary separation of the two groups, possibly caused by a deletion in a common ancestral gene. In this

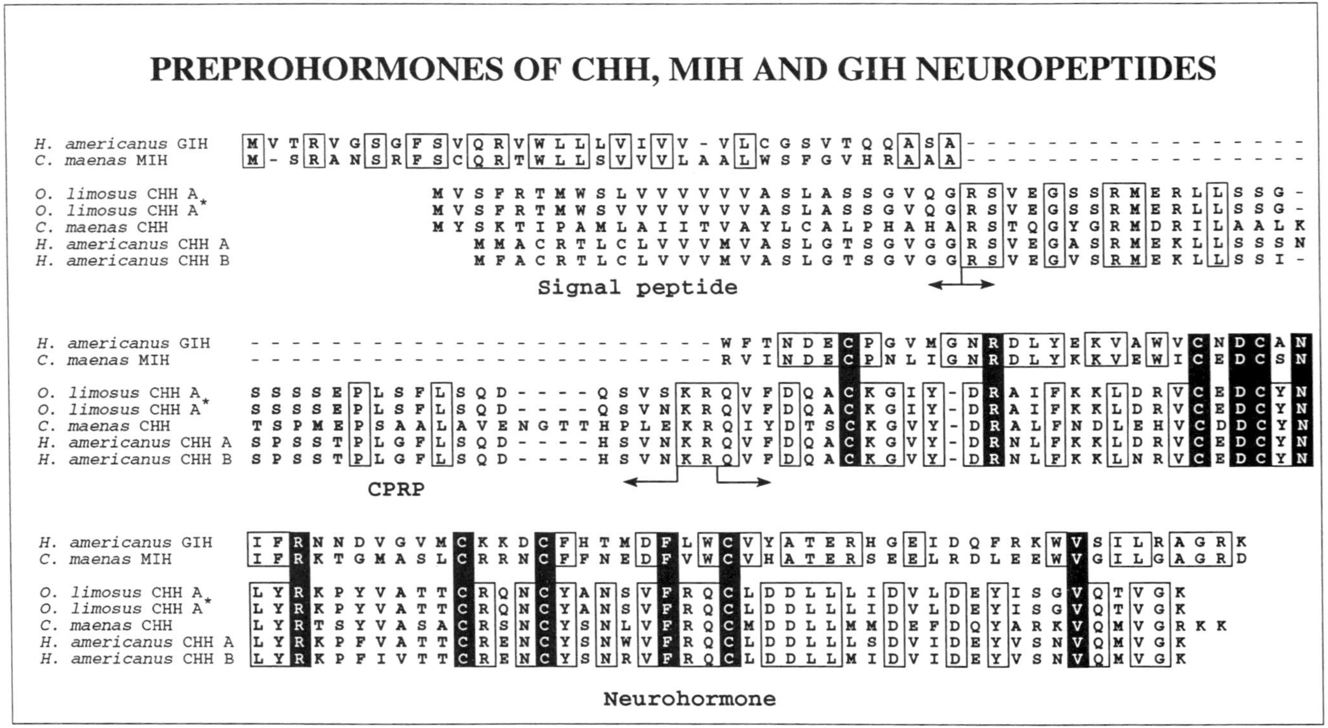

PREPROHORMONES OF CHH, MIH AND GIH NEUROPEPTIDES

```
H. americanus GIH   M V T R V G S G F S V Q R V W L L L V I V V - V L C G S V T Q Q A S A - - - - - - - - - - - - - - - - - - - - - -
C. maenas MIH       M - S R A N S R F S C Q R T W L L S V V V L A A L W S F G V H R A A A - - - - - - - - - - - - - - - - - - - - - -

O. limosus CHH A                M V S F R T M W S L V V V V V V A S L A S S G V Q G R S V E G S S R M E R L L S S G -
O. limosus CHH A*               M V S F R T M W S V V V V V V V A S L A S S G V Q G R S V E G S S R M E R L L S S G -
C. maenas CHH                   M Y S K T I P A M L A I I T V A Y L C A L P H A H A R S T Q G Y G R M D R I L A A L K
H. americanus CHH A             M M A C R T L C L V V V M V A S L G T S G V G G R S V E G A S R M E K L L S S S N
H. americanus CHH B             M F A C R T L C L V V V M V A S L G T S G V G G R S V E G V S R M E K L L S S I -
                                                     Signal peptide

H. americanus GIH   - - - - - - - - - - - - - - - - - - - - - - - - - - - W F T N D E C P G V M G N R D L Y E K V A W V C N D C A N N
C. maenas MIH       - - - - - - - - - - - - - - - - - - - - - - - - - - - R V I N D E C P N L I G N R D L Y K K V E W I C E D C S N N

O. limosus CHH A    S S S S E P L S F L S Q D - - - - Q S V S K R Q V F D Q A C K G I Y - D R A I F K K L D R V C E D C Y N N
O. limosus CHH A*   S S S S E P L S F L S Q D - - - - Q S V N K R Q V F D Q A C K G I Y - D R A I F K K L D R V C E D C Y N N
C. maenas CHH       T S P M E P S A A L A V E N G T T H P L E K R Q I Y D T S C K G V Y - D R A L F N D L E H V C D D C Y Y N
H. americanus CHH A S P S S T P L G F L S Q D - - - - H S V N K R Q V F D Q A C K G V Y - D R N L F K K L D R V C E D C Y N N
H. americanus CHH B S P S S T P L G F L S Q D - - - - H S V N K R Q V F D Q A C K G V Y - D R N L F K K L N R V C E D C Y N N
                                    CPRP

H. americanus GIH   I F R N N D V G V M C K K D C F H T M D F L W C V Y A T E R H G E I D Q F R K W V S I L R A G R K
C. maenas MIH       I F R K T G M A S L C R R N C F F N E D F V W C V H A T E R S E E L R D L E E W V G I L G A G R D

O. limosus CHH A    L Y R K P Y V A T T C R Q N C Y A N S V F R Q C L D D L L L I D V L D E Y I S G V Q T V G K
O. limosus CHH A*   L Y R K P Y V A T T C R Q N C Y A N S V F R Q C L D D L L L I D V L D E Y I S G V V Q T V G K
C. maenas CHH       L Y R T S Y V A S A C R S N C Y S N L W V F R Q C M D D L L M M D E F D Q Y A R K V V Q M V G R K K
H. americanus CHH A L Y R K P F V A T T C R E N C Y S N R V F R Q C L D D L L L S D V I D E Y V S N V V Q M V G K
H. americanus CHH B L Y R K P F I V T T C R E N C Y S N R V F R Q C L D D L L M I D V I D E Y V S N V Q M V G K
                                           Neurohormone
```

context, the putative CPRP sequence in the partially elucidated pro-hormone from a MIH-like neuropeptide described in *P. vannamei* by Sun (1994), shows that this MIH-like prawn peptide has a higher structural identity with the CHH preprohormones than with the GIH/MIH subgroup of the family.

The presence of several mRNAs encoding CHH preprohormones in a single species confirms the polymorphism described in the aforementioned biochemical studies. To date, two preproCHH-encoding mRNAs have been found in the crayfish *O. limosus* (De Kleijn *et al.*, 1994a), and in the lobster *H. americanus* (De Kleijn *et al.*, 1995a). The identification of only two CHH preprohormones and four isoforms of the respective neuropeptides in the lobster and only one CHH preprohormone and two isoforms in the crayfish points to a post-translational modification phenomenon. Figure 2 summarises the hypothesis that the polymorphism for CHH in the lobster is reached after post-translational modification: genes *A* and *B* are responsible for expression of mRNAs encoding the CHH A and CHH B precursor. These preprohormones are then translated to the neuropeptides CHH A1 and CHH B1, both characterised by an L-Phe on the third position. Afterwards, a post-translational modification takes place in some of the molecules resulting in the formation of CHH A2 and CHH B2 containing the enantiomer, D-Phe, at that position. The overall result is that the four isoforms accumulate in, and are released from, the sinus gland.

Some years ago, it was shown that sinus gland extract decreases the secretion rate of methyl farnesoate (MF) by the mandibular organs (MO) *in vitro* and *in vivo* (for a review, see Laufer *et al.*, 1993). The nature of the active compound(s) from the sinus glands has remained unknown for a long time but, recently, two studies have shown that peptides belonging to the CHH family are responsible for the inhibition of MF secretion *in vitro*. Working with the crab *Cancer pagurus*, Wainwright *et al.* (1996) characterised two peptides consisting of 78 amino acid residues and differing only by one amino acid. These peptides have striking similarities with crab MIH and lobster GIH. In contrast, three peptides inhibiting MF synthesis were purified from sinus glands of the crab *Libinia emarginata* (Liu & Laufer, 1996). These (72–76 residue) neuropeptides have molecular masses ranging between 8398 and 8474 Da and are considered as isoforms of one neuropeptide

---

Fig. 1. Sequence comparison of preprohormones from the different members of the CHH neuropeptide family. Amino acid residues identical between all preprohormones are indicated in black boxes and those identical in both subfamilies are shown in open frames. For abbreviations, see the text. (Modified versions from De Kleijn & Van Herp (1995.)

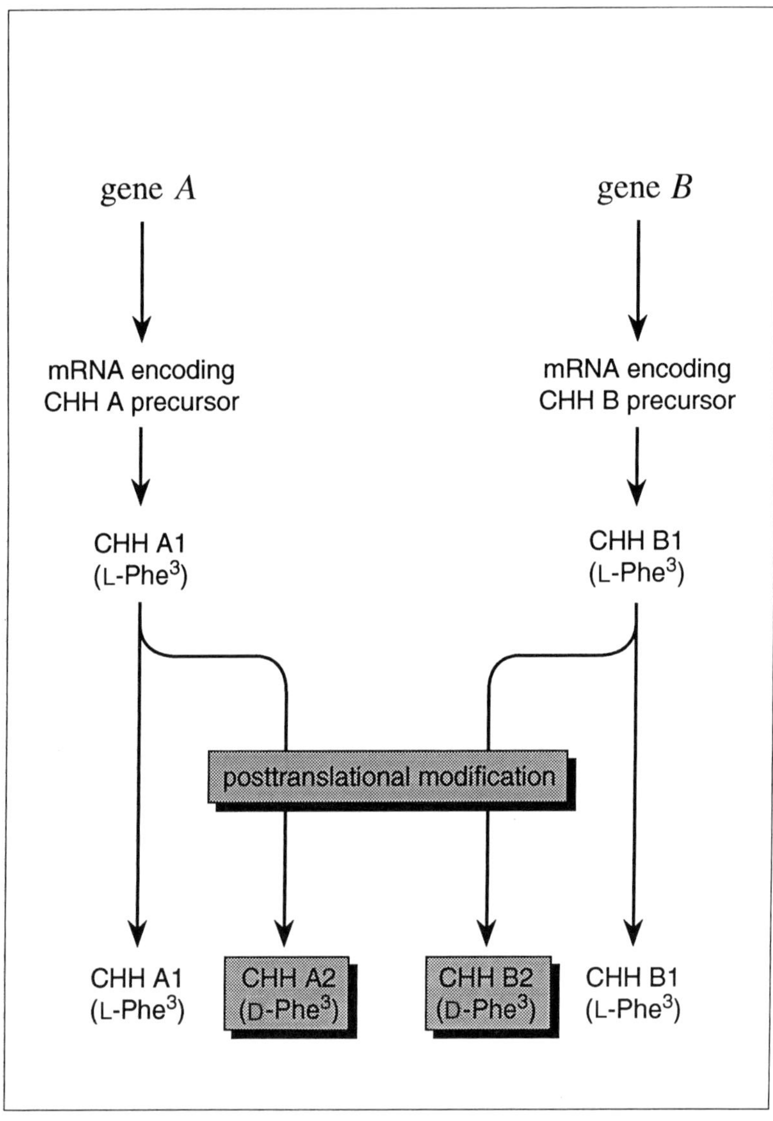

Fig. 2. Polymorphism of CHH in lobster derived from the posttranslational modification of the translated forms CHH A1 and CHH B1 into CHH A2 and CHH B2.

family and exhibit hyperglycemic activity. A partial sequence of one of these MOIHs resembles CHH from the shore crab, *Carcinus maenas*.

In conclusion, all four neurohormones CHH, GIH, MIH and MOIH belong to a peptide family which has thus far been found mostly in crustaceans. Nevertheless it is of interest to note that CHH-like peptides have been found in some other arthropods (Audsley, McIntosh & Phillips, 1992; Gasparini *et al.*, 1994), and in particular, a CHH-like peptide, ion transport peptide (ITP) which stimulates Cl⁻ transport in the locust ileum, has been fully characterised by cDNA cloning (Meredith *et al.*, 1996). (For a comprehensive review, see Phillips *et al.*, this volume.)

Cytophysiological aspects of the neuroendocrine cells synthesising neuropeptides belonging to the CHH family

Early attempts to identify neuroendocrine centres in crustaceans were, by necessity, based upon 'classical' extirpation/reimplantation experiments. However, during the last few years the availability of specific antisera and cDNA probes have facilitated the precise cellular localisation and expression of many of the CHH group neuropeptides in both adult and larval crustaceans.

Several immunocytochemical studies carried out in crab, crayfish, lobster and prawn (for a review, see Van Herp & Kallen, 1991) have shown that CHH, MIH and GIH are synthesised in the same regions of the X-organ sinus gland complex of the eyestalk. CHH was the first crustacean neuropeptide that was localised by immunocytochemistry in the crayfish *Astacus leptodactylus* (Van Herp & Van Buggenum, 1979) and the crab *C. maenas* (Jaros & Keller, 1979). MIH has been localised in the X-organ sinus gland complex of the crab *C. maenas* (Dircksen, Webster & Keller, 1988) while GIH was visualised in the same complex of the lobster, *H. americanus* (Kallen & Meusy, 1989). From both studies it became clear that cellular co-localisation is found for CHH and GIH in *H. americanus*, but not for CHH and MIH in *C. maenas*. Immunocytological studies of larvae and embryos indicated that CHH and GIH are already present at the metanauplius stage of the lobster (Rotllant *et al.*, 1995) while MIH appears at the first zoeal stage in crabs (Webster & Dircksen, 1991).

Non-radioactively labelled cRNA probes have been successfully used to study the expression of the preprohormones. Such investigations have been carried out for CHH in *O. limosus* (Tensen, Coenen & Van Herp, 1991a), GIH (Laverdure *et al.*, 1992) and GIH in combination with CHH in *H. americanus* (De Kleijn *et al.*, 1992) and MIH in *C. maenas* (Klein *et al.*, 1993b). These investigations confirmed the results of the immunocytochemical localisation studies. Moreover, *in situ* hybridisation in combination with

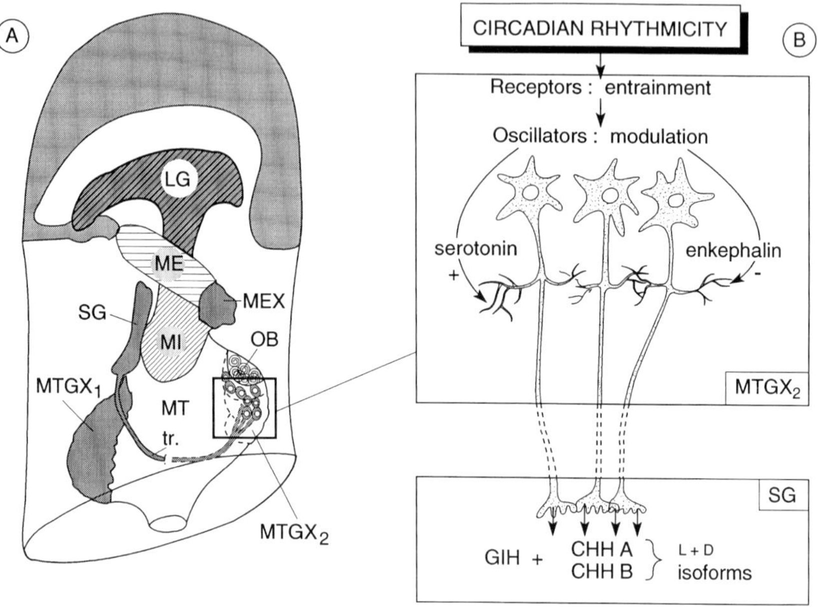

Fig. 3. Schematic representation of some cytophysiological aspects of the neuroendocrine cells synthesising CHH and GIH in lobster. (A) Longitudinal section of a lobster eyestalk showing the neuroendocrine X-organ–sinus gland complex. Abbreviations: LG, lamina ganglionaris; ME, medulla externa; MEX, medulla externa X-organ; MI, medulla interna; MT, medulla terminalis; $MTGX_1$ and $MTGX_2$, medula terminalis ganglionic X-organs; OB, organ of Bellonci; SG, sinus gland; tr. X-organ sinus gland tractus. (B) Dynamics in the cellular activity of the neuroendocrine cells according to circadian rhythmicity. (Fig. 3A from Charmantier, Charmantier-Daures & Van Herp, 1997).

immunocytochemical staining made it possible to examine the secretory activity of the X-organ–sinus gland complex at the mRNA and protein levels, as shown in the combined studies on the neuroendocrine cells producing CHH and GIH in the adult lobster (De Kleijn *et al.*, 1992) and in their larvae (Rotllant *et al.*, 1993). Interestingly, the study in the adult lobster revealed that GIH is also present in males, an observation which should stimulate research towards identifying roles for this neurohormone in male crustaceans.

Figure 3 gives an impression of the production sites of the CHH neurohormone family in the eyestalk of the lobster and summarises the dynamics of the cellular activity of the neuroendocrine cells synthesising the neuropeptides in general. As illustrated in Fig. 3A, the neuroendocrine cells synthesising both CHH and GIH neuropeptides are localised in the most important part of the X-organ, the medulla terminalis X-organ ($MTGX_2$). They are located distally on the latero-ventral side of the $MTGX_2$, and the perikarya can be subdivided into two subpopulations according to size. The neuropeptides are synthesised in the perikarya, transported along the axons, grouped in the X-organ sinus gland tractus, and stored and released from the sinus gland.

In relation to the presence of several CHH isoforms, characterised by a L-Phe or D-Phe at the third place of the amino acid sequence, the neuroendocrine cells in the $MTGX_2$ of the lobster producing CHH as well as GIH, have very recently been studied using specific antisera against the L- and D-isoforms of CHH (D. Soyez *et al.*, unpublished results). From the preliminary results of this study, it seems that most of the larger CHH/GIH-producing perikarya have the capacity to isomerise L-Phe CHH to D-Phe CHH, while this post-translational modification phenomenon, as shown in Fig. 2, seems to be absent in the group of smaller CHH/GIH-producing cells. The results obtained by double immunolabelling at the electron microscopical level, demonstrate that both isoforms are present in the same axon terminals and secretory granules, suggesting that the sorting of both forms during transport from the cell perikarya to the neurohaemal region does not occur. Additionally, results from a recent study, using specific probes for the mRNAs encoding lobster CHH A and B for *in situ* hybridisation, indicate that all CHH/GIH perikarya simultaneously express both mRNAs (D. P. V. De Kleijn & F. Van Herp, unpublished observations). In conclusion, all these immunocytochemical and *in situ* hybridisation results show that individual neuroendocrine cells in the X-organ of the eyestalk have the capacity to synthesise several neuropeptides and must, therefore, be able to tune their secretary activity in response to variations in external parameters.

Cytophysiological and regulative aspects of the CHH-producing cells have been extensively studied in the crayfish *A. leptodactylus* by Kallen and colleagues and are reviewed by Van Herp & Kallen (1991). The most important points are summarised in Fig. 3B. Studies on the secretory activity of the CHH-producing cells in relation to circadian rhythmicity (Gorgels-Kallen & Voorter, 1985; Kallen, Abrahamse & Van Herp, 1990) and surgical treatments on different optic areas on the entrainment of that rhythm (Kallen, Rigiani & Trompenaars, 1988) have indicated an endogenous circadian rhythm of CHH release entrained by a light–dark schedule, and shown that the biological clock must be located within the medulla terminalis.

There is good evidence that 5-hydroxytryptamine (5-HT) and Leu-enkephalin modulate, in an antagonistic manner, the secretory activity of the CHH perikarya in the $MTGX_2$. Physiological experiments investigating the effects of injections of several neurotransmitters and neuropeptides in

Crustacea have been carried out in several species. The results indicate that 5-HT provokes the release of CHH from the sinus gland resulting in hyperglycemia (Keller & Beyer, 1968; Strolenberg & Van Herp, 1977; Martin, 1978; Kallen, 1988), and that Leu-enkephalin inhibits CHH release and thus decreases blood glucose level indirectly (Kallen, 1988; Rothe *et al.*, 1991). 5-HT immunoreactive neurones have been demonstrated in the X-organ of the prawn *Palaemon serratus* (Bellon-Humbert & Van Herp, 1988), and ultrastructural studies have demonstrated that axonal ramifications of the CHH-producing cells in the X-organ of the crayfish *A. leptodactylus* are innervated by serotonergic synaptic structures (Van Herp & Kallen, 1991). Leu-enkephalin has been isolated from the eyestalk of the crab *C. maenas* and was visualised by immunocytochemistry (Rothe *et al.*, 1991). Modulation of the release of neuropeptides belonging to the CHH family by 5-HT has been reported many times: Fingerman (1995) has recently reviewed possible modulatory roles of neurotransmitters in controlling reproduction, and it has been reported that 5-HT mediates the release of MIH from isolated crab eyestalk ganglia (Mattson & Spaziani, 1985).

The tissue distribution of mRNAs encoding crustacean neurohormones have recently been studied using northern blot analysis and RNase protection assays. Although northern blot analysis demonstrated that preproCHH mRNA in *O. limosus* (De Kleijn *et al.*, 1994a) and preproGIH mRNA in *H. americanus* (De Kleijn *et al.*, 1994b) are expressed only in the X-organ sinus gland complex, in the lobster *H. americanus* significant expression of CHH-encoding mRNA was observed in the ventral nervous system (De Kleijn *et al.*, 1995a). The presence of CHH in the thoracic ganglia is very interesting in relation to the earlier studies, which demonstrated that some CHH isoforms have a gonad-stimulating effect (Tensen, Janssen & Van Herp, 1989; Van Herp, 1992). Since very early studies suggested that the thoracic ganglia of crustaceans possess so-called vitellogenesis-stimulating hormone (VSH) or gonad-stimulating hormone (GSH) (Otsu, 1963; Gomez & Nayar, 1965) activities, it is entirely possible that CHH may have a stimulatory role in reproduction, and this possibility deserves further attention.

Recently, RNase protection assays have been used to examine whether prepro-CHH gene expression in the eyestalk of *O. limosus* is regulated at the transcriptional or translational level (De Kleijn *et al.*, 1994a). In this study, levels of the CHH A and CHH A* mRNAs, (both encoding identical CHH peptides) from the X-organ were quantified together with HPLC measurement of both CHH isoforms in sinus glands of individual crayfish. Although levels of both mRNAs and peptides varied between individuals, the ratio between the two CHH peptides remained constant, in contrast to the ratio between the CHH and CHH A* mRNAs which showed considerable individual variation. Therefore, De Kleijn *et al.* (1994a) suggested the existence of a regulatory mechanism at the translational or post-translational level. Clearly, detailed studies on individual animals at defined stages of moult and development, using the molecular techniques briefly described here are now timely, and should prove fruitful.

Physiological aspects of the members of the CHH neuropeptide family

The primary physiological roles of CHH, GIH and MIH are now recognised: CHH provokes hyperglycemia in the haemolymph, GIH inhibits vitellogenesis in female crustaceans and MIH inhibits the moulting process. Nevertheless, since it is now quite clear that CHH exists in several isoforms in single species, it is evident that a critical reassessment of the roles of these peptides is of some importance! In this context, it has already been shown that CHH has a multi-functional role in crustaceans: it plays a central role in carbohydrate metabolism, but also has an inhibitory effect on ecdysteroid and MF synthesis, and may have a stimulatory effect on oocyte growth. As regards the lobster CHHs, Tensen *et al.* (1989) found that the two CHH B isoforms have a stimulatory effect on the growth of non-vitellogenic oocytes from *Palaemonetes varians* females. Both CHH B isoforms also stimulate the endocytotic activity of isolated oocytes from the same species when measured in a homologous *in vitro* assay (Van Herp, 1992). A recent study on the expression, storage and release of CHH during the reproductive cycle of the female lobster demonstrated that the CHH isoforms may have a differential effect on the onset of vitellogenesis and maturation (De Kleijn *et al.*, 1995b). Chang *et al.* (1990) characterised MIH in the lobster and noted that this neuropeptide has not only an inhibitory effect on ecdysteroid synthesis by Y-organs *in vitro*, but also significant hyperglycemic activity. Recently, Chang (1997) suggested that lobster MIH is identical to CHH A. Additionally, there is some evidence to suggest that the lobster CHH isoforms may be involved in osmoregulatory processes (Charmantier-Daures *et al.*, 1994). While these observations clearly demonstrate that CHH is a physiologically important neuropeptide, detailed studies investigating the differential biological activities of the CHH isoforms are still necessary in other crustacean species. In this context, it is interesting to note that CHH isoforms from the crayfish *Procambarus clarkii* have an inhibitory effect on MF synthesis in the MOs (Laufer, Liu & Van Herp, 1994). Similarly, in the crab *Libinia emarginata*, the modulation of MF production seems be due to CHH isoforms (Liu & Laufer, 1996).

Screening of GIH activity in heterologous assays revealed an inhibitory effect of lobster GIH on the growth of oocytes in the prawn *Palaemonetes varians* (Soyez *et al.*, 1987), and a decrease of the vitellin synthesis in cultured ovaries of the prawn *Penaeus vannamei* by crayfish GIH (Aguilar *et al.*,

1992). These results suggest that the biological activity of GIH seems not to be strictly species specific. The mode of action of GIH is different according to the species. In lobster, for example, GIH seems to inhibit the incorporation of vitellogenin into the oocytes (Jugan & Soyez, 1985; Van Herp & Payen, 1991). In other species, such as penaeid shrimps or the crab *Callinectes sapidus*, egg vitellins appear to arise mainly from endogenous synthesis in the oocyte (Fainzilber *et al.*, 1992). In such species, GIH may act by repressing protein and vitellin synthesis (Quackenbush, 1989; Lee & Watson, 1995). As mentioned earlier, GIH has been found not only in female but also in male crustaceans. Therefore, this hormone may also play an important, but as yet undefined role in the hormonal control of male crustaceans. This finding puts the name VIH into question, and it might be better to use the name GIH in order to generalise the role of this neuropeptide in the control of reproduction in both sexes. Details on the hormonal control of reproduction have recently been reviewed by Van Herp & Soyez (1997).

MIH acts by inhibiting the synthesis of ecdysteroids by the Y-organ. Current research is focused on the study of the cellular mechanisms by which MIH modulates ecdysteroid production by the Y-organs. Furthermore, there are indications that MIH may also be involved in the moulting process of crustacean larvae before metamorphosis. Recent information concerning the role of MIH in growth (and reproduction) can be found in the reviews by Chang (1996) and Charmantier, Charmantier-Daures & Van Herp (1997).

The neuroendocrine control of the MOs has been a point of discussion for several years. A stimulatory effect of RPCH and an inhibition by PDH on the activity of the MOs and the synthesis of MF has been discussed by Landau, Laufer & Homola (1989). But, as already mentioned, CHH may also modulate MF production in the MOs (Laufer *et al.*, 1994; Liu & Laufer, 1996). The recent characterisation of several so-called MOIH neuropeptides by Wainwright *et al.* (1996) and Laufer (1996) are the first contributions that may result in a clearer understanding of neuroendocrine regulation of the MOs by the different members of the CHH peptide family.

The schematic presentation in Fig. 4 gives an overview of the roles of the members of the CHH family in the regulation of different physiological processes in crustaceans. However, since these peptides influence some of the most fundamental physiological processes, many interendocrine interactions, and indeed further roles, for this hormone family are to be expected.

Acknowledgements

I am grateful to the following colleagues for their enthusiastic cooperation: Daniel Soyez for the studies on the CHH isoforms; Susan Waddy and colleagues for the study of reproduction in lobster; my colleagues in our laboratory Tony

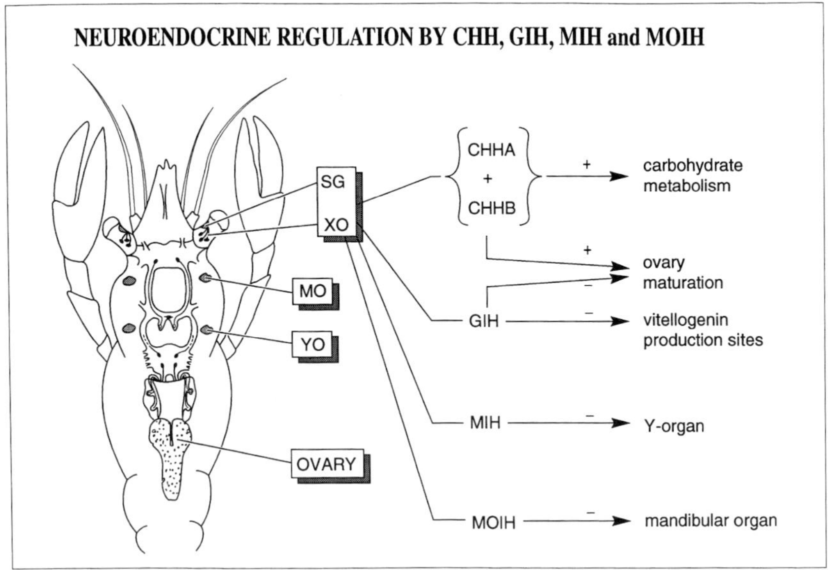

Fig. 4. Summary of the regulative role of the members of the CHH neuropeptide family in the control of different physiological processes in crustaceans. XO, X-organ; YO, Y-organ MO, mandibular organ.

Coenen, Dominique De Kleijn, Karel Janssen, Janine Kallen and Gerard Martens for the molecular biological, biochemical and cytological work.

References

Aguilar, M. B., Quackenbush, L. S., Hunt, D. T., Shabnowitz, J. & Huberman, A. (1992). Identification, purification and initial characterization of the vitellogenesis-inhibiting hormone from the Mexican crayfish *Procambarus bouvieri*. *Comparative Biochemistry and Physiology* **102B**, 491–498.

Aguilar, M. B., Soyez, D., Falchetto, R., Arnott, D., Shabanowitz, J., Hunts, D. F. & Huberman, A. (1995). Amino acid sequence of the minor isomorph of the crustacean hyperglycemic hormone (CHH-II) of the Mexican crayfish *Procambarus bouvieri* (Ortmann): presence of a D-amino acid. *Peptides* **16**, 1375–1381.

Audsley, N., McIntosh, C. & Phillips, J. E. (1992). Isolation of a neuropeptide from locust corpus cardiacum which influence ileal transport. *Journal of Experimental Biology* **173**, 261–274.

Bellon-Humbert, C. & Van Herp, F. (1988). Localization of serotin-like immunoreactivity in the eyestalk of the prawn *Palaemon serratus* (Crustacea, Decapoda, Natantia). *Journal of Morphology* **196**, 307–320.

Chang, E. S. (1997). Chemistry of crustacean hormones that regulate growth and reproduction. In *Recent Advances in Marine Biotechnology*, ed. M. Fingerman, R. Nagabhushanan & M. F. Thompson, pp. 163–178. New Dehli: Oxford and IBH Publishing Co.

Chang, E. S., Prestwich, G. D. & Bruce, M. J. (1990). Amino acid sequence of a peptide with both moult-inhibiting and hyperglycemic activities in the lobster, *Homarus americanus*. *Biochemical and Biophysical Research Communications* **171**, 818–826.

Charmantier, G., Charmantier-Daures, M. & Van Herp, F. (1997). Hormonal regulation of growth and reproduction in crustaceans. In *Recent Advances in Marine Biotechnology*, ed. M. Fingerman, R. Nagabhushanan & M. F. Thompson pp. 109–161. New Delhi: Oxford and IBH Published Co.

Charmantier-Daures, M., Charmantier, G., Janssen, K. P. C., Aiken, D. E. & Van Herp, F. (1994). Involvement of eyestalk factors in the neuroendocrine control of osmoregulation in adult American lobster *Homarus americanus*. *General and Comparative Endocrinology* **94**, 281–293.

De Kleijn, D. P. V., Coenen, T., Laverdure, A. M., Tensen, C. P. & Van Herp, F. (1992). Localization of mRNAs encoding the crustacean hyperglycemic hormone (CHH) and gonad-inhibiting hormone (GIH) in the X-organs sinus gland complex of the lobster *Homarus americanus*. *Neuroscience* **51**, 121–128.

De Kleijn, D. P. V., De Leeuw, E. P. H., Van Den Berg, M. C., Martens, G. J. M. & Van Herp, F. (1995a). Cloning and expression of two mRNAs encoding structurally different crustacean hyperglycemic hormone precursors in the lobster *Homarus americanus*. *Biochemica et Biophysica Acta* **1260**, 62–66.

De Kleijn, D. P. V., Janssen, K. P. C., Martens, G. J. M. & Van Herp, F. (1994a). Cloning and expression of two crustacean hyperglycemic hormone (CHH) mRNAs in the eyestalk of the crayfish *Orconectes limosus*. *European Journal of Biochemistry* **224**, 623–628.

De Kleijn, D. P. V., Janssen, K. P. C., Waddy, S. L., Hegeman, R., Lai, W. Y., Martens, G. J. M. & Van Herp, F. (1995b). Characterization and expression of lobster preprohormones involved in metabolism, moulting and reproduction. *Netherlands Journal of Zoology* **45**, 238–240.

De Kleijn, D. P. V., Sleutels, F. J. G. T., Martens, G. J. M. & Van Herp, F. (1994b). Cloning and expression of mRNA encoding prepro-gonad-inhibiting hormone (GIH) in the lobster *Homarus americanus*. *FEBS Letters* **353**, 255–258.

De Kleijn, D. P. V. & Van Herp, F. (1995). Review: molecular biology of neurohormone precursors in the eyestalk of Crustacea. *Comparative Biochemistry and Physiology* **112B**, 573–579.

Dircksen, H., Webster, S. G. & Keller, R. (1988). Immunocytochemical demonstration of the neurosecretory systems containing putative moult inhibiting hormone and hyperglycemic hormone in the eyestalk of brachyuran crustaceans. *Cell and Tissue Research* **251**, 3–12.

Fainzilber, M., Tom, M., Shafir, S., Applebaum, S. W. & Lubzens, E. (1992). Is there extraovarian synthesis of vitellogenin in penaeid shrimp? *Biological Bulletin* **183**, 233–241.

Fingerman, M. (1995). Endocrine mechanisms in crayfish, with emphasis on reproduction and neurotransmitter regulation of hormone release. *American Zoologist* **35**, 68–78.

Gasparini, S., Kiyatkin, N., Drevet, P., Boulain, J.-C., Tacnet, F., Ripoche, P., Forest, E., Grishin, E. & Menez, A. (1994). The low molecular weight protein which co-purifies with a latrotoxin is structurally related to crustacean hyperglycemic hormone. *Journal of Biological Chemistry* **269**, 19 803–19 809.

Gomez, R. & Nayar, K. K. (1965). Certain endocrine influences in the reproduction of the crab *Parathelphusa hydrodromous. Zoologische Jahrbücher, Abteilung für Allgemeine Zoologie und Physiologie der Tiere* **71**, 694–701.

Gorgels-Kallen, J. L. & Voorter, C. E. M. (1985). The secretory dynamics of the CHH-producing cell group in the eyestalk of the crayfish *Astacus leptodactylus*, in the course of a day/night cycle. *Cell and Tissue Research* **241**, 361–366.

Huberman, A., Aguilar, M. B., Brew, K., Shabanowitz, J. & Hunt, D. F. (1993). Primary structure of the major isomorph of the crustacean hyperglycemic hormone (CHH-1) from the sinus gland of the mexican crayfish *Procambarus bouvieri* (Ortmann): interspecies comparison. *Peptides* **14**, 7–16.

Jaros, P. P. & Keller, R. (1979). Immunocytochemical identification of the hyperglycemic hormone-producing cells in the eyestalk of *Carcinus maenas. Cell and Tissue Research* **204**, 379–385.

Jugan, P. & Soyez, D. (1985). Démonstration in vitro de l'inhibition de l'endocytose ovocytaire par un extrait de glande de sinus chez la crevette *Macrobrachium rosenbergii. Comptes Rendus de l'Académie des Sciences, Paris* **300**, 705–709.

Kallen, J. L. (1988). Quelques aspects de la régulation du système neuroendocrine produisant la CHH et de la relation entre le rythme circadien et la glycémie. In *Aspects récents de la biologie des Crustacés. Actes de Colloques, 8*, ed. Y. Le Gal & A. Van Wormhoudt, pp. 105–108. Plouzané: IFREMER.

Kallen, J. L., Abrahamse, S. L. & Van Herp, F. (1990). Circadian rhythmicity of the crustacean hyperglycemic hormone (CHH) in the hemolymph of the crayfish. *Biological Bulletin* **179**, 351–357.

Kallen, J. L. & Meusy, J. J. (1989). Do the neurohormones VIH (vitellogenesis inhibiting hormone) and CHH (crustacean hyperglycemic hormone) of crustaceans have a common precursor? Immunolocalization of VIH and CHH in the X-organ sinus gland complex of the lobster *Homarus americanus. Invertebrate Reproduction and Development* **16**, 43–53.

Kallen, J. L., Rigiani, N. R. & Trompenaars, H. J. A. J. (1988). Aspects of entrainment of CHH cell activity and hemolymph glucose levels in the crayfish. *Biological Bulletin* **175**, 137–143.

Kegel, G., Reichwein, B., Tensen, C. P. & Keller, R. (1991). Amino acid sequence of crustacean hyperglycemic hormone (CHH) from the crayfish *Orconectes limosus*. Emergence of a novel neuropeptide family. *Peptides* **12**, 909–913.

Kegel, G., Reichwein, B., Weese, S., Gaus, G., Peter-Katalinic, J. & Keller, R. (1989). Amino acid sequence of the crustacean hyperglycemic hormone (CHH) from the shore crab, *Carcinus maenas. FEBS Letters* **255**, 10–14.

Keller, R. (1992). Crustacean neuropeptides: Structures, functions and comparative aspects. *Experientia* **48**, 439–448.

Keller, R. & Beyer, J. (1968). Zur hyperglykämischen Wirkung von Serotonin und Augenstielextract beim Flusskrebs *Orconectes limosus. Zeitschrift für Vergleichende Physiologie* **59**, 78–85.

Klein, J. M., De Kleijn, D. P. V., Hunemeyer, G., Keller, R. & Weidemann, W. M. (1993a). Demonstration of the cellular expression of genes for moulting inhibiting and crustacean hyperglycemic hormone in the eyestalk of the shore crab *Carcinus maenas. Cell and Tissue Research* **274**, 515–519.

Klein, J. M., Mangerich, S., De Kleijn, D. P. V., Keller, R. & Weidemann, W. M. (1993b). Molecular cloning of crustacean putative moult-inhibiting hormone (MIH) precursor. *FEBS Letters* **334**, 139–142.

Landau, M., Laufer, H. & Homola, E. (1989). Control of methyl farnesoate synthesis in the mandibular organ of the crayfish *Procambarus clarkii*: evidence for peptide neurohormones with dual functions. *Invertebrate Reproduction and Development* **16**, 165–168.

Laufer, H., Liu, L. & Van Herp, F. (1994). A neuropeptide family that inhibits the mandibular organ of Crustacea and may regulate reproduction. In *Insect Neurochemistry and Neuropharmacology 1993*, ed. A. B. Borkovec & M. J. Loeb, pp. 203–206. Boca Raton: CRC Press.

Laufer, H., Wainwright, G., Young, N. J., Sagi, A., Ahl, J. S. B. & Rees, H. H. (1993). Ecdysteroids and juvenoid in two male morphotypes of *Libinia emarginata. Insect Biochemistry and Molecular Biology* **23**, 171–174.

Laverdure, A. M., Breuzet, M., Soyez, D. & Becker, J. (1992). Detection of the mRNA encoding vitellogenesis inhibiting hormone in neurosecretory cells of the X-organ in *Homarus americanus. General and Comparative Endocrinology* **87**, 443–450.

Lee, C. Y. & Watson, R. D. (1995). In vitro study of vitellogenesis in the blue crab (*Callinectes sapidus*): site and control of vitellin synthesis. *Journal of Experimental Zoology* **271**, 364–372.

Liu, L. & Laufer, H. (1996). Isolation and characterization of sinus gland neuropeptides with both mandibular organ-inhibiting and hyperglycemic effects from the spider crab, *Libinia emarginata. Archives of Insect Biochemistry and Physiology* **32**, 375–385.

Martin, G. (1978). Action de la sérotonine sur la glycémie et sur la libération des neurosécrétions contenues dans la glande du sinus de *Porcellio dilatatus* Brandt (Crustacé, Isopode, Oniscoide). *Comptes Rendus de la Société Biologique* **172**, 303–308.

Martin, G., Sorokine, O. & Van Dorsselaer, A. (1993). Isolation and molecular characterisation of a hyperglycemic neuropeptide from the sinus gland of the terrestrial isopod *Armadillidium vulgare* (Crustacea). *European Journal of Biochemistry* **211**, 601–607.

Mattson, M. P. & Spaziani, E. (1985). 5-Hydroxytryptamine mediates release of moult-inhibiting hormone activity from isolated crab eyestalk ganglia. *Biological Bulletin* **169**, 246–255.

Otsu, T. (1963). Bihormonal control of sexual cycle in the freshwater crab *Potamon dehaani*. *Embryologica* **8**, 1–20.

Quackenbush, L. S. (1989). Vitellogenesis in the shrimp, *Penaeus vannamei: in vitro* studies of the isolated hepatopancreas and ovary. *Comparative Biochemistry and Physiology* **94**, 253–261.

Rothe, H., Lüschen, W., Asken, A., Willig, A. & Jaros, P. P. (1991). Purified crustacean enkephalin inhibits release of hyperglycemic hormone in the crab *Carcinus maenas*. *Comparative Biochemistry and Physiology* **99C**, 57–62.

Rotllant, G., Charmantier-Daures, M., De Kleijn, D. P. V., Charmantier, G. & Van Herp, F. (1995). Ontogeny of neuroendocrine centers in the eyestalk of *Homarus americanus* embryos: an anatomical and hormonal approach. *Invertebrate Reproduction and Development* **27**, 233–245.

Rotllant, G., De Kleijn, D., Charmantier-Daures, M., Charmantier, G. & Van Herp, F. (1993). Localization of crustacean hyperglycemic hormone (CHH) and gonad-inhibiting hormone (GIH) in the eyestalk of *Homarus gammarus* larvae by immunocytochemistry and *in situ* hybridization. *Cell and Tissue Research* **271**, 507–512.

Soyez, D., Le Caer, J. P., Noël, P. Y. & Rossier, J. (1991). Primary structure of two isoforms of the vitellogenesis inhibiting hormone from the lobster *Homarus americanus*. *Neuropeptides* **20**, 25–32.

Soyez, D., Van Deijnen, J. E. & Martin, M. (1987). Isolation and characterization of a vitellogenesis-inhibiting factor from sinus glands of the lobster *Homarus americanus*. *Journal of Experimental Zoology* **244**, 479–484.

Soyez, D., Van Herp, F., Rossier, J., Le Caer, J.-P., Tensen, C. P. & Lafont, R. (1994). Evidence for a conformational polymorphism of invertebrate neurohormones. *Journal of Biological Chemistry* **269**, 18 295–18 298.

Strolenberg, G. E. C. M. & Van Herp, F. (1977). Mise en évidence du phénomène d'exocytose dans la glande du sinus *d'Astacus leptodactylus* (Nordmann) sous l'influence d'injection de sérotonine. *Comptes Rendus de l'Académie des Sciences* **D 284**, 57–60.

Sun, P. S. (1994). Molecular cloning and sequence analysis of a cDNA encoding a moult-inhibiting hormone-like neuropeptide from the white shrimp *Penaeus vanamei*. *Molecular Marine Biology and Biotechnology* **3**, 1–6.

Tensen, C. P., Coenen, T. & Van Herp, F. (1991a). Detection of mRNA encoding crustacean hyperglycemic hormone (CHH) in the eyestalk of the crayfish *Orconectes limosus* using non-radioactive *in situ* hybridization. *Neuroscience Letters* **124**, 178–182.

Tensen, C. P., De Kleijn, D. P. V. & Van Herp, F. (1991b). Cloning and sequence analysis of DNAs encoding two crustacean hyperglycemic hormones in the lobster *Homarus americanus*. *European Journal of Biochemistry* **200**, 103–106.

Tensen, C. P., Janssen, K. P. C. & Van Herp, F. (1989). Isolation, characterization and physiological specificity of the crustacean hyperglycemic factors from the sinus gland of the lobster *Homarus americanus* (Milne-Edwards). *Invertebrate Reproduction and Development* **16**, 155–164.

Van Herp, F. (1992). Inhibiting and stimulating neuropeptides controlling reproduction in Crustacea. *Invertebrate Reproduction and Development* **22**, 21–30.

Van Herp, F. & Kallen, J. L. (1991). Neuropeptides and neurotransmitters in the X-organ sinus gland complex, an important neuroendocrine integration center in the eyestalk of Crustacea. In *Comparative Aspects of Neuropeptides*, ed. G. B. Stefano & E. Florey), pp. 211–221. Manchester: Manchester University Press.

Van Herp, F. & Payen, G. G. (1991). Crustacean neuroendocrinology: perspectives for the control of reproduction in aquacultural systems. *Bulletin of the Institute of Zoology, Academia Sinica (Monograph)* **16**, 513–539.

Van Herp, F. & Soyez, D. (1997). 8. Arthropoda – Crustacea. In *Progress in Reproductive Endocrinology*, Part A, Vol. X, ed. T. S. Adams (Series: Reproductive Biology in Invertebrates). New Delhi: Oxford and IBH Publishing Co., in press.

Van Herp, F. & Van Buggenum, H. J. M. (1979). Immunocytochemical localization of hypergycemic (HGH) in the neurosecretary system of the eyestalk of the crayfish *Astacus leptodactylus*. *Experientia* **35**, 1527–1528.

Wainwright, G., Webster, S. G., Wilkinson, M. C., Chung, J. S. & Rees, H. H. (1996). Structure and significance of mandibular organ-inhibiting hormone in the crab, *Cancer pagurus*; involvement in multihormonal regulation of growth and reproduction. *Journal of Biological Chemistry* **127**, 12 749–12 574.

Webster, S. G. (1991). Amino acid sequence of putative moult-inhibiting hormone from the crab *Carcinus maenas*. *Proceedings of the Royal Society of London* **244B**, 247–252.

Webster, S. G. & Dircksen, H. (1991). Putative moult inhibiting hormone in larvae of the shore crab *Carcinus maenas* L. An immunocytochemical approach. *Biological Bulletin* **180**, 65–71.

Weidemann, W., Gromoll, J. & Keller, R. (1989). Cloning and sequence analysis of cDNA for precursor of a crustacean hyperglycemic hormone. *FEBS Letters* **257**, 31–34.

Yasuda, A., Yasuda, Y., Fujita, T. & Naya, Y. (1994). Characterization of crustacean hyperglycemic hormone from the crayfish (*Procambarus clarkii*): multiplicity of molecular stereoversion and diverse functions. *General and Comparative Endocrinology* **95**, 387–398.

XAVIER BELLÉS

Endocrine effectors in insect vitellogenesis

Introduction: a formidable diversity

With almost one million species described, and some nine millions still to be named, insects are among the most successful creatures on earth. This formidable diversity is apparent not only at the morphological level, but also from a functional point of view. For example, insects have evolved a fascinating diversity of reproductive strategies to ensure the continuance of the species. Such strategies range from oviparity to viviparity, and from bisexual reproduction to functional hermaphroditism. This is further complicated with cases of parthenogenesis, paedogenesis and heterogony.

Reproduction and, in particular, vitellogenesis and egg maturation, are under hormonal control, and the diversity of reproductive strategies promises a parellel multiplicity of endocrine mechanisms of regulation. The present contribution deals with this multiplicity, and it has been written with the general purpose of discerning common themes, and to fit diverse observations into an orderly pattern.

The subject is too broad to be covered in a short review, therefore the bounds of the problem have been set in a rather restrictive manner. Firstly, only formally identified hormones that directly act on the vitellogenic tissues have been considered. This means that discussion of unidentified factors ('food factors', for example), or 'long-loop' mechanisms of regulation have been avoided. Secondly, the general idea has been to discuss only the essential facts and recent findings, and to display the information bearing in mind the phylogenetic position of the insect groups considered.

The above clarifications already suggest that this contribution is by no means an encyclopaedia, and makes no claim to completeness. Although objective considerations determined the problems to be treated, the choice and the differential emphasis was ultimately subjective. In any case, many comprehensive reviews (for examples, see Engelmann, 1983; Hagedorn, 1985; Koeppe *et al.*, 1985; Valle, 1993) provide the bulk of the information and the historical background that could not be included here.

Vitellogenic proteins, tissues and hormones

In most insect species, vitellogenins are produced by the female fat body under the action of juvenile hormone, released into the haemolymph, and then incorporated into developing oocytes (Engelmann, 1983). However, not all insects synthesise the same type of vitellogenic protein; there are vitellogenic tissues other than the fat body; and there are hormones other than juvenile hormone involved in vitellogenesis.

Vitellogenins and yolk proteins

Vitellogenins are the precursors of yolk proteins in insects and other oviparous animals. In most insects, the primary vitellogenin gene product, with a molecular mass of more than 200 kDa, is processed into large (about 200 kDa) and small (about 50 kDa) units. This is the case in many orthopterans, heteropterans, lepidopterans and lower dipterans (Valle, 1993; Yano *et al.*, 1994a,b; Hiremath, Lehtoma & Nagarajan, 1994). In bees and wasps, the primary gene product, with a molecular mass of some 180 kDa, is secreted without processing (Valle, 1993; Kageyama *et al.*, 1994).

In contrast, the yolk proteins of higher Diptera are quite different from those of other insects. For example, in *Drosophila melanogaster* there are three major yolk proteins of around 40 kDa, each encoded by single-copy genes located on the X-chromosome (Bownes *et al.*, 1993). Similar examples are found in other higher Diptera such as the fruit fly *Ceratitis capitata* (Rina & Savakis, 1991) or the blowfly *Calliphora erythrocephala* (Martínez & Bownes, 1994).

In addition, the molecular analysis of vitellogenic proteins for which genomic or cDNA sequences are available (Table 1), has afforded new, more consistent information on the evolutionary relationships between different vitellogenic proteins. These analyses indicate that the yolk proteins of higher Diptera are not homologous to the vitellogenins of other insects (Romans *et al.*, 1995). Therefore, it seems that during their evolution flies shifted from the primitive type of vitellogenin to another protein to fulfil the role of nutrient reserve for the embryo.

Vitellogenic tissues

The fat body is the exclusive site of vitellogenin synthesis in the great majority of insects (Valle, 1993). During vitellogenesis fat body cells undergo dramatic changes and produce huge amounts of protein in a relatively short time (Keeley, 1985). However, in the Zygentoma (*Thermobia domestica*), in some Coleoptera (*Coccinella septempunctata, Leptinotarsa decemlineata*), and in most higher Diptera (*Dacus oleae, Musca domestica, Calliphora vicina, C. erythrocephala, Sarcophaga argyrostoma, D. melanogaster*) vitellogenesis occurs

Table 1 *Species where genomic or cDNA sequences of vitellogenin (Vg) or yolk proteins (YP) are available*

Species (Order)	Type of protein	Type of sequence available	References
Locusta migratoria (Orthoptera)	Vg	Genomic (partial)	Locke *et al.*, 1987
Anthonomus grandis (Coleoptera)	Vg	Genomic	Trewitt *et al.*, 1992
Aedes aegypti (Diptera)	Vg	Genomic	Romans *et al.*, 1995
		cDNA	Chen *et al.*, 1994
Drosophila melanogaster (Diptera)	YP	Genomic	Hung & Wensink, 1993; Garabedian *et al.*, 1987
Ceratitis capitata (Diptera)	YP	Genomic	Rina & Savakis, 1991
Calliphora erythrocephala (Diptera)	YP	Genomic	Martínez & Bownes, 1994
Bombyx mori (Lepidoptera)	Vg	Genomic cDNA	Yano *et al.*, 1994a Yano *et al.*, 1994b
Lymantria dispar (Lepidoptera)	Vg	cDNA (partial)	Hiremath *et al.*, 1994
Athalia rosae (Hymenoptera)	Vg	cDNA (partial)	Kageyama *et al.*, 1994

both in the ovaries and in the fat body. An extreme case is represented by the stable fly *Stomoxys calcitrans*, where yolk protein synthesis occurs exclusively in the ovary (Valle, 1993; Rousset & Bitsch, 1993; Martínez & Bownes, 1994). The fact that in the primitive insect *T. domestica* vitellogenins are produced by the ovary suggests this is a primitive feature that has been lost in more modified insect groups that have panoistic ovaries (like most orthopteroids). Conversely, the production of vitellogenin by the ovary in beetles and flies would have been a later development in meroistic ovaries.

Hormones

Identified hormones involved in vitellogenesis belong to the three classic families of insect endocrines: juvenile hormones, ecdysteroids, and peptides and proteins.

The best-known hormones regulating vitellogenesis are juvenile hormones (Engelmann, 1983). These are sesquiterpenoids that are produced mainly in the corpora allata, part of the retrocerebral complex. Juvenile hormones were initially identified from their juvenilising action, prolonging the larval stage. Juvenile hormone III (JH III), methyl $(2E,6E)$-10,11-epoxy-3,7,11-trimethyl-2,6-dodecadienoate, is the JH homologue most frequently involved in vitellogenesis, and it is present in the great majority of insect orders. In adult females of higher Diptera, the bisepoxy homologue, methyl $(2E)$-6,7;10,11-bis-epoxy-3,7,11-trimethyl-$(2E)$-dodecenoate also seems to play a role in vitellogenesis (Kelly, 1994).

Ecdysteroids stimulate vitellogenesis in several insect species (Hagedorn, 1985), although they can also inhibit this process in others (see below). They were first known as moulting hormones, and in juvenile stages they are produced mainly by the prothoracic glands. In the adult, the principal ecdysteroidogenic organ is the ovary, and the ecdysteroid which acts most frequently in vitellogenesis is 20-hydroxyecdysone.

With regard to peptides, adipokinetic hormones and allatostatins may be involved in the regulation of vitellogenesis. The most typical action of adipokinetic hormones is to regulate energy metabolism by mobilising fat body nutrient reserves. Locust adipokinetic hormone I (AKH I) (pGlu-Leu-Asn-Phe-Thr-Pro-Asn-Trp-Gly-Thr-NH$_2$) has been shown to inhibit vitellogenin production in *Locusta migratoria*. Finally, allatostatins were initially described as inhibitors of juvenile hormone biosynthesis (Stay, Tobe & Bendena, 1994). However, it has recently been shown that these compounds can participate in the termination of vitellogenesis in cockroaches (Martín, Piulachs & Bellés, 1996b).

Hormonal control of initiation and maintenance of vitellogenesis

The fact that there is a formidable bulk of data on the initiation and maintenance of vitellogenesis (Engelmann, 1983; Hagedorn, 1985; Koeppe *et al.*, 1985; Valle, 1993) invites a systematic approach. A classification of the different cases into four groups has therefore been attempted. Only those hormones truly sustaining vitellogenesis have been considered here; that is, juvenile hormones and ecdysteroids.

The co-ordination of moulting and reproduction in the Zygentoma

Within the Apterygota, the order Zygentoma deserves special attention not only because it is one of the most primitive among insects, but also because

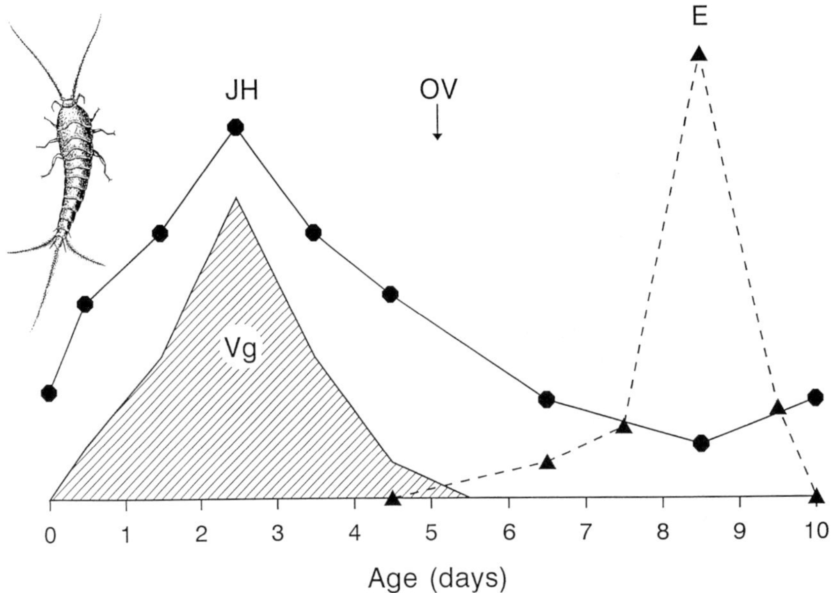

Fig. 1. Patterns of juvenile hormone (JH), ecdysteroids (E) and vitellogenin (Vg, shaded area) during the first days of adult life in the firebrat *Thermobia domestica*. OV, oviposition. Data from Rousset & Bitsch (1993) and references therein.

its representatives continue to moult during adult life. Most of the data available have been obtained with the firebrat, *T. domestica*. In this species, allatectomy prevents the vitellogenic phase of oocyte development, and treatment with allatotoxins (precocenes) also inhibits vitellogenesis. This indicates that juvenile hormone stimulates vitellogenesis (Rousset & Bitsch, 1993, and references therein).

In the firebrat, there are two large vitellogenins processed from large precursors and produced by both the fat body and the ovaries (Rousset & Bitsch, 1993). Interestingly, in this species moulting in the female (regulated by ecdysteroids) is interspersed with gonadotropic cycles (regulated by juvenile hormone) (Fig. 1), which suggests that moult and reproductive cycles are regulated in a co-ordinated manner.

Vitellogenesis directed by juvenile hormone

The most thorough studies on the vitellogenic action of juvenile hormones, including data on molecular action, have been carried out on the orthopteran *L. migratoria*. In the female of this species, juvenile hormone

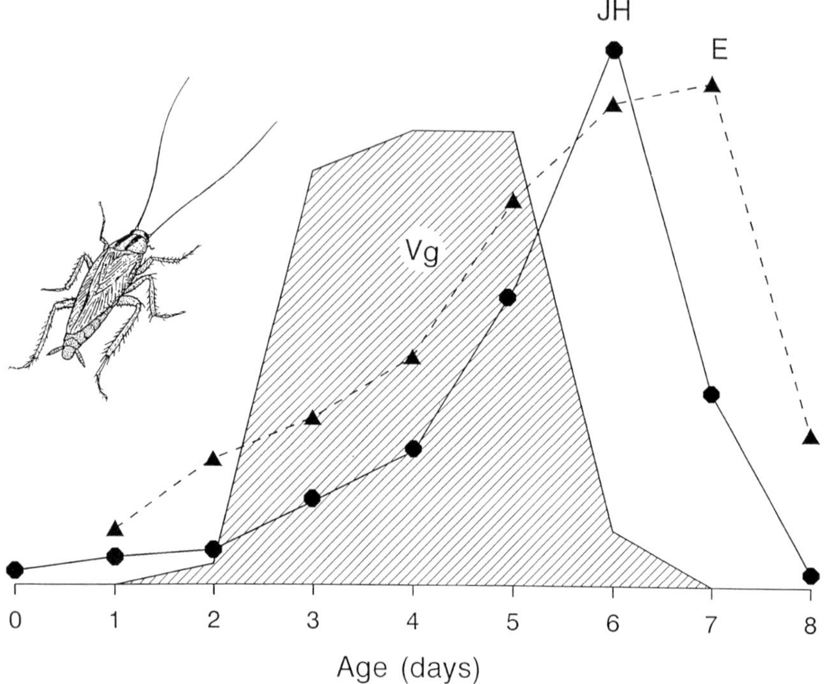

Fig. 2. Patterns of juvenile hormone (JH), ecdysteroids (E) and vitellogenin (Vg, shaded area) during the first days of adult life in the German cockroach, *Blattella germanica*. Data from Bellés *et al.*, 1987 (JH); Romañá *et al.*, 1995 (E); and Martín *et al.*, 1995 (Vg).

regulates vitellogenin synthesis in the fat body, vitellogenin mRNAs have been detected in this organ, and it has been shown that juvenile hormone influences their levels (Zhang, McCracken & Wyatt, 1993, and references therein). More recently, Glinka & Wyatt (1996) have shown that a juvenile hormone analogue activates vitellogenin gene transcription in the fat body.

As in other orthopteroids, it is also juvenile hormone that directs the synthesis of vitellogenin in the fat body of cockroaches. Most of the information has been obtained in the oviparous *Periplaneta americana*, the ovoviviparous *Blattella germanica*, the pseudoviviparous *Leucophaea maderae* and *Nauphoeta cinerea*, and the viviparous *Diploptera punctata* (Engelmann, 1983). In viviparous, pseudoviviparous or ovoviviparous species, in which vitellogenesis follows discrete cycles (separated by periods of egg transport when the endocrine system is quiescent), juvenile hormone production rates and circulating vitellogenin levels show cyclic patterns (Fig. 2). Conversely, in oviparous species, where vitellogenesis is sustained throughout successive phases of oocyte maturation, juvenile hormone production and haemolymph levels of vitellogenin are rather constant (Martín *et al.*, 1995, and references therein).

Within Heteroptera most data have been obtained from the bloodsucking bug *Rhodnius prolixus* (Davey, 1993), in which juvenile hormone stimulates vitellogenin synthesis in the fat body. Similar data have been reported for other heteropteran species, both haematophagous, such as *Triatoma protracta*, and phytophagous, for example *Oncopeltus fasciatus* or *Pyrrhocoris apterus* (Engelmann, 1983).

In Coleoptera, the influence of the corpora allata and juvenile hormone on vitellogenin synthesis has been thoroughly studied in *L. decemlineata*. Juvenile hormone also directs vitellogenesis in *Tenebrio molitor, Dendroctonus pseudotsugae* or *Pterostichus nigrita* (Engelmann, 1983).

The flexibility of Lepidoptera

Within the Lepidoptera it is important to distinguish between species that start vitellogenesis after adult emergence, and those in which vitellogenin production begins before adult emergence (Fig. 3).

In species that begin vitellogenesis after adult emergence, the influence of corpora allata and juvenile hormone on oocyte growth has been reported in many cases (Engelmann, 1983), and direct evidence that juvenile hormone (or juvenile hormone analogues) induce vitellogenin production has been found in *Danaus plexippus, Nymphalis antiopa, Helicoverpa zea* and *Pseudaletia unipuncta* (Cusson *et al.*, 1994, and references therein). In *P. unipuncta*, a correlation has been observed between ovarian growth, juvenile hormone release rates *in vitro*, and vitellogenin synthesis (Cusson *et al.*, 1994).

Species in which vitellogenin production begins in the late pharate adult stage, such as *Plodia interpunctella* and *Manduca sexta*, constitute a special case. In *P. interpunctella*, vitellogenesis coincides with the decline in ecdysteroid titre which occurs at the pharate adult stage, and treatment with 20-hydroxyecdysone inhibits vitellogenin production (Shirk, Bean & Brookes, 1990). In *M. sexta*, vitellogenesis starts three to four days before adult emergence and proceeds in the absence of the pupal corpora allata. Nevertheless, juvenile hormone is necessary to complete oocyte growth, although it does not seem to be essential for vitellogenin production (Imboden & Law, 1983, and references therein). In summary, vitellogenin production by the fat body in *M. sexta* appears to be stimulated by juvenile hormone and ecdysteroids.

In *Hyalophora cecropia*, vitellogenin production begins in the prepupal stage, whereas in *Bombyx mori* it starts during early pupal development. In these two species, juvenile hormone does not appear to be involved in

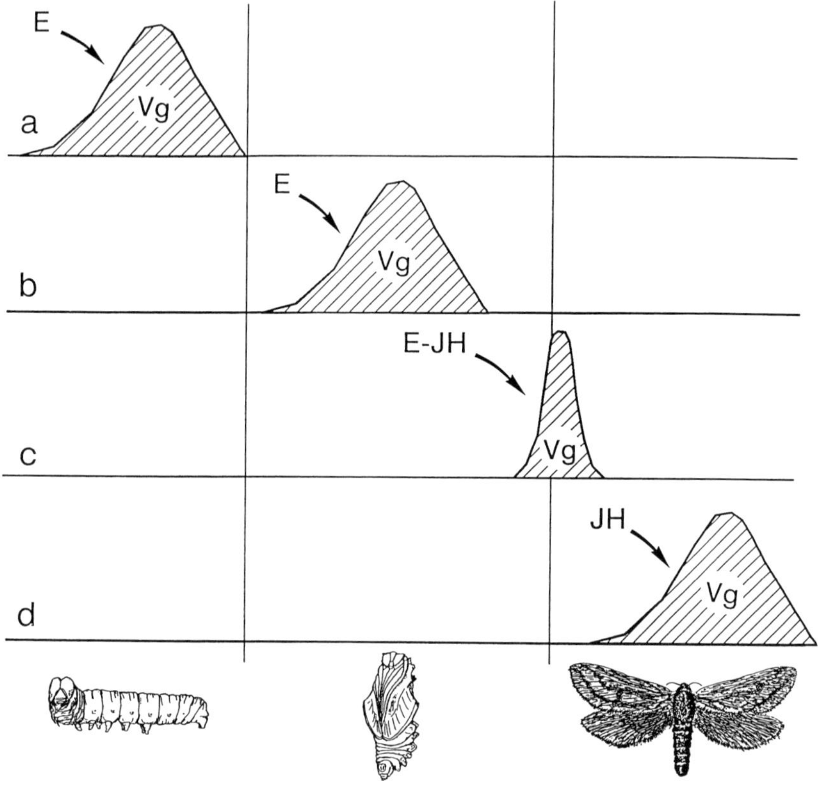

Fig. 3. In Lepidoptera, vitellogenesis (Vg, represented schematically by a shaded area) is sustained by ecdysteroids (E) and/or juvenile hormone (JH) depending upon whether it takes place in last larval stage (a), in the pupae (b), in late pharate adult stage (c) or after adult emergence (d). See the text for references.

ovarian development, whereas ecdysteroids have been shown to stimulate this process. The beginning of vitellogenin synthesis and uptake in *B. mori* coincides with a peak in ecdysteroid titre, and administration of 20-hydroxyecdysone to brainless pupae stimulates vitellogenin production (Tsuchida, Nagata & Suzuki, 1987).

A possibly similar case occurs in *Lymantria dispar*, in which vitellogenin production begins during the last larval instar. The treatment of larvae of this species with a juvenile hormone analogue suppresses vitellogenin production (Fescemyer *et al.*, 1992) and vitellogenin mRNA in the fat body (Hiremath & Jones, 1992). This suggests that a low or declining titre of juvenile hormone is necessary for vitellogenin accumulation in *L. dispar*.

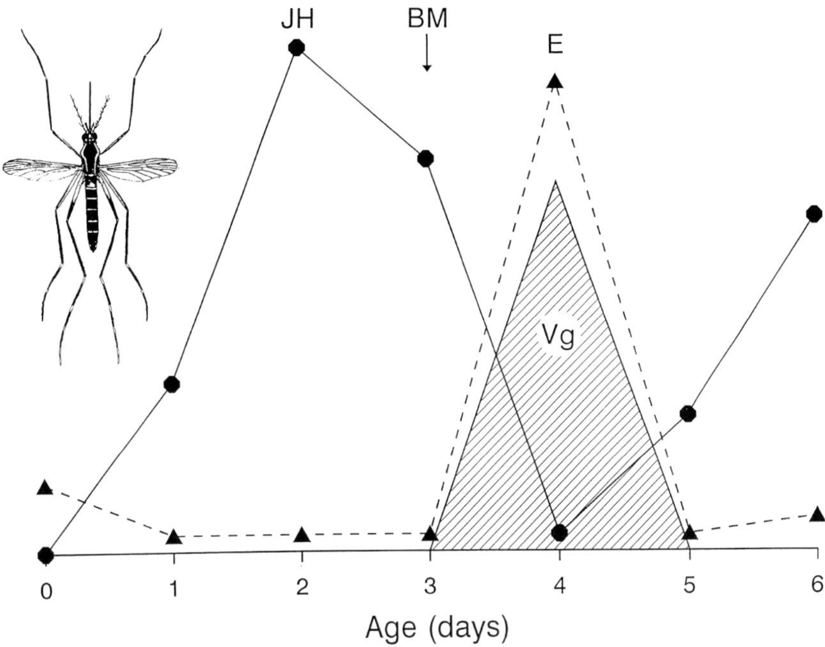

Fig. 4. Patterns of juvenile hormone (JH), ecdysteroids (E) and vitellogenin (Vg, shaded area) during the first days of adult life in the yellow fever mosquito, *Aedes aegypti*. BM, blood meal. Data are from diverse sources (e.g. Hagedorn, 1985; Raikhel, 1992).

The shift from juvenile hormone to ecdysteroids in Diptera

Within the Diptera, a distinction should be made between orthorraphid (lower) and cyclorrhaphid (higher) species, because there are substantial differences between these two groups.

With regard to the Orthorrhapha, most of the data available come from research on the mosquito *Aedes aegypti* (Hagedorn, 1985; Raikhel, 1992; Dhadialla & Raikhel, 1994). In this species, Hagedorn and co-workers proposed 20–hydroxyecdysone as the factor directing vitellogenin synthesis in the fat body, and that the ovary was the source of ecdysteroids. Data supporting this view have also been reported for other mosquitoes: *Culex pipiens, Aedes atropalpus* or *Anopheles stephensi* (Hagedorn, 1985).

In *A. aegypti*, it has also been shown that juvenile hormone potentiates the action of 20-hydroxyecdysone, and that juvenile hormone may induce fat body capacitation in previtellogenesis, whereas ecdysteroids would direct the vitellogenic phase (Fig. 4). In addition, the importance of a blood meal and of midgut factors to the full development of vitellogenesis in response to physiological doses of 20-hydroxyecdysone has also been stressed (Dhadialla & Raikhel, 1994).

A number of groups have described the action of ecdysteroids at a molecular level on vitellogenin production in *A. aegypti* (Dhadialla & Raikhel, 1994, and references therein). The production and, particularly, the accumulation of a high-molecular weight vitellogenin precursor (provitellogenin) is stimulated by 20-hydroxyecdysone *in vitro*, which suggests that the hormone triggers the transcription of vitellogenin genes. More recent studies have shown that cycloheximide inhibits the vitellogenic response induced by 20-hydroxyecdysone (Deitsch *et al.*, 1995), which suggests that the action of this hormone is indirect, and is mediated by transcriptional factors.

Among the Cyclorrhapha, the species that has been studied most thoroughly is *D. melanogaster*. As stated above, yolk proteins of *D. melanogaster*, and presumably of higher Diptera in general, are peculiar and different from vitellogenins of other insects. In addition, yolk proteins are produced not only by the fat body but also by the ovary. The sex of the adults is the primary factor for correct expression of yolk proteins genes in *D. melanogaster* (Bownes *et al.*, 1993). Once the genes are active in the female, the level of expression is modulated by 20-hydroxyecdysone, juvenile hormone and neurosecretory factors. The endocrine regulation of yolk proteins in *D. melanogaster* has recently been reviewed by Kelly (1994). Both juvenile hormone and ecdysteroids stimulate yolk protein synthesis in the fat body and in the ovary and juvenile hormone induces yolk protein uptake in the ovary. Significant data are also available in the housefly, *M. domestica*. In the adult female, 20-hydroxyecdysone and juvenile hormone can induce yolk protein synthesis, and the two hormones have an additive effect (Adams & Filipi, 1988). Both the fat body and the ovary are stimulated by 20-hydroxyecdysone, whereas juvenile hormone induces a greater response in the ovary than in the fat body (Agui *et al.*, 1991). 20-Hydroxyecdysone induces lower production of yolk proteins in houseflies allatectomised before juvenile hormone was released (Adams & Filipi, 1988), and in specimens from which both corpora allata and ovaries had been removed (Adams & Gerst, 1992). Taken together, the results indicate that both ecdysteroids and juvenile hormone are necessary for vitellogenesis, and the data of Agui *et al.* (1991) suggest that juvenile hormone could direct the synthesis of yolk proteins in the ovary whereas 20-hydroxyecdysone could do this in the fat body.

Ecdysteroids also induce vitellogenesis in *C. erythrocephala*, *C. vicina*, *Neobellieria* (= *Sarcophaga*) *bullata*, *Protophormia terraenovae*, *Phormia regina* and *Lucilia caesar* (Yin & Stoffolano, 1994).

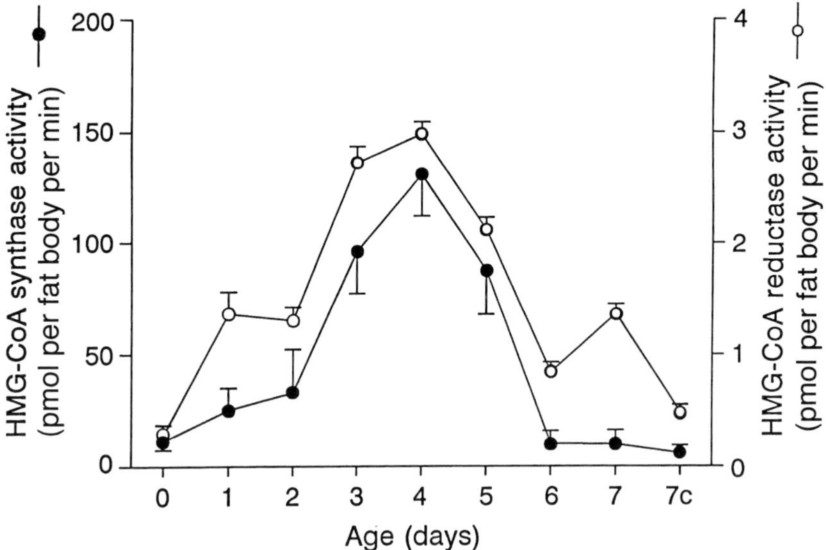

Fig. 5. Pattern of activity of the enzymes 3-hydroxy-3-methylglutaryl-CoA (HMG-CoA) synthase and HMG-CoA reductase in the fat body of *Blattella germanica*. Data from Casals *et al.* (1996).

The problems of cyclicity

The brief outline of vitellogenesis in different insect groups presented above shows that cyclicity may be related to physiological constraints (vitellogenesis co-ordinated with moulting in Zygentoma, Fig. 1, or in some Lepidoptera, Fig. 3) critically connected to trophic factors. For example, vitellogenic cycles are associated with the timing of blood meals in most haematophagous insects (Fig. 4), or may be interspersed with discrete oviposition pulses as in locusts.

In addition to being a by-product, so to speak, of trophic or physiological constraints, cyclicity might also be a genuine evolutionary step towards viviparity. In this context, cockroaches offer a paradigm of a continuum from oviparous species, with almost constant production of vitellogenin (*P. americana*), to ovoviviparous or viviparous species, with cycles of vitellogenesis separated by periods of ootheca or embryo transport equivalent to pregnancy (*D. punctata, N. cinerea* or *B. germanica* (Martín *et al.*, 1995, and references therein) (Fig. 2).

Furthermore, cyclicity is observed not only in vitellogenin production, but also in other parameters related to vitellogenesis. A recent example is the cyclic pattern of expression and activity of the enzymes 3-hydroxy-3-methylgutaryl-CoA (HMG-CoA) synthase and HMG-CoA reductase observed in the fat body of *B. germanica* (Casals *et al.*, 1996). These enzymes regulate the mevalonic acid pathway, which leads, among other end-products, to dolichol, a donor of oligosaccharide residues in the process of glycosylation of proteins. In the fat body of *B. germanica*, the cyclic expression of these enzymes (Fig. 5) is related to the glycosylation of vitellogenin.

These cases suggest not only that vitellogenesis is hormonally induced but also that it is terminated in a highly integrated manner.

Regulation by a short-loop feedback

The simplest mechanism to regulate vitellogenesis would be a short inhibitory loop resulting from the increasing levels of circulating vitellogenin. Indeed, this mechanism has been proposed for mosquito vitellogenesis (Raikhel, 1992). In connection with this, it has been shown that lysosomal activity in the fat body contributes to the termination of vitellogenesis in *A. aegypti*, and also that a high concentration of circulating vitellogenin in ovariectomized vitellogenic mosquitoes provides feedback regulation of this lysosomal activity (Raikhel, 1992).

In orthopteroids, the huge amounts of vitellogenin found in the haemolymph of ovariectomized females seems to suggest that the possible inhibitory action of vitellogenin upon its own synthesis would not operate under physiological conditions. Instead, these cases point to the ovarian ecdysteroids as candidates for terminating vitellogenesis.

Ecdysteroids as candidates for terminating vitellogenesis

Experiments involving ovariectomy have been useful in elucidating homeostatic relationships between the ovary and the fat body during vitellogenesis (Maestro *et al.*, 1994, and references therein). A frequent consequence of ovariectomy is the continuous synthesis of vitellogenin and its marked accumulation in the haemolymph. This has been well described in primitive insects such as locusts (*L. migratoria*) and cockroaches (*Byrsotria fumigata, N. cinerea, D. punctata, B. germanica*) (Martín *et al.*, 1996a, and references therein), and suggests that the ovary is involved in the termination of the vitellogenic cycle.

It has also been shown that the ovaries of these species produce considerable amounts of ecdysteroids, and these hormones have been detected in the haemolymph. In both ovary and haemolymph, ecdysteroids reach maximal levels towards the end of vitellogenesis, and haemolymph ecdysteroid levels decrease after ovariectomy (*Blaberus craniifer, N. cinerea, B. germanica, P. americana*) (Romañá, Pascual & Bellés, 1995, and references therein). It seems, therefore, that some of the ecdysteroids produced by the ovary are released to the haemolymph, and they could eventually act on the fat body. Experiments with the fat body of *B. germanica* incubated *in vitro* (D. Martín, N. Pascual, M. D. Piulachs & X. Bellés, unpublished observations) showed that physiological doses of 20-hydroxyecdysone added to the incubation medium inhibited the production of vitellogenin, which supports the above hypothesis.

This inhibitory action of ecdysteroids could be extended to other insects in which vitellogenesis is directed by juvenile hormone. In the case of Zygentoma, the increase in ecdysteroid levels at the end of the reproductive cycle induces a subsequent moult, but could also be related to the termination of vitellogenesis. In Heteroptera, Davey (1993) has suggested that ovarian ecdysteroids in *R. prolixus* might inhibit vitellogenin production. With regard to beetles, ecdysteroids have been detected in the haemolymph and ovaries of *L. decemlineata* (Hagedorn, 1985), although the possible role of ecdysteroids in vitellogenin synthesis has not been investigated.

The contribution of peptides

The first clue to the involvement of an identified peptide in the negative control of vitellogenesis came from the experiments of Carlisle and Loughton in the late 1970s (cited in Kodrík & Goldsworthy, 1995), demonstrating that locust adipokinetic hormone I (AKH I) (pGlu-Leu-Asn-Phe-Thr-Pro-Asn-Trp-Gly-Thr-NH$_2$) inhibited the synthesis of proteins in the fat body of *L. migratoria*. Furthermore, Moshitzky & Applebaum (1990) stated that vitellogenesis by the fat body of *L. migratoria* incubated *in vitro* is inhibited when locust adipokinetic hormone I is added to the incubation medium. This, together with the observation that titres of locust AKH I markedly increase in the haemolymph at the end of egg maturation (Moshitzky & Applebaum, 1990), seemingly implicate AKHs in the negative control of vitellogenesis of *L. migratoria*. More recently, Kodrík & Goldsworthy (1995) have shown that locust AKHs inhibit RNA synthesis *in vitro* in the fat body of *L. migratoria*.

The inhibitory effect of locust AKH I on protein production in *L. migratoria* fat body contrasts with reports describing the stimulatory effect of hypertrehalosaemic hormone (HTH), which belongs to the same peptide family as locust adipokinetic hormone, on protein synthesis in the fat body of the cockroach *Blaberus discoidalis* under certain experimental conditions (Keeley *et al.*, 1991). This suggests that hormonal regulation may be different in locusts and cockroaches.

Allatostatins belonging to the Tyr-Xaa-Phe-Gly-Lys-NH$_2$ family were identified on the basis of their inhibitory action on insect juvenile hormone synthesis (Stay *et al.*, 1994) (see Weaver *et al.*, this volume), but their ubiquity

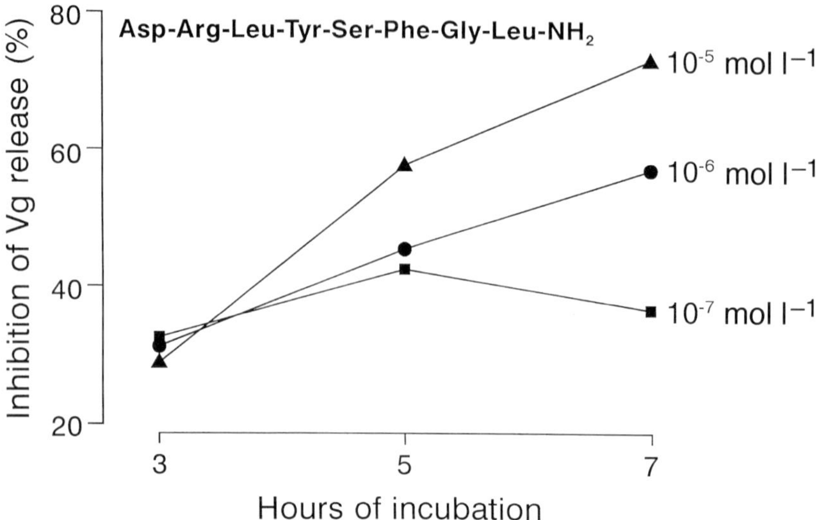

Fig. 6. The inhibition of vitellogenin (Vg) release from fat bodies incubated *in vitro* produced by allatostatin 2 of *Blattella germanica*. Periovaric fat bodies from 4-day-old females of *B. germanica* were used. The inhibition was dose- and incubation time-dependent. Data from Martín *et al.* (1996b).

in the central and peripheral nervous system and gut suggests that they have a broad spectrum of functions (see Duve *et al.*, this volume). In a search for new functions for this peptide family, the effect of one of the allatostatins of *B. germanica* (BLAST-2: Asp-Arg-Leu-Tyr-Ser-Phe-Gly-Leu-NH$_2$, Bellés *et al.*, 1994) on vitellogenin production by fat body incubated *in vitro* (Martín *et al.*, 1996b) was investigated. The results (Fig. 6) showed that the allatostatin inhibited vitellogenin production *in vitro* by fat bodies from four-day-old females, an age which precedes the decline of vitellogenin production (see Fig. 2). Furthermore, results from enhanced chemiluminescence labelling (ECL) western blot analysis suggested that the allatostatin had inhibited the glycosylation of vitellogenin, which would account for the reduction in release of vitellogenin previously observed. Martín *et al.* (1996b) suggested that the allatostatin might have inhibited the mevalonic acid pathway and the synthesis of dolichol, an intermediate necessary to vitellogenin glycosylation. In line with this hypothesis, it was shown that mevalonolactone, used as precursor of the mevalonic acid pathway, was able to restore the normal levels of vitellogenin release in allatostatin-treated fat bodies. Taken together, these results suggest that allatostatins contribute to the termination of the vitellogenic cycle in the cockroach *B. germanica*, by inhibiting the glycosylation, and thus interrupting the release, of vitellogenin.

Epilogue: tailored solutions

The first conclusions that can be drawn from this overview is that hormonal systems regulating vitellogenesis can vary greatly from group to group. However, this is not surprising given the great diversity at all levels (morphological, physiological, ecological and behavioural) not only within the class Insecta, but also within a single order or even a family.

Results from the Zygentoma, the least modified group studied so far, suggest that the primitive trophic hormone in insect vitellogenesis is juvenile hormone, and that ecdysteroids would 'frame' each reproductive cycle. The fact that the most primitive groups within the Pterygota (all the orthopteroids, for example) also use juvenile hormone to direct vitellogenesis supports this hypothesis. It also appears that in these primitive groups, where vitellogenesis is directed by juvenile hormone, ecdysteroids may be involved in terminating the process, occasionally in conjunction with peptide hormones.

The Lepidoptera are a paradigm of plasticity. Juvenile hormone directs vitellogenin synthesis in those species in which oocyte growth is strictly a post-emergence phenomenon. Conversely, in those which develop oocytes in late larval or in pupal stages, stimulation by juvenile hormone would be in conflict with the regulation of metamorphosis, which requires low titres of this hormone. In these lepidopterans, ecdysteroids seem to play the role of gonadotrophic hormone, in the context of the hormonal events of moulting and metamorphosis, which require a high titre of this hormone. Species in which vitellogenin production is initiated a few days before adult eclosion would represent an intermediate case, where both juvenile hormone (perhaps as a primary, capacitatory, factor acting on the fat body and/or the ovary) and ecdysteroids (acting as gonadotrophic hormone to induce the synthesis of vitellogenin in the fat body) may be required for vitellogenesis and oocyte development.

In Diptera, two important developments are found. One, accomplished by the higher dipterans (Cyclorrhapha), is the change from the primitive type of vitellogenin to another protein to fulfil the role of nutrient reserve for the embryo. The other is the shift from juvenile hormone to ecdysteroids to direct vitellogenesis. In dipterans, juvenile hormone continues to participate in the gonadotropic process, but ecdysteroids are the main factor directing the production of vitellogenic proteins.

The shift from juvenile hormone to ecdysteroids to direct vitellogenesis is an intriguing issue. Although the nature of the juvenile hormone receptor remains elusive, it is plausible that belongs to the nuclear receptor superfamily (for a review, see Mangelsdorf *et al.*, 1995), as does the ecdysone receptor. If this were true, then it would open many possibilities of 'interchangeability', so to speak, of both juvenile hormone and ecdysteroids to induce or repress vitellogenin transcription, with discrete changes, either at the ligand- or DNA-binding sites in the receptor, at the site of regulatory elements in vitellogenic protein genes, or even through different combinations with the heterodimer partner (see also Thummel, 1995). In addition, evidence on the induction of transcription of vitellogenic proteins suggests that the mode of action of these hormones is indirect, and is mediated by transcriptional factors (Wyatt, 1991; Deitsch *et al.*, 1995), which opens further possibilities to explain the interchangeability of both kinds of hormone.

Taken together, the evidence suggests that endocrine strategies do not follow strict phylogenetic lines, nor do the hormones have rigid actions along these lines. Rather it seems that the evolution and constraints of the reproductive strategies have modelled hormonal systems of regulation *ad hoc*, and that this has been achieved by using (co-opting, if you wish) the arsenal of hormonal molecules available in a quite flexible manner. Therefore, the most certain conclusion is that stated in the first sentence of this epilogue, and we are just seeing the tip of the iceberg. Every new species studied brings new surprises, and there are more than nine million of them yet to be investigated!

Acknowledgements

I wish to express my sincere appreciation to the DGICYT, Spain, for its sustained support of investigation on 'The reproductive biology in cockroaches'. Thanks are also due to those who contributed to this research program, and especially to M. D. Piulachs, N. Paswal, J. L. Maestro, D. Martin and L. I. Vilaplana.

References

Adams, T. S. & Filipi, P. A. (1988). Interaction between juvenile hormone, 20-hydroxyecdysone, the corpus cardiacum–allatum complex, and the ovaries in regulating vitellogenin levels in the housefly, *Musca domestica. Journal of Insect Physiology* **34**, 11–19.

Adams, T. S. & Gerst, J. W. (1992). Interaction between diet and hormones on vitellogenin levels in the housefly, *Musca domestica. Invertebrate Reproduction and Development* **21**, 91–98.

Agui, N., Shimada, T., Izumi, S. & Tomino, S. (1991). Hormonal control of vitellogenin mRNA levels in the male and female housefly, *Musca domestica. Journal of Insect Physiology* **37**, 383–390.

Bellés, X., Casas, J., Messeguer, A. & Piulachs, M. D. (1987). *In vitro* biosynthesis of JH III by the corpora allata of adult females of *Blattella germanica* (L.). *Insect Biochemistry* **17**, 1007–1010.

Bellés, X., Maestro, J. L., Johnson, A. H., Duve, H. & Thorpe, A. (1994). Allatostatic neuropeptides from the cockroach *Blattella germanica* (L.) (Dictyoptera, Blattellidae). Identification, immunolocalization and activity. *Regulatory Peptides* **53**, 237–247.

Bownes, M., Ronaldson, E., Mauchline, D. & Martínez, A. (1993). Regulation of vitellogenesis in *Drosophila*. *International Journal of Insect Morphology and Embryology* **22**, 349–367.

Casals, N., Buesa, C., Piulachs, M. D., Cabañó, J., Marrero, P. F., Bellés, X. & Hegardt, F. G. (1996). Coordinated expression and activity of 3-hydroxy-3-methylglutaryl coenzyme A synthase and reductase in the fat body of *Blattella germanica* (L.) during vitellogenesis. *Insect Biochemistry and Molecular Biology* **26**, 837–843.

Chen, J.-S., Cho, W.-L. & Raikhel, A. S. (1994). Analysis of mosquito vitellogenin cDNA, similarity with vertebrate phosvitins and arthropod serum proteins. *Journal of Molecular Biology* **237**, 641–647.

Cusson, M., Yu, C. G., Carruthers, K., Wyatt, G. R., Tobe, S. S. & McNeil, J. N. (1994). Regulation of vitellogenin production in Armyworm moths, *Pseudaletia unipuncta*. *Journal of Insect Physiology* **40**, 129–136.

Davey, K. G. (1993). Hormonal control of egg production in *Rhodnius prolixus*. *American Zoologist* **33**, 397–402.

Deitsch, K. W., Dittmer, N., Kapitskaya, M. Z., Chen, J. S., Cho, W.-L. & Raikhel, A. S. (1995). Regulation of gene expression by 20-hydroxy-ecdysone in the fat body of *Aedes aegypti* (Diptera: Culicidae). *European Journal of Entomology* **92**, 237–244.

Dhadialla, T. & Rhaikel, A. (1994). Endocrinology of mosquito vitellogenesis. In *Perspectives in Comparative Endocrinology*, ed. K. G. Davey, R. E. Peter & S. S. Tobe, pp. 275–281. Ottawa: National Research Council of Canada.

Engelmann, F. (1983). Vitellogenesis controlled by juvenile hormone. In *Endocrinology of Insects*, ed. R. G. H. Downer & H. Laufer, pp. 259–270. New York: Alan R. Liss.

Fescemyer, H. W., Masler, E. P., Davis, R. E. & Kelly, T. J. (1992). Vitellogenin synthesis in female larvae of the gypsy moth, *Lymantria dispar* (L.): suppression by juvenile hormone. *Comparative Biochemistry and Physiology* **103B**, 533–542.

Garabedian, M. J., Shirras, A. D., Bownes, M. & Wensink, P. C. (1987). The nucleotide sequence of the gene coding for *Drosophila melanogaster* yolk protein 3. *Gene* **55**, 1–8.

Glinka, A. V. & Wyatt, G. R. (1996). Juvenile hormone activation of gene transcription in locust fat body. *Insect Biochemistry and Molecular Biology* **26**, 13–18.

Hagedorn, H. (1985). The role of ecdysteroids in reproduction. In *Comprehensive Insect Physiology, Biochemistry and Pharmacology*, ed. G. A. Kerkut & L. I. Gilbert, Vol. 8, pp. 205–261. Oxford: Pergamon Press.

Hiremath, S. & Jones, D. (1992). Juvenile hormone regulation of vitellogenin in the gypsy moth, *Lymantria dispar*: Suppression of vitellogenin mRNA in the fat body. *Journal of Insect Physiology* **6**, 461–474.

Hiremath, S., Lehtoma, K. & Nagarajan, M. (1994). cDNA cloning and expression of the large subunit of juvenile hormone-regulated vitellogenin from *Lymantria dispar*. *Journal of Insect Physiology* **40**, 813–821.

Hung, M.-C. & Wensink, P. C. (1983). Sequence and structure conservation in yolk proteins and their genes. *Journal of Molecular Biology* **164**, 481–492.

Imboden, H. & Law, J. H. (1983). Heterogeneity of vitellins and vitellogenins of the tobacco hornworm, *Manduca sexta* L. Time course of vitellogenin appearance in the haemolymph of the adult female. *Insect Biochemistry* **13**, 151–162.

Kageyama, Y., Kinoshita, T., Umesono, Y., Hatakeyama, M. & Oishi, K. (1994). Cloning of cDNA for vitellogenin of *Athalia rosae* (Hymenoptera) and characterization of the vitellogenin gene expression. *Insect Biochemistry and Molecular Biology* **24**, 599–605.

Keeley, L. L. (1985). Physiology and biochemistry of the fat body. In *Comprehensive Insect Physiology, Biochemistry and Pharmacology*, ed. G. A. Kerkut & L. I. Gilbert, Vol. 3, pp. 211–248. Oxford: Pergamon Press.

Keeley, L. L., Hayes, T. K., Bradfield, J. Y. & Sowa, S. M. (1991). Physiological actions by hypertrehalosemic hormone and adipokinetic peptides in adult *Blaberus discoidalis* cockroaches. *Insect Biochemistry* **21**, 121–129.

Kelly, T. J. (1994). Endocrinology of vitellogenesis in *Drosophila melanogaster*. In *Perspectives in Comparative Endocrinology*, ed. K. G. Davey, R. E. Peter & S. S. Tobe, pp. 282–290. Ottawa: National Research Council of Canada.

Kodrík, D. & Goldsworthy, G. J. (1995). Inhibition of RNA synthesis by adipokinetic hormones and brain factor(s) in adult fat body of *Locusta migratoria*. *Journal of Insect Physiology* **41**, 127–133.

Koeppe, J. K., Fuchs, M., Chen, T. T., Hunt, L.-M., Kovalick, G. E. & Briers, T. (1985). The role of juvenile hormone in reproduction. In *Comprehensive Insect Physiology, Biochemistry and Pharmacology*, ed. G. A. Kerkut & L. I. Gilbert, Vol. 8, pp. 165–203. Oxford: Pergamon Press.

Locke, J., White, B. N. & Wyatt, G. R. (1987). Cloning and 5′ end nucleotide sequences of two juvenile hormone-inducible vitellogenin genes of the African migratory locust. *DNA* **6**, 331–342.

Maestro, J. L., Danés, M. D., Piulachs, M. D., Cassier, P. & Bellés, X. (1994). Juvenile hormone inhibition in corpora allata from ovariectomized *Blattella germanica*. *Physiological Entomology* **19**, 342–348.

Mangelsdorf, D. J., Thummel, C., Beato, M., Herrlich, P., Schütz, G., Umesono, K., Blumberg, B., Kastner, P., Mark, M., Chambon, P. & Evans, R. M. (1995). The nuclear receptor superfamily: The second decade. *Cell* **83**, 835–839.

Martín, D., Piulachs, M. D. & Bellés, X. (1995). Patterns of haemolymph vitellogenin and ovarian vitellin in the German cockroach, and the role of juvenile hormone. *Physiological Entomology* **20**, 59–65.

Martín, D., Piulachs, M. D. & Bellés, X. (1996a). Production and extraovarian processing of vitellogenin in ovariectomized *Blattella germanica* (L.) (Dictyoptera, Blattellidae). *Journal of Insect Physiology* **42**, 101–105.

Martín, D., Piulachs, M. D. & Bellés, X. (1996b). Inhibition of vitellogenin production by allatostatin in the German cockroach. *Molecular and Cellular Endocrinology*, **121** 191–196.

Martínez, A. & Bownes, M. (1994). The sequence and expression pattern of the *Calliphora erythrocephala* yolk protein A and B genes. *Journal of Molecular Evolution*, **38**, 336–351.

Moshitzky, P. & Applebaum, S. W. (1990). The role of adipokinetic hormone in the control of vitellogenesis in locusts. *Journal of Insect Physiology* **20**, 319–323.

Raikhel, A. (1992). Vitellogenesis in mosquitoes, *Advances in Disease Vector Research* **9**, 1–39.

Rina, M. & Savakis, C. (1991). A cluster of vitellogenin genes in the Mediterranean fruit fly, *Ceratitis capitata*: sequence and structural conservation in dipteran yolk proteins and their genes. *Genetics* **127**, 769–780.

Romañá, I., Pascual, N. & Bellés, X. (1995). The ovary is a source of circulating ecdysteroids in *Blattella germanica* (L.) (Dictyoptera, Blattellidae). *European Journal of Entomology* **92**, 93–103.

Romans, P., Tu, Z., Ke, Z. & Hagedorn, H. (1995). Analysis of a vitellogenin gene of the mosquito, *Aedes aegypti* and comparisons to vitellogenins from other organisms. *Insect Biochemistry and Molecular Biology* **25**, 939–958.

Rousset, A. & Bitsch, C. (1993). Comparison between endogenous and exogenous yolk proteins along an ovarian cycle in the firebrat *Thermobia domestica* (Insecta, Thysanura). *Comparative Biochemistry and Physiology* **104B**, 33–44.

Shirk, P. D., Bean, D. W. & Brookes, V. J. (1990). Ecdysteroids control vitellogenesis and egg maturation in pharate adult females of the Indian meal moth, *Plodia interpunctella*. *Archives of Insect Biochemistry and Physiology* **15**, 183–199.

Stay, B., Tobe, S. S. & Bendena, W. G. (1994). Allatostatins: identification, primary structures, functions and distribution. *Advances in Insect Physiology* **25**, 267–337.

Thummel, C. S. (1995). From embryogenesis to metamorphosis: the regulation and function of *Drosophila* nuclear receptor superfamily members. *Cell* **83**, 871–877.

Trewitt, P. M., Heilmann, L. J., Degrugillier, S. S. & Kumaran, A. K. (1992). The boll weevil vitellogenin gene: nucleotide sequence, structure, and evolutionary relationship to nematode and vertebrate vitellogenin genes. *Journal of Molecular Biology* **34**, 478–492.

Tsuchida, K., Nagata, M. & Suzuki, A. (1987). Hormonal control of ovarian development in the silkworm, *Bombyx mori*. *Archives of Insect Biochemistry and Physiology* **5**, 167–177.

Valle, D. (1993). Vitellogenesis in insects and other groups. A review. *Memorias del Instituto Oswaldo Cruz* **88**, 1–26.

Wyatt, G. R. (1991). Gene regulation in insect reproduction. *Invertebrate Reproduction and Development* **20**, 1–35.

Yano, K., Sakurai, M. T., Izumi, S. & Tomino, S. (1994a). Vitellogenin gene of the silkworm, *Bombyx mori*: structure and sex-dependent expression. *FEBS Letters* **356**, 207–211.

Yano, K., Sakurai, M. T., Watabe, S., Izumi, S. & Tomino, S. (1994b). Structure and expression of mRNA for vitellogenin in *Bombyx mori*. *Biochimica et Biophysica Acta* **1218**, 1–10.

Yin, Ch.-M. & Stoffoloano, J. G. (1994). Endocrinology of vitellogenesis in blow flies. In *Perspectives in Comparative Endocrinology*, ed. K. G. Davey, R. E. Peter & S. S. Tobe, pp. 291–298. Ottawa: National Research Council of Canada.

Zhang, J., McCracken, A. & Wyatt, G. R. (1993). Properties and sequence of a female-specific, juvenile hormone induced protein from locust hemolymph. *Journal of Biological Chemistry* **268**, 3282–3288.

LEE O. LOMAS and HUW H. REES

Endocrine regulation of development and reproduction in acarines

Introduction

Our knowledge and understanding of developmental hormones in insects and crustaceans has progressed substantially since the demonstration of the hormonal basis of moulting in insects (Kopec, 1917; Wigglesworth, 1934) and crustaceans (Gabe, 1953; Echalier, 1955). Moulting, juvenile, and peptide hormones have been identified and characterised in these classes and, in some instances, the complex interactions between these hormones have been elucidated (for a review, see Gupta, 1990; Kerkut & Gilbert, 1985a,b). Similar information concerning the identification and/or functional significance of hormones in other arthropod groups is, however, fragmented or non-existent. This chapter deals with endocrinological events in arachnids, with specific reference to ticks (acarines).

Ticks

Ticks (Acari: Ixodidea) are obligate blood-feeding parasites of birds, reptiles and mammals that evolved during the mid-late Palaeozoic (Savory, 1977; Hoogstraal, 1983, 1985). Approximately 600 species are distributed within the families Ixodidae (ixodid or hard ticks; approximately 450 species) and Argasidae (argasid or soft ticks; approximately 140 species), with a third family, Nuttalliellidae, represented by only one species with characteristics intermediate between Ixodidae and Argasidae.

Both ixodid and argasid ticks undergo ametabolous development. However, there are fundamental differences in their behavioural and physiological strategies that can make direct endocrinological comparisons between the two families difficult (see Table 1). Where appropriate, such distinctions are highlighted in the text.

Role of the blood meal in ixodid ticks

Female ixodid ticks typically require 7–10 days to fully engorge and increase their body weight approximately 100-fold. According to Balashov (1972) the

Table 1 *Comparison of ixodid and argasid ticks*

	Ixodid	Argasid
Developmental stages	Egg, larvae, nymph (1 instar), adult	Egg, larvae, nymph (2–8 instars), adult
Feeding	Feeding period lasts several days to weeks	Feeding period lasts several minutes to hours
	One blood meal taken during each free-living stage	Several blood meals taken during each free-living stage
	Feeds to 100× unfed weight	Feeds to 3–5× unfed weight
	Secretes excess fluid back into the host via the salivary glands	Secretes excess fluid to the external environment via the coxal glands
Reproduction	Gametogenesis is not completed until after feeding begins (exception is *Ixodes* species)	Gametogenesis is completed during the nymphal–adult moult before feeding begins
	Completes a single gonotrophic cycle, then dies	Completes multiple gonotrophic cycles
	Lays thousands of eggs	Lays tens–hundreds of eggs

first 48 h comprise the preparative feeding phase where the female attaches to the host and the feeding lesion is prepared. During the next five to seven days, the female completes a slow feeding phase where the fed-to-unfed weight ratio increases by a factor of 10. Within this period, the salivary glands become competent to secrete fluid at maximal rates (Kaufman, 1976), undifferentiated 'stem' cells within the gut develop into digestive and secretory cells (for a review, see Coons, Rosell-Davis & Tarnowski, 1986), and oogenesis and spermatogenesis are reinitiated. The final phase of feeding, the rapid engorgement phase, requires an additional 24 h and is characterised by a further 10-fold increase in body weight. The blood meal taken during this time provides the bulk of the nutrient needed to produce the egg mass.

The transition between the slow feeding phase and the rapid engorgement phase (defined as the critical weight or approximately 10 times the unfed weight; Harris & Kaufman, 1984; Lindsay & Kaufman, 1988) appears to be a critical control point for the regulation of many endocrinological events, such as salivary gland degeneration, vitellogenesis and egg laying (see below). Females prematurely removed from a host below this critical weight retain a host-seeking strategy and will reattach to a host if given the opportunity, do not undergo salivary gland degeneration, and will not lay eggs. If females are prematurely removed above the critical weight, they are unable to reattach to a host, their salivary glands degenerate, and they will lay as many eggs as the acquired blood meal will allow.

Tick feeding beyond the critical weight is also dependent on the mating status of the female (Show, 1969; Pappas & Oliver, 1971; Aeschlimann & Grandjean, 1973). If a female has mated (usually within the first three to five days of feeding), she will readily feed beyond the critical weight. If the female remains virgin, feeding is halted at a point just prior to the critical weight. Pappas & Oliver (1972) demonstrated that, in the ixodid *Dermacentor variabilis*, the stimulus for rapid engorgement was due to a chemical factor rather than a volatile pheromone or mechanical stimulation associated with spermatophore transfer. The relationship between mating and feeding in ticks may be similar to that seen in many insects, where chemical signals from the male, usually in the form of 'fecundity-enhancing substances' and 'receptivity-inhibiting substances' promote overall egg production (for a review, see Gillot, 1988).

Overview of insect and crustacean developmental hormone systems

As alluded to earlier, of all the arthropod developmental hormone systems, those of the insects and crustaceans are probably best understood. At this juncture, it will be helpful to provide a brief overview of the major features of these systems for comparative purposes (for a review, see Gupta, 1990).

In insects, brain neurosecretory cells produce the neuropeptide, prothoracicotropic hormone (PTTH), which is released from the neurohaemal organs, corpora cardiaca and in some cases from the corpora allata (Watson, Spaziani & Bollenbacher, 1989). PTTH stimulates ecdysone production (in most species) by prothoracic glands in immature stages. Ecdysone undergoes 20-hydroxylation to the generally more active 20-hydroxyecdysone in several peripheral tissues (Smith, 1985). Similarly, production of juvenile hormones (JHs) by corpora allata may be regulated by two groups of neuropeptides, the allatostatins (Kramer *et al.*, 1991; Bendena *et al.*, 1994) and allatotropins (Kataoka *et al.*, 1989; see Weaver *et al.*, this volume). In immature stages, JHs prevent metamorphosis (Riddiford, 1994) and also inhibit PTTH release (Bollenbacher & Granger, 1985).

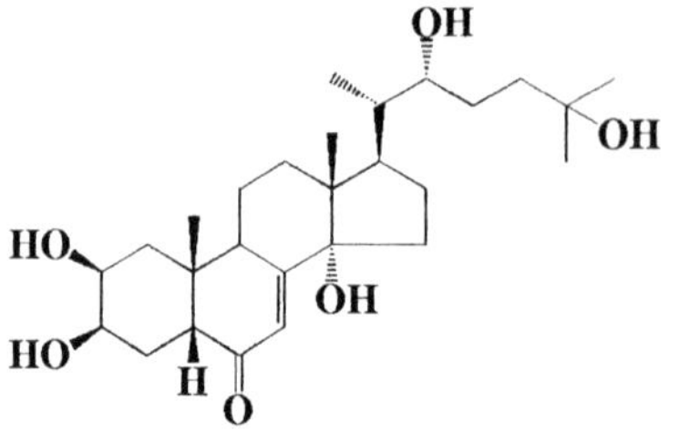

Ecdysone

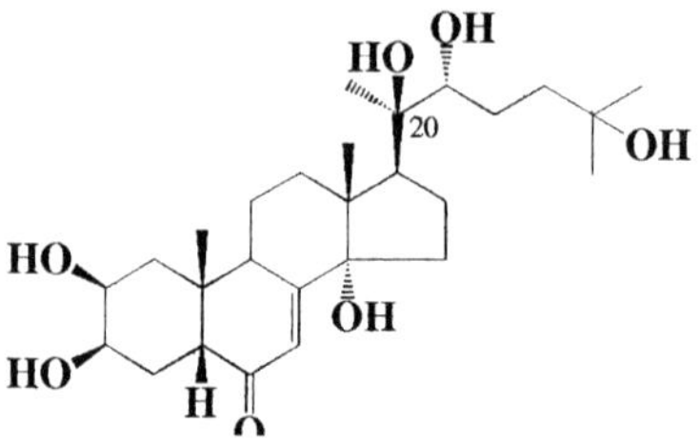

20-Hydroxyecdysone

In adults, JHs function in regulating aspects of reproduction, e.g. vitellogenin synthesis and uptake by ovaries; in Diptera, ecdysteroids are also involved in regulation of vitellogenesis (Koeppe *et al.*, 1985; see Bellés, this volume). The prothoracic glands generally degenerate by the adult stage, when ecdysteroid synthesis primarily occurs in females in the ovarian follicle cells, under the stimulation of brain neuropeptide(s). In males, there is evidence that the testes are a source of ecdysteroids (Hagedorn, 1985).

In Crustacea, the X-organ (part of the X-organ–sinus gland complex of the eyestalk) is a source of neuropeptides, which are stored in the sinus gland prior to release. A major difference compared to insects in the crustacean developmental hormone system is that moult-inhibiting hormone (MIH) negatively regulates production of ecdysteroid by the Y-organs (Webster, 1986, 1991; see Webster, this volume), which can produce 25-deoxyecdysone as well as ecdysone (Lachaise, 1990). As in insects, 20-hydroxylation occurs in certain peripheral tissues.

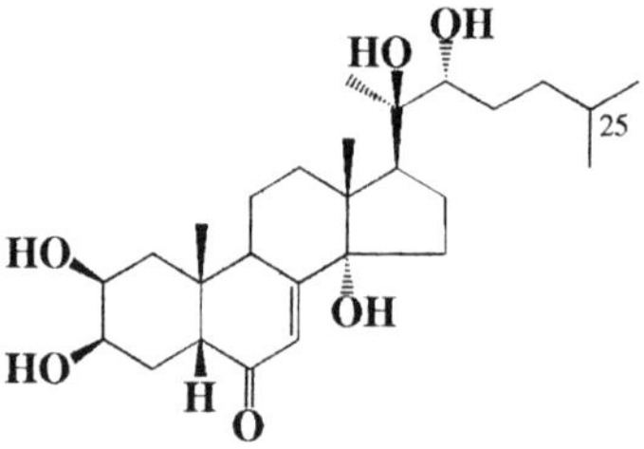

25-Deoxyecdysone

Current evidence indicates that methyl farnesoate (MF) is the crustacean equivalent of the JHs of insects. This putative hormone is produced in the mandibular organs and is negatively regulated by a neuropeptide, mandibular organ-inhibiting hormone (MOIH), produced and secreted by the X-organ–sinus gland complex (Wainwright *et al.*, 1996). Interestingly, MF appears to promote ecdysteroid synthesis by Y-organs (Tamone & Chang, 1993). For a fuller discussion of the crustacean hormones see Webster (this volume).

In common with many other invertebrate phyla, arthropods do not appear capable of sterol synthesis *de novo* and rely upon a dietary source (Pennock, 1977). Work on ecdysteroid biosynthesis has been undertaken primarily in insects and crustaceans (for reviews, see Grieneisen, 1994; Rees, 1995).

Ecdysteroids

Within the limits of the available data, the roles of ecdysteroids in ticks appear largely similar to those in insects and crustaceans. Evidence for this is based mostly on a combination of the observed effects of exogenously administered hormones and correlations of changes in ecdysteroid titres with physiological events. There is still a lack of information on direct effects of the hormones on specific genes. Excellent reviews on the physiological roles of ecdysteroids are available (Diehl, Connat & Dotson, 1986; Sonenshine, 1991; Oliver & Dotson, 1993).

Immature stages

Physiological roles

Regulation of moulting
In vitro feeding of nymphs of the argasid *Ornithodoros porcinus*, with porcine blood containing 20-hydroxyecdysone, accelerated moulting (cited in Solomon, Mango & Obenchain, 1982). However, similar treatment of nymphs of *O. parkeri* with 20-hydroxyecdysone (Campbell & Oliver, 1984) or of *O. moubata* (Diehl *et al.*, 1986) with 22,25-dideoxyecdysone did not affect the moulting period, although the ticks underwent additional moults without taking another blood meal. As with *O. procinus*, in nymphs of the ixodid *Hyalomma dromedarii*, topical application of 20-hydroxyecdysone speeded up moulting (Khalil *et al.*, 1984). As would be expected, the sensitivity of ticks is highly dependent upon the precise developmental time, method of exposure, and dose of ecdysteroid administered (see Sonenshine, 1991). A dose-dependent induction of apolysis and cuticle formation by injection of 20-hydroxyecdysone into female *O. parkeri* has been demonstrated (Pound, Oliver & Andrews, 1984).

There is a close temporal correlation between haemolymph ecdysteroid titres and the moulting process. In fifth-stage nymphs of the argasid *O. moubata*, haemolymph and whole-body ecdysteroid titres have been correlated with the moulting processes, which are induced by the blood meal (Germond, Diehl & Morici, 1982). Between days 2 and 3 post-feeding, a few procuticle lamellae were deposited and the mitotic period initiated.

Ecdysteroid titres were low during these first three days, but then began to increase (days 3–4), with the mitotic period ending on day 4. The titres then rose sharply concomitant with apolysis and formation of the exuvial space (days 4–5), with highest ecdysteroid titres being measured during the deposition of the new epicuticle (days 5–6). Concomitant with the beginning of procuticle deposition and digestion of the old cuticle (day 6), the titres began to decrease, reaching low values shortly before ecdysis (day 9–10). A similar correlation has been observed in nymphs of *O. parkeri* (Zhu *et al.*, 1994) and the ixodids *Amblyomma hebraeum* (Diehl, Germond & Morici, 1982) and *A. variegatum* (Stauffer & Connat, 1990), with evidence for a comparable situation in larvae of *O. moubata* (cited in Diehl *et al.*, 1986; Dotson, Connat & Diehl, 1991). In a careful study on well-synchronised *A. variegatum*, all the integumental events were realised along an anterior–posterior gradient. Furthermore, a small peak of ecdysteroid preceded the major peak of hormone during this nymphal–adult moulting cycle. In addition to many structural changes observed in the integument, salivary glands and dermal glands during this moulting cycle, the muscles of the mouthparts and the pharynx undergo a temporary dedifferentiation at the end of the first small peak of ecdysteroids. At the end of the second peak, when the ecdysteroid titre drops to basal levels, these muscles reappear in an adult differentiated state (Stauffer & Connat, 1990). The occurrence of two ecdysteroid peaks, the first of which appears important in many morphogenetic processes, is reminiscent of the larval–pupal moult in many holometabolous insect species. In all tick species examined, 20-hydroxyecdysone, the presumed major active hormone, accompanied by ecdysone, occurred at high titres (also see Delbecque, Diehl & O'Connor, 1978; Dees, Sonenshine & Breidling, 1984b). Transformation of ecdysone into 20-hydroxyecdysone has widespread tissue distribution, including the midgut (cited in Diehl *et al.*, 1986) and fat body (Zhu, Oliver & Dotson, 1991b).

Although adult ticks do not normally moult, certain argasids, but not ixodids, have been induced experimentally to supermoult by administration of ecdysteroids (for references, see Diehl *et al.*, 1986). Apparently, sensitivity to induction of supermoulting depends on the hormone, the type and timing of application, and the species of tick. In *O. porcinus*, supermoulting resulted in healthy specimens which could subsequently feed, supermoult, and oviposit (Mango, Odhiambo & Galun, 1976). Interestingly, in this case, the speculation was made that supermoulting could occur in nature when the host feeds on phytoecdysteroid-rich plants. On the other hand, supermoulting in *O. moubata* and *O. parkeri* frequently resulted in failure of ecdysis, which rendered subsequent feeding impossible (Connat *et al.*, 1983; Campbell & Oliver, 1984).

Termination of larval diapause

Termination of larval diapause by ecdysteroids has been demonstrated in the ixodids *Dermacentor albipictus* and *Rhipicephalus sanquieneus* (Wright, 1969; Sannasi & Subramonian, 1972). However, the significance of this to the situation *in vivo* remains to be elucidated.

Inactivation

Ecdysteroid inactivation/metabolism in nymphs has been primarily investigated in *O. moubata*. Metabolism of injected [^{3}H]ecdysone has been studied in last instar nymphs of this species at different stages of the moult cycle (Bouvier, Diehl & Morici, 1982). Ecdysone was efficiently converted into 20-hydroxyecdysone, the transformation having wide tissue distribution, being especially active in midgut (cited in Diehl *et al.*, 1986). 20-Hydroxyecdysone is further metabolised to putative 20,26-dihydroxyecdysone. All three ecdysteroids are then converted into polar compounds of unknown structures. This polar pathway was particularly active when endogenous ecdysteroid titres were decreasing, leading to an accumulation of polar products in the digestive tract, suggesting that it is an important inactivation mechanism for endogenous nymphal ecdysteroids. It was not operative with ingested ecdysteroids and also not in females (see Adult females, pp. 100–105).

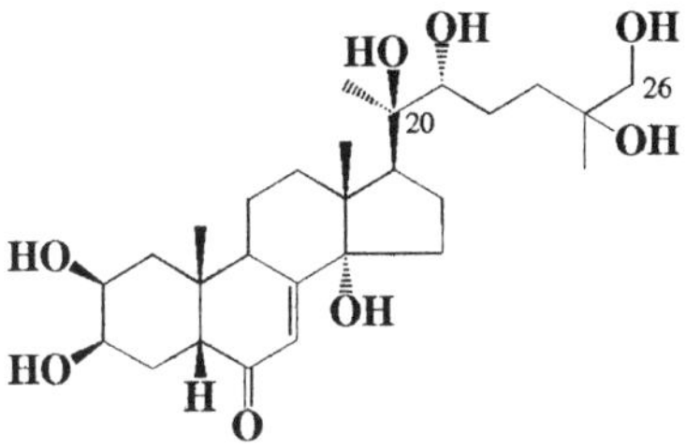

20,26-Dihydroxyecdysone

Injected ecdysone, 20-hydroxyecdysone, or putative 20,26-dihydroxyecdysone were also rapidly metabolised via another pathway into apolar metabolites. Such metabolites were even more rapidly produced after ingestion of ecdysteroids (Connat, Diehl & Thompson, 1986) and have been identified as 22-long-chain fatty acyl esters (Diehl *et al.*, 1985). Subsequently, such apolar conjugates were gradually converted into uncharacterised slightly more polar conjugates. Most ingested hormone was retained in the digestive tract, where it was transformed by the midgut cells into the latter conjugates which accumulated in the midgut lumen (Diehl *et al.*, 1986). Although the physiological significance of the apolar pathway is unclear, it may serve to inactivate ecdysteroids which could be ingested with blood from hosts that feed on plants containing phytoecdysteroids (Diehl *et al.*, 1985).

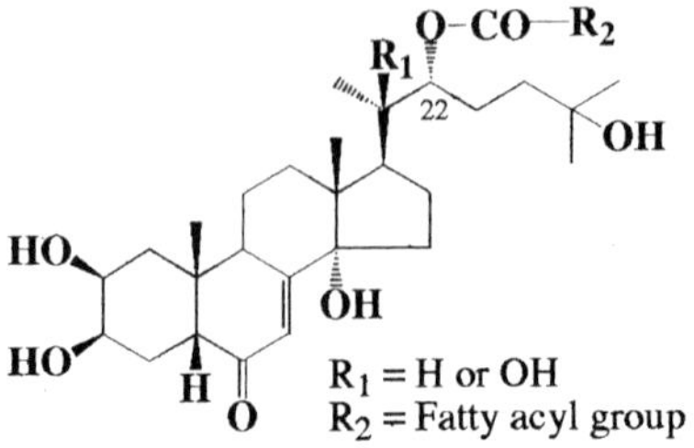

$$R_1 = H \text{ or } OH$$
$$R_2 = \text{Fatty acyl group}$$

22-Long chain fatty acyl ester

The apolar pathway could also inactivate endogenous ecdysteroids. Metabolic pathways of ecdysone in fed nymphs of the ixodid *A hebraeum* were very similar to those observed in *O. moubata* (Diehl *et al.*, 1985).

Clearly, esterification of ecdysteroids at C-22 with fatty acids is a prominent metabolic route in ticks and preliminary evidence suggests that formation of such esters may be widespread amongst arthropods (Connat & Diehl, 1986). In Lepidoptera, ingested ecdysteroids are also inactivated by 22-fatty acylation in the gut (Kubo *et al.*, 1987; Robinson *et al.*, 1987). It is significant that ingested 22,25-dideoxyecdysone was particularly effective in causing supermoulting in *O. moubata*, presumably because it lacks a 22-hydroxyl group and, hence, evades the efficient inactivation by 22-fatty acylation in the midgut (Connat *et al.*, 1983; Connat *et al.*, 1986a).

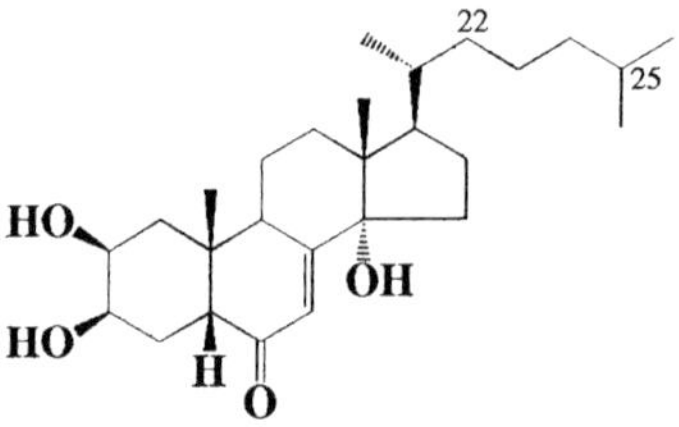

22,25-Dideoxyecdysone

Adult male ticks

Sperm maturation

In male ticks, sperm production and maturation occurs in two phases: spermatogenesis (the development of haploid spermatids through meiotic and mitotic division) and spermiogenesis (the final growth and maturation of spermatids; reviewed in Sonenshine, 1991). Spermatogenesis begins during the nymphal–adult moult, but with the exception of *Ixodes* sp., *Amblyomma triguttatum* (Guglielmone & Moorhouse, 1983), *Aponomma hydrosauri, A. concolor* and *A. varanensis* (Oliver, 1989), stops after the development of secondary spermatogonia at the time of adult ecdysis. Sperm development then remains arrested until after the adult male begins to feed. In the noted exceptions, spermatogenesis proceeds to completion during the nymphal–adult moult, and unfed males are equipped with mature prospermia and are competent to mate.

The mechanism responsible for reinitiation of sperm development is unclear. However, there is evidence to suggest that ecdysteroids are involved. Ecdysteroids are present in males (Oliver, 1986a; Oliver & Dotson, 1993) and injection of 20-hydroxyecdysone into unfed males had potent effects on germ cell DNA synthesis; injection of a crude synganglial extract could mimic this effect (Oliver, 1986a). The biphasic accumulation of differentiated spermatocytes in *D. variabilis* nymphs parallels ecdysteroid titres described in *A. hebraeum* (Dumser & Oliver, 1981; Diehl *et al.*, 1982). Unlike the situation in some insects where there is evidence for ecdysteroid synthesis in testis (Loeb *et al.*, 1982, 1988; Gelman *et al.*, 1989; Jarvis, Earley & Rees, 1994), possibly under the influence of a brain peptide (Loeb, Brandt & Woods, 1986; Loeb *et al.*, 1987), the testes of ticks have not been identified hitherto as a source of ecdysteroids.

Spermiogenesis (the final growth and maturation of spermatids) is further divided into a growth and elongation phase (producing prospermia) and a capacitation phase (sperm maturation; El-Said *et al.*, 1981). Sperm growth and elongation in males is continuous with the completion of spermatogenesis (three to four days after feeding begins), with mature prospermia being found throughout the testis after five days of feeding (Khalil, 1970). Capacitation, the final phase of sperm development, requires 24 hours and is initiated by a 12.5 kDa protein in the male accessory gland (El-Said *et al.*, 1981; Shepherd, Oliver & Hall, 1982).

Non-steroidal mating factors

Mating in ixodid ticks is essential in co-ordinating aspects of the feeding and/or the reproductive cycle (for a detailed overview, see Kaufman & Lomas, 1996). If mating does not occur in feeding ixodid ticks, engorgement is prevented and most females do not feed beyond the critical weight. The stimulus associated with mating is chemical rather than mechanical (Pappas & Oliver, 1972; Oliver, Murphy & Obenchain, 1975) and, to date, three such copulatory factors have been identified. These include a chemical factor derived from within the endospermatophore that initiates the rapid expansion of feeding in *D. variabilis* (Pappas & Oliver, 1972), a high molecular weight protein factor (>400 kDa) secreted by the prospermia after capacitation that is responsible for stimulating oviposition (Germond & Aeschlimann, 1977; Connat *et al.*, 1986b), and a chemical factor derived from the male genital tract which indirectly promotes salivary gland degeneration in engorged female *A. hebraeum* (Lomas & Kaufman, 1992). Although the oviposition stimulating factor and the salivary gland degenerating factor have been established as peptidic in nature, further characterisation has yet to be completed.

Adult females

Regulation of salivary glands

The salivary glands of ixodids comprise three acinar cell types, denoted type I, II and III, in females (four cell types in males) and function primarily in osmotic regulation. During periods of dehydration, type I acini secrete a hygroscopic liquid onto the mouthparts that can absorb water from sub-saturated atmospheres (Rudolph & Knulle, 1974). This hydrated liquid is then ingested and acts as a source of water to rehydrate the tick. During times of feeding, the salivary glands secrete a variety of compounds, such as attachment cements (Moorhouse, 1969), anti-coagulants (Waxman *et al.*, 1990; Vlasuk, 1993; Chinery, 1981) and prostaglandins (Ribeiro *et al.*, 1992), that maintain the flow of blood to the feeding lesion while inhibiting host immune responses directed against the feeding ticks (for reviews, see Sonenshine, 1991; Sauer & Hair, 1986).

The salivary glands are also responsible for secreting the excess water generated by concentration of the blood meal in the gut, back into the host (Kaufman, 1989). Development of the secretory function of the salivary glands is dependent on the stage of feeding. Although adult males secrete little fluid throughout the feeding cycle (Kaufman, 1990), the salivary glands of females develop a high fluid secretory capacity within the first seven days of feeding (Kaufman, 1978). The mechanism for development of a high secretory capacity is poorly understood. However, it has been demonstrated that this development of secretory capacity is under the influence of haemolymph derived factor(s) (Coons & Kaufman, 1988). Once females engorge, they detach from the host and the salivary glands degenerate within four days (Harris & Kaufman, 1984). Using implantation experiments, Harris & Kaufman (1981) demonstrated that this degeneration is triggered by a haemolymph-borne factor, which they called 'tick salivary gland degeneration factor' (TSGDF). Release of TSGDF depends on the degree of engorgement of the female and is probably mediated by abdominal stretch, as severing the neural connections between the abdomen and the synganglion (see p. 110) prevents salivary gland degeneration (Harris & Kaufman, 1984). Considerable evidence has accumulated to suggest that TSGDF is an ecdysteroid. Both ecdysone and 20-hydroxyecdysone mimic the physiological effects of TSGDF when glands are exposed to these compounds *in vivo* (Harris & Kaufman, 1985; Lindsay & Kaufman, 1988).

20-Hydroxyecdysone induces the formation of autophagic vacuoles in the salivary glands of partially fed females, a characteristic indicator of salivary gland degeneration (Lindsay & Kaufman, 1988). The haemolymph ecdysteroid titre in *A. hebraeum* increases markedly during the first few days post engorgement (Connat *et al.*, 1985; Kaufman, 1991), and correlates well with degeneration of the salivary glands (Kaufman, 1991). Ecdysteroid receptors have been characterised in the salivary glands and the number of ecdysteroid receptors increases just prior to salivary gland degeneration (Mao, McBlain & Kaufman, 1995).

The functional significance of salivary gland degeneration is not known, but several possibilities exist. Ecdysteroid receptors in salivary glands may be a 'carryover' from immature stages and salivary gland degeneration may simply be a consequence of the increasing ecdysteroid titre, which is important for aspects of reproduction. More likely, however, is that salivary gland degeneration during oviposition is important for water conservation, preventing the secretion of fluid when such resources are needed for egg production.

Regulation of sex pheromone production

As alluded to earlier, the completion of a successful gonadotrophic cycle in ixodid ticks relies on co-ordinating host location, feeding and mating. This can be somewhat problematic given the free-living lifestyle of ixodids, and consequently, the dependence on 'chance' encounters with suitable hosts and mates. A number of pheromonal systems have thus evolved to ensure that appropriate hosts and mates are found. These include assembly pheromones, aggregation-attachment pheromones and sex pheromones (for reviews, see Sonenshine, 1986, 1991).

Assembly pheromones have been reported in at least 14 argasid and six ixodid tick species (for a review, see Sonenshine, 1986) and evidence suggests that guanine may be the active compound (Otieno *et al.*, 1985). Release of this pheromone causes clustering of unfed ticks, usually in cracks or crevices, and may represent a signal to indicate a suitable microclimate while the tick searches for a suitable host (for discussion, see Sonenshine, 1991).

Aggregation–attachment pheromones have been identified only in *Amblyomma* species and are produced by feeding males. They consist of mixtures of *o*-nitrophenol, methyl salicylate, and nonanonic acid (Schöni *et al.*, 1984) and promote both the excitation of unfed ticks on nearby vegetation, and attachment of unfed males and females on the host. Presumably, these compounds are used by unfed ticks to discriminate tick-infested from uninfested hosts (may be an indication of host suitability), and to promote attachment of males and females in close proximity to facilitate mating.

In ixodid ticks, courtship is a complex process involving a hierarchy of discrete behavioural events co-ordinated by sex pheromones (for reviews, see Sonenshine, 1986, 1991). This process involves attractant sex pheromones (2,6-dichlorophenol; Berger, 1972), mounting sex pheromones (identified as cholesteryl oleate in *D. variabilis*; Hamilton, Sonenshine & Lusby, 1989) and a genital sex pheromone. The genital sex pheromone consists of long-chain saturated fatty acids (Allan *et al.*, 1988). Taylor, Sonenshine & Phillips (1991) provided evidence to suggest that ecdysteroids are also important in genital sex pheromone activity. Ecdysteroids were identified in vaginal tract washings and on the external body surfaces surrounding the genital pore in both *D. andersoni* and *D. variabilis*. If females were neutered by severing the vaginal tract from the genital pore, ecdysteroids could no longer be detected on the external surface and males would not readily copulate. Ecdysone and 20-hydroxyecdysone were found to be highly stimulatory to sexually excited males. When they were artificially introduced into the gential pore of neutered females, males would readily copulate. Taylor *et al.* (1991) also demonstrated that males would sometimes misidentify wound scars as the female gential pore if the wound scars were treated with ecdysone and 20-hydroxyecdysone.

Ecdysteroids may also regulate aspects of sex pheromone production in some ticks. Production of the attractant sex pheromone, 2,6-dichlorophenol, by the foveal glands begins shortly after the nymphal–adult moult. Dees, Sonenshine & Breidling (1984a, 1985) demonstrated an increase in 2,6-dichlorophenol production in unfed *H. dromedarii* females when stimulated by exogenous 20-hydroxyecdysone prior to moulting. Furthermore, Jaffe *et al.* (1986) demonstrated a similar increase in unfed females and unfed males when treated with 22,25-dideoxyecdysone.

Oogenesis and oviposition

As alluded to earlier, in most ticks, feeding and mating are necessary for oogenesis and oviposition to occur. However, there is a paucity of information on the factors involved in regulation of oogenesis and oviposition, with interactions between nervous, endocrine and reproductive systems being involved (for reviews, see Connat *et al.*, 1986; Oliver, 1986b; Oliver & Dotson, 1993). Surprisingly, even endocrine control of vitellogenesis is poorly understood (Oliver & Dotson, 1993). Although ecdysteroids are known to play an important role in oogenesis in some insect species (Hagedorn, 1985), their role in such a process in ticks is uncertain.

Difficulties have been experienced, for some unknown reason, in analysis of ecdysteroids in females of the argasid *O. moubata*, but relatively low amounts of immunoreactive ecdysteroids were observed in haemolymph of vitellogenic females and in eggs (Diehl *et al.*, 1986). In line with eggs of several tick species, *O. moubata* eggs contain maternally derived apolar conjugates of ecdysteroids (Connat & Dotson, 1988). Administration of exogenous ecdysteroids to argasid females not only induces supermoulting, but inhibits oogenesis and causes resorption of previously formed eggs (Diehl *et al.*, 1986). A working hypothesis for possible control of the gonotrophic cycle in *O. moubata* has been suggested, involving action of ecdysteroids on the ovary to induce vitellus resorption (Connat *et al.*, 1987). It has been further suggested that since argasid epidermis is apparently very sensitive to ecdysteroids, it seems unlikely that such steroids play a significant role in oogenesis unless, perhaps, tissues associated with oogenesis are more sensitive to small amounts of them than are epidermal cells (Oliver & Dotson, 1993). Such inhibition of oogenesis has also been reported in ixodids, although there are conflicting results as well (see Diehl *et al.*, 1986).

A detailed study of the ecdysteroid profile during the gonotrophic cycle of an ixodid tick has been undertaken in *A. hebraeum* (Connat *et al.*, 1985). The salivary glands degenerate four days post-engorgement, when the haemolymph and whole-body ecdysteroid titre have increased drastically. This is consistent with another study supporting release of ecdysteroid into the haemolymph when the tick exceeds a 'critical weight', the hormone then triggering salivary gland degeneration (Kaufman, 1991). Interestingly, salivary gland degeneration occurs in parallel with vitellogenesis, which is initiated after day 3, although there is no evidence implicating a role for ecdysteroids in regulation of this process. The haemolymph ecdysteroid titre continues rising to a peak one day before the beginning of oviposition. Although the significance of this peak is uncertain, it could provide a signal for egg laying, analogous to the indirect effect of ecdysteroid in triggering ovulation in *Rhodnius* (Ruegg *et al.*, 1981). In contrast to the haemolymph profile, whole-body ecdysteroid titre continued to increase throughout the beginning of oviposition (Connat *et al.*, 1985). Appreciable amounts of ecdysteroids accumulated in the ovaries and were also passed in the free form into the eggs. Throughout, the major ecdysteroid was 20-hydroxyecdysone accompanied by ecdysone.

Apart from induction of salivary gland degeneration, the significance of the free ecdysteroids in *A. hebraeum* is unclear. Other possibilities include roles in vitellogenesis, egg chorion deposition, or maturation of Gene's organ (involved in egg waxing; Diehl *et al.*, 1986). In fact, in *Ixodes scapularis*, there is preliminary evidence that 20-hydroxyecdysone may induce vitellogenin uptake into the ovary (cited in Oliver & Dotson, 1993). In certain insect and crustacean species, ecdysteroids are involved in meiotic reinitiation in developing oocytes (Lanot *et al.*, 1989; Lanot & Cledon, 1989), but there is no information on such a role of the hormones in ticks.

Significantly, in *Rh. appendiculatus* (ixodid), a distinct peak in free ecdysteroid titre is observed a few days post engorgement and just preceding oviposition (Magee, Jones & Rees, 1996), analogous to the situation in *A. hebraeum*. In *Rh. appendiculatus*, apolar esters were present throughout oviposition and were incorporated, with free ecdysteroids, into the eggs. The situation in *Boophilus microplus* is different in that a major peak of ecdysteroids in whole ticks was observed just before full engorgement (Wigglesworth, Lewis & Rees, 1985). At peak titre, ecdysone was the major hormone accompanied by smaller amounts of 20-hydroxyecdysone. This decrease in ecdysteroid titre before repletion may be accounted for by intense esterification of the hormones at C-22 with fatty acids (Crosby *et al.*, 1986). The esters accumulate in the eggs, together with much lower amounts of free ecdysteroids. As in *A. hebraeum*, a peak of immunoreactive ecdysteroids has been observed in the haemolymph near the beginning of oviposition, but in this case, much of it was apparently esterified (Diehl *et al.*, 1986). Whether there is sufficient free hormone present to influence salivary gland degeneration is unclear. Although the exact significance of the peak in whole-tick free ecdysteroid titre before engorgement is uncertain, it is unlikely to affect gland degeneration at such an early stage. Preliminary results suggest that the ecdysteroid titre changes in *D. variabilis* females show some features analogous to those in *A. hebraeum* (Dees *et al.*, 1984b).

Metabolism and inactivation of ecdysteroids

Ovaries of female *O. moubata* incubated *in vitro* with [^{3}H]ecdysone yield 20-hydroxyecdysone and apolar esters of similar properties to the 22-fatty acyl esters identified in nymphs of this species (Connat, Diehl & Morici, 1984). Such esters accumulated in eggs of females injected with [^{3}H]ecdysone or 20-hydroxyecdysone during vitellogenesis, but remained unchanged during embryonic development (Connat, Dotson & Diehl, 1988). Similarly, during the moulting cycle of larvae, the esters remained unchanged. These observations led to the conclusion that in *O. moubata*, apolar ecdysteroid conjugates are inactivation metabolites that are not reutilized during development of the animal. However, in the ixodid *B. microplus*, the situation appears different (see Embryogenesis). In this species, several tissues, including Malpighian tubules, ovaries, gut, and fat body, could produce these esters from [^{3}H]ecdysone *in vitro* (Wigglesworth *et al.*, 1985). In *A. hebraeum* as well, formation of putative ecdysone 22-fatty acyl esters occurred rapidly in all tissues incubated with [^{3}H]ecdysone *in vitro* (Connat, Lafont & Diehl, 1986c). Although only low amounts of 20-hydroxyecdysone were formed, Malpighian tubules and gut had the highest 20-hydroxylase activity. Surprisingly, a massive conversion of ecdysone into 3-epiecdysone was observed in ovaries, with detection of the expected intermediate, 3-dehydroecdysone, as well. Additionally, a low level of epimerization was observed

3-Epiecdysone　　　　　　　　　*3-Dehydroecdysone*

in the carcass. The extensive formation of apolar esters in these ovary incubations is an enigma in view of the large accumulation of only free hormones (ecdysone and 20-hydroxyecdysone) in the case of endogenous ecdysteroids (Connat *et al.*, 1985), and of *in vivo* labelling experiments, which also demonstrated incorporation of only free ecdysteroids into the newly laid eggs (Connat *et al.*, 1987). It was suggested (Connat *et al.*, 1987) that *in vivo* the ecdysteroids may be bound to vitellogenins/vitellins, as suggested for *Rh. appendiculatus* (Whitehead *et al.*, 1986), and thus be protected from metabolism. Similarly, 3-epiecdysteroids were not detected during studies on [³H]ecdysone metabolism *in vivo* (Connat *et al.*, 1987). It could be that the epimerization enzymes occur in the oocytes for utilisation *in vivo* during embryogenesis, when 3-epiecdysteroid conjugates are formed (cited in Connat *et al.*, 1986c). The latter results for apolar ester and 3-epiecdysteroid formation illustrate that experiments *in vitro* may not always reflect the *in vivo* situation.

Taking the foregoing studies together, as well as a comparative study of ecdysteroids in newly laid eggs of ticks (Connat & Dotson, 1988), it appears that the ecdysteroids occur either in the free form (*A. hebraeum* and *A. variegatum*), or as apolar conjugates (*O. moubata* and *B. microplus*), or a mixture of both forms (*Rh. appendiculatus* and *H. dromedarii*). The significance of incorporation of the free ecdysteroids into eggs of *Amblyomma* species is unknown. Following identification of ecdysteroid 22-fatty acyl esters in newly laid eggs of ticks, such esters were also detected in eggs of certain insect species (for a review, see Isaac & Slinger, 1989). These esters may possibly serve as storage forms of maternal ecdysteroids releasing free hormone during embryogenesis (Slinger & Isaac, 1988; Whiting, Sparks & Dinan, 1993), as was the case for 22-phosphate conjugates in locusts (Rees & Isaac, 1984).

Embryogenesis

During embryogenesis in the ixodid *B. microplus*, electron microscopic evidence was obtained of the successive formation of three embryonic membranes/cuticles before the larval cuticle deposition (Crosby, Lewis & Rees, 1987). Although broad peaks in the free ecdysteroid titre were observed, it was difficult to correlate these with specific cuticular events, presumably owing to lack of synchronisation in the developing embryos. None the less, there is preliminary evidence to suggest that the maternal ecdysteroid 22-fatty acyl esters incorporated into the eggs are utilised in early embryogenesis (T. Crosby, H. H. Rees & L. O. Lomas, unpublished results) and can at least serve as a source of free ecdysteroids until the embryos have differentiated the synthetic machinery. This does not preclude the existence of other sources of hormone. During embryogenesis, there is successive formation of a new class of apolar esters, more polar on reversed-phase high-performance liquid chromatograph (HPLC) than the 22-fatty acyl esters, together with ecdysteroid 26-oic acids, both types of compound being presumed hormone inactivation products. Although the complete structure of these apolar esters is not known, evidence suggests that they may be fatty acyl derivatives of the 26-oic acids (Crosby *et al.*, 1987).

Ecdysteroid 26-oic acid

Interestingly, in *A. hebraeum*, where free maternal ecdysteroids are incorporated into the eggs, the embryonic cuticular events appear similar to *B. microplus*, with no significant peaks of hormone being detected (Dotson, Connat & Diehl, 1995). During embryogenesis, maternally incorporated [³H]ecdysteroid was converted into 22-fatty acyl esters and 3-epimers of the latter, both types of metabolites being regarded as hormone inactivation products. The free ecdysteroids appear to be completely inactivated by the time of hatching. Although the functions of the maternal free ecdysteroids is uncertain, various possibilities have been suggested (Dotson *et al.*, 1995): they may be involved in events during oocyte maturation, such as a role in eggshell production; an involvement in production of the first two embryonic membranes/cuticles is also possible, before the free hormones are inactivated; the possibility exists that the ecdysteroids act as a feeding deterrent to predators.

In an extensive study of embryogenesis and larval development in *O. moubata*, formation of three embryonic 'cuticles'/envelopes was observed with the larval cuticulin starting on day 8 of embryonic development and procuticle deposition continuing after hatching until apolysis of the larval cuticle (Dotson *et al.*, 1991). Although no titre peaks of immunoreactive ecdysteroids were detected during deposition of the three envelopes, a very small peak coincided with the shortening of the germ band and a second distinct peak coincided with deposition of the larval epicuticle. The embryos apparently did not hydrolyse the maternal ecdysteroid apolar esters during development, but, on the contrary, they apparently synthesised ecdysteroids at the times of the peaks and these then seemed to be esterified (Dotson *et al.*, 1991). In fact, it was shown that hanging-drop cultures of embryos could metabolise [³H]ecdysone into 22-fatty acyl esters, 20,26-dihydroxyecdysone and 20-hydroxyecdysone. 20-Hydroxylation first became evident in two-day-old embryos with highest activity occurring during increasing endogenous ecdysteroid titres (Dotson *et al.*, 1993). Whereas esterification occurred in all stages, it was more pronounced during periods of low endogenous ecdysteroid titres. Although the significance of the apolar ecdysteroid esters in *O. moubata* eggs is uncertain, it has been suggested that they may represent a detoxification mechanism (Dotson *et al.*, 1993).

Juvenoids

In insects, juvenile hormones (JHs; juvenoids) function in immature stages in preventing metamorphosis (Riddiford, 1985) and in adults regulate aspects of reproduction (Koeppe *et al.*, 1985). In females, the process of vitellogenin synthesis, secretion into the haemolymph, and uptake into the oocyte where it becomes vitellin, is controlled by JH, and by ecdysteroid in some dipterans (see Bellés, this volume). Although JH has not been identified in ticks, there is much indirect evidence suggesting the occurrence of a juvenoid or JH-like compound in regulating their development and reproduction. A comprehensive coverage of the earlier evidence is beyond the scope of this article and has been reviewed (Solomon, Mango & Obenchain, 1982). More recent reviews are also available (Sonenshine, 1991; Oliver & Dotson, 1993).

The true significance of conflicting reports suggesting effects of juvenoids on moulting in ticks is difficult to assess (Solomon *et al.*, 1982). For example, JH I delayed moulting of nymphs in *O. porcinus* (cited in Solomon *et al.*, 1982) and *H. dromedarii* (Khalil *et al.*, 1984). Interestingly, JH III led to a

Juvenile hormone I

Juvenile hormone III

dose-dependent reversal of 20-hydroxyecdysone-induced supermoulting in mated females of *O. porcinus* and promotion of normal oviposition (Obenchain & Mango, 1980).

Appreciable evidence indicates that ticks possess a gonadotropin of somewhat similar function to insect JH. That administration of synganglion homogenate from fed, mated females induced oviposition in *O. moubata* (Aeschlimann, 1968) is in concordance with ligation experiments implicating the synganglion as the source of gonadotropin (Shanbaky & Khalil, 1975). Other workers have also suggested that the synganglion-lateral organ complex is the putative site of JH/gonadotropin production (Binnington, 1981; Marzouk, Mohamed & Khalil, 1985). However, in these cases, neurosecretory products as well as hormones may be involved in the control processes under study.

Furthermore, experiments aimed at interfering with production of endogenous JH or examining the effects of applying exogenous JHs have furnished further indirect evidence suggesting a role for JH-like compounds in reproduction of ticks. Application of precocene-2, a compound which suppresses corpora allata function and JH production in insects, sterilised the argasids, *Argas persicus, O. coriaceus* and the ixodid *Rh. sanguineus* (Leahy & Booth, 1980). Significantly, such precocene-2-induced sterility was partially reversed by JH III application in *O. parkeri* (Pound & Oliver, 1979). However, in another study of the effects of various anti-JH agents, clear-cut effects of lower doses of precocene analogues on fecundity in *O. moubata* were not observed (Connat & Nepa, 1990).

Topical application of a JH mimic led to termination of reproductive diapause in female *A. arboreus* (Bassal & Roshdy, 1974), whereas similar application of JH I and JH III induced maturation and oviposition in fed, virgin female *O. porcinus* (Obenchain & Mango, 1980). In contrast, a role of JH in egg development in the ixodid *A. hebraeum* could not be demonstrated (Lunke & Kaufman, 1993).

In fed, virgin female *O. moubata*, JH analogues, JH I or JH III induced vitellogenesis and oviposition (Connat, Ducommun & Diehl, 1983). In contrast, in other studies with unfed *O. moubata* or *O. parkeri*, JH and analogues did not stimulate vitellogenesis (Chinzei, Taylor & Ando, 1991; Taylor *et al.*, 1991). Since many of the foregoing studies used fairly high doses of JHs and analogues, the observed effects may not necessarily be direct, but taken together, they are suggestive of the occurrence of a JH-related hormone in ticks. However, known insect JHs could not be detected in haemolymph from vitellogenic *O. moubata* by gas chromatography–mass spectrometry (Connat *et al.*, 1986b).

Indirect evidence supporting the occurrence of JH-like compounds in ixodid ticks has been based largely on biochemical evidence rather than on biological treatments. In insects, a major route of JH inactivation involves hydrolysis by a haemolymph JH-specific esterase, with tissue epoxide hydrolase-catalysed conversion to JH-diol also occurring (Hammock, 1985). In adult female *D. variabilis*, incubation of JHs with haemolymph yielded primarily JH-acid, demonstrating the occurrence of esterase activity, whereas appreciable epoxide hydrolase activity as well was observed with whole body homogenates (Venkatesh *et al.*, 1990). Significantly, these enzymatic activities were influenced by feeding and mating, with different profiles being observed for the JH ester hydrolysis and general esterase activity. This is consistent with a conceivable role for these JH-metabolising enzymes in the control of JH titre during adult development and reproduction.

When haemolymph from adult female *D. variabilis* and *H. dromedarii* was subjected to HPLC with collection of fractions for JH-radioimmunoassay, peaks of immunoreactivity were observed corresponding to the position of JH III (Sonenshine *et al.*, 1989). However, as indicated by Sonenshine *et al.*, owing to likely cross-reactivity of the antiserum with other compounds related to JH III, no conclusions can be drawn regarding the identity of the immunoreactive material.

In insect haemolymph, specific JH-binding proteins (JHPs) are involved in JH transport. Using [^{3}H]EFDA, a photoreactive analogue of JH III, [^{3}H]EFDA-labelled proteins were demonstrated in haemolymph of mated, vitellogenic females, but not of virgin females of *D. variabilis*. The JH III-displacement of the [^{3}H]EFDA binding suggested that the binding proteins recognised JH-like substances in vitellogenic female ticks (Kulcsar, Prestwich & Sonenshine, 1989).

Although the foregoing studies, taken together, are highly suggestive of the occurrence of a juvenoid in ticks, final confirmation awaits definitive identification of such a functional compound.

Sites of hormone synthesis

Nervous and neuroendocrine systems

Structure of the nervous system

The structure of the tick nervous system has previously been extensively described (Obenchain, 1974a,b; Obenchain & Oliver, 1975; Binnington, 1981, 1986, 1987; Sonenshine, 1991). Unlike other arthropods, where segmentation of the body has preserved discrete neuronal segments in the form of

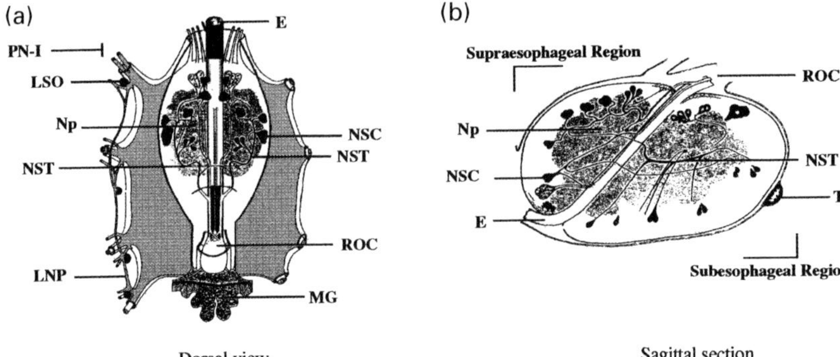

Fig. 1. Diagrammatic reconstruction of the synganglion of the ixodid tick *Dermacentor variabilis* showing a portion of the neurosecretory centres and connecting neurosecretory tracts. (a) Dorsal view. (b) Sagittal section. PN-I, pedal nerve I; LSO, lateral segmental organ; Np, neuropile; NST, neurosecretory tract; LNP, lateral nerve plexus; E, oesophagus; NSC, neurosecretory centre; ROC, retrocerebral organ complex; MG, midgut epithelium; TR, trachea. (Modified from original figure by Obenchain, F. D. & Oliver, J. H. Jr (1975) *Journal of Morphology* **145**, 269–294, Copyright © Journal of Morphology. Reprinted by permission of Wiley-Liss Inc., a subsidiary of John Wiley & Sons, Inc.)

segmental ganglia, ticks have evolved a highly condensed, fused nerve mass in which the cerebral ganglia and ventral nerve cord (with its associated segmental ganglia) have coalesced into a perioesophageal synganglion (Fig. 1; Obenchain & Oliver, 1975; Binnington, 1987). This tissue can be divided into the supra-oesophageal region (equivalent to the protocerebrum, cheliceral, palpal and optic ganglia) and the suboesophageal region (contains the pedal and posterior opisthosomal ganglia; Binnington, 1987). The entire synganglia is ensheathed by an acellular neurilemma and by perineurial glial cells (Binnington, 1987). Tick peripheral nerves, comprising axons surrounded by glial cells, branch laterally from the synganglia, and innervate all organs throughout the body. Associated with the synganglia and peripheral nerves are paired discrete neurohaemal/endocrine organs of unknown function; the lateral segmental organs (lying along the pedal nerve trunks) and the retrocerebral organ complex (situated adjacent to the supra-oesophageal region).

The neuroendocrine system

Neurosecretory and neurohaemal centres

Identification of neurosecretory centres within the synganglia have relied primarily on histological sections based on paraldehyde fuchsin or vital dye (such as methylene blue) staining (Obenchain, 1974a,b; Obenchain & Oliver, 1975). Eighteen such centres have been identified throughout the cortex of the synganglion of *D. variabilis*, with comparable centres identified in the argasid tick, *O. parkeri* (Pound & Oliver, 1984). A complex system of neurosecretory tracts extends between these identified neurosecretory cells, between the supra-oesophageal and suboesophageal regions, and between the retrocerebral organ complex (Obenchain & Oliver, 1975; for a review, see Sonenshine, 1991). In addition, the neurilemma is also innervated by axons from the neurosecretory cells that terminate in a diffuse network of neurosecretory terminals. This may be suggestive of a general release of neurohormones from the neurilemma, and is supported by the observation that, unlike insects, the tick neurilemma/perineurial membrane does not function as a blood–brain barrier and macromolecules can readily penetrate through these layers into the haemolymph (Binnington & Lane, 1980; Hart, Beadle & Botham, 1980).

During periods of desiccation (unfed stages) and after engorgement, secretory products accumulate along the axonal pathways within the supra-oesophageal and suboesophageal regions of the synganglia and extend to the perineural layers, including the retrocerebral organ complex (Obenchain & Oliver, 1975; Binnington & Oliver, 1982). The significance of these accumulations of secretory products is unknown as none of the identified neurosecretory centres have been ascribed functions. The most intriguing observation relating to functional significance came from Zhu *et al.* (1991), who showed that three neurosecretory centres stain strongly with an antibody raised against bovine insulin. This may be significant, since the neuropeptide bombyxin (a prothoracicotropic hormone that stimulates ecdysteroid synthesis in prothoracic glands) from the silkworm, *Bombyx mori*, and bovine insulin have a high degree of amino acid sequence identity (Iwami *et al.*, 1989).

Previously, no neuropeptide has been identified from the tick synganglion that has a demonstrated function in ticks. However, a synganglial peptide(s) has recently been identified in the ixodid tick *A. hebraeum* that is important in the regulation of ecdysteroid synthesis (Lomas, Turner & Rees, 1997). As in immature stages of ixodid and argasid ticks (Zhu *et al.*, 1991a), the epidermal tissue of adult female *A. hebraeum* is competent to synthesise ecdysteroids (Lomas *et al.*, 1997). In *A. hebraeum* females, this synthesis is dependent on the presence of a synganglial peptide(s). Epidermal tissue incubated in the absence of a synganglial extract did not synthesise immunoreactive ecdysteroids, whereas epidermal tissue incubated in the presence of synganglial extracts did synthesise such ecdysteroids. The importance of synganglial factors in stimulating ecdysteroid synthesis in *O. parkeri* nymphs (argasid) has previously been explored using ligation experiments (described in Oliver & Dotson, 1993); however, the results were inconclusive.

Nymphal cuticle formation is stimulated by ecdysteroids. When *O. parkeri* nymphs were ligated in such a manner as to exclude synganglial involvement in the anterior region of the tick, cuticle formation in this region also failed to occur. Endogenous application of 20-hydroxyecdysone to the anterior cuticle was, however, also ineffective in stimulating cuticle formation, and Oliver & Dotson (1993) suggested that continuous stimulation by the synganglion may be necessary for cuticle formation.

As mentioned previously, a peptide factor has recently been identified from the synganglion of *A. hebraeum* that stimulates ecdysteroidogenesis. This neuropeptide–endocrine axis shows several similarities to the pro-thoracicotropic hormone (PTTH)-prothoracic gland interaction of insects. Both PTTH and the synganglial peptide identified in *A. hebraeum* appear to stimulate ecdysteroid synthesis in both a dose- and time-dependent manner. Involvement of cAMP in the action of PTTH is well established (Smith *et al.*, 1996). At least the effect of synganglial peptide on epidermal ecdysteroidogenesis can be mimicked either by cAMP analogues or experimental elevation of endogenous cAMP concentration (Lomas *et al.*, 1997). Thus, there is evidence in *A. hebraeum* for the occurrence of a novel neuro-peptide–second messenger–epidermal ecdysteroidogenesis endocrine axis.

Retrocerebral organ complex

The retrocerebral organ complexes of argasids (*A. arboreus*, Roshdy & Marzouk cited in Sonenshine, 1991; Obenchain & Oliver, 1975) and ixodids (*D. variabilis*, Obenchain & Oliver, 1975; *H. dromedarii*, Marzouk *et al.*, 1985) are believed to resemble the eyestalk neurosecretory systems of crustaceans and the pars intercerebralis–corpora cardiaca–corpora allata system of insects (for comparisons, see Obenchain & Oliver, 1975). Although light and electron microscopy indicate that the retrocerebral organ complex is innervated by a pair of compound nerves that end in putative neurosecretory terminals, a detailed ultrastructural study of this complex in *B. microplus* did not show aggregations of neuroendocrine terminals typical of neurohemal organs (Binnington, 1983). Currently, no known function has been associated with the retrocerebral organ complex. In a recent review, Sonenshine (1991) suggested that the retrocerebral organ complex may be a prime candidate as a site of production of juvenile hormone and gonadotrophic hormone, although there is no evidence for this.

Lateral segmental organs

Associated with the pedal nerves at the level of the synganglial perigan-glionic sheath are two to four pairs of putative endocrine glands collectively called the lateral segmental organs (Binnington, 1981). Ultrastructural studies by Binnington (1981) showed that the lateral segmental organs contain small clusters of cells rich in smooth endoplasmic reticulum, a characteristic of steroid hormone-secreting tissues. The lateral segmental organs were subsequently suggested to be a candidate tissue for ecdysteroid synthesis; however, no direct evidence for such a function has been furnished.

Acknowledgement

We thank The Wellcome Trust for financial support of work from our laboratory.

References

Aeschlimann, A. A. (1968). La ponte chez *Ornithodoros moubata* Murray (Ixodoidea: Argasidae). *Revue Suisse Zoologie* **75**, 1033–1039.

Aeschlimann, A. A. & Grandjean, O. (1973). Influence of natural and artificial mating on feeding, digestion, vitellogenesis and oviposition in ticks (Ixodoidea). *Folia Parasitology* **20**, 67–74.

Allan, S. A., Phillips, J. S., Taylor, D. & Sonenshine, D. E. (1988). Genital sex pheromones of ixodid ticks: evidence for the role of fatty acids from the anterior reproductive tract in mating of *Dermacentor variabilis* and *Dermacentor andersoni. Journal of Insect Physiology* **34**, 315–323.

Balashov, Y. S. (1972). Bloodsucking ticks (Ixodoidea) – Vectors of disease of man and animals. (English translation), ed. H. Hoogsraal & R. J. Tatchell. *Miscellaneous Publications of the Entomological Society of America* **8**, 161–376.

Bassal, T. T. M. & Roshdy, M. A. (1974). *Argas (Persicargas) arboreus*: juvenile hormone analog termination of diapause and oviposition control. *Experimental Parasitology* **36**, 34–39.

Bendena, W. G., Donly, B. C., Ding, Q. & Tobe, S. S. (1994). Molecular characterisation of cockroach allatostatins. In *Perspectives in Comparative Endocrinology*, ed. K. G. Davey, R. E. Peter & S. S. Tobe, pp. 373–377. Ottawa: National Research Council of Canada.

Berger, R. S. (1972). 2,6-Dichlorophenol, sex pheromone of the lone star tick. *Science* **177**, 704–705.

Binnington, K. C. (1981). Ultrastructural evidence for the endocrine nature of the lateral organs of the cattle tick *Boophilus microplus. Tissue and Cell* **13**, 475–490.

Binnington, K. C. (1983). Ultrastructural identification of neurohaemal sites in a tick: evidence that the dorsal complex may be a true endocrine gland. *Tissue and Cell* **15**, 317–327.

Binnington, K. C. (1986). Ultrastructure of the tick neuroendocrine system. In *Morphology, Physiology, and Behavioral Biology of Ticks* ed. J. R. Sauer & J. A. Hair, pp. 152–164. Chichester: Ellis Horwood.

Binnington, K. C. (1987). Histology and ultrastructure of the acarine synganglion. In *Arthropod Brain: Its Evolution, Development, Structure, and Functions*, ed. A. P. Gupta, pp. 95–109. Oxford: Pergamon Press.

Binnington, K. C. & Lane, N. J. (1980). Perineural and glial cells in the tick *Boophilus microplus* (Acarina: Ixodidae): freeze-fracture and tracer studies. *Journal of Neurocytology* **9**, 343–362.

Binnington, K. C. & Oliver, J. H. Jr (1982). Structure and function of the circulatory, nervous and neuroendocrine systems of ticks. In *Physiology of Ticks* ed. F. D. Obenchain & R. Galun, pp. 351–398. Oxford: Pergamon Press.

Bollenbacher, W. E. & Granger, N. A. (1985). Endocrinology of the prothoracicotropic hormone. In *Comprehensive Insect Physiology, Biochemistry and Pharmacology*, Vol. 7, ed. G. A. Kerkut & L. I. Gilbert, pp. 109–151. Oxford: Pergamon Press.

Bouvier, J., Diehl, P. A. & Morici, M. (1982). Ecdysone metabolism in the tick *Ornithodoros moubata* (Argasidae, Ixodoidea). *Revue Suisse de Zoologie* **89**, 967–976.

Campbell, J. D. & Oliver, J. H. Jr (1984). Membrane feeding and developmental effects of ingested β-ecdysone on *Ornithodoros parkeri* (Acari: Argasidae). In *Acarology VI*, Vol. 1, ed. D. A. Griffiths & C. E. Bowman, pp. 393–399. Chichester: Ellis Horwood.

Chinery, W. A. (1981). Observation on the saliva and salivary gland extract of *Haemaphysalis spinigera* and *Rhipicephalus sanguineus*. *Journal of Parasitology* **67**, 15–19.

Chinzei, Y., Taylor, D. & Ando, K. (1991). Effects of juvenile hormone and its analogs on vitellogenin synthesis and ovarian development in *Ornithodoros moubata* (Acari: argasidae). *Journal of Medical Entomology* **28**, 506–513.

Connat, J.-L. & Diehl, P. A. (1986). Probable occurrence of ecdysteroid fatty acid esters in different classes of arthropods. *Insect Biochemistry* **16**, 91–97.

Connat, J.-L., Diehl, P. A., Dumont, N., Carminati, S. & Thompson, M. J. (1983). Effects of exogenous ecdysteroids on the female tick, *Ornithodoros moubata*: Induction of supermolting and influence on oogenesis. *Zeitschrift für Angewandte Entomologie* **96**, 520–530.

Connat, J.-L., Diehl, P. A., Gfeller, H. & Morici, M. (1985). Ecdysteroids in females and eggs of the ixodid tick *Amblyomma hebraeum*. *International Journal of Invertebrate Reproduction and Development* **8**, 103–116.

Connat, J.-L., Diehl, P. A. & Morici, M. (1984). Metabolism of ecdysteroids during the vitellogenesis of the tick *Ornithodoros moubata* (Ixodoidea: Argasidae): accumulation of apolar metabolites in the eggs. *General and Comparative Endocrinology* **56**, 100–110.

Connat, J.-L., Diehl, P. A. & Thompson, M. J. (1986a). Possible inactivation of ingested ecdysteroids by conjugation with long-chain fatty acids in the female tick *Ornithodoros moubata* (Acarina: Argasidae). *Archives of Insect Biochemistry and Physiology* **3**, 235–252.

Connat, J.-L. & Dotson, E. M. (1988). Comparative investigation of the egg ecdysteroids of ticks using radioimmunoassay and metabolic studies. *Journal of Insect Physiology* **34**, 639–645.

Connat, J.-L., Dotson, E. M. & Diehl, P. A. (1987). Metabolism of ecdysteroids in the female tick *Amblyomma hebraeum* (Ixodoidea: Ixodidae): accumulation of free ecdysone and 20-hydroxyecdysone in the eggs. *Journal of Comparative Physiology B* **157**, 689–699.

Connat, J.-L., Dotson, E. M. & Diehl, P. A. (1988). Apolar conjugates of ecdysteroids are not used as a storage form of molting hormone in the argasid tick *Ornithodoros moubata*. *Archives of Insect Biochemistry and Physiology* **9**, 221–235.

Connat, J.-L., Ducommun, J. & Diehl, P. A. (1983). Juvenile hormone-like substances can induce vitellogenesis in the tick *Ornithodoros moubata* (Acarina: Argasidae). *International Journal of Invertebrate Reproduction* **6**, 285–294.

Connat, J.-L., Ducommun, J., Diehl, P. A. & Aeschlimann, A. (1986b). Some aspects of the control of the gonotrophic cycle in the tick *Ornithodoros moubata* (Ixodoidea, Argasidae). In *Morphology, Physiology, and Behavioral Biology of Ticks*. ed. J. R. Sauer & J. A. Hair, pp. 194–216. Chichester, Ellis Horwood.

Connat, J.-L., Lafont, R. & Diehl, P. A. (1986c). Metabolism of [³H]ecdysone by isolated tissues of the female ixodid tick *Amblyomma hebraeum* (Ixodoidea: Ixodidae). *Molecular and Cellular Endocrinology* **47**, 257–267.

Connat, J.-L. & Nepa, M.-C. (1990). Effects of different anti-juvenile hormone agents on the fecundity of the female tick *Ornithodoros moubata*. *Pesticide Biochemistry and Physiology* **37**, 266–274.

Coons, L. B. & Kaufman, W. R. (1988). Evidence that developmental changes in type III acini in the tick *Amblyomma hebraeum* (Acari: Ixodidae) are initiated by a haemolymph borne factor. *Experimental and Applied Acarology* **4**, 117–139.

Coons, L. B., Rosell-Davis, R. & Tarnowski, B. I. (1986). Bloodmeal digestion in ticks. In *Morphology, Physiology and Behavioral Biology of Ticks*, ed. J. R. Sauer & J. A. Hair, pp. 248–279. Chichester: Ellis Horwood.

Crosby, T., Evershed, R. P., Lewis, D., Wigglesworth, K. P. & Rees, H. H. (1986). Identification of ecdysone 22-long-chain fatty acyl esters in newly laid eggs of the cattle tick *Boophilus microplus*. *Biochemical Journal* **240**, 131–138.

Crosby, T., Lewis, D. & Rees, H. H. (1987). Ecdysteroids in adult and embryonic development of the cattle tick, *Boophilus microplus*. VIIIth ecdysone Workshop, Marburg, Germany, Poster 16, p. 41.

Dees, W. H., Sonenshine, D. E. & Breidling, E. (1984a). Ecdysteroids in *Hyalomma dromedarii* and *Dermacentor variabilis* and their effects on sex pheromone activity. In *Acarology VI*, ed. D. A. Griffiths & C. E. Bowman, pp. 405–413. Chichester: Ellis Horwood.

Dees, W. H., Sonenshine, D. E. & Breidling, E. (1984b). Ecdysteroids in the American dog tick, *Dermacentor variabilis* (Acari: Ixodidae), during different periods of tick development. *Journal of Medical Entomology* **21**, 514–523.

Dees, W. H., Sonenshine, D. E. & Breidling, E. (1985). Ecdysteroids in the camel tick, *Hyalomma dromedarii* (Acari: Ixodidae) and comparison with sex pheromone activity. *Journal of Medical Entomology* **22**, 22–27.

Delbecque, J. P., Diehl, P. A. & O'Connor, J. D. (1978). Presence of ecdysone and ecdysterone in the tick *Amblyomma hebraeum* Koch. *Experientia* **34**, 1379–1381.

Diehl, P. A., Connat, J.-L., Girault, J. P. & Lafont, R. (1985). A new class of apolar ecdysteroid conjugates: esters of 20-hydroxy-ecdysone with long-chain fatty acids in ticks. *International Journal of Invertebrate Reproduction and Development* **8**, 1–13.

Diehl, P. A., Connat, J. L. & Dotson, E. M. (1986). Chemistry, function, and metabolism of tick ecdysteroids. In *Morphology, Physiology and Behavioral Biology of Ticks*, ed. J. R. Sauer & J. H. Hair, pp. 165–193. Chichester: Ellis Horwood.

Diehl, P. A., Germond, J. E. & Morici, M. (1982). Correlations between ecdysteroid titres and integument structure in nymphs of the tick, *Amblyomma hebraeum* Koch (Acarina: Ixodidae). *Revue Suisse de Zoologie* **89**, 859–868.

Dotson, E. M., Connat, J.-L. & Diehl, P. A. (1991). Cuticle deposition and ecdysteroid titres during embryonic and larval development of the argasid tick *Ornithodoros moubata*. *General and Comparative Endocrinology* **82**, 386–400.

Dotson, E. M., Connat, J.-L. & Diehl, P. A. (1993). Metabolism of [^{3}H]ecdysone in embryos and larvae of the tick *Ornithodoros moubata*. *Archives of Insect Biochemistry and Physiology* **23**, 67–78.

Dotson, E. M., Connat, J.-L. & Diehl, P. A. (1995). Ecdysteroid titre and metabolism and cuticle deposition during embryogenesis of the ixodid tick *Amblyomma hebraeum* (Koch). *Comparative Biochemistry and Physiology* **110B**, 155–166.

Dumser, J. B. and Oliver, J. H. Jr (1981). Kinetics of spermatogenesis, cell-cycle analysis, and testis development in nymphs of the tick *Dermacentor variabilis*. *Journal of Insect Physiology* **27**, 743–753.

Echalier, G. (1955). Rôle de l'organe Y dans le déterminisme de la mue de *Carcinides (Carcinus) mænas* L. (Crustacés Décapodes): expériences d'implantation. *Comptes Rendus de l'Académie des Sciences, Paris, Série D* **240**, 1581–1583.

El-Said, A., Swiderski, Z., Aeschlimann, A. and Diehl, P. A. (1981). Fine structure of spermiogenesis in the tick *Amblyomma hebraeum* (Acari: Ixodidae): Late stages of differentiation and structure of the mature spermatozoon. *Journal of Medical Entomology* **18**, 464–476.

Gabe, M. (1953). Sur l'existence, chez quelques Crustacés Malacostracés, d'un organe comparable à la grande de mue des Insects. *Comptes Rendus de l'Académie des Sciences, Paris, Série D* **237**, 1111–1113.

Gelman, D. B., Woods, C. W., Loeb, M. J. & Borkovec, A. B. (1989). Ecdysteroid synthesis by testes of 5th instar and pupae of the European corn borer, *Ostrinia nubilalis* (Hubner). *Invertebrate Reproduction and Development* **15**, 177–184.

Germond, J.-E. & Aeschlimann, A. A. (1977). Influence of copulation on vitellogenesis and egg-laying in *Ornithodoros moubata* Murry (Ixodoidea: Argasidae). In *Advances in Inveretebrate Reproduction*, ed. K. G. Adiyodi & R. G. Adiyodi, pp. 308–318. Amsterdam: Elsevier.

Germond, J.-E., Diehl, P. A. & Morici, M. (1982). Correlations between integument structure and ecdysteroid titres in fifth-stage nymphs of the tick, *Ornithodoros moubata*. *General and Comparative Endocrinology* **46**, 255–266.

Gillot, C. (1988). Arthropoda – Insecta. In *Reproductive Biology of Invertebrates*, Vol. III. *Accessory Sex Glands* ed. K. G. Adiyodi & R. G. Adiyodi pp. 319–471. Chichester: John Wiley and sons.

Grieneisen, M. L. (1994). Recent advances in our knowledge of ecdysteroid biosynthesis in insects and crustaceans. *Insect Biochemistry and Molecular Biology* **24**, 115–132.

Guglielmone, A. A. & Moorhouse, D. E. (1983). Copulation and successful insemination by unfed *Amblyomma triguttatum triguttatum* Koch. *Journal of Parasitology* **69**, 786–787.

Gupta, A. P. (ed.) (1990). *Morphogenetic Hormones of Arthropods*. New Brunswick: Rutgers University Press.

Hagedorn, H. H. (1985). The role of ecdysteroids in reproduction. In *Comprehensive Insect Physiology, Biochemistry and Pharmacology*, Vol. 8, ed. G. A. Kerkut & L. I. Gilbert, pp. 205–262. Oxford: Pergamon Press.

Hamilton, J. G. C., Sonenshine, D. E. & Lusby, W. R. (1989). Cholesteryl oleate; mounting sex pheromone of the hard tick, *Dermacentor variabilis* (Say) (Acari: Ixodidae). *Journal of Insect Physiology* **35**, 873–879.

Hammock, B. D. (1985). Regulation of juvenile hormone titer: degradation. In *Comprehensive Insect Physiology, Biochemistry and Pharmacology*, Vol. 7, ed. G. A. Kerkut & L. I. Gilbert, pp. 431–472. Oxford: Pergamon.

Harris, R. A. & Kaufman, W. R. (1981). Hormonal control of salivary gland degeneration in the ixodid tick *Amblyomma hebraeum*. *Journal of Insect Physiology* **27**, 241–243.

Harris, R. A. & Kaufman, W. R. (1984). Neural involvement in the control of salivary gland degeneration in the ixodid tick *Amblyomma hebraeum*. *Journal of Experimental Biology* **109**, 281–290.

Harris, R. A. & Kaufman, W. R. (1985). Ecdysteroids: possible candidates for the hormone which triggers salivary gland degeneration in the ixodid tick *Amblyomma hebraeum*. *Experientia* **41**, 740–742.

Hart, R. J., Beadle, D. J. & Botham, R. P. (1980). The penetration of ionic lanthanum into the central nervous system of the tick *Amblyomma variegatum*. *Physiological Entomology* **5**, 401–405.

Hoogstraal, H. (1983). Ticks. In *World Animal Science B: Disciplinary Approach II: Parasites, Pests, and Predators*, ed. S. M. Gaafar, W. E. Howard & R. E. Marsh, pp. 347–370. Amsterdam: Elsevier.

Hoogstraal, H. (1985). Argasid and nuttalliellid ticks as parasites and vectors. *Advances in Parasitology* **24**, 135–238.

Isaac, R. E. & Slinger, A. J. (1989). Storage and excretion of ecdysteroids. In *Ecdysone*, ed. J. Koolman, pp. 250–253. Stuttgart: G. Thieme.

Iwami, M., Kawakami, A., Ishizaki, H., Takahasi, S. Y., Adachi, T., Suzuki, Y., Hagawasa, H. & Suzuki, A. (1989). Cloning of a gene encoding bombyxin, an insulin-like brain secretory peptide of the silkmoth *Bombyx mori* with prothoracicotropic activity. *Developmental Growth and Differentiation* **31**, 31–37.

Jaffe, H., Hayes, K. K., Sonenshine, D. E., Dees, W. H., Beveridge, M. & Thompson, M. J. (1986). Controlled release reservoirs system for the delivery of insect steroid analogues against ticks. *Journal of Medical Entomology* **23**, 685–691.

Jarvis, T. D., Earley, F. G. & Rees, H. H. (1994). Ecdysteroid biosynthesis in larval testes of *Spodoptera littoralis*. *Insect Biochemistry and Molecular Biology* **24**, 531–537.

Kataoka, H., Toschi, A., Li, J. P., Carney, R. L., Schooley, D. A. & Kramer, S. G. (1989). Identification of an allatotropin from adult *Manduca sexta*. *Science* **243**, 1481–1483.

Kaufman, W. R. (1976). The influence of various factors on fluid secretion by *in vitro* salivary glands of ixodid ticks. *Journal of Experimental Biology* **64**, 727–742.

Kaufman, W. R. (1978). Actions of some transmitters and their antagonists on salivary secretion in a tick. *American Journal of Physiology* **235**, R76–R81.

Kaufman, W. R. (1989). Tick–host interaction: a synthesis of current concepts. *Parasitology Today* **5**, 47–56.

Kaufman, W. R. (1990). Effect of 20-hydroxyecdysone on the salivary glands of the male tick, *Amblyomma hebraeum*. *Experimental and Applied Acarology* **9**, 87–95.

Kaufman, W. R. (1991). Correlation between haemolymph ecdysteroid titre, salivary gland degeneration and ovarian development in the ixodid tick, *Amblyomma hebraeum* Koch. *Journal of Insect Physiology* **37**, 95–99.

Kaufman, W. R. & Lomas, L. O. (1996). Male factors in ticks: their role in feeding and egg development. *Invertebrate Reproduction and Development*, **30**, 191–198.

Kerkut, G. A. & Gilbert, L. I. (eds.) (1985a). *Comprehensive Insect Physiology, Biochemistry and Pharmacology*, Vol. 7. Oxford: Pergamon Press.

Kerkut, G. A. & Gilbert, L. I. (eds.) (1985b). *Comprehensive Insect Physiology, Biochemistry and Pharmacology*, Vol. 8. Oxford: Pergamon Press.

Khalil, G. M. (1970). Biochemical and physiological studies on certain ticks (Ixodoidea). Gonad development and gametogenesis in *Hyalomma* (H.) *anatolicum excavatum* Koch (Ixodidae). *Journal of Parasitology* **56**, 596–610.

Khalil, G. M., Shaarawy, A. A. A., Sonenshine, D. E. & Gad, S. M. (1984). β-Ecdysone effects on the camel tick, *Hyalomma dromedarii* (Acari: Ixodidae). *Journal of Medical Entomology* **21**, 188–193.

Khalil, G. M., Sonenshine, D. E., Hanafy, H. A. & Abdelmonem, A. E. (1984). Juvenile hormone I effects on the camel tick, *Hyalomma dromedarii* Acari: Ixodidae). *Journal of Medical Entomology* **21**, 561–566.

Koeppe, J. K., Fuchs, M., Chen, T. T., Hunt, L.-M., Kovalich, G. E. & Bries, T. (1985). The role of juvenile hormone in reproduction. In *Comprehensive Insect Physiology, Biochemistry and Pharmacology*, Vol. 8, ed. G. A. Kerkut & L. I. Gilbert, pp. 165–203. Oxford: Pergamon Press.

Kopec, S. (1917). Experiments on metamorphosis of insects. *Bulletin International Academy Cracovie B*, 57–60.

Kramer, S. J., Toschi, A., Miller, C. A., Kataoka, H., Quistad, G. B., Li, J. P., Carmey, R. L. & Schooley, D. A. (1991). Identification of an allatostatin from the tobacco hornworm, *Manduca sexta*. *Proceedings of the National Academy of Sciences USA* **88**, 9458–9462.

Kubo, I., Komatsu, S., Asaka, Y. & De Boer, G. (1987). Isolation and identification of apolar metabolites of ingested 20-hydroxyecdysone in frass of *Heliothis virescens* larvae. *Journal of Chemical Ecology* **13**, 785–794.

Kulcsar, P., Prestwich, G. G. & Sonenshine, D. E. (1989). Detection of binding proteins for juvenile hormone-like substances in ticks by photoaffinity labeling. In *Host Regulated Developmental Mechanisms in Vector Arthropods*, ed. D. Borovsky & A. Spielman, pp. 18–23. Vero Beach: University of Florida Press, IFAS.

Lachaise, F. (1990). Synthesis, metabolism, and effects on molting of ecdysteroids in Crustacea, Chelicerata, and Myriapoda. In *Morphogenetic Hormones of Arthropods*, ed. A. P. Gupta, pp. 276–323. New Brunswick: Rutgers University Press.

Lanot, R., Dorn, A., Gunster, B., Thiebold, J., Lagueux, M. & Hoffmann, J. A. (1989). Functions of ecdysteroids in oocyte maturation and embryonic development of insects. In *Ecdysone*, ed. J. Koolman, pp. 262–270. Stuttgart: George Thieme Verlag.

Lanot, R. & Cledon, P. (1989). Ecdysteroids and meiotic reinitiation in *Palaemon serratus* (Crustacea Decapoda Natantia) and in *Locusta migratoria* (Insecta Orthoptera). A comparative study. *Invertebrate Reproduction and Development* **16**, 169–175.

Leahy, M. G. & Booth, K. S. (1980). Precocene induction of tick sterility and ecdysis failure. *Journal of Medical Entomology* **17**, 18–21.

Lindsay, P. J. & Kaufman, W. R. (1988). Action of some steroids on salivary gland degeneration in the ixodid tick *A. americanum*. *Journal of Insect Physiology* **34**, 351–359.

Loeb, M. J., Brandt, E. P. & Woods, C. W. (1986). Effects of exogenous ecdysteroid titer on endogenous ecdysteroid production *in vitro* by testes of the tobacco budworm, *Heliothis virescens*. *Journal of Experimental Zoology* **240**, 75–82.

Loeb, M. J., Brandt, E. P., Woods, C. W. & Bell, R. A. (1988). Secretion of ecdysteroid by sheets of testes of the gypsy moth, *Lymantria dispar*, and its regulation by testis ecdysiotropin. *Journal of Experimental Zoology* **248**, 94–100.

Loeb, M. J., Brandt, E. P., Woods, C. W. & Borkovec, A. B. (1987). An ecdysiotropic factor from brains of *Heliothis virescens* induces testes to produce immunodetectable ecdysteroid *in vitro*. *Journal of Experimental Zoology* **243**, 275–282.

Loeb, M. J., Woods, C. W., Brandt, E. P. & Borkovec, A. B. (1982). Larval testes of the tobacco budworm. A new source of insect ecdysteroids. *Science* **218**, 896–898.

Lomas, L. O. & Kaufman, W. R. (1992). The influence of a factor from the male genital tract on salivary gland degeneration in the female ixodid tick *Amblyomma hebraeum*. *Journal of Insect Physiology* **38**, 595–601.

Lomas, L. O., Turner, P. C. & Rees, H. H. (1997). A novel neuropeptide–endocrine interaction controlling ecdysteroid production in ixodid ticks. *Proceedings of the Royal Society of London B* **264**, 589–596.

Lunke, M. D. & Kaufman, W. R. (1993). Hormonal control of ovarian development in the tick *Amblyomma hebraeum* Koch (Acari: Ixodidae). *Invertebrate Reproduction and Development* **23**, 25–38.

Magee, R. M., Jones, L. D. & Rees, H. H. (1996). Ecdysteroids in relation to adult development and reproduction in female *Rhipicephalus appendiculatus* (Acari: Ixodidae). *Archives of Insect Biochemistry and Physiology* **31**, 197–206.

Mango, C., Odhiambo, T. R. & Galun, R. (1976). Ecdysone and the super tick. *Nature* **260**, 318–319.

Mao, H., McBlain, W. A. & Kaufman, W. R. (1995). Some properties of the ecdysteroid receptor in the salivary gland of the ixodid tick, *Amblyomma hebraeum*. *General and Comparative Endocrinology* **99**, 340–348.

Marzouk, A. S., Mohamed, F. S. A. & Khalil, G. M. (1985). Neurohemal-endocrine organs in the camel tick, *Hyalomma dromedarii* (Acari: Ixodoidea: Ixodidae). *Journal of Medical Entomology* **22**, 385–391.

Moorhouse, D. E. (1969). The attachment of some ixodid ticks to their natural hosts. *Proceedings of the 2nd International Congress of Acarology* (Budapest) 319–327.

Obenchain, F. D. (1974a). Structure and anatomical relationships of the synganglion in the American dog tick, *Dermacentor variabilis* (Acari: Ixodidae). *Journal of Morphology* **142**, 205–223.

Obenchain, F. D. (1974b). Neurosecretory system of the American dog tick,

Dermacentor variabilis (Acari: Ixodidae). I. Diversity of cell types. *Journal of Morphology* **142**, 433–446.

Obenchain, F. D. & Mango, C. K. A. (1980). Effects of exogenous ecdysteroids and juvenile hormones on female reproductive development in *Ornithodoros p. porcinus. American Zoologist* **20**, Abstract No. 1192.

Obenchain, F. D. & Oliver, J. H. Jr (1975). Neurosecretory system of the American dog tick, *Dermacentor variabilis* (Acari: Ixodidae). II. Distribution of secretory cell types, axonal pathways and putative neuro-hemal-neuroendocrine associations: comparative histological and anatomical implications. *Journal of Morphology* **145**, 269–294.

Oliver, J. H. Jr (1986a). Relationship among feeding, gametogenesis, mating and syngamy in ticks. In *Host Regulated Development Mechanisms in Vector Arthropods*, ed. D. Borovsky & A. Spielman, pp. 93–99. Vero Beach: University of Florida Press, IFAS.

Oliver, J. H. Jr (1986b). Induction of oogenesis and oviposition in ticks. *In Morphology, Physiology and Behavioral Biology of Ticks*, ed. J. R. Sauer & J. A. Hair, pp. 233–247. Chichester: Ellis Horwood.

Oliver, J. H. Jr (1989). Ticks (Acari: Ixodidae). *Annual Review of Ecology Systematics* **20**, 397–430.

Oliver, J. H. Jr & Dotson, E. M. (1993). Hormonal control of molting and reproduction in ticks. *American Zoologist* **33**, 384–396.

Oliver, J. H. Jr, Murphy, R. W. & Obenchain, F. D. (1975). Reproduction in ticks (Acari: Ixodoidea). 4. Effects of mechanical and chemical stimulation on oocyte development in *Amblyomma americanum. Journal of Parasitology* **61**, 782–784.

Otieno, D. A., Hassanali, A., Obenchain, F. D., Sternbert, A. & Galun, R. (1985). Identification of guanine as an assembly pheromone of ticks. *Insect Science Applications* **6**, 667–670.

Pappas, P. J. and Oliver, J. H. J. (1971). Mating necessary for complete feeding of female *Dermacentor variabilis* (Acari: Ixodidae). *Journal of the Georgia Entomological Society* **6**, 122–124.

Pappas, P. J. and Oliver, J. H. Jr (1972). Reproduction in ticks (Acari: Ixodoidea). Analysis of the stimulus for rapid and complete feeding of *Dermacentor variabilis* (Say). *Journal of Medical Entomology* **9**, 47–50.

Pennock, J. F. (1977). Terpenoids in marine invertebrates. In *International Reviews of Biochemistry*, Vol. 14, *Biochemistry of Lipids*, ed. T. W. Goodwin, pp. 153–213. Baltimore: University Park Press.

Pound, J. M. & Oliver, J. H. Jr (1979). Juvenile hormone: evidence of its role in the reproduction of ticks. *Science* **206**, 355–357.

Pound, J. M. & Oliver, J. H. Jr (1984). Synganglial and neurosecretory morphology of female *Ornithodoros parkeri* (Cooley) (Acari: Argasidae). *Journal of Morphology* **173**, 159–177.

Pound, J. M., Oliver, J. H. Jr & Andrews, R. H. (1984). Induction of apolysis and cuticle formation in female *Ornithodoros parkeri* (Acari: Argasidae) by hemocoelic injections of β-ecdysone. *Journal of Medical Entomology* **21**, 612–614.

Rees, H. H. (1995). Ecdysteroid biosynthesis and inactivation in relation to function. *European Journal of Entomology* **92**, 9–39.

Rees, H. H. & Isaac, R. E. (1984). Biosynthesis of ovarian ecdysteroid phosphates and their metabolic fate during embryogenesis in *Schistocerca gregaria*. In *Biosynthesis, Metabolism and Mode of Action of Invertebrate Hormones*, ed. J. Hoffmann & M. Porchet, pp. 181–195. Berlin: Springer-Verlag.

Ribeiro, J. M. C., Evans, P. M., MacSwain, J. L. & Sauer, J. (1992). *Amblyomma americanum*: Characterization of salivary prostaglandins E_2 and $F_{2\alpha}$ by RP-HPLC/bioassay and gas chromatograph-mass spectrometry. *Experimental Parasitology* **74**, 112–116.

Riddiford, L. M. (1985). Hormone action at the cellular level. In *Comprehensive Insect Physiology, Biochemistry and Pharmacology*, Vol. 8, ed. G. A. Kerkut & L. I. Gilbert, pp. 37–84. Oxford: Pergamon.

Riddiford, L. M. (1994). Cellular and molecular actions of juvenile hormone. I. General considerations and premetamorphic actions. *Advances in Insect Physiology* **24**, 213–274.

Robinson, P. D., Morgan, E. D., Wilson, Y. D. & Lafont, R. (1987). The metabolism of ingested and injected [³H]ecdysone by final instar larvae of *Heliothis armigera*. *Physiological Entomology* **12**, 321–330.

Rudolph, D. & Knulle, W. (1974). Site and mechanism of water vapour uptake from the atmosphere in ixodid ticks. *Nature* **249**, 84–85.

Ruegg, R. P., Kriger, F. L., Davey, K. G. & Steel, C. G. H. (1981). Ovarian ecdysone elicits release of a myotropic ovulation hormone in *Rhodnius* (Insecta: Hemiptera). *International Journal of Invertebrate Reproduction* **3**, 357–361.

Sannasi, A. & Subramoniam, T. (1972). Hormonal rupture of larval diapause in the tick *Rhipicephalus sanguineus* (Lat.). *Experientia* **28**, 666–667.

Sauer, J. R. & Hair, J. A. (eds.) (1986). *Morphology, Physiology, and Behavioral Biology of Ticks*. Chichester: Ellis Horwood.

Savory, T. (1977). *Arachnida*. New York: Academic Press.

Schöni, R., Hess, E., Blum, W. & Ramstein, K. (1984). The aggregation-attachment pheromone of the bont tick *Amblyomma variegatum* Fabricius (Acari: Ixodidae). Isolation, identification, and action of its active components. *Journal of Insect Physiology* **30**, 613–618.

Shanbaky, N. M. & Khalil, G. M. (1975). The subgenus *Persicargas* (Ixodoidea: Argasidae: *Argas*). 22. The effect of feeding on hormonal control of egg development in *Argas (Persicargas) arboreus*. *Experimental Parasitology* **37**, 361–366.

Shepherd, J., Oliver, J. H. Jr & Hall, J. D. (1982). A polypeptide from male accessory glands which triggers maturation of tick spermatozoa. *International Journal of Invertebrate Reproduction* **5**, 129–137.

Slinger, A. J. & Isaac, R. E. (1988). Ecdysteroid titers during embryogenesis of the cockroach, *Periplaneta americana*. *Journal of Insect Physiology* **34**, 1119–1125.

Smith, S. L. (1985). Regulation of ecdysteroid titer: synthesis. In

Comprehensive Insect Physiology, Biochemistry and Pharmacology, Vol. 7, ed. G. A. Kerkut & L. I. Gilbert, pp. 295–341. Oxford: Pergamon Press.

Smith, W. A., Varghese, A. H., Healy, M. S. & Lou, K. J. (1996). Cyclic AMP is a prerequisite messenger in the action of big PTTH in the prothoracic glands of pupal *Manduca sexta. Insect Biochemistry and Molecular Biology* **26**, 161–170.

Snow, K. R. (1969). Life history of *Hyalomma anatolicum anatolicum* Koch 1844 (Ixodoidea: Ixodidae) under laboratory conditions. *Parasitology* **59**, 105–122.

Solomon, K. R., Mango, C. K. A. & Obenchain, F. D. (1982). Endocrine mechanisms in ticks: effects of insect hormones and their mimics on development and reproduction. In *Physiology of Ticks*, ed. F. D. Obenchain & R. Galun, pp. 399–438. Oxford: Pergamon.

Sonenshine, D. E. (1986). Tick pheromones: an overview. In *Morphology, Physiology, and Behavioral Biology of Ticks* ed. J. R. Sauer & J. A. Hair, pp. 342–360. Chichester: Ellis Horwood Ltd.

Sonenshine, D. E. (1991). *Biology of Ticks*, Vol. 1. New York: Oxford University Press.

Sonenshine, D. E., Roe, R. M., Venkatesh, K., Apperson, C., Winder, B., Schriefer, M. E. & Baehr, J. C. (1989). Biochemical evidence of the occurrence of a juvenoid in ixodid ticks. In *Host Regulated Developmental Mechanisms In Vector Arthropods*, ed. D. Borovsky & A. Spielman, pp. 9–17. Vero Beach: University of Florida Press, IFAS.

Stauffer, A. & Connat, J.-L. (1990). Anterioposterior gradient during nymphal–adult moulting cycle of the tropical bond tick, *Amblyomma variegatum* (Acarina: Ixodidae). Correlations between ecdysteroid titers and integument structure. *Roux's Archives of Developmental Biology* **198**, 309–321.

Tamone, S. L. & Chang, E. S. (1993). Methyl farnesoate stimulates ecdysteroid secretion from crab Y-organs *in vitro. General and Comparative Endocrinology* **89**, 425–432.

Taylor, De M., Chinzei, Y., Muria, K. & Ando, K. (1991). Vitellogenin synthesis, processing and hormonal regulation in the tick, *Ornithodoros parkeri* (Acari: Argasidae). *Insect Biochemistry* **21**, 723–733.

Taylor, D., Sonenshine, D. E. & Phillips, J. S. (1991). Ecdysteroids as a - component of the genital sex pheromone in two species of hard ticks, *Dermacentor variabilis* (Say) and *Dermacentor andersoni* Stiles (Acari: Ixodidae). *Experimental and Applied Acarology* **12**, 275–296.

Venkatesh, K., Roe, R. M., Apperson, C. S., Sonenshine, D. E., Schriefer, M. E. & Boland, L. M. (1990). Metabolism of juvenile hormone during adult development of *Dermacentor variabilis* (Acari: Ixodidae). *Journal of Medical Entomology* **27**, 36–42.

Vlasuk, G. P. (1993). Structural and functional characterization of tick and anticoagulant peptide (TAP): a potent and selective inhibitor of blood coagulation factor Xa. *Thrombone Haemostasis* **70**, 212–216.

Wainwright, G., Webster, S. G., Wilkinson, M. C., Chung, J. S. & Rees, H. H. (1996). Structure and significance of mandibular organ-inhibiting hormone in the crab, *Cancer pagurus* – involvement in multihormonal regulation of growth and reproduction. *Journal of Biological Chemistry* **271**, 12749–12754.

Watson, R. D., Spaziani, E. & Bollenbacher, W. E. (1989). Regulation of ecdysone biosynthesis in insects and crustaceans: a comparison. In *Ecdysone*, ed. J. Koolman, pp. 188–203. Stuttgart: G. Thieme.

Waxman, L., Smith, D. E., Arcuri, K. E. & Vlasuk, G. P. (1990). Tick anti-coagulant peptide (TAP) is a novel inhibitor of blood coagulation factor Xa. *Science* **248**, 593–596.

Webster, S. G. (1986). Neurohormonal control of ecdysteroid biosynthesis by *Carcinus maenas* Y-organs *in vitro*, and preliminary characterisation of the putative moult-inhibiting hormone. *General and Comparative Endocrinology* **61**, 237–247.

Webster, S. G. (1991). Amino acid sequence of putative moult-inhibiting hormone from the crab, *Carcinus maenas. Proceedings of the Royal Society of London B* **244**, 247–252.

Whitehead, D. L., Osir, E. W., Obenchain, F. D. & Thomas, L. S. (1986). Evidence for the presence of ecdysteroids and preliminary characterization of their carrier proteins in the eggs of the brown ear tick *Rhipicephalus appendiculatus* (Neumann). *Insect Biochemistry* **19**, 112–133.

Whiting, P., Sparks, S. & Dinan, L. (1993). Ecdysteroids during embryogenesis of the house cricket, *Acheta domesticus*: occurrence of novel ecdysteroid conjugates in developing eggs. *Insect Biochemistry* **23**, 319–329.

Wigglesworth, K. P., Lewis, D. & Rees, H. H. (1985). Ecdysteroid titre and metabolism to novel apolar derivatives in adult female *Boophilus microplus* (Ixodidae). *Archives of Insect Biochemistry and Physiology* **2**, 39–54.

Wigglesworth, V. B. (1934). The physiology of ecdysis in *Rhodinus prolixus* (Hemiptera). II. Factors controlling moulting and 'metamorphosis'. *Quarterly Journal of Microscrope Science* **77**, 191–222.

Wright, J. E. (1969). Hormonal termination of larval diapause in *Dermacentor albipictus. Science* **163**, 390–391.

Zhu, X. X., Oliver, J. H. Jr & Dotson, E. M. (1991a). Epidermis as the source of ecdysone in an argasid tick. *Proceedings of the National Academy of Sciences USA* **88**, 3744–3747.

Zhu, X. X., Oliver, J. H. & Dotson, E. M. (1991b). Immunocytochemical localization of an insulin-like substance in the synganglion of the tick, *Ornithodoros parkeri* (Acari: Argasidae). *Experimental and Applied Acarology* **13**, 153–159.

Zhu, X. X., Oliver, J. H. Jr, Dotson, E. M. & Ren, H. L. (1994). Correlation between ecdysteroids and cuticulogenesis in nymphs of the tick *Ornithodoros parkeri* (Acri: Argasidae). *Journal of Medical Entomology* **31**, 479–485.

DIETRICH SEDLMEIER
and ALEXANDRA SEINSCHE

Ecdysteroid synthesis in the crustacean Y-organ: role of cyclic nucleotides and Ca^{2+}

Introduction

Crustaceans are forced to shed their exoskeleton in order to grow. Moulting is induced by ecdysteroids which are produced in, and released from, the so-called Y-organs, paired glands of ectodermal origin (for reviews, see Spaziani, 1991; Lachaise *et al.*, 1993). According to a dogma in arthropod endocrinology, these animals are unable to synthesise cholesterol. Ecdysteroids are produced from cholesterol taken up with the diet in a series of so far mostly unknown steps. The first step seems to be the conversion of cholesterol to 7-dehydrocholesterol (Warren *et al.*, 1988; Lachaise *et al.*, 1989; Grieneisen, Sakurai & Gilbert, 1991). The next compound, which is produced in a series of unknown reactions, is 5β-ketodiol. This compound is converted into ecdysone by three successive hydroxylation steps taking place at positions C25, C22, and C2 (Meister *et al.*, 1985; Lachaise *et al.*, 1986). To add to the complexity of the system, it has been repeatedly demonstrated in recent years that not only is ecdysone produced by the Y-organ as suggested by *in vitro* studies on some crustacean species (Chang & O'Connor, 1977; Willig & Keller, 1976; Keller & Schmid, 1979), but also 25-deoxyecdysone (25dE) as demonstrated in the shore crab, *Carcinus maenas* (Lachaise *et al.*, 1986), or 3-dehydroecdysone demonstrated in the rock crab, *Cancer antennarius* (Spaziani *et al.*, 1989), *Menippe mercenaria* (Rudolph & Spaziani, 1992), the white shrimp, *Penaeus vannamei* (Blais *et al.*, 1994), the red swamp crayfish, *Procambarus clarkii* (Sonobe *et al.*, 1991) and the crayfish *Orconectes limosus* (Böcking *et al.*, 1993).

Of central importance to the regulation of circulating ecdysteroids levels are those factors that influence the secretory activity of the Y-organ. The moult-inhibiting hormone (MIH) is regarded as the prime regulator of ecdysteroid production in crustaceans (for further details, see Van Herp, this volume, and Webster, this volume).

To gain some insight into the possible mechanisms by which these reactions might be regulated, it seemed logical to start looking at the mode of action of the moult-inhibiting hormone. The ecdysteroid biosynthetic pathway could be regulated either: (a) by influencing synthesis of the enzymes involved in ecdysteroid biosynthesis; or (b) via direct influence on the synthesising enzymes by phosphorylation/dephosphorylation procedures. Here, data in support of these mechanisms are summarised.

Protein synthesis

In *O. limosus*, a significant increase in *de novo* protein synthesis has been demonstrated during the premoult stage as ecdysteroid production is elevated (Dauphin-Villemant, Böcking & Sedlmeier, 1995). The qualitative analysis of newly synthesised Y-organ proteins suggested that in the premoult stage the overall protein synthesis is enhanced and synthesis of specific proteins is also stimulated. In order to determine whether the presence of these newly synthesised proteins is required to permit the large increase in ecdysteroid production in early premoult, the effect of the translation inhibitor cycloheximide was studied at this moulting stage. As a result, incorporation of [^{14}C]leucine was totally inhibited within 2 h, and at the same time ecdysteroid production was rapidly inhibited (70% reduction within 4 h). This effect was corroborated in the crab *C. antennarius*, where leucine incorporation was inhibited to 10% of controls, whereas the maximal effect on ecdysteroid synthesis was an inhibition of 60% after 24 h ((Mattson & Spaziani, 1986a). Actinomycin D depressed RNA synthesis in *C. antennarius* Y-organs, but did not affect ecdysteroid synthesis.

The regulatory neuropeptide MIH is also involved in short-term regulation of ecdysteroid biosynthesis. Treatment with sinus gland extract rapidly inhibited ecdysteroid production *in vitro*, and also induced reproducible methionine incorporation into proteins within a few hours (Dauphin-Villemant *et al.*, 1995). Similar results were observed in studies on crabs (Soumoff & O'Connor, 1982; Mattson & Spaziani, 1986a).

Protein synthesis appears to be involved in both acute (possible effect on cholesterol or early ecdysteroid precursor supply to rate limiting enzymes) and chronic (possible effect on enzyme level) regulation of ecdysteroid production.

Role of cyclic nucleotides
Involvement of cyclic AMP and cyclic GMP

The first experiments concerning this subject were performed with the crab *C. antennarius*. Using an *in vitro* assay developed by Soumoff & O'Connor (1982), Mattson & Spaziani (1985) demonstrated an increase of Y-organ cyclic AMP levels of about 70% above controls in glands treated with eyestalk extract. At the same time, the release of ecdysteroid was decreased. However, in this study a rather large dose (4 eyestalk equivalents) was needed to achieve maximal increase in cyclic AMP levels. Rather surprisingly, when these experiments were repeated using Y-organs from the crayfish *O. limosus*, exposure to rather more physiologically relevant doses of effector (for example, using culture medium in which sinus glands had previously been incubated, diluted to give an equivalent of 0.1 sinus glands per Y-organ) did not result in increases in cyclic AMP, but cyclic GMP values were enhanced by around six-fold within 2 h, and levels increased for about 9 h. (Sedlmeier & Fenrich, 1993).

In the green crab, *C. maenas*, pure MIH (Webster, 1991) was used. Y-organs of premolt animals responded to application of MIH (10 nmol 1^{-1}) with a modest and transient increase of cyclic AMP (two-fold) 1 min after addition of the hormone, which was maintained for about 4 min only. In contrast, cyclic GMP levels increased steadily, with a 10-fold increase after 3 min and reaching a maximum value of 360 pmol Y-organ^{-1} after 30 min, a 60-fold increase above controls (Saïdi *et al.*, 1994). In intermoult animals, there was no effect on cyclic AMP levels, whereas the effect on cyclic GMP levels was comparable to premoult animals.

In view of the facts that *C. maenas* Y-organ membranes display both MIH-receptors and CHH-receptors (Webster, 1993; see also Webster, this volume), that both hormones belong to the same peptide family and show a certain sequence similarity, and that eyestalk extracts or sinus gland material contain both hormones, one has to regard the experiments performed in *O. limosus* and *C. antennarius* with some caution. The effect of high-performance liquid chromatography (HPLC)-purified CHH on cyclic AMP and cyclic GMP levels has been measured in *O. limosus*. High doses (0.11 μg=content of 1 sinus gland) of CHH elevated cyclic GMP levels to a degree that is comparable to the effect of sinus gland extract (0.05 sinus gland equivalents) containing MIH, but no increases in cyclic AMP levels were observed.

Protein kinase activity

The hormonal message represented at the level of cyclic nucleotides is transferred to the relevant protein kinases, in this case cyclic AMP-dependent protein kinase (PKA) and cyclic GMP-dependent protein kinase (PKG). Both enzymes are present in *O. limosus* Y-organs. Eyestalk removal, which removes the source of MIH and CHH peptides, decreases the activity ratio (the ratio of basal versus stimulated enzyme activity) of PKG from 0.8 to 0.5. Incubation of such Y-organs with MIH or sinus gland extract reverses this ratio to the value in intact animals (von Gliscynski & Sedlmeier, 1993). PKA activity is not influenced by eyestalk removal, reflecting the situation at the level of cyclic nucleotides. Baghdassarian *et al.* (1996) have shown that the predominant relevant protein kinase in *C. maenas* Y-organs is PKG. This enzyme has an activation constant for cyclic GMP of about 30 nmol l^{-1}, but is also activated by much higher concentrations of cyclic AMP ($K_a = 1$ μmol l^{-1}). Incubation of premolt Y-organs with MIH for 1 h (which produces a large increase in cyclic GMP levels) increases the activity ratio for PKG about 3.5-fold. From these studies it is not yet clear whether PKG works alone in steroidogenesis or in combination with PKA. Data on the activities of PKA and PKG in *C. antennarius* are not available.

Phosphorylation pattern

If phosphorylation/dephosphorylation mechanisms do indeed play a role in the regulation of ecdysteroid synthesis, phosphorylation of certain proteins due to the action of MIH should take place, and experiments involving ^{32}P radio-labelling followed by electrophoresis might be expected to show some heavily labelled protein bands in the intermoult stage, which then should diminish in early premoult. Indeed, while certain protein bands in *O. limosus* intermoult Y-organs (17, 68, 90 kDa) appear to show decreased labelling in premoult, a 95 kDa phosphoprotein increased in abundance as moulting proceeded, suggesting the synthesis of proteins in premoult which for some reason have to be phos-phorylated (Böcking & Sedlmeier, 1994). Unfortunately, it has so far not been possible to identify these proteins. From vertebrates, it is known that cytochromes P450 are involved in steroid synthesis, and it is also known that some of these enzymes are subject to phosphorylation/dephosphorylation mechanisms. However, with regard to crustacean ecdysteroid synthesis we are still awaiting confirmation of the assumption that one or more of the synthesising enzymes in ecdysteroid biosynthesis pathway can be regulated (may it be an inhibitory or stimulatory effect) through phosphorylation/dephosphorylation. In this context, the recent results of Lachaise *et al.* (1996; see also Lachaise & Sommé, this volume), are of particular interest. These workers have shown that activity of a transaldolase, which is highly expressed by the Y-organ, is strongly correlated to the ecdysteroidogenic capacity of *C. maenas* Y-organs. It is entirely possible that one action of MIH would be to decrease the activity or synthesis of transaldolase, one of the enzymes of the non-oxidative part of the pentose phosphate pathway involved in generation of NADPH, because cytochromes P450 are NADPH dependent. Indeed, evidence has been presented showing a reduction in transaldolase activity in MIH-treated Y-organs (Lachaise *et al.*, 1996; see Lachaise & Sommé, this volume).

Role of calcium

Mattson & Spaziani (1986b) observed enhanced ecdysteroid synthesis in *C. antennarius* Y-organs after elevation of extracellular calcium. In addition,

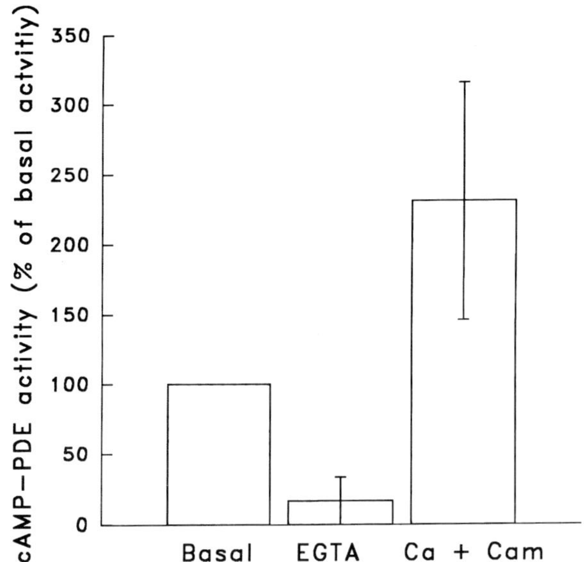

Fig. 1. Cyclic AMP-phosphodiesterase activity in the Y-organ of the cray-fish *O. limosus*. Phosphodiesterase activity was estimated according to the method of Thompson *et al.* (1979). Basal activity was measured in the presence of 1 μmol cyclic AMP 1^{-1}. EGTA, activity in the presence of 1 μmol cyclic AMP 1^{-1} +5 mmol EGTA 1^{-1}; Ca-Cam, activity in the presence of 1 μmol cyclic AMP 1^{-1}+0.8 mmol Ca^{2+} 1^{-1}+2 μg calmodulin. Results are presented as the means±SD of six experiments.

ecdysteroid production was stimulated by application of the calcium ionophore A23187. The fact that A23187 coincidentally decreased basal and stimulated Y-organ cyclic AMP levels led to the conclusion that calcium counteracts the activity of MIH by stimulating a Ca^{2+}-calmodulin-dependent phosphodiesterase (PDE), thus decreasing MIH-induced cyclic AMP-levels. The evidence for a calmodulin-dependent PDE is based upon the observation that the enzyme is inhibited by the calmodulin-inhibitor tri-fluoperazine by about 38% (Mattson & Spaziani, 1986b).

In *O. limosus*, both a cyclic AMP-PDE and a cyclic GMP-PDE have been found. The cyclic AMP-PDE seems to be a Ca^{2+}-calmodulin-dependent enzyme, because the presence of the Ca^{2+}-chelator EGTA almost completely abolished the activity of the enzyme. Addition of calcium plus calmodulin increase activity above basal levels (Fig. 1). In contrast, the cyclic GMP-PDE was completely insensitive to calcium-calmodulin (Fig. 2). Thus, in *O. limosus*, where cyclic GMP levels are enhanced by MIH, a link between cyclic GMP levels and calcium via the appropriate PDE seems not to exist.

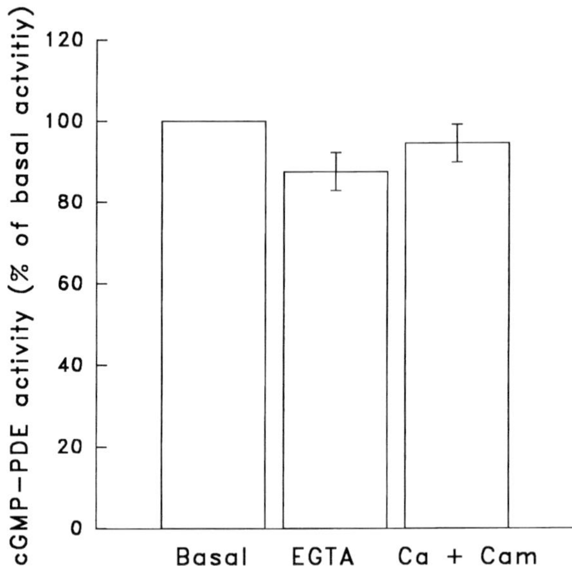

Fig. 2. Cyclic GMP-phosphodiesterase activity of *O. limosus* Y-organs. Basal activity was estimated in the presence of 1 μmol cyclic GMP l^{-1}; EGTA, activity in the presence of 1 μmol cyclic GMP l^{-1}+5 mmol EGTA l^{-1}; Ca-Cam, enzyme activity in the presence of 1 μmol cyclic GMP l^{-1}+0.8 mmol Ca^{2+} l^{-1}+1 μg calmodulin. Results are presented as the means±SD of six experiments.

As in *C. antennarius* (Mattson & Spaziani 1986b), in *O. limosus* Y-organ cells an increase of ecdysteroid levels due to increases in extracellular calcium was observed. Moreover, addition of the ionophore A23187 augmented ecdysteroid production with maximal values at 10 μmol A23187 l^{-1}. A most significant observation was that the presence of MIH or sinus gland material reduced the stimulatory effect to the same extent as MIH in control incubations (Fig. 3), suggesting that MIH action is not counteracted by calcium.

To assess the role of calcium influx upon ecdysteroid synthesis, different calcium channel blockers were used. Verapamil (an L-type Ca^{2+} channel blocker) and ω-conotoxin (an N-type Ca^{2+} channel blocker) did not have any effect. Nimodipine (an L-type Ca^{2+} channel blocker), flunarizine (a T-type Ca^{2+} channel blocker) and pimozide (a T-type Ca^{2+} channel blocker) reduced ecdysteroid synthesis by about 50% (Fig. 4). This suggests an involvement of L- and T-type calcium channels in the stimulation of ecdysteroid synthesis. [45]Ca studies in *C. antennarius* dispersed Y-organ cells showed no effect of MIH on calcium influx, whereas calcium efflux was increased by about 30% (Mattson & Spaziani 1986b).

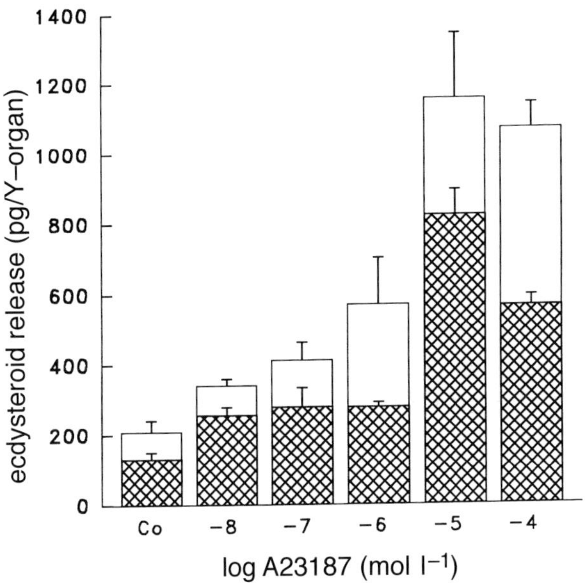

Fig. 3. Ecdysteroid release from *O. limosus* Y-organ cells (prepared according to the method of Toullec & Dauphin-Villemant, 1994) in relation to the ionophore A23187. The extracellular Ca^{2+} concentration was 10 mmol l^{-1}. Crossed bars represent experiments in the presence of MIH (1 sinus gland equivalent). Results are presented as the means$\pm$SD of eight experiments.

Apart from a calcium influx-related enhancement of ecdysteroid synthesis, depletion of intracellular stores by the Ca^{2+}-ATPase inhibitor thapsigargin (Thastrup *et al.*, 1990) resulted in a profound increase in ecdysteroid release (Fig. 5). Again, the presence of MIH reduced the degree of stimulation to an extent that is seen for MIH inhibition in untreated cells, suggesting an action of MIH independent of calcium concentrations.

The action of a many hormones is mediated by the second messenger inositol 1,4,5-trisphosphate (IP_3), resulting in release of calcium from intracellular stores. Incubation of Y-organ cells with various concentrations of IP_3 produced a three-fold increase in ecdysteroid release (Fig. 6). Additionally, the other compound originating in the action of phospholipase C, diacylglycerol (DAG), augmented ecdysteroid production, suggesting a role for protein kinase C (PKC) in the ecdysteroid production pathway (Fig. 7). The presence of a PKC in *C. antennarius* Y-organs was demonstrated by Mattson & Spaziani (1987). These authors assumed that this enzyme is not connected to the cyclic AMP system.

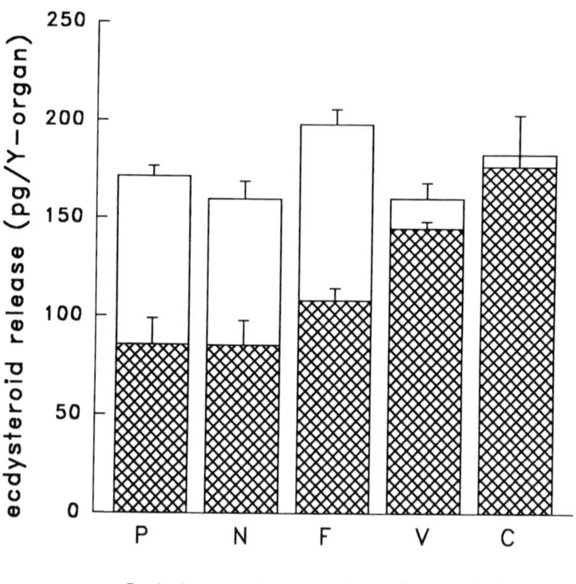

Fig. 4. Ecdysteroid release from *O. limosus* Y-organ cells in the presence of various Ca^{2+} channel blockers. The channel blockers were used at a concentration of $10\ \mu mol\ l^{-1}$. P, pimozide; N, nimodipine; F, flunarizine; V, verapamil; C, ω-conotoxin. Open bars represent controls, crossed bars show cells treated with channel blockers. Results are presented as the means$\pm$SD of eight experiments.

Conclusions

It is quite obvious that there is still an enormous lack of information concerning the regulation of moulting in Crustacea. One of the problems is the fact that we do not know most of the steps in the biosynthetic pathway of ecdysteroids, neither the compounds nor the enzymes involved. Once we do have this information, it will be only a short step to unravel the regulated pathways and the mechanisms involved.

From the results available to date, it is clear that MIH has an inhibitory action on ecdysteroid production (for reviews, see Watson, Spaziani & Bollenbacher, 1989; Smith & Sedlmeier, 1990). It also seems clear, that cyclic nucleotides are involved in the mechanism of action of MIH. Both cyclic nucleotides, cyclic AMP and cyclic GMP, have been shown to be affected by MIH, even if cyclic GMP levels in three species (*O. limosus*, *C. maenas*, and *Callinectes sapidus*) were increased to a much larger degree. In the opinion of the authors, this has to be re-evaluated in view of the fact that *C. maenas* Y-

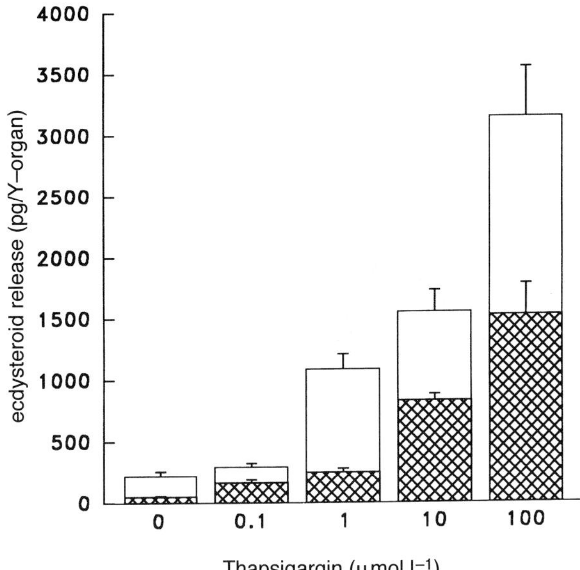

Fig. 5. Ecdysteroid release from *O. limosus* Y-organ cells in the presence of the Ca^{2+} ATPase inhibitor thapsigargin. Open bars represent experiments with various concentrations of thapsigargin, whereas the crossed bars show the effect of simultaneous MIH (1 sinus gland equivalent) treatment. Results are presented as the means±SD of eight experiments.

organs obviously display both MIH- and CHH-receptors (Webster, 1993), and in most of the above-mentioned experiments sinus gland extracts or eyestalk extracts were used. In addition, to date, MIH sequences exist only from crabs (Webster, 1991; Webster, this volume). Thus, it is possible that crayfish (in which CHHs appear to act as MIH) use second messenger systems different from those of crabs, which possess separate CHH and MIH neuropeptides.

Concerning the influence of calcium on ecdysteroid production, it seems clear that this divalent cation has a stimulatory role. Both possibilities of increasing intracellular calcium concentrations seem to be realized. On the one hand, influx from the extracellar space can be demonstrated by the use of Ca^{2+} ionophores, Ca^{2+}-channel blockers and ^{45}Ca. On the other hand, release from intracellular stores can be demonstrated by the application of thapsigargin and by the action of IP_3. The involvement of IP_3 and the effect of DAG point to a stimulatory factor (hormone) that is acting antagonistically to MIH on ecdysteroid production. To what degree both these pathways are interconnected is still an open question. Calcium does not appear to

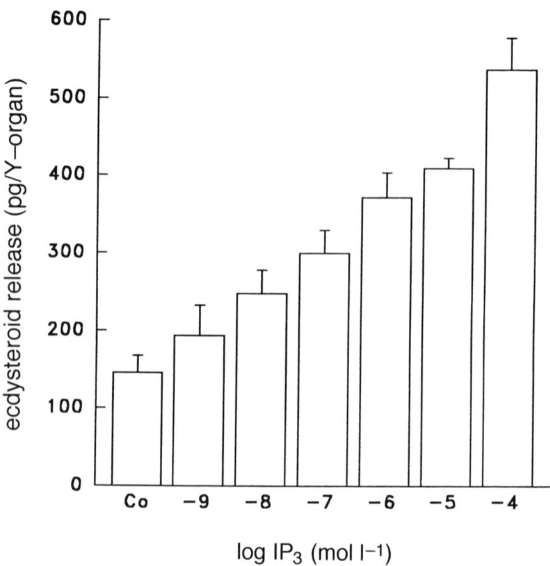

Fig. 6. Ecdysteroid release of *O. limosus* Y-organ cells after treatment with various concentrations of inositol 1,4,5-trisphosphate (IP_3). Results are presented as the means$\pm$SD of eight experiments.

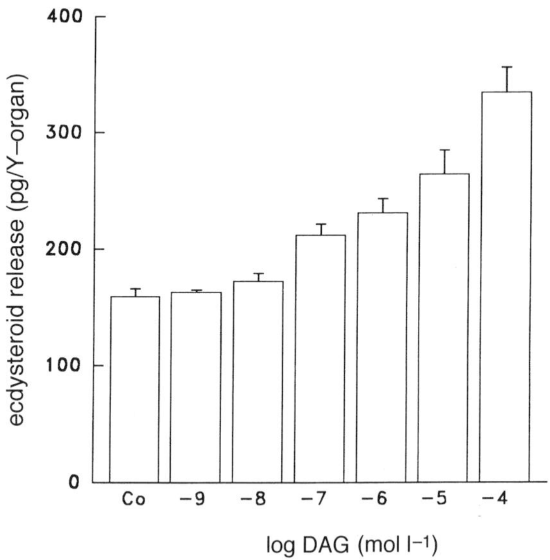

Fig. 7. Release of ecdysteroids from *O. limosus* Y-organ cells by treatment with various concentrations of diacylglycerol (DAG). Results are presented as the means$\pm$SD of eight experiments.

have any effect on the action of MIH, which persists independent of the extracellular Ca^{2+} concentration, in the presence of Ca^{2+}-channel blockers, and is not influenced by the intracellular Ca^{2+} concentration. On the contrary, co-treatment with MIH prevented stimulation of ecdysteroid production by calcium.

References

Baghdassarian, D., De Bessé, N., Saïdi, B., Sommé, G. & Lachaise, F. (1996). Neuropeptide-induced inhibition of steroidogenesis in crab moulting glands: involvement of cGMP-dependent protein kinase. *General and Comparative Endocrinology* **104**, 41–51.

Blais, C., Sefiani, M., Toullec, J.-Y. & Soyez, D. (1994). In vitro production of ecdysteroids by Y-organs of *Penaeus vannamei* (Crustacea, Decapoda). Correlation with hemolymph titers. *Invertebrate Reproduction and Development* **26**, 3–12.

Böcking, D., Dauphin-Villemant, C., Sedlmeier, D., Blais, C. & Lafont, R. (1993). Ecdysteroid biosynthesis in moulting glands of the crayfish *Orconectes limosus*: evidence for the synthesis of 3-dehydroecdysone by in vitro synthesis and conversion studies. *Insect Biochemistry and Molecular Biology* **23**, 57–63.

Böcking, D. & Sedlmeier, D. (1994). Protein phosphorylation in the moulting gland of the crayfish, *Orconectes limosus*: Role of cyclic nucleotides, calcium, and moult inhibiting hormone (MIH). *Invertebrate Reproduction and Development* **26**, 237–245.

Chang, E. S. & O'Connor, J. D. (1977). Secretion of α-ecdysone by crab Y-organs in vitro. *Proceedings of the National Academy of Sciences USA* **74**, 615–618.

Dauphin-Villemant, C., Böcking, D. & Sedlmeier, D. (1995). Regulation of steroidogenesis in crayfish molting glands: involvement of protein synthesis. *Molecular and Cellular Endocrinology* **109**, 97–103.

Grieneisen, M. L., Sakurai, S. & Gilbert, L. I. (1991). A putative route to ecdysteroids: metabolism of cholesterol in vitro by mildly disrupted prothoracic glands of *Manduca sexta*. *Insect Biochemistry* **21**, 41–51.

Keller, R. & Schmid, E. (1979). In vitro secretion of ecdysteroids by Y-organs and lack of secretion by mandibular organs of the crayfish following moult induction. *Journal of Comparative Physiology B* **130**, 347–353.

Lachaise, F., Meister, M. F., Hetru, C. & Lafont, R. (1986). Studies on the biosynthesis of ecdysone by the Y-organs of *Carcinus maenas*. *Molecular and Cellular Endocrinology* **45**, 235–262.

Lachaise, F., Carpentier, G., Sommé, G., Colardeau, J. & Beydon, P. (1989). Ecdysteroid synthesis by crab Y-organs. *Journal of Experimental Zoology* **252**, 283–292.

Lachaise, F., Le Roux, A., Hubert, M. & Lafont, R. (1993). The molting gland of crustaceans: localization, activity, and endocrine control (a review). *Journal of Crustacean Biology* **13**, 198–234.

Lachaise, F., Sommé, G., Carpentier, G., Granjeon, E., Webster, S. G. & Baghdassarian, D. (1996). A transaldolase: an enzyme implicated in crab steroidogenesis. *Endocrine* **5**, 23–32.

Mattson, M. P. & Spaziani, E. (1985). Cyclic AMP mediates the negative regulation of Y-organ ecdysteroid production. *Molecular and Cellular Endocrinology* **42**, 185–189.

Mattson, M. P. & Spaziani, E. (1986a). Regulation of Y-organ ecdysteroidogenesis by molt-inhibiting hormone in crabs: involvement of cyclic AMP-mediated protein synthesis. *General and Comparative Endocrinology* **63**, 414–423.

Mattson, M. P. & Spaziani, E. (1986b). Regulation of crab Y-organ steroidogenesis in vitro: evidence that ecdysteroid production increases through activation of cAMP-phosphodiesterase by calcium-calmodulin. *Molecular and Cellular Endocrinology* **48**, 135–151.

Mattson, M. P. & Spaziani, E. (1987). Demonstration of protein kinase C activity in crustacean Y-organs, and partial definition of its role in regulation of ecdysteroidogenesis. *Molecular and Cellular Endocrinology* **49**, 159–171.

Meister, M., Dimarcq, J. L., Kappler, C., Hétru, C., Lagueux, M., Lanot, R., Luu, B. & Hoffmann, J. A. (1985). Conversion of a radiolabelled ecdysone precursor, 2,22,25-trideoxyecdysone, by embryonic and larval tissues of *Locusta migratoria*. *Molecular and Cellular Endocrinology* **41**, 27–44.

Rudolph, P. H. & Spaziani, E. (1992). Formation of ecdysteroids by the Y-organs of the crab, *Menippe mercenaria*. II. Incorporation of cholesterol into 7-dehydrocholesterol and secretion products in vitro. *General and Comparative Endocrinology* **88**, 235–242.

Saïdi, B., de Bessé, N., Webster, S. G., Sedlmeier, D. & Lachaise, F. (1994). Involvement of cAMP and cGMP in the mode of action of molt-inhibiting hormone (MIH), a neuropeptide which inhibits steroidogenesis in a crab. *Molecular and Cellular Endocrinology* **102**, 53–61.

Sedlmeier, D. & Fenrich, R. (1993). Regulation of ecdysteroid biosynthesis in crayfish Y-organs. I. Role of cyclic nucleotides. *Journal of Experimental Zoology* **265**, 448–453.

Smith, W. A. & Sedlmeier, D. (1990). Neurohormonal control of ecdysone production: comparison of insects and crustaceans. *Invertebrate Reproduction and Development* **18**, 77–89.

Sonobe, H., Kamba, M., Ohta, K., Ikeda, M. & Naya, Y. (1991). In vitro secretion of ecdysteroids by Y-organs of the crayfish, *Procambarus clarkii*. *Experientia* **47**, 948–952.

Soumoff, C. & O'Connor, J. D. (1982). Repression of Y-organ secretory activity by molt-inhibiting hormone in the crab, *Pachygrapsus crassipes*. *General and Comparative Endocrinology*, **48**, 432–439.

Spaziani, E. (1991). Morphology, histology, and ultrastructure of the ecdysial gland (Y-organ) in Crustacea. In *Morphogenetic Hormones of Arthropods: Embryonic and Postembryonic Sources*, Vol. I, ed. A. P. Gupta, pp. 233–267. New Brunswick: Rutgers University Press.

Spaziani, E., Rees, H. H., Wang, W. L. & Watson, R. D. (1989). Evidence that Y-organs of the crab *Cancer antennarius* secrete 3-dehydroecdysone. *Molecular and Cellular Endocrinology* **66**, 17–25.

Thastrup, O., Cullen, P. J., Drobak, B. K., Hanley, M. R. & Dawson, A. P. (1990). Thapsigargin, a tumor promotor, discharges intracellular Ca^{2+} stores by specific inhibition of the endoplasmic reticulum Ca^{2+}-ATPase. *Proceedings of the National Academy of Sciences USA* **87**, 2466–2470.

Thompson, W. J., Terasaki, W. L., Eckstein, P. M. & Strada, S. J. (1979). Assay of cyclic nucleotide phosphodiesterase and resolution of molecular forms of the enzyme. In *Advances in Cyclic Nucleotide Research*, Vol. 10, ed. G. Brooker, P. Greengard & G. A. Robison, pp. 69–92. New York: Raven Press.

Toullec, J.-Y. & Dauphin-Villemant, C. (1994). Dissociated cell suspensions of *Carcinus maenas* Y-organs as a tool to study ecdysteroid production and its regulation. *Experientia* **50**, 153–158.

von Gliscynski, U. & Sedlmeier, D. (1993). Regulation of ecdysteroid biosynthesis in crayfish Y-organs: II. Role of cyclic nucleotide-dependent protein kinases. *Journal of Experimental Zoology* **265**, 454–458.

Warren, J. T., Sakurai, S., Rountree, D. B. & Gilbert, L. I. (1988). Synthesis and secretion of ecdysteroids by the prothoracic glands of *Manduca sexta*. *Journal of Insect Physiology* **34**, 571–576.

Watson, R. D., Spaziani, E. & Bollenbacher, W. E. (1989). Regulation of ecdysone biosynthesis in insects and Crustaceans: a comparison. In *Ecdysone. From Chemistry to Mode of Action*, ed. J. Koolman, pp. 188–203. Stuttgart: G. Thieme.

Webster, S. G. (1991). Amino acid sequence of putative moult-inhibiting hormone from the crab *Carcinus maenas*. *Proceedings of the Royal Society of London B* **244**, 247–252.

Webster, S. G. (1993). High-affinity binding of putative moult-inhibiting hormone (MIH) and crustacean hyperglycaemic hormone (CHH) to membrane-bound receptors on the Y-organ of the shore crab *Carcinus maenas*. *Proceedings of the Royal Society of London B* **251**, 53–59.

Willig, A. & Keller, R. (1976). Biosynthesis of α- and β-ecdysone by the crayfish *Orconectes limosus* in vivo and by its Y-organs in vitro. *Experientia* **3**, 936–937.

FABIENNE LACHAISE
and GHISLAINE SOMMÉ

Regulation of steroidogenesis: role of transaldolase in crab moulting glands

Regulation of steroidogenesis – principles established in vertebrates

It is now generally accepted that regulatory mechanisms involved in steroid hormone biosynthesis may influence this process over very brief time scales of a few minutes, or may involve long-term processes leading to maintenance of steroidogenic capacity in the steroidogenic tissues, and that such regulation generally involves cyclic AMP as a second messenger (Waterman & Simpson, 1989). Short-term regulation involves cholesterol mobilisation from lipid stores within steroidogenic cells to the inner mitochondrial membrane, where it is converted to pregnenolone by cytochrome $P450_{scc}$ (scc, side chain cleavage; Hanukoglu, 1992). This acute regulation involves various types of protein, which are either constitutive, for example sterol carrier protein 2 (SCP2; Pfeifer *et al.*, 1993), diazepam binding inhibitor (DBI; Garnier *et al.*, 1994) or labile, for example steroidogenic activator peptide (SAP) in conjunction with GTP (Xu *et al.*, 1991) and the mitochondrial phosphoproteins belonging to the PP30 family, such as steroidogenic acute regulatory protein (StAR; Clark *et al.*, 1994). The synthesis of these proteins is stimulated in the presence of cyclic AMP and, for the PP30 family, phosphorylated by protein kinase A. Long-term regulation operates at the level of steroidogenic enzyme synthesis, the expression of which varies depending on tissues (tissue specificity) and during development. Such long-term regulation occurs at the transcriptional level via regulatory proteins such as the steroid hydroxylase-inducing proteins (SHIPs; Simpson & Waterman, 1988) and cyclic AMP-responsive element-binding protein (CREB; Payne & Youngblood, 1995). These proteins are activated by the couple cyclic AMP/protein kinase A and bind to a promoting region of the hydroxylase gene.

There is growing evidence that regulation of steroidogenesis does not result solely from protein regulation of cholesterol translocation and expression of the enzymes of steroid biosynthesis. Other less direct, but equally important, regulatory mechanisms may exist, acting for example on the mitochondrial and microsomal electron transport pathways necessary to support steroid hydroxylase activities in all steroidogenic tissues. For example, in rat adrenal glands the existence of a 43 kDa protein capable of stimulating steroid synthesis has been reported (Paz *et al.*, 1994). Moreover, a flavoprotein necessary for P450 activity was shown to be post-transcriptionally regulated by cyclic AMP, the second messenger in hormonal stimulation of steroidogenesis (Hum & Miller, 1993). In addition, for steroidogenic tissues of vertebrates, it has long been shown that gonadotropins (McKerns & Ryschkewitsch, 1976) and adrenocorticotropic hormone (ACTH) (Criss & McKerns, 1969) stimulate G6PDH (glucose 6-phosphate dehydrogenase: EC 1.1.149), a key enzyme of the pentose phosphate pathway (Schaaff, Hohmann & Zimmerman, 1990). Since the activity of this enzyme was correlated with steroidogenic activity, the assumption has been made that G6PDH plays a role in regulation of steroidogenesis through an increase in NADPH (McKerns, 1966; McKerns & Ryschkewitsch, 1976; Bentley *et al.*, 1990), which is required for activation of P450 cytochrome enzymes necessary for steroidogenesis.

Steroidogenesis in crustaceans

In arthropods, as in vertebrates, the biosynthesis of ecdysteroids from cholesterol by steroidogenic glands (Y-organs in crustaceans) has been shown to involve P450 cytochromes (Greineisen, 1994) which are NADPH-dependent enzymes (Waterman & Simpson, 1989). For crustaceans, in contrast to insects, the hormones regulating steroidogenesis (ecdysteroid synthesis) are uniquely inhibitory: neuropeptides belonging to the so-called crustacean hyperglycemic hormone family, namely moult-inhibiting hormone (MIH) and hyperglycemic hormone (CHH), inhibit ecdysteroid synthesis by the Y-organ in many decapod crustaceans (see chapters by Webster and Sedlmeier & Seinsche, this volume). For insects, it has long been established that the prothoracicotropic hormone (PTTH) stimulates ecdysteroidogenesis. Although the precise mechanisms by which these neuropeptides regulate ecdysteroid production are by no means fully understood in either group, it is clear that biosynthesis of ecdysteroids by the Y-organ requires the continuous synthesis of labile proteins as it is profoundly reduced by protein synthesis inhibitors such as cycloheximide (Mattson & Spaziani, 1986; Dauphin-Villemant, Böcking & Sedlmeier, 1995), and MIH accordingly reduces protein synthesis (Mattson & Spaziani, 1985; Sedlmeier & Fenrich, 1993; Saïdi *et al.*, 1994; Dauphin-Villemant *et al.*, 1995). With regard to the second messengers involved in MIH signal transduction, for crabs, and in particular for the shore crab, *Carcinus maenas*, it is clear that MIH causes rapid (within a few minutes) and sustained (60-fold above basal levels) increases in cyclic GMP and rather transient and less pronounced (two-fold above basal levels) increases in cyclic AMP (Saïdi *et al.*, 1994). Furthermore, in this species it has recently been found that incubation of Y-organs in the presence of 10 nmol MIH l^{-1} leads to a three- to four-fold increase in cyclic GMP-dependent protein kinase activity and subsequent phosphorylation of endogenous protein(s) in the Y-organ (Baghdassarian *et al.*, 1996). These observations clearly suggest that cyclic GMP is the relevant second messenger involved in MIH-mediated inhibition of ecdysteroid synthesis in crabs.

Transaldolase: a key protein in the regulation of steroidogenesis in *C. maenas* Y-organs

Since MIH reduces protein synthesis by the Y-organs, and in view of the observation that protein synthesis is dramatically increased during premoult, any proteins which are (a) abundantly expressed by the Y-organ, and (b) differentially expressed during the moult cycle, in relation to invariantly expressed proteins (for example actin), might be suspected of being involved in steroidogenesis, and thus, the synthesis and/or activity of such proteins might be ultimately controlled by MIH.

Computer analysis of 2D-PAGE (two-dimensional polyacrylamide gel electrophoresis) of Y-organs has revealed that the staining intensity of certain proteins varies during the moulting cycle in accordance with the steroidogenic activity of the tissue (Fig. 1). Among protein spots which are differentially expressed during the moulting cycle, a cytosolic 36.2 kDa protein was characterized as a transaldolase (EC: 2.2.1.2) (Fig. 1, Table 1). Transaldolase was invariably detectable in Y-organs regardless of the moulting stage, and its synthesis and enzymatic activity increased with steroidogenic activity (Table 2). In Y-organs, transaldolase enzymatic activity is considerably higher (10–100-fold) than in any other crab tissue and is in the range found in vertebrate tissues (Kuhn & Brand, 1972).

Transaldolase synthesis was inhibited by 10 nmol MIH l^{-1}, but this is only measurable after an incubation period of about 6 h and cannot account for the rapid inhibition of steroidogenesis, which is measurable after a few minutes' exposure to MIH (Saïdi *et al.*, 1994). Nevertheless, after 18 h of incubation, a significant reduction in synthesis of transaldolase in MIH-treated Y-organs was observed (Table 3). Previous work has shown that MIH increases levels of cyclic GMP in the Y-organ about 60-fold in *Carcinus* Y-organs, suggesting that cyclic GMP is the second messenger in MIH action (Saïdi *et al.*, 1994). A slight cyclic GMP-dependent phosphorylation of transaldolase was observed in *Carcinus* Y-organ homogenates previously incubated for a few minutes in the presence of $[\gamma^{32}P]ATP$ (Table 1). After 1 h of incubation with 10 nmol MIH l^{-1}, transaldolase enzymatic activity is signifi-

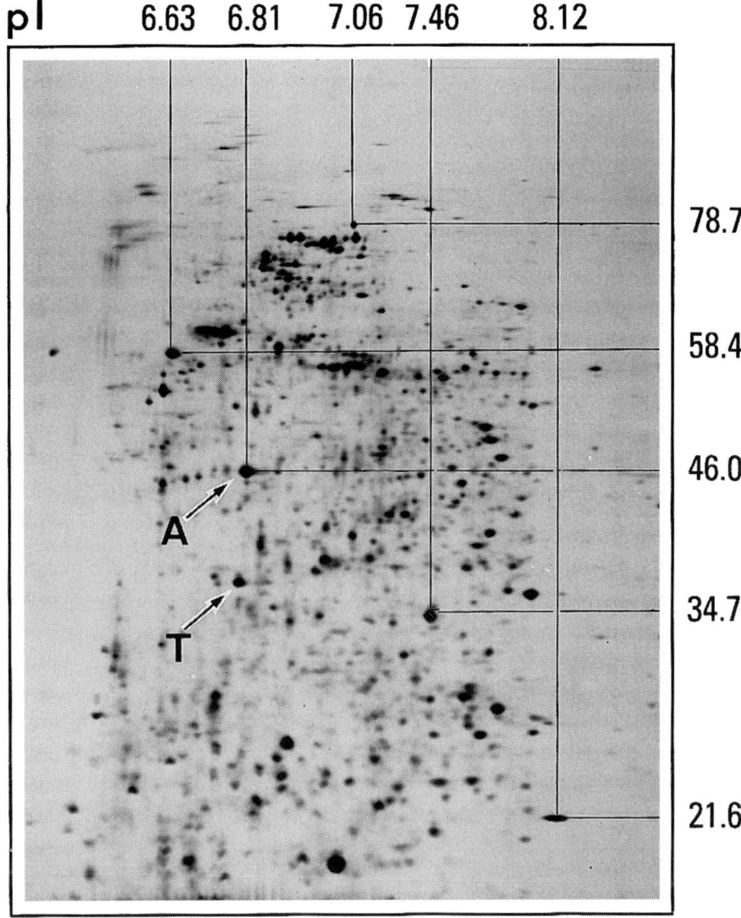

Fig. 1. Silver-stained two-dimensional polyacrylamide gel electrophoresis (2D-PAGE) of crab Y-organ homogenates exhibiting high steroidogenic activity. Molecular mass (m: kDa) and isoelectric point (pI) are matched for reference. A, actin; T, transaldolase.

cantly decreased (Table 3). These combined results suggest that acute inhibition of transaldolase enzymatic activity in the presence of MIH might be consequent upon increases in cyclic GMP. Consequently, cyclic GMP-dependent phosphorylation of transaldolase could be associated with its inactivation (Fig. 2).

An important question to address is how transaldolase can play a part in steroidogenesis and in its inhibition. Although transaldolase is one of the

Table 1 *Characteristics of crab transaldolase*

Tissue specificity: Y-organs
Subcellular localization: Cytosol

Molecular mass: 36 200 Da
Isoelectric point: 6.8

Phosphorylation in presence of cGMP
100% sequence identity with signature peptide[a]

Notes:
Using 2D-PAGE, transalodase was localized, its biochemical characteristics were measured and partial amino acid sequences were obtained. See Lachaise *et al.* (1996) for details.
[a] Reizer, Reizer & Saier (1995).

Table 2 *Relationship between transaldolase expression and enzymatic activity of transaldolase as a function of steroidogenic activity of crab Y-organs*

	Steroidogenic activity	
Transaldolase	Low	Med
Intensity of protein	15±2% of actin spot	45±9% of actin spot
Enzymatic activity	0.36±.01 U mg^{-1}	0.9±0.4 U mg^{-1}

Note:
% of transaldolase intensity of protein was determined using invariant actin spot as reference (see Lachaise *et al.*, 1996; Figs. 3, 7). Transaldolase enzymatic activity was expressed in units of proteins mg^{-1}, where one unit is defined as the amount of enzyme catalyzing the formation of 1 µmol of glyceraldehyde 3-phosphate.

Table 3 *Inhibition of synthesis and enzymatic activity of transaldolase by 10 nmol MIH l⁻¹*

Transaldolase enzyme activity (U (mg protein)$^{-1}$)
Control
    1h   0.6±0.1 (n=6)
    18 h 0.64±0.1 (n=7)
MIH
    1 h   0.4±0.1 (n=6)
    18 h 0.4±0.1 (n=7)

% inhibition of transaldolase synthesis
6–18 h 20–55% (t=3.9; df=8; P=0.0005)

Note:
Incorporation of labelled methionine into transaldolase was inhibited by MIH in long-term incubation. Transaldolase activity is inhibited by MIH after short (1 h) and long (18 h) term incubation (see Lachaise *et al.*, 1996: 'MIH inhibits incorporation of [^{35}S]methionine into P36' and Fig. 8).

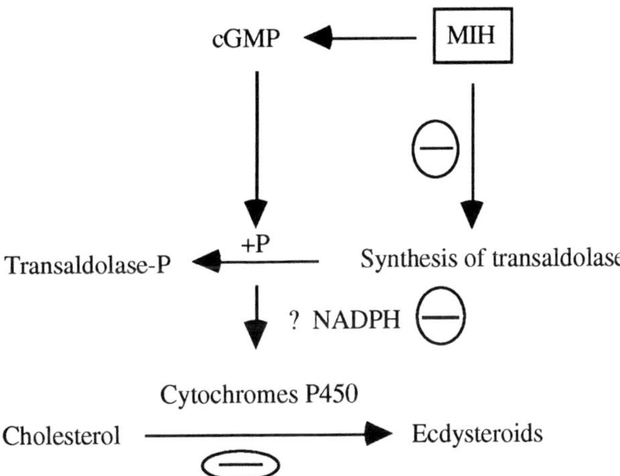

Fig. 2. Scheme summarising the possible role of transaldolase in the inhibition (minus signs) of ecdysteroid synthesis by MIH (moult-inhibiting hormone) in crab steroidogenic glands. cGMP, cyclic GMP.

enzymes of the non-oxidative part of the pentose phosphate pathway this enzyme plays an important role for the balance of metabolites in this pathway, which is known to generate NADPH. As NADPH is necessary for P450 cytochrome activation, the role of this transaldolase would most likely to be in inhibiting the conversion of sterols or steroids via NADPH reduction (Fig. 2). Transaldolase genes have been identified recently in bacteria (Yura *et al.*, 1992), yeast (Jacoby *et al.*, 1993; Schaaff *et al.* 1990) and human myelomonocytic leukemia cells (Banki, Halladay & Perl, 1994). These enzymes are highly conserved and show significant structural identity. However, the *Carcinus* transaldolase is closer to the enzyme characterised in human cells.

In view of these observations, it can be hypothesized that, in crustaceans, one of the inhibition pathways of steroidogenesis by MIH could proceed via transaldolase: in the short term by its phosphorylation resulting in its inactivation; in the long term by inhibition of its synthesis. It would be of interest to test whether the crab model of transaldolase implicated in steroidogenesis and its regulation can be found in other groups of animals, arthropodian invertebrates or vertebrates.

References

Banki, K., Halladay, D. & Perl, A. (1994). Cloning and expression of the human gene for transaldolase, a novel highly repetitive element constitutes an integral part of the coding sequence. *Journal of Biological Chemistry* **269**, 2847–2851.

Baghdassarian, D., de Bessé, N., Saïdi, B., Sommé, G. & Lachaise, F. (1996). Neuropeptide induced inhibition of steroidogenesis in crab molting glands: involvement of cGMP-dependent protein kinase. *General and Comparative Endocrinology* **104**, 41–51.

Bentley, J. D., Jentoft, J. E., Foreman, D. & Ambrose, D. (1990). ^{31}P nuclear magnetic resonance (NMR) identification of sugar phosphates in isolated rat ovarian follicular granulosa cells and the effects of follicle-stimulating hormone. *Molecular and Cellular Endocrino*logy **73**, 179–185.

Clark, B. J., Wells, J., King, S. R. & Stocco, D. M. (1994). The purification, cloning, and expression of a novel luteinizing hormone induced mitochondrial protein in MA-10 mouse Leydig tumor cells. *Journal of Biological Chemistry* **269**, 28 314–28 322.

Criss, W. E. & McKerns, K. W. (1969). Studies on ACTH-induced changes at the NADP-binding site of bovine adrenal glucose-6-phosphate deshydrogenease. *Archives of Biochemistry and Biophysics* **135**, 118–125.

Dauphin-Villemant, C., Böcking, D. & Sedlmeier, D. (1995). Regulation of steroidogenesis in crayfish molting glands: involvement of protein synthesis. *Molecular and Cellular Endrocinology* **109**, 97–103.

Garnier, M., Boujrad, N., Ogwuegbu, S. O., Hudson, J. R. & Papadopoulos, V. (1994). The polypeptide diazepam-binding inhibitor and higher affinity mitochondrial peripheral-type benzodiazepine receptor sustain constitutive steroidogenesis in the R2C Leydig tumor cell line. *Journal of Biological Chemistry* **269**, 22 105–22 112.

Grieneisen, M. L. (1994). Recent advances in our knowledge of ecdysteroid biosynthesis in insects and crustaceans. *Insect Biochemistry and Molecular Biology* **24**, 115–132.

Hanukoglu, I. (1992). Steroidogenic enzymes: structure, function, and role in regulation of steroid hormone biosynthesis. *Journal of Steroid Biochemistry and Molecular Biology* **43**, 779–804.

Hum, D. W. & Miller, W. L. (1993). Transcriptional regulation of human genes for steroidogenic enzymes. *Clinical Chemistry* **39**, 333–340.

Jacoby, J., Hollenberg, C. P. & Heinisch, J. J. (1993). Transaldolase mutants in the yeast *Kluyveromyces lactis* provide evidence that glucose can be metabolized through the pentose phosphate pathway. *Molecular Microbiology* **10**, 867–876.

Kuhn, E. & Brand, K. (1972). Purification and properties of transaldolase from bovine mammary gland. *Biochemistry* **11**, 1767–1772.

Lachaise, F., Sommé, G., Carpentier, G., Granjeon, E., Webster, S. & Baghdassarian, D. (1996). A transaldolase: an enzyme implicated in crab steroidogenesis. *Endocrine* **5**, 23–32.

Mattson, M. P. & Spaziani, E. (1985). Cyclic-AMP mediates the negative regulation of Y-organ ecdysteroid production. *Molecular and Cellular Endocrinology* **42**, 185–189.

Mattson, M. P. & Spaziani, E. (1986). Regulation of Y-organ ecdysteroidogenesis by moult inhibiting hormone in crabs: involvement of cyclic AMP-mediated protein synthesis. *General and Comparative Endocrinology* **63**, 414–423.

McKerns, K. W. (1966). Hormone regulation of genetic potential through the pentose phosphate pathway. *Biochimica et Biophysica Acta* **121**, 207–209.

McKerns, K. W. & Ryschkewitsch, W. (1976). Regulation of RNA synthesis in corpora lutea nuclei by luteinizing hormone and derivates. *Biochimica et Biophysica Acta* **454**, 51–54.

Payne, A. H. & Youngblood, G. L. (1995). Regulation of expression of steroidogenic enzymes in Leydig cells. *Biology of Reproduction* **52**, 217–225.

Paz, C., Dada, L. A., Cornejo Maciel, M. F., Mele, P. G., Cymering, C. B., Neuman, I., Mendez, C. F., Finkelstein, C. V., Solano, A. R., Park, M., Fischer, W. H., Towbin, H., Scartazzini, R. & Podesta, E. (1994). Purification of a novel 43-kDa protein (p43) intermediary in the activation of steroidogenesis from rat adrenal gland. *European Journal of Biochemistry* **224**, 709–716.

Pfeifer, S. M., Furth, E. E., Ohba, T., Chang, Y. J., Rennert, H., Sakuragi, N., Billheimer, J. T. & Strauss III, J. F. (1993). Sterol carrier protein 2: role in steroid hormone synthesis. *Journal of Steroid Biochemistry and Molecular Biology* **47**, 167–192.

Reizer, J., Reizer, A. & Saier, M. H. Jr (1995). Novel phosphotransferase system genes revealed by bacterial genome analysis – A gene cluster encoding a unique enzyme I and the proteins of a fructose-like permease system. *Microbiology* **141**, 961–971.

Saïdi, B., de Bessé, N., Webster, S. G., Sedlmeier, D. & Lachaise, F. (1994). Involvement of cAMP and cGMP in the mode of action of moult-inhibiting hormone (MIH), a neuropeptide which inhibits steroidogenesis in a crab. *Molecular and Cellular Endocrinology* **102**, 53–61.

Schaaff, I., Hohmann, S. & Zimmermann, F. K. (1990). Molecular analysis of the structural gene for yeast transaldolase. *European Journal of Biochemistry* **188**, 587–603.

Sedlmeier, D. & Fenrich, R. (1993). Regulation of ecdysteroid biosynthesis in crayfish Y-organs: role of cyclic nucleotides. *Journal of Experimental Zoology* **265**, 448–453.

Simpson, E. R. & Waterman, M. R. (1988). Regulation of the synthesis of steroidogenic enzymes in adrenal cortical cells by ACTH. *Annual Review of Physiology* **50**, 427–440.

Waterman, M. R. & Simpson, E. R. (1989). Regulation of steroid hydroxylase gene expression is multifactorial in nature. *Recent Progress in Hormone Research* **45**, 533–566.

Xu, T., Bowman, E. P., Glass, D. B. & Lambeth, J. D. (1991). Stimulation of adrenal mitochondrial cholesterol side chain cleavage by GTP, steroidogenesis activator polypeptide (SAP), and sterol carrier protein$_2$. *Journal of Biological Chemistry* **266**, 6801–6807.

Yura, T., Mori, H., Nagai, H., Nagata, T., Ishihama, A., Fujita, N., Isono, K., Mizobuchi, K. & Nakata, A. (1992). Systematic sequencing of the *Escherichia coli* genome: analysis of the 0–2.4 min region. *Nucleic Acids Research* **20**, 3305–3308.

Part II

Control of intermediary metabolism, and ion and water balance

MICHAEL J. LEE
and GRAHAM J. GOLDSWORTHY

New perspectives on the structures, assays and actions of locust adipokinetic hormones

Introduction

The adipokinetic hormones (AKHs) of insects exert a wide range of effects, many of which are analogous to those of vertebrate glucagons (see Goldsworthy, 1994). In addition to hypertrehalosaemia or hyperlipaemia, responses in locusts include the inhibition of protein synthesis (Carlisle & Loughton, 1979), the activation of glycogen phosphorylase (Van Marrewijk, Van den Broek & Beenakkers, 1980, 1986; Goldsworthy, Mallison & Wheeler, 1986a), inhibition of lipid synthesis (Gokuldas, Hunt & Candy, 1988), and inhibition of uridine incorporation into RNA (Kodrík & Goldsworthy, 1995; Michalik *et al.*, 1995). The influences of AKH on lipid and carbohydrate metabolism during flight are of obvious physiological significance to a flying insect. However, the inhibition of protein, lipid and RNA synthesis is of less certain physiological relevance, although the extremely low concentrations at which these latter actions are exerted suggest that they are likely to be of physiological significance, and their exact roles remain to be determined.

The first insect AKH to be characterised was the decapeptide AKH-I isolated from the locusts *Schistocerca gregaria* and *Locusta migratoria* (Stone *et al.*, 1976). Its similarity in sequence to a peptide hormone, red pigment-concentrating hormone (RPCH), isolated previously from prawns by Fernlund & Josefsson (1972), led to these two peptides being recognized eventually as the first members of an AKH/RPCH family of arthropod hormones. This family comprises >30 members isolated from 30 or more insect species (Gäde, Reynolds & Beeching, 1994). Figure 1 is an analysis of 32 members of the family (see Gäde *et al.*, 1994), and shows the frequency of particular residues at each position of the consensus sequence, that of AKH-I.

The relationships between structures and activities

The molecular conformations of locust adipokinetic peptides

Soon after its primary structure was elucidated, Stone *et al.* (1978) predicted from its linear sequence that AKH-I had the potential to adopt a β-turn

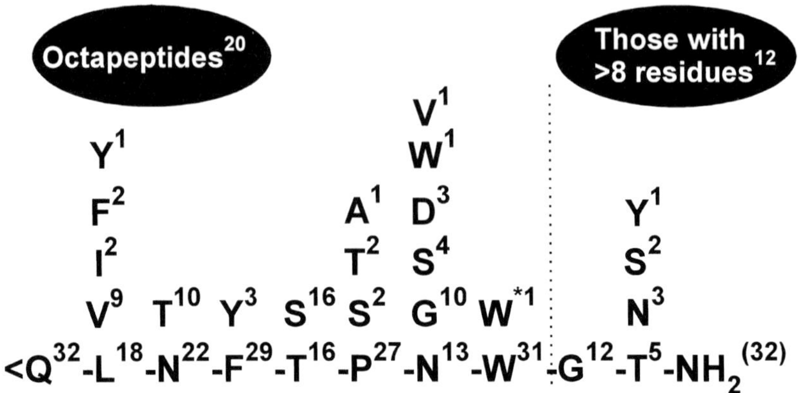

Fig. 1. The frequencies of the natural variations of the AKH/RPCH family. A sample of 32 peptides was analysed from the data given by Gäde *et al.* (1994) and Gäde (1996, 1997a,b): the indices refer to the frequency of occurrence of each amino acid signified by the single-letter code at each position. The tryptophan marked with an asterisk (W*) is probably a glycosylated form (see Gäde *et al.*, 1994). Note that the consensus sequence is actually that of AKH-1.

conformation in solution, but it was more than 10 years before this hypothesis was put to the test. Goldsworthy *et al.* (1990, 1992) used circular dichroism (CD) spectroscopy in different solvents to examine the secondary structure of a number of AKHs. The CD spectra of AKH-I and II at room temperature in water are characteristic of disordered conformations, but such spectra could arise from the superposition of a set of different complementary conformations which 'cancel each other out' to give a rather featureless difference spectrum (see Fig. 2, spectrum 1). The different component conformations can be studied by using different solvents or different temperatures. Thus, for AKH-I in SDS (Goldsworthy *et al.*, 1990) its CD spectrum undergoes changes that are typical of a β-structure (Fig. 2, spectrum 2), but the CD spectrum of AKH-II (Goldsworthy *et al.*, 1990) is relatively unaffected by SDS (Fig. 2, spectrum 7). This apparent inability of AKH-II to form a β-structure may correlate with the low activity of this peptide compared with the other two natural AKHs in *L. migratoria*, in both the lipid mobilisation (*in vivo*) and the acetate uptake (*in vitro*) assays (Lee & Goldsworthy, 1995a).

At low temperatures, the CD spectra of both AKH-I and II change dramatically (Fig. 2): as the temperature decreases, an increasingly positive dichroism at 217 nm emerges (Goldsworthy *et al.*, 1992) that stabilizes below −75 °C: a feature typical of the left-handed extended helix (P_{II}) conformation found for example in polyproline II and many other proteins (Siligardi & Drake, 1995). This ordered low temperature state is not perturbed by the presence of 4 mol urea 1^{-1} (Fig. 2, spectrum 5), confirming that intramolecular hydrogen bonds are not involved in its maintenance; again, this is a feature characteristic of the extended P_{II} conformation (Siligardi & Drake, 1995). Nuclear magnetic resonance (NMR) studies of *Emp*-AKH, an octapeptide from a praying mantis, *Empusa pennata*, suggest the presence of a β-turn between residues 5 and 8, with evidence for a β-sheet conformation for residues 1–5 (Zubrzycki & Gäde, 1994). Thus, data from both NMR and CD studies suggest a β-structure in at least some AKHs, and part of the molecule can adopt an extended conformation.

The publication of the sequence of the third *Locusta*-AKH, AKH-III (Oudejans *et al.*, 1991), prompted a study of its secondary structure. Unfortunately, the 'double tryptophan' motif at residues 7 and 8 dominates the CD spectra (Lee, Drake & Goldsworthy, 1996). This is clear from SDS and low temperature CD spectra of AKH-III itself (Fig. 2, spectra 11–13). The strong CD in the 210–230 nm region is clearly related to the indole chromophore and does not convey directly information about the peptide backbone conformation. The bi-signate nature of this tryptophan-based CD at lower temperatures or in SDS indicates a defined indole/indole orientation. These CD data suggest that AKH-III can also adopt two limiting conformations. There is one at low temperatures with a negative/positive CD couplet around 215 nm, and another conformation is characterised by the positive/negative couplet at about 225 nm. This suggests that, in a membrane-like environment, AKH-I and III may adopt a conformation quite different from that expected from the thermodynamically preferred conformation prevailing at low temperature. The precise nature of these conformations for AKH-III, however, remain to be identified. From the SDS and low temperature studies, the water spectrum of AKH-I can be interpreted as the 'difference spectrum' of the ordered states which predominate under these two conditions. However, the lack of sensitivity of the CD spectrum of AKH-II to SDS means that, if the P_{II} conformation of this peptide contributes to its water spectrum at room temperature, the identity of the AKH-II 'complementary' conformation again remains to be determined.

If the 'active conformations' at the receptor site were to correlate with those shown in the presence of SDS micelles, then the activity of AKH peptides would be influenced by their capacity to assume, for example, a β-structure. Data from a study of natural AKHs tested in the lipid mobilisation assay broadly support this hypothesis (Goldsworthy *et al.*, 1990), but biological activity data from an assay measuring the inhibition of acetate uptake into fat body tissue do not fit in with that interpretation of the CD data presented here. The significance of this is unclear, but has been used to argue for

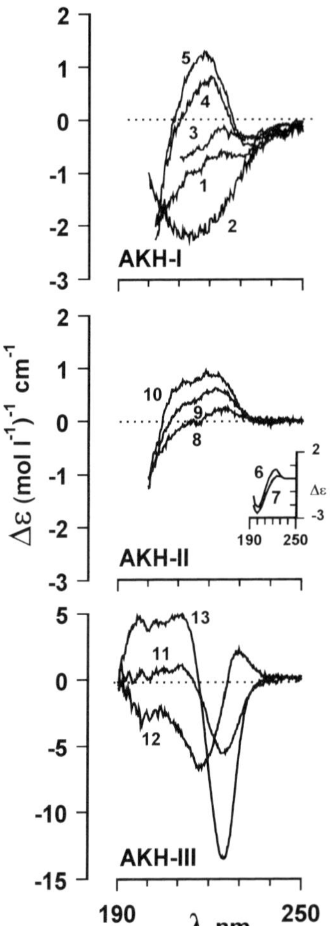

Fig. 2. Circular dichroism of AKH-I, II and III. Spectra are shown of AKH-I in water (spectrum 1) and 0.6% SDS (2) at room temperature; and at –30 (3), –90 (4), and at –90 °C in the presence of 4 mol urea l^{-1} (5). CD spectra are shown of AKH-II in water (6) and 0.6% SDS (7), and in ethane-diol/water at 0.5 (8), –36 (9), and –85 °C (10). CD spectra of AKH-III are shown as a function of temperature (11 and 13) and in 0.6% SDS (12).

Except for spectra 6 and 7 (from a J40CS, digitized, and reproduced by hand: taken from Goldsworthy *et al.*, 1992), all the CD data were recorded on a JASCO J600 spectropolarimeter. The latter, more recent unpublished data for AKH-I and II, substantiate earlier reports (see Goldsworthy *et al.*, 1992) except for spectrum 5, which has not been described previously (G. J. Goldsworthy, O. Cusinato & A. F. Drake, unpublished observations). The peptide concentrations were in the range 0.1–0.3 mg ml^{-1}, in a cell of qualitative differences in the receptors involved in the two responses (Goldsworthy, Lee & Luswata, 1995). In contrast, the potencies of four AKH analogues each containing two tryptophan residues, assayed in the acetate uptake assay or in the lipid mobilisation assay, correlate well with the magnitude of the CD signal at 225 nm at 25 °C; the similarity of this correlation in both assays (Lee *et al.*, 1996) indicates that the hormone receptors involved in the two responses are similar, if not identical. Full resolution of this problem awaits direct studies of receptor-binding for the different AKHs of locusts.

Development of new functional assays

Lipid mobilisation

The lipid mobilisation response of locusts forms a robust assay *in vivo* that has been used both to assess the release of AKHs during flight (Cheeseman, Jutsum & Goldsworthy, 1976; Cheeseman & Goldsworthy, 1979; Orchard & Lange, 1983) and to provide indirect information about their receptors (Goldsworthy *et al.*, 1986b). This assay detects large differences in the potencies of the natural AKHs when locusts are used as the recipients, with AKH-I being the most potent peptide assayed so far. However, when conducting assays *in vivo* it is difficult or impossible to control for secondary actions or any removal (or endogenous release) of hormone, or the presence of antagonists. In an attempt to prevent possible effects of tissue other than the fat body (the presumed primary target) on the apparent potency of AKHs, lipid mobilisation assays *in vitro* have been developed in several laboratories (see Goldsworthy, 1983), but the absolute requirement for acceptor lipoprotein *in vitro* (see Tietz, 1962; Peled & Tietz, 1973, 1975; Spencer & Candy, 1976; Orchard *et al.*, 1982) makes them technically cumbersome.

Acetate uptake

To overcome the problems described above, a new assay for AKHs *in vitro* (Lee & Goldsworthy, 1995a,b) was developed, based on the AKH-induced inhibition of fatty acid synthesis in locusts described by Gokuldas *et al.*

Fig. 2. (*cont.*)

0.05 cm pathlength. The concentration of each peptide was determined initially by weight, and then corrected by means of the Trp absorbance at 280 nm. SDS was added to the aqueous samples at 0.6%. Low temperature studies were carried out in a 2:1 (v/v) ethanediol/water mixture. Values are quoted as $M^{-1} cm^{-1}$ per residue.

(1988). Weighed pieces of fat body are incubated with radiolabelled acetate and test hormone in a microfiltration plate. After incubation at 40°C for 1 h, the medium is removed by vacuum filtration, the tissues washed, and then the filter disc (with trapped tissue) punched out into a scintillation vial to determine the radioactivity retained. The assay is quick, incubation conditions (such as time, pH, temperature, osmolarity) can be controlled individually, and the initial hormone concentration is known precisely. Of all AKHs tested, AKH-III is the most potent inhibitor of acetate uptake with an EC_{50} of about 0.1 nmol l^{-1} (Lee & Goldsworthy, 1995a,b), and the response occurs at a threshold concentration of less than 10 pmol l^{-1}; AKH-I and II are equally effective, but somewhat less potent. The acetate uptake assay shows good specificity for AKHs; none of the other hormones tested from insect (for example, proctolin, octopamine, 5-hydroxytryptamine) or other unrelated sources (for example, oxytocin, adrenaline) significantly altered the rates of acetate uptake (Lee & Goldsworthy, 1995a). In spite of its speed and specificity, the acetate uptake assay does have a number of drawbacks. First, the basic cost of raw materials (microfiltration plates, particularly) make it more expensive than the lipid mobilisation assay. Second, acetate uptake is inhibited by low concentrations of organic solvent (see Goldsworthy *et al.*, 1997), so for peptides which have a limited solubility in aqueous solution and/or have a low potency, it is more difficult to use than the lipid mobilisation assay. Third, although this assay is considerably faster than the lipid mobilisation assay, it involves weighing around 90 pieces of fat body per plate, a procedure which takes around 40 min. Finally, like the lipid mobilisation assay, it is subject to high variability in the response between replicates, which necessitates at least seven or eight observations; this means that only one dose response curve per plate can be determined.

The use of dispersed fat body cells in the acetate uptake assay circumvents most of these problems: reducing time and labour involved in preparing dose–response curves to different test peptides, and increasing the precision of the estimate of potency. As a consequence of the reduced variability, fewer replicates per dose are required and as many as eight dose–response curves can be constructed per plate (although for routine use so many is not recommended); this therefore reduces quite considerably both the cost of the assay and the time taken per peptide. The availability of a routine method for preparing viable dispersed fat body cells also facilitates studies of AKH actions by other methods (for example, the determination of changes in intracellular calcium in response to hormone; see below and see Goldsworthy *et al.*, 1997).

Collagenase is a frequently used enzyme to prepare dispersed cell populations and it has been used successfully to disperse fat body from *Periplaneta americana* (Steele & Ireland, 1994) and *Locusta migratoria* (Asher *et al.*,

1984). In our laboratory, commercially available collagenase preparations did not produce good yields of viable fat body cells from *L. migratoria*, but bovine chymotrypsin, in conjunction with a small amount of DNase to prevent cell clumping, gives an excellent yield of cells compared with other enzymes (Lee & Goldsworthy, 1995b). Following the final wash to remove as much as possible of any protease, it is still necessary to add a chymotrypsin inhibitor (for example SBTI, soybean trypsin inhibitor) to prevent degradation of test peptides during subsequent incubations. Aliquots of cells may then be dispensed into the wells of a 96-well microfiltration plate using a multichannel pipettor.

Chymotrypsin has also been used successfully to disperse fat body from *Sarcophaga* (Kurata, Saito & Natori, 1990) where the natural enzyme produced during pupation is a cysteine protease (Kurata, *et al.*, 1992a,b; Takahashi, Kurata & Natori, 1993). An attempt was made to improve the uptake of the dispersed fat body preparation further by mimicking the action of this enzyme using papain, a broad specificity cysteine protease with an approximately neutral pH optimum. While papain disperses the fat body optimally at 0.1% (w/v) (a slightly lower concentration than the equivalent optimum for chymotrypsin), and although the fat body appeared both well-dispersed and similar under the light microscope, the resulting cell preparations took up acetate at only about half the rate of preparations dispersed using chymotrypsin (Lee and Goldsworthy, unpublished observations). None the less, it is possible that other proteases with broad specificity (such as bromelain, dispase, thermolysin) may be sufficiently effective and potent to be useful for fat body dispersal, although the problem of inactivating any residual non-serine protease with an effective, but economical, non-toxic and water-soluble (to avoid organic solvents) inhibitor would still remain. In this laboratory, chymotrypsin used together with soybean trypsin inhibitor remains the best approach for preparing dispersed cell preparations.

The uptake of acetate into fat body cells or pieces of tissue has been determined at 40 °C. This is a rather higher temperature than is often used for short-term culture *in vitro*, but both *L. migratoria* fat body cells and tissue are tolerant of this temperature, and higher temperatures increase the rates of uptake of radioactive substrates, thus reducing the length of incubation time required for adequate levels of uptake. Indeed, the temperature optima for both cells and tissue is in the region of 45–50 °C (M. J. Lee & G. J. Goldsworthy, unpublished observations), but at higher temperatures the cells are killed quickly. There appear to be few, if any studies, on the effect of temperature on AKHs and their potencies. Would AKHs be expected to have different potencies at different temperatures, either *in vivo* or *in vitro*?

Second messengers and the mechanism of action

Both the lipid mobilisation and phosphorylase activation responses in locusts are mediated, at least in part, through increases in the concentration of intracellular cyclic AMP (Gäde & Holwerda, 1976; Spencer & Candy, 1976; Goldsworthy *et al.*, 1986a; Vroemen, Van Marrewijk & Van der Horst, 1995a: see Van Marrewijk & Van der Horst, this volume). However, this does not appear to be the case for the inhibition of acetate uptake (see Lee & Goldsworthy, 1995a). Cyclic AMP analogues are either inactive or only weakly active in moderating the rate of uptake of acetate into locust fat body, indicating that cyclic AMP does not play a major role in the transduction of this particular effect of AKH. Similarly, Orr *et al.* (1985) and Lee & Keeley (1994) concluded that cyclic AMP is not involved in the intracellular transduction mechanism for the stimulation of trehalose synthesis by hypertrehalosaemic hormones (HrTHs) in the fat body of the cockroaches *P. americana* and *Blaberus discoidalis*, respectively. A lack of involvement of cyclic AMP in the transduction mechanism for the inhibition of RNA synthesis by AKH in locust fat body has also been suggested (Kodrík & Goldsworthy, 1995). By contrast, Vroemen *et al.* (1995a) have shown that the activation of phosphorylase by AKHs is strongly attenuated by GDPβS and mimicked by cholera toxin, indicating that the AKH receptor(s) for this response are coupled to cyclic AMP formation and glycogen phosphorylase activation via a stimulatory guanine nucleotide-binding protein (G_s: see Van Marrewijk & Van der Horst, this volume).

The AKHs are known to activate the inositol trisphosphate (IP_3) pathway in locust fat body (Pancholi *et al.*, 1991), but a number of inhibitors of protein kinase C (PKC), such as bisindolyl maleimide (Lee & Goldsworthy, 1995a), cheleryethrine, phloretin, quercetin, and staurosporine (M. J. Lee & G. J. Goldsworthy, unpublished data), are ineffective at preventing the inhibition of acetate uptake by AKHs, although staurosporine is an inhibitor of acetate uptake in its own right. In addition, an activator of protein kinase C, phorbol myristate acetate, does not mimic the effects of AKH (Lee & Goldsworthy, 1995a), Inhibitors of phospholipase C, such as 2-nitro-4-carboxyphenyl-*N*,*N*-diphenylcarbamate (NCDC; Walenga, Vanderhoek & Feinstein, 1980) and neomycin (Vergara, Tsien & Delay, 1985) do not show any antagonistic activity to AKH action on acetate uptake, although neomycin inhibits acetate uptake (Lee & Goldsworthy, unpublished data), possibly via Ca^{2+} mobilisation from intracellular stores (see Nakashima *et al.*, 1987). Thimerosal, which sensitizes inositol-1,4,5-trisphosphate receptors of the endoplasmic reticulum for calcium release (Bootman, Taylor & Berridge, 1992) inhibits acetate uptake at micromolar concentrations; at present, therefore, it is not possible to substantiate or rule out IP_3 as a second messenger in the action of AKH on which the acetate uptake assay is based.

As with other AKH-mediated responses of locust fat body (see Spencer & Candy, 1976; Van Marrewijk *et al.*, 1991; Vroemen *et al.*, 1995a; Kodrík & Goldsworthy, 1995), inhibition of acetate uptake by AKH-I is dependent on the presence of calcium ions in the external medium (Lee & Goldsworthy, 1995a). The mobile ion carriers, ionomycin and A23187, inhibit acetate uptake (EC_{50}=2.2 and 10.1 μmol 1^{-1}, respectively; M. J. Lee & G. J. Goldsworthy, unpublished observations). A23187 has been shown to inhibit lipid synthesis in the fat body of *S. gregaria* (Gokuldas, 1989), to activate fat body phosphorylase (Van Marrewijk *et al.*, 1991), and to increase $[Ca^{2+}]_i$ in cultured haemocytes from a lepidopteran, *Malacosoma diastria* (Jahagirdar *et al.*, 1987). On the other hand, neither 10 μmol A23187 1^{-1} nor ionomycin mimic HrTH effects in the fat body of *B. discoidalis* (Lee & Keeley, 1994; Keeley & Hesson, 1995). It is not clear precisely how these ion carriers mimic AKHs in the acetate uptake assay – either by enabling external calcium ions to enter the cell and/or allowing movement of ions from intracellular calcium stores to calcium-sensitive regions of the cell. It does, however, appear that in *L. migratoria*, AKHs cause a calcium influx (see also Van der Horst, Vroemen & Van Marrewijk, 1997; Vroemen *et al.*, 1995b: see Van Marrewijk & Van der Horst, this volume), although Keeley & Hesson (1995) suggest that the intracellular calcium release is more significant than the extracellular calcium entry for transduction of the HrTH response in the cockroach. Indeed, release of calcium from intracellular stores does mimic AKH action in the locust: thapsigargin, an inhibitor of the endoplasmic reticulum Ca^{2+} ATPase pump (Thastrup *et al.*, 1990), reduces acetate uptake in dispersed fat cells from locusts (Lee & Goldsworthy, unpublished observations), suggesting that these may play a role in AKH signal transduction, at least for its action on acetate uptake.

The ability to produce viable hormone-sensitive dispersed fat body cells, has enabled fluorimetric measurements of changes in intracellular calcium in response to AKH. After a short lag phase there is a rapid six- to seven-fold increase in $[Ca^{2+}]_i$, maintained for different times depending on the applied peptide (Fig. 3). It is significant that individual cells are able to respond to each of the three hormones. It has been shown that another AKH, neurohormone D, from *P. americana* induces an increase in an N-type Ca^{2+} current in cockroach neurones (Wicher, Walther & Penzlin, 1995). A number of different calcium channel modulators have been tested to determine whether a specific calcium channel could be linked to the transduction of the AKH signal in reducing acetate uptake (M. J. Lee & G. J. Goldsworthy, unpublished observations). The addition of calcium channel blockers does not prevent the AKH-induced inhibition of acetate uptake: it appears that

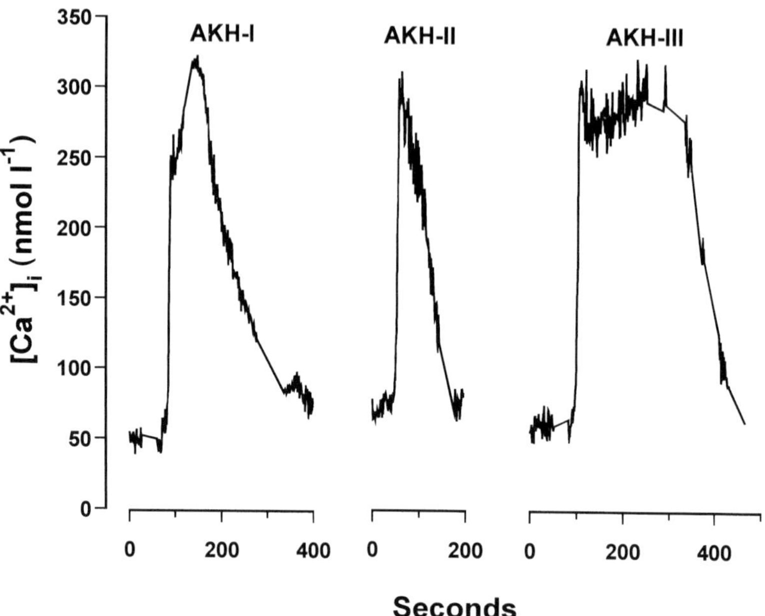

Fig. 3. Changes in $[Ca^{2+}]_i$ in a single *Locusta* fat body cell in response to each of the three *Locusta* AKHs. Dispersed cells (see Lee & Goldsworthy, 1995b) were allowed to attach themselves to a coverslip, and loaded with Fura-2 by incubation at room temperature for up to 80 min in saline containing 10 μmol Fura-2 AM l^{-1}. The coverslip with Fura-2-loaded cells was then placed in an experimental chamber where any dye not trapped in the cells, together with cell debris and any remaining chymotrypsin (from the cell preparation), was washed off in a continuous flow of Ringer solution. Hormone was added and removed via this perfusion system. A perfusion rate of 2 ml min^{-1} was used, and this completely exchanged the volume of Ringer solution surrounding the cells (about 300 μl) within 10 s. Isolated cells were exposed alternatively to ultraviolet light at 380 and 340 nm over a 1 s period. The ratio (R) of the two fluorescence intensities was calculated each second, and used to monitor $[Ca^{2+}]_i$ over time; an increase in the ratio corresponded to an increase in concentration; the $[Ca^{2+}]_i$ (in nmol l^{-1}) was calculated from R using the equation $[Ca^{2+}]_i = 450(R-0.7)(5)/(9-R)$ (Grynkiewicz, Poenie & Tsien, 1985; Williams & Fay, 1990).

Following a 10 s pulse of hormone at a concentration of 66 nmol l^{-1} (20 pmol per 300 μl), there was a rapid increase in $[Ca^{2+}]_i$ which then declined relatively slowly. The responses to the *Locusta*-AKHs last for 90, 32 and 250 s (AKH-I, II and III, respectively), suggesting a relationship between the persistence of increased $[Ca^{2+}]_i$ and the potencies of the hormones in the acetate uptake assay (Lee & Goldsworthy, 1995b). The significance of this remains to be determined.

N-type (inhibited by ω-conotoxin GVIA), T-type (inhibited by flunarizine) and L-type (inhibited by nifedipine and verapamil) calcium channels are not involved in the AKH transduction mechanism. Other non-specific calcium channel blockers, La^{3+}, Cd^{2+} and Gd^{3+} also fail to prevent AKH action.

Structure activity studies

Attempts at indirect characterisations of the receptor(s) for AKHs have made use of both the variety of naturally occurring analogues which are known, and the chemical synthesis of analogues (Goldsworthy *et al.*, 1986a,b; Ford, Hayes & Keeley, 1988; Gäde, 1990; Hayes & Keeley, 1990; Ziegler *et al.*, 1991; Lee & Goldsworthy, 1995a). It is clear from such studies that the pGlu[1], Phe[4] and Trp[8] residues are the most important for biological activity of the adipokinetic hormones. For example, a study on *Blaberus* HrTH has shown that replacement of Phe[4] or Trp[8] either with alanine or D-amino acids results in biologically inactive analogues (Ford *et al.*, 1988), although D-Phe[4] in HrTH is tolerated, but with lower potency (Hayes, Nails & Ford, 1994). The relative intolerance of the receptor to modifications of the hormone at these positions is not unexpected when the natural variability of the AKHs discovered to date is considered (see Fig. 1 and Goldsworthy *et al.*, 1997; Gäde *et al.*, 1994): the pGlu[1] and Trp[8] residues of AKHs are invariate, and all but three of 32 structural variants have Phe[4].

N-terminal modifications to AKH-I

While pGlu[1] is present in all AKHs found to date, it is not itself essential for biological activity; its deletion from AKH-I results in a peptide (des-pGlu-AKH-I) which is essentially inactive (Gäde, 1990), but acetylation of the N-terminal leucine restores some activity in the lipid mobilisation and acetate uptake assays (Lee *et al.*, 1997). In AKH-I, pGlu[1] can also be replaced by hydroxyphenyl propionate (HPP), with only a minor reduction in potency, particularly in the assay *in vitro* (Lee *et al.*, 1997), but HPP-[Pro[1]]-*Blaberus*-HrTH is virtually inactive (Hayes *et al.*, 1994). Unblocked amino acids may also be substituted at this position in *Locusta*-AKH-I, so [Ala[1]]-, [Pro[1]]- or [Glu[1]]-AKH-I do have some activity in the lipid mobilisation assay *in vivo* and the acetate uptake assay *in vitro* (unpublished observations). Thus, while [Ala[1]]-AKH-I is only very weakly active *in vivo*, perhaps due to rapid break-down (via an aminopeptidase?) and removal rather than poor receptor binding per se, [AcAla[1]]-AKH-I is very much more potent *in vivo* than is [Ala[1]]-AKH-I (Lee *et al.*, 1997). Ziegler *et al.* (1991) found that [Gly[1]]- and [Tyr[1]]-*Manduca* AKH were unable to activate glycogen phosphorylase in fat body from *Manduca sexta* larvae, but as is the case with locust AKH-I, the acetylated peptides [AcGly[1]]-*Manduca*-AKH (Ziegler *et al.*, 1991) and [AcAla[1]]-*Blaberus*-HrTh (Ford *et al.*, 1988) retain activity.

C-terminal modifications

Modifications of Thr[10] in AKH-I

The importance of the C-terminal carboxyamide of AKH-I in influencing potency has been investigated (Poulos *et al.*, 1995; Lee *et al.*, 1996). Replacement of this amide by the free acid causes a severe loss in potency in both lipid mobilisation and acetate uptake assays (100-fold *in vivo* and 230-fold *in vitro*), as noted for other AKHs (Ford *et al.*, 1988; Gäde, 1990). Replacement of the C-terminal carboxyamide with a methyl ester was tolerated best of all the carboxyamide modifications studied, with only a 12-fold loss of potency in the lipid mobilisation assay, and only a five-fold loss in the acetate uptake assay. Replacement of one carboxyamide hydrogen by a methyl or a phenyl group, or both carboxyamide hydrogens by methyl groups results in very dramatic reductions of potency, although these modifications are better tolerated in the acetate uptake assay (60–160-fold reduction in potency) than the lipid mobilisation assay (580–1050-fold reduction in potency). It appears that the C-terminal carboxyamide group of AKH-I is important for potency in both the lipid and acetate uptake assays. However, given that the losses in potency are more marked in the lipid mobilisation assay *in vivo*, higher rates of removal of the analogues *in vivo* cannot be excluded, perhaps by a mechanism that has a high affinity for peptides which are more hydrophobic; unfortunately, data on the effects of C-terminal modifications on the rate of removal or inactivation of AKHs in the blood are not available to support this hypothesis – the differential rates of breakdown for *Locusta* AKHs themselves were only known recently (Oudejans *et al.*, 1996: see Van Marrewijk & Van der Horst, this volume).

Attempts to understand why AKH-III is so potent in the acetate uptake assay

AKH-III is the most potent of the three *Locusta* AKH peptides at inhibiting uptake of acetate into *L. migratoria* fat body *in vitro* (Lee & Goldsworthy, 1995a), inhibiting RNA synthesis (Kodrík & Goldsworthy, 1995), stimulating cyclic AMP production, activating glycogen phosphorylase (Vroemen *et al.*, 1995a), and inducing Ca^{2+} efflux (Vroemen *et al.*, 1995b). This high potency is surprising in comparison with its moderate potency in the lipid mobilisation assay *in vivo*. An attempt has been made to investigate the basis for these findings by synthesizing novel analogues of AKHs with Trp[7] either included or removed (Lee & Goldsworthy, 1996), because the 'double-tryptophan' Trp[7]-Trp[8] motif of AKH-III appears to be unique amongst the naturally occurring AKHs. Thus, the decapeptide [Trp[7]]-AKH-I, and the octapeptides [des-Gly[9]-Thr[10]]-AKH-III, [Trp[7]]-AKH-II and [Trp[7]]-*Acheta*-AKH were tested in both the acetate uptake and the lipid mobilisation assays. With the exception of [Trp[7]]-AKH-I, which has potency similar to that of AKH-I when measured in the acetate uptake assay, each of these analogues is less potent than its respective parent in each assay. However, the acetate uptake assay appears more tolerant of AKH analogues with the sequence Trp[7]-Trp[8], whereas inclusion of this sequence markedly reduces potency in the lipid mobilisation assay. The apparent differences in potencies of the peptides between assays may be consequent on different binding preferences of distinct AKH receptors, or alternatively, they may be due to different rates of breakdown or removal; AKH-III is inactivated very much more rapidly *in vivo* than either AKH-I or AKH-II (Oudejans *et al.*, 1996: see Van Marrewijk & Van der Horst, this volume). Presumably rates of breakdown or removal for any particular peptide would be higher in the intact insect, but will vary also between peptides, depending on their primary structure. Thus, apparent changes in potency could be due to differential rates of breakdown or removal rather than to qualitative variations in the peptide–receptor interaction. Such a phenomenon would be more pronounced in its effect on potencies determined *in vivo* particularly, but could be significant *in vitro*, although to a lesser degree. However, the potency of AKH-III tested *in vivo* (lipid mobilisation assay) is unaffected (M. J. Lee & G. J. Goldsworthy, unpublished observations) by either pre-injection and/or concommittant injection of phosphoramidon (200 μmol l^{-1} final concentration in the haemolymph), an inhibitor of the insect endopeptidase 24.11 which is thought to be the protease involved in the initial inactivation of neuropeptides, including AKHs (Lamango & Isaac, 1993). Unfortunately, it is not certain that phosphoramidon is effective *in vivo*; the general role of proteases in determining the potency of neuropeptides remains uncertain (see also Yagi *et al.*, 1992).

Substances which stimulate acetate uptake by *Locusta* fat body tissue

The acetate uptake assay determines the ability of adipokinetic hormones to *inhibit* the rate of fatty acid synthesis *in vitro* – an essentially glucagon-like property. Many biological processes are controlled by push–pull hormonal mechanisms and, in vertebrate systems, insulin is one hormone which is antagonistic to glucagon. Molecules with insulin-like activities or structures related to vertebrate insulin are known in insects (Thorpe & Duve, 1984), and the cDNA and deduced primary structure is known for a *Locusta* insulin (Kromer & Lageux, 1994). Hardly anything is known of the physiology of insulin in insects, but such a molecule, if it has biological activities in parallel with its structural relatedness to vertebrate insulin, might be expected to stimulate acetate uptake.

As part of a search for factors which might antagonise AKHs, it was observed that $N\alpha$-p-tosyl-L-lysine chloromethyl ketone (TLCK) has an insulin-like effect on acetate uptake (that is increased fatty acid synthesis) in *L. migratoria* fat body. Incubation of *L. migratoria* fat body with TLCK for 1 h prior to the addition of radiolabelled acetate causes a three-fold increase in the rate of uptake compared with the controls, with an optimum at 0.5 mmol 1^{-1} and an EC_{50} value of 0.27 mmol 1^{-1} (Fig. 4A). TLCK does not prevent the inhibitory action of AKH-I on acetate uptake although, because the level of acetate uptake is very high in TLCK-treated fat body, the inhibition by AKH-I represents only a small percentage change. Further, the addition of 10 pmol AKH-I to the fat body incubation medium does not prevent the stimulatory effects of TLCK on acetate uptake (Fig. 5), and the EC_{50} value remains unaffected. These data would suggest that AKHs and TLCK are acting at different sites.

Structural requirements of chloromethyl ketones for stimulatory activity

A number of simple organic chloromethyl ketones, all lacking an amino acid, together with amino acid-linked chloromethyl ketones have been tested *in vitro*. None of the simple chloromethyl ketones are effective at stimulating acetate uptake and one, chloroacetone, is inhibitory (Lee & Goldsworthy, unpublished data). Chloroacetyl-L-leucine is also ineffective (data not shown). Of several amino acid-containing chloromethyl ketones tested, acetyl-L-lysine chloromethyl ketone (ALCK), phenylalanine chloromethyl

---

Fig. 4. The effect of various amino acid linked chloromethyl ketones on acetate uptake (A) Pre-incubation of the fat body with TLCK for 1 h prior to the addition of radiolabelled acetate causes a three-fold increase in the rate of uptake compared with the controls, with an optimum at 0.5 mmol 1^{-1}. (B) Pre-incubation of up to 0.2 mmol TPCK 1^{-1} (concentration limited by the intolerance of the assay to high concentrations of organic solvent) for up to 4 h fails to increase the rate of acetate uptake above the control. (C and D) PheCK and LeuCK pre-incubated with fat body in the presence of radiolabelled acetate for 1 h causes a doubling in the rate of uptake compared to controls, with an optimum between 0.2 and 0.6, and 0.2 and 0.4 mmol 1^{-1}, respectively.

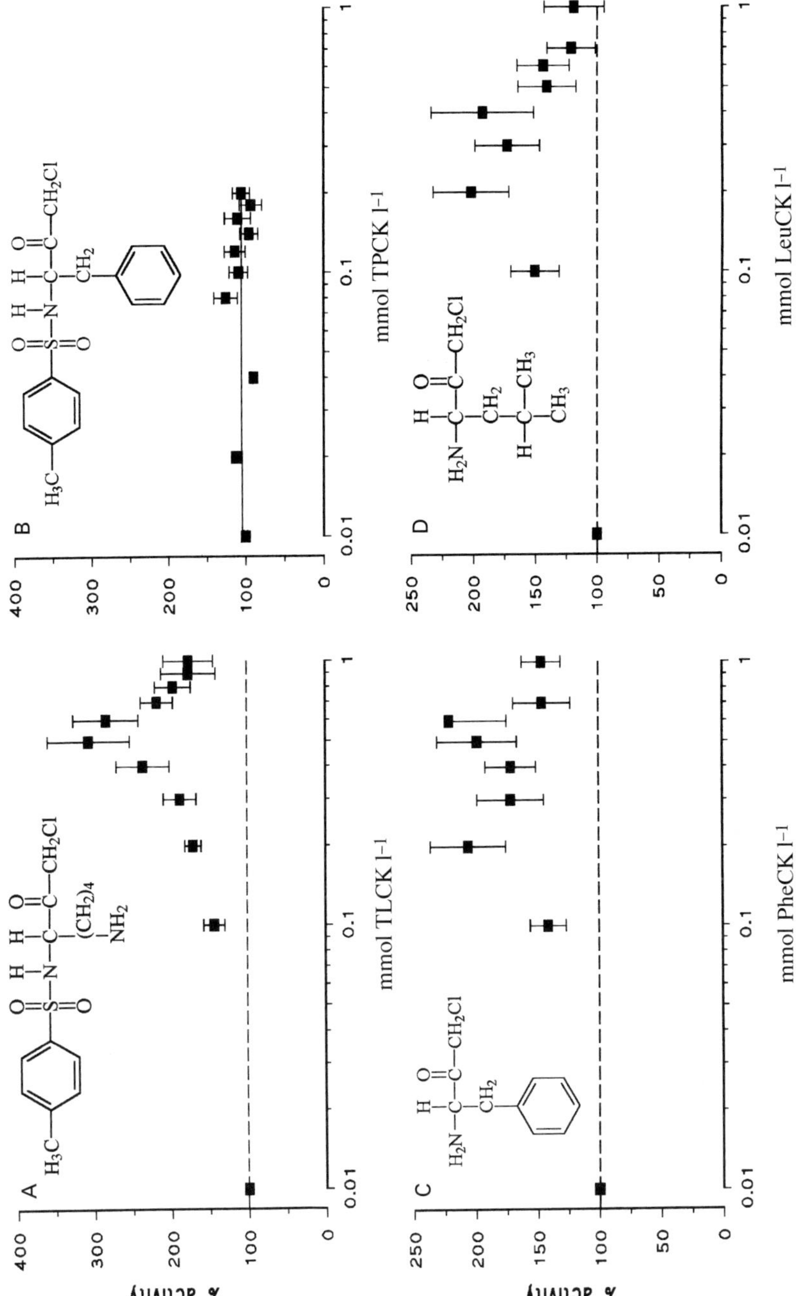

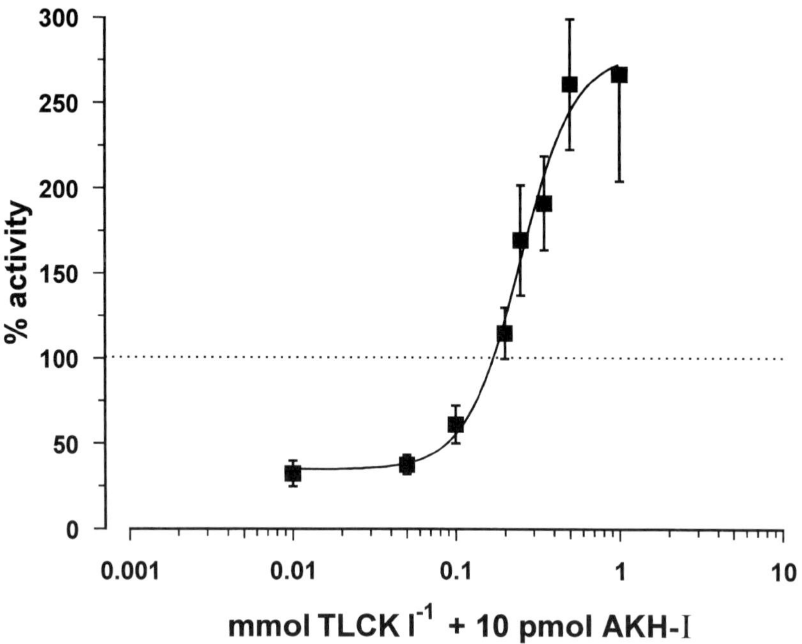

Fig. 5. The addition of 10 pmol AKH-I to the fat body incubation medium does not prevent the stimulatory effects of TLCK on acetate uptake. The EC_{50} values for the stimulation by TLCK in either the presence or absence of AKH remain unaffected, at about $0.25 \, \text{mmol} \, l^{-1}$.

ketone (PheCK) and leucine chloromethyl ketone (LeuCK) are all able to stimulate the rate of uptake, but with a less obvious optimal concentration (see Fig. 4C and D; data for ALCK not shown). However, N-tosyl-L-phenylalanine chloromethyl ketone (TPCK) does not elicit an insulin-like response even when fat body is pre-incubated in its presence for up to 4 h (Fig. 4B). These data suggest that, for stimulatory activity, the chloromethyl ketone moiety must be linked to an amino acid with a free amino group (either primary in PheCK and LeuCK or secondary in TLCK and ALCK), because TPCK (an amino acid-containing chloromethyl ketone without a free amine) and chloroacetamide (a chloromethyl ketone with a free amine, but no amino acid) are inactive.

Mechanism of action of chloromethyl ketones

Chloromethyl ketones are known primarily for their irreversible inhibition of serine proteases (Schoellmann & Shaw, 1963), but they also inhibit other classes of protease (see e.g. Kettner, Glover & Prescott, 1974; Barrett & Kirschke, 1981) and other enzymes, particularly the kinases, for example protein kinase A (Kupfer *et al.*, 1979), protein kinase C (Solomon, O'Brien & Weinstein, 1985), and tyrosine kinase (Navarro, Abdel Ghany & Racker, 1982). In an attempt to determine their mechanism of action in stimulating acetate uptake in the fat body of the locust, specific inhibitors of proteases and kinases have been incubated with fat body tissue to test whether they could stimulate acetate uptake. Thus far, all such inhibitors are ineffective at mimicking the effects of the chloromethyl ketones (M. J. Lee & G. J. Goldsworthy, unpublished observations); the mechanism of action of the chloromethyl ketones in stimulating acetate uptake therefore remains unknown.

Conclusions

The relationships between the activities of the AKHs and their primary and secondary structures is not well understood, and the prospect of a quantitatively exact predictive model of these relationships in various assays *in vitro* and *in vivo* is some way off; the possibility that breakdown and removal of the hormones affects their apparent potency clearly needs further investigation. Similarly, the interplay of the various second-messenger systems involved in the transduction of the AKH signal is gradually being teased apart, but is by no means complete (see Van Marrewijk & Van der Horst, this volume). Finally, the insulin-like stimulation of acetate uptake effected by the chloromethyl ketones is of particular interest, especially in view of the limited data relating to biologically active insulin-like materials in insects. It indicates that fat body has the cellular machinery necessary to increase acetate uptake and may suggest that biologically active insulin-like endogenous factors are present in *L. migratoria*. Indeed, preliminary experiments (M. Owusu, K. Siegert, W. Mordue & G. Goldsworthy, unpublished observations) support such a view.

Acknowledgements

We are grateful to Mary Lightfoot and Gita Panchal for technical assistance. Work described in this review is supported by grants to GJG from the SERC Invertebrate Neuroscience Initiative and the BBSRC.

References

Asher, C., Moshitzky, P., Ramachandran, J. & Applebaum, S. W. (1984). The effects of synthetic locust adipokinetic hormone on dispersed locust fat body cell preparations: cAMP induction, lipid mobilization, and inhibition of protein synthesis. *General and Comparative Endocrinology* **55**, 167–173.

Barrett, A. J. & Kirschke, H. (1981). Cathepsin B, cathepsin H, and cathepsin L. *Methods In Enzymology* **80**, 535–561.

Bootman, M. D., Taylor, C. W. & Berridge, M. J. (1992). The thiol reagent, thimerosal, evokes Ca^{2+} spikes in HeLa cells by sensitizing the inositol 1,4,5-trisphosphate receptor. *Journal of Biological Chemistry* **267**, 25113–25119.

Carlisle, J. A. & Loughton, B. G. (1979). Adipokinetic hormone inhibits protein synthesis in *Locusta. Nature* **282**, 420–421.

Cheeseman, P. & Goldsworthy, G. J. (1979). The release of adipokinetic hormone during flight and starvation in *Locusta. General and Comparative Endocrinology* **37**, 35–43.

Cheeseman, P., Jutsum, A. R. & Goldsworthy, G. J. (1976). Quantitative studies on the release of locust adipokinetic hormone. *Physiological Entomology* **1**, 115–121.

Fernlund, P. & Josefsson, L. (1972). Crustacean color-change hormone: amino acid sequence and chemical synthesis. *Science* **177**, 173–175.

Ford, M. M., Hayes, T. K. & Keeley, L. L. (1988). Structure–activity relationships for insect hypertrehalosemic hormone: the importance of side chains and termini. In *Peptides. Chemistry and Biology*, ed. G. M. Marshall, pp. 653–655. Leiden, The Netherlands, ESCOM.

Gäde, G. (1990). Structure–function studies on hypertrehalosemic and adipokinetic hormones: activity of naturally occurring analogues and some N- and C-terminal modified analogues. *Physiological Entomology* **15**, 299–316.

Gäde, G. (1996). The revolution in insect neuropeptides illustrated by the adipokinetic hormone/red pigment-concentrating hormone family of peptides. Zeitschrift für Naturforschung **51c**, 607–617.

Gäde, G. (1997a). Hyperprolinaemia caused by novel members if the adipokinetic/red pigment-concentrating family of peptides isolated from corpora cardiaca of onitine beetles. *Biochemical Journal* **321**, 201–206.

Gäde, G. (1997b). Distinct sequences of AKH/RPCH family members in beetle (*Scarabaeus*-species) corpus cardiacum contain three aromatic amino acid residues. *Biochemical and Biophysical Research Communications* **230**, 16–21.

Gäde, G. & Holwerda, D. A. (1976). Involvement of adenosine 3'-5'cyclic monophosphate in lipid mobilization in *Locusta migratoria. Insect Biochemistry* **6**, 535–540.

Gäde, G., Reynolds, S. E. & Beeching, J. R. (1994). Molecular evolution of peptides of the AKH-RPCH family. In *Perspectives in Comparative Endocrinology*, ed. K. G. Davey, R. E. Peter & S. S. Tobe, pp. 119–128. Ottawa, National Research Council of Canada.

Gokuldas, M. (1989). Hormonal regulation of lipid synthesis in the locust *Schistocerca gregaria*. Ph.D. thesis, University of Birmingham, England.

Gokuldas, M., Hunt, P. A. & Candy, D. J. (1988). The inhibition of lipid synthesis *in vitro* in the locust, *Schistocerca gregaria*, by factors from the corpora cardiaca. *Physiological Entomology* **13**, 43–48.

Goldsworthy, G. J. (1983). The endocrine control of flight metabolism in locusts. *Advances in Insect Physiology* **17**, 149–204.

Goldsworthy, G. J. (1994). The adipokinetic hormones of insects: are they the insect glucagons? In *Perspectives in Comparative Endocrinology* ed. K. G. Davey, R. E. Peter & S. S. Tobe, pp. 486–492. Ottawa: National Research Council of Canada.

Goldsworthy, G. J., Coast, G., Wheeler, C., Cusinato, O., Kay, I. & Khambay, I. (1992). The structure and functional activity of neuropeptides. In *Insect Molecular Biology; 16th Symposium of the Royal Entomological Society of London, 12–13 September 1991*, ed. J. M. Crampton & P. Eggleston, pp. 205–225. London: Academic Press.

Goldsworthy, G. J., Lee, M. J. & Luswata, R. (1995). Adipokinetic hormones: interassay variations in potencies as clues to hormone–receptor interaction in the locust. In *Insects: Chemical, Physiological and Environmental Aspects 1994*, ed. D. Konopinska, pp. 17–27. Poland: University of Wroclaw.

Goldsworthy, G. J., Lee, M. J., Luswata, R., Drake, A. & Hyde, D. (1997). Structures, assays and receptors for locust adipokinetic hormones. *Comparative Biochemistry and Physiology* **117B**, 483–496.

Goldsworthy, G. J., Mallison, K. & Wheeler, C. H. (1986a). The relative potencies of two known locust adipokinetic hormones. *Journal of Insect Physiology* **32**, 95–101.

Goldsworthy, G. J., Mallison, K., Wheeler, C. H. & Gäde, G. (1986b). Relative adipokinetic activities of members of the adipokinetic hormone/red pigment concentrating hormone family. *Journal of Insect Physiology* **32**, 433–438.

Goldsworthy, G. J., Wheeler, C. H., Cusinato, O. & Wilmot, C. M. (1990). Adipokinetic hormones: structures and functions. In *Progress in Comparative Endocrinology*, ed. A. Epple, C. G. Scanes & M. H. Stetson, pp. 28–34. New York: Alan R. Liss. Inc.

Grynkiewicz, G., Poenie, M. & Tsien, R. Y. (1985). A new generation of Ca^{2+} indicators with greatly improved fluorescence properties. *Journal of Biological Chemistry* **260**, 3440–3450.

Hayes, T. K. & Keeley, L. L. (1990). Structure–activity relationships on hyperglycaemia by representatives of the adipokinetic/hyperglycaemic hormone family in *Blaberus discoidalis* cockroaches. *Journal of Comparative Physiology B* **160**, 187–194.

Hayes, T. K., Nails, F. L. & Ford, M. M. (1994). Probing the steric tolerance of the insect hypertrehalosemic hormone receptor to develop an effective photoaffinity probe. In *Peptides: Chemistry, structure and biology*, ed. R. Hodges & J. Smith, pp. 666–668. Leiden, The Netherlands: ESCOM.

Jahagirdar, A. P., Milton, G., Viswanatha, T. & Downer, R. G. H. (1987). Calcium involvement in mediating the action of octopamine and hypertrehalosemic peptides on insect haemocytes. *FEBS Letters* **219**, 83–87.

Keeley, L. L. & Hesson, A. S. (1995). Calcium-dependent signal transduction by the hypertrehalosemic hormone in the cockroach fat body. *General and Comparative Endocrinology* **99**, 373–381.

Kettner, C., Glover, G. I. & Prescott, J. M. (1974). Kinetics of inhibition of *Aeromonas* aminopeptidase by leucine methyl ketone derivatives. *Archives of Biochemistry and Biophysics* **165**, 739–743.

Kodrík, D. & Goldsworthy, G. J. (1995). Inhibition of RNA synthesis by adipokinetic hormones and brain factor(s) in adult fat body of *Locusta migratoria*. *Journal of Insect Physiology* **41**, 127–133.

Kromer, E. & Lageux, M. (1994). Studies on *Locusta* insulin related peptides. In: *Perspectives in Comparative Endocrinology*. ed. K. G. Davey, R. E. Peter & S. S. Tobe, pp. 220–225. Ottawa: National Research Council of Canada.

Kupfer, A., Gani, V., Jiménez, J. S. & Shaltiel, S. (1979). Affinity labelling of the catalytic subunit of cyclic AMP-dependent protein kinase by N^α-tosyl-L-lysine chloromethyl ketone. *Proceedings of the National Academy of Sciences USA* **76**, 3073–3077.

Kurata, S., Saito, H. & Natori, S. (1990). Participation of hemocyte proteinase in dissociation of the fat body on pupation of *Sarcophaga peregrina* (flesh fly). *Insect Biochemistry* **20**, 461–465.

Kurata, S., Saito, H. & Natori, S. (1992a). The 29-kDa hemocyte proteinase dissociates fat body at metamorphosis of *Sarcophaga*. *Developmental Biology* **153**, 115–121.

Kurata, S., Saito, H. & Natori, S. (1992b). Purification of a 29-kDa hemocyte proteinase of *Sarcophaga peregrina*. *European Journal of Biochemistry* **204**, 911–914.

Lamango, N. S. & Isaac, R. E. (1993). Metabolism of insect neuropeptides: properties of a membrane-bound endopeptidase from heads of *Musca domestica*. *Insect Biochemistry and Molecular Biology* **23**, 801–808.

Lee, M. J., Cusinato, O., Luswata, R., Wheeler, C. H. & Goldworthy, G. J. (1997). N-terminal modifications to AKH-I from *Locusta migratoria*: assessment of biological potencies *in vivo* and *in vitro*. *Regulatory Peptides* **69**, 69–76.

Lee, M. J., Drake, A. F. & Goldsworthy, G. J. (1996). *Locusta*-AKH-III and related peptides containing two tryptophan residues have unusual CD spectra. *Biochemical and Biophysical Research Communications* **226**, 407–412.

Lee, M. J. & Goldsworthy, G. J. (1995a). Acetate uptake test; the basis of a rapid method for determining potencies of adipokinetic peptides for structure-activity studies. *Journal of Insect Physiology* **41**, 163–170.

Lee, M. J. & Goldsworthy, G. J. (1995b). The preparation and use of dispersed cells from fat body of *Locusta migratoria* in a filtration plate assay for adipokinetic peptides. *Analytical Biochemistry* **228**, 155–161.

Lee, M. J. & Goldsworthy, G. J. (1996). Modified adipokinetic peptides containing two tryptophan residues and their activities *in vitro* and *in vivo* in *Locusta*. *Journal of Comparative Physiology B* **166**, 61–67.

Lee, M. J., Goldsworthy, G. J., Poulos, C. P. & Velentza, A. (1997). Synthesis and biological activity of adipokinetic hormone analogues modified at the C-terminus. *Peptides* **17**, 1285–1290.

Lee, Y.-H. & Keeley, L. L. (1994). Intracellular transduction of trehalose synthesis by hypertrehalosemic hormone in the fat body of the tropical cockroach, *Blaberus discoidalis. Insect Biochemistry and Molecular Biology* **24**, 473–480.

Michalik, J., Szolajska, E., Rosinski, G., Lombarska-Sliwinska, D. & Konopinska, D. (1995). Some physiological effects of AKH-RPCH peptides on *Tenebrio molitor* fat body in vitro. In *Insects: Chemical, Physiological and Environmental Aspects 1994*, ed. D. Konopinska, pp. 235–241. Poland: University of Wroclaw.

Nakashima, S., Tohmatsu, T., Shirato, L., Takenaka, A. & Nozawa, Y. (1987). Neomycin is a potent agent for arachidonic acid release in human platelets. *Biochemical and Biophysical Research Communications* **146**, 820–826.

Navarro, J., Abdel Ghany, M. & Racker, E. (1982). Inhibition of tyrosine protein kinases by halomethyl ketones. *Biochemistry* **21**, 6138–6144.

Orchard, I., Carlisle, J. A., Loughton, B. G., Gole, J. W. D. & Downer, R. G. H. (1982). *In vitro* studies on the effects of octopamine on locust fat body. *General and Comparative Endocrinology* **48**, 7–13.

Orchard, I. & Lange, A. B. (1983). Release of identified adipokinetic hormones during flight and following neural stimulation in *Locusta migratoria. Journal of Insect Physiology* **29**, 425–429.

Orr, G. L., Gole, J. W. D., Jahagirdar, A. P., Downer, R. G. H. & Steele, J. E. (1985). Cyclic AMP does not mediate the action of synthetic hypertrehalosemic peptides from the corpus cardiacum of *Periplaneta americana. Insect Biochemistry* **15**, 703–709.

Oudejans, R. C. H. M., Kooiman, F. P., Heerma, W., Versluis, C., Slotboom, A. J. & Beenakkers, A. M. T. (1991). Isolation and structure elucidation of a novel adipokinetic hormone (Lom AKH-III) from the glandular lobes of the corpus cardiacum of the migratory locust, *Locusta migratoria. European Journal of Biochemistry* **195**, 351–359.

Oudejans, R. C. H. M., Vroemen, S. F., Jansen, R. F. R. & Van der Horst, D. J. (1996). Locust adipokinetic hormones: carrier-independent transport and differential inactivation at physiological concentrations during rest and flight. *Proceedings of the National Academy of Sciences USA* **93**, 8654–8659.

Pancholi, S., Barker, C. J., Candy, D. J., Gokuldas, M. & Kirk, C. J. (1991). Effects of adipokinetic hormones on inositol phosphate metabolism in locust fat body. *Biochemical Society Transactions* **19**, 1045.

Peled, Y. & Tietz, A. (1973). Fat transport in the locust *Locusta migratoria*: the role of protein synthesis. *Biochimica et Biphysica Acta* **296**, 499–509.

Peled, Y. & Tietz, A. (1975). Isolation and properties of a lipoprotein from the haemolymph of the locust, *Locusta migratoria. Insect Biochemistry* **5**, 61–72.

Poulos, C., Velentza, A., Lee, M. J. & Goldsworthy, G. J. (1995). Synthesis and biological activity of AKH-I analogues with modifications at the threonine residue in position 10. In *Insects: Chemical, Physiological and*

Environmental Aspects 1994, ed. D. Konopinska, pp. 228–230. Poland: University of Wroclaw.

Schoellmann, G. & Shaw, E. (1963). Direct evidence for the presence of histidine in the active center of chymotrypsin. *Biochemistry* **2**, 252–255.

Siligardi, G. & Drake, A. F. (1995). The importance of extended conformations and in particular the P_{II} conformation for the molecular recognition of peptides. *Biopolymers (Peptide Science)* **37**, 281–292.

Solomon, D. H., O'Brian, C. A. & Weinstein, I. B. (1985). *N*-α-tosyl-L-lysine chloromethyl ketone and *N*-α-tosyl-L-phenylalanine chloromethyl ketone inhibit protein kinase C. *FEBS Letters* **190**, 342–344.

Spencer, I. M. & Candy, D. J. (1976). Hormonal control of diacyl glycerol mobilization from fat body of the desert locust *Schistocerca gregaria*. *Insect Biochemistry* **6**, 289–296.

Steele, J. E. & Ireland, R. (1994). The preparation of trophocytes from disaggregated fat body of the cockroach *Periplaneta americana*). *Comparative Biochemistry and Physiology* **107A**, 517–522.

Stone, J. V., Mordue, W., Batley, K. E. & Morris, H. R. (1976). Structure of locust adipokinetic hormone, a neurohormone that regulates lipid utilization during flight. *Nature* **263**, 207–211.

Stone, J. V., Mordue, W., Broomfield, C. E. & Hardy, P. M. (1978). Structure–activity relationships for the lipid-mobilising action of locust adipokinetic hormone. Synthesis and activity of a series of hormone analogues. *European Journal of Biochemistry* **89**, 195–202.

Takahashi, N., Kurata, S. & Natori, S. (1993). Molecular cloning of cDNA for the 29 kDa proteinase participating in decomposition of the larval fat body during metamorphosis of *Sarcophaga peregrina* (flesh fly). *FEBS Letters* **334**, 153–157.

Thastrup, O., Cullen, P. J., Drøbak, B. K., Hanley, M. R. & Dawson, A. P. (1990). Thapsigargin, a tumor promoter, discharges intracellular Ca^{2+} stores by specific inhibition of the endoplasmic reticulum Ca^{2+}-ATPase. *Proceedings of the National Academy of Sciences USA* **87**, 2466–2470.

Thorpe, A. & Duve, H. (1984). Insulin- and glucagon-like peptides in insects and molluscs. *Molecular Physiology* **5**, 235–260.

Tietz, A. (1962). Fat transport in the locust. *Journal of Lipid Research* **3**, 421–426.

Van der Horst, D. J., Vroemen, S. F. & Van Marrewijk, W. J. A. (1997). Metabolism of stored reserves in insect fat body: hormonal signal transduction implicated in glycogen mobilization, and biosynthesis of the lipophorin system. *Comparative Biochemistry and Physiology* **117B**, 463–474.

Van Marrewijk, W. J. A., Van Den Broek, A. Th. M. & Beenakkers, A. M. (1980). Regulation of gluconeogenesis in the fat body during flight. *Insect Biochemistry* **10**, 675–679.

Van Marrewijk, W. J. A., Van Den Broek, A. Th. M. & Beenakkers, A. M. Th. (1986). Hormonal control of fat body glycogen mobilization for locust flight. *General and Comparative Endocrinology* **64**, 136–142.

Van Marrewijk, W. J. A., Van den Broek, A. T. M. & Beenakkers, A. M. T. (1991). Adipokinetic hormone is dependent on extracellular Ca^{2+} for its stimulatory action on the glycogenolytic pathway in locust fat body *in vitro*. *Insect Biochemistry* **21**, 375–380.

Vergara, J., Tsien, R. & Delay, M. (1985). Inositol 1,4,5-trisphosphate: a possible chemical link in excitation–contraction coupling in muscle. *Proceedings of the National Academy of Sciences USA* **82**, 6352–6356.

Vroemen, S. F., Van Marrewijk, W. J. A. & Van der Horst, D. J. (1995a). Stimulation of glycogenolysis by three locust adipokinetic hormones involves G_s and cAMP. *Molecular and Cellular Endocrinology* **107**, 165–171.

Vroemen, S. F., Van Marrewijk, W. J. A., Schepers, C. C. J. & Van der Horst, D. J. (1995b). Signal transduction of adipokinetic hormones involves Ca^{2+} fluxes and depends on extracellular Ca^{2+} to potentiate cAMP-induced activation of glycogen phosphorylase. *Cell Calcium* **17**, 459–467.

Walenga, R., Vanderhoek, J. Y. & Feinstein, M. B. (1980). Serine esterase inhibitors block stimulus-induced mobilization of arachidonic acid and phosphatidylinositide-specific phospholipase C activity in platelets. *Journal of Biological Chemistry* **255**, 6024–6027.

Wicher, D., Walther, Ch. & Penzlin, H. (1995). The neurohormone D-induced increase of a N-type Ca^{2+} current leads to an enhanced after-hyperpolarization in central cockroach neurones. In *Insects: Chemical, Physiological and Environmental Aspects 1994*, ed. D. Konopinska, pp. 146–151. Poland: University of Wroclaw.

Williams, D. A. & Fay, F. S. (1990). Intracellular calibration of the fluorescent calcium indicator fura-2. *Cell Calcium* **11**, 75–83.

Yagi, K. J., Yu, C. G. & Tobe, S. S. (1992). Phosphoramidon enhances allatostatin-mediated inhibition of juvenile hormone biosynthesis in the coprora allata of the cockroach, *Diploptera punctata*. *Experientia* **48**, 758–761.

Ziegler, R., Eckart, K., Jasensky, R. D. & Law, J. H. (1991). Structure–activity studies on adipokinetic hormones in *Manduca sexta*. *Archives of Insect Biochemistry and Physiology* **18**, 229–237.

Zubrzycki, I. Z. & Gäde, G. (1994). Conformational study on an insect neuropeptide of the AKH/RPCH family by combined ^{1}H-NMR spectroscopy and molecular mechanics. *Biochemical and Biophysical Research Communications* **198**, 228–235.

WIL J. A. VAN MARREWIJK
and DICK J. VAN DER HORST

Signal transduction of adipokinetic hormone

Introduction

Flight activity of insects provides an attractive model system to study regulation of metabolic processes during physical exercise. Several long-distance flying insects mobilise both carbohydrate and lipid reserves as fuels for migratory flight from the fat body, which combines many of the properties and functions of vertebrate liver and adipose tissue, and plays a fundamental role in energy metabolism. This chapter focuses on current knowledge on a few aspects of metabolic processes in the fat body, employing particularly the migratory locust, *Locusta migratoria*, as an insect model. The mobilisation of energy substrates from the fat body is controlled by adipokinetic hormones (AKHs), peptide neurohormones which upon flight are released by the corpus cardiacum, a neuroendocrine gland connected with the brain. Recent data on the involvement of the distinct AKHs identified in locust corpora cardiaca in flight-related processes is discussed, with special emphasis on their role in substrate mobilisation. However, the major focus of this chapter is on the mechanism through which the AKH signal is transduced intracellularly. The structural features of the fat body, which is capable of functioning *in vitro*, render this metabolically highly active tissue a suitable model for studies on signal perception and transduction in an intact animal system. Therefore data on AKH signal transduction presented in this chapter result primarily from studies performed on fat body *in vitro*.

Adipokinetic hormones and substrate mobilisation

Insect flight activity stimulates the adipokinetic cells in the glandular lobes of the corpus cardiacum to release the adipokinetic hormones. These are N- and C-terminal blocked small peptides which are structurally related; in glandular lobes of the migratory locust, a decapeptide AKH-I is present in addition to two octapeptides (AKH-II and III), the molar proportions of AKH-I:AKH-II:AKH-III being approximately 14:2:1 (Oudejans *et al.*, 1993). All three AKHs are synthesised as preprohormones; recent data using *in situ* hybridisation showed the mRNA signals encoding the three different AKH precursors to be co-localised in cell bodies of the glandular lobes of the corpus cardiacum (Bogerd *et al.*, 1995). Additionally, expression of the genes for all three AKH precursors is stimulated by flight activity. These data suggest that all three AKHs are involved in flight-related processes. A physiologically relevant question therefore is why these three related hormones occur simultaneously.

The action of the hormones on the insect fat body results ultimately in the mobilisation of stored reserves as fuels for flight. Carbohydrate (trehalose) is mobilised from glycogen reserves, implying hormonal activation, by phosphorylation, of the key enzyme glycogen phosphorylase. Similarly, lipid (*sn*-1,2-diacylglycerol, DAG) is mobilised from stored triacylglycerol, which implies hormonal activation of the fat body triacylglycerol lipase (for a review, see Beenakkers, Van der Horst & Van Marrewijk, 1985). The carbohydrate and lipid substrates are transported in the haemolymph to the working flight muscles. Carbohydrate provides most of the energy for the initial period of flight; thereafter, in long-term flying insects like locusts, the concentration of lipid substrate in the blood is increased and gradually takes over, so the principal substrate for long-term flight is lipid. Although glycogen phosphorylase in the locust fat body is converted into its phosphorylated active form within a few minutes of flight activity, it takes almost half an hour before newly synthesised trehalose is released from the fat body (Beenakkers *et al.*, 1985). A possible explanation for this lag period between the phosphorylation of glycogen phosphorylase and the release of newly synthesised trehalose may be that the allosteric enzyme is kept in its inactive T (tense) state as long as the concentration of trehalose is high, and will switch to its active R (relaxed) state only after a substantial reduction of trehalose levels in the fat body, as may occur during flight activity. A similar mechanism has been described for the inhibition of phosphorylase *a* by glucose in vertebrate liver (Johnson, 1992).

In a bioassay, all three AKHs are able to activate both glycogen phosphorylase and triacylglycerol lipase, although there may be differences in potency, as shown in Fig. 1. Recent preliminary results obtained with combinations of two or three AKHs, which are likely to occur together in insect haemolymph under physiological conditions *in vivo*, revealed that the maximal responses for both the carbohydrate- and the lipid-mobilising effects were much lower than the theoretically calculated responses that were based on dose–response curves for the individual hormones. In the lower (probably physiological) range, however, combinations of the AKHs were more effective or at least as effective as the theoretical values calculated from the responses to the individual hormones (Oudejans *et al.*, 1992). In addition to their different potencies in eliciting substrate mobilisation, all three AKHs

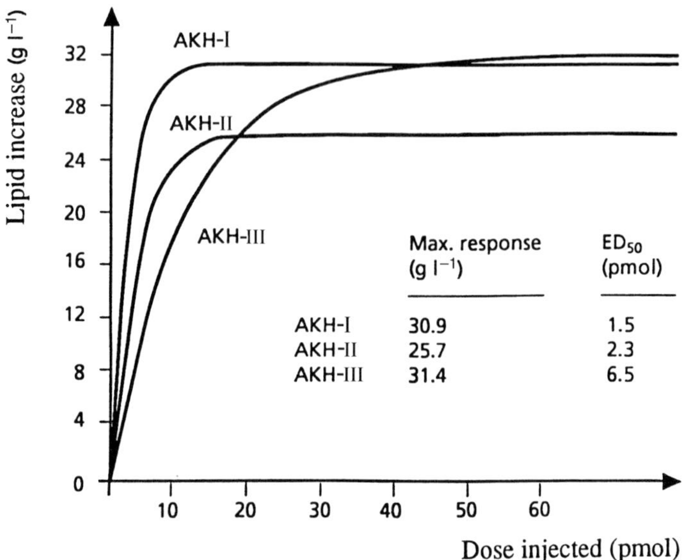

Fig. 1. Dose–response curves for the lipid mobilising effects of AKH-I, II and III in *L. migratoria*. Adult male locusts were injected with different doses of AKH or, in the controls, with saline, and after 120 min the lipid content in haemolymph was determined. Responses represent increase in haemolymph lipid in AKH-injected locusts, expressed as g l⁻¹. For clarity, data points and standard error bars have been omitted. ED_{50}, effective dose eliciting 50% of the maximum response (after Oudejans *et al.*, 1992).

inhibit RNA synthesis in the locust fat body *in vitro* (Kodrík & Goldsworthy, 1995). In males, the responses were dose-dependent, with their potencies decreasing in the order AKH-III>AKH-II>AKH-I. All three AKHs were equally efficacious. These data may point to the occurrence of different receptors for the AKHs or, alternatively, the observed differences in effectiveness of the three AKHs may be due to differences in the mechanism of hormonal signal transduction. However, a decisive factor for the ability to elicit physiological responses is the relative abundance of each of the hormones, which is determined by their rates of synthesis and release, and by their rate of inactivation. Studies of hormone dynamics in intact locusts injected with physiological amounts of tritiated AKH (1 pmol) have demonstrated for all three AKHs a decrease in total radioactivity (non-degraded hormone plus radioactive degradation products) with time, both at rest and during flight (Fig. 2A). The rate of decrease was quite different for each of the hormones (AKH-III>>>AKH-II>AKH-I), and was faster during flight than at rest (Oudejans *et al.*, 1996). Although the first step in the

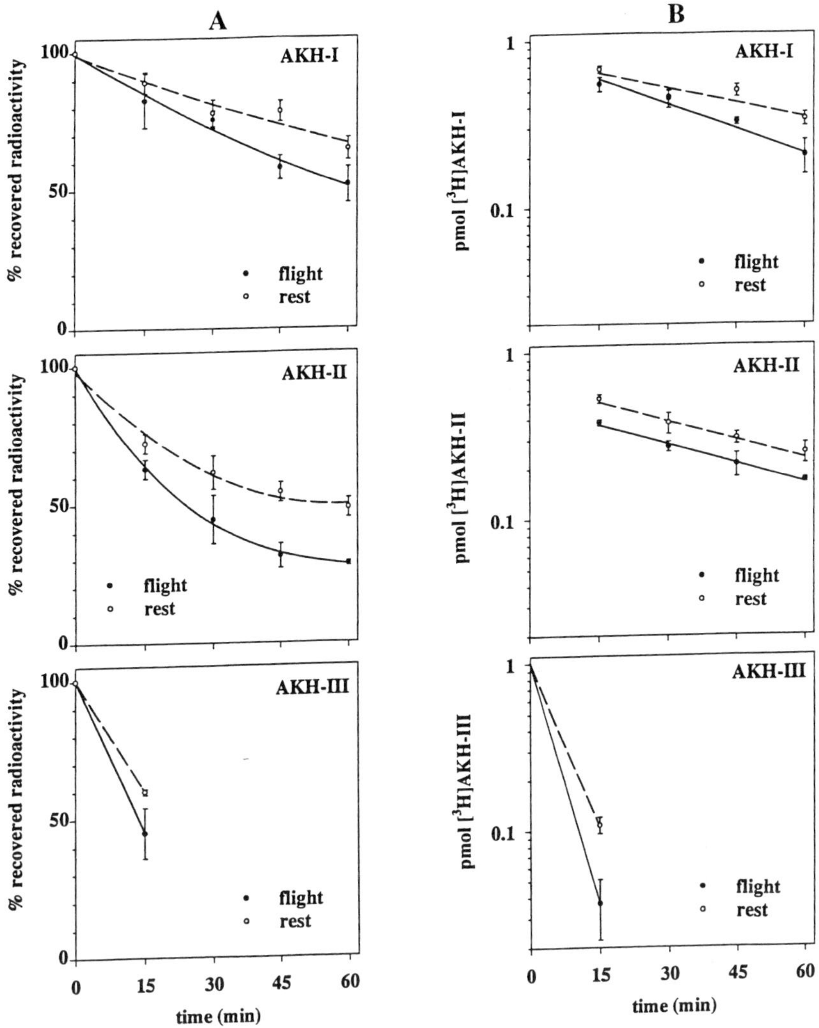

Fig. 2. Differential inactivation of the three AKHs in *L. migratoria* under rest and flight conditions. Locusts were injected with 1.0 pmol of tritiated AKH-I, II or III, and at the times indicated haemolymph was sampled for measurement of radioactivity and high-performance liquid chromatography analysis ($n=6$ for rest and $n=3$ for flight). (A) Time course of proportional radioactivity recovered in haemolymph (non-degraded AKH plus radioactive degradation products). (B) Semi-logarithmic plots of AKH remaining in the haemolymph versus time (after Oudejans *et al.*, 1996).

inactivation of the hormones occurs by an endopeptidase cleaving the N-terminal tripeptide pGlu-Leu-Asn which they have in common, their half-lives were different and were changed by flight activity, as estimated from semi-logarithmic plots of remaining AKH in haemolymph versus time (Fig. 2B): in resting locusts, half-lives were 51, 40 and 5 min for AKH-I, II and III, respectively, whereas during flight values of 35, 37 and 3 min were obtained. This rapid and differential degradation of the three AKHs will lead to changes from the proportions in which they were released, and therefore has important consequences for concerted hormone action at the level of the target organ receptor(s), suggesting that each of the AKHs may play its own biological role in the overall syndrome of locust flight.

Hormonal signal transduction in insect fat body

Role of extracellular calcium in phosphorylase stimulation by AKH

As to the mechanism of transduction of the three AKH signals – which additionally might provide some possible clues as to why in *L. migratoria* three different AKHs occur – the activation of glycogen phosphorylase was monitored in fat body *in vitro*, which provides a convenient and very sensitive test system. Stimulation of fat body phosphorylase by all three AKHs is dependent on the presence of extracellular Ca^{2+} (Vroemen *et al.*, 1995a). At low concentrations of calcium ions in the medium (<0.2 mmol l^{-1}) no significant phosphorylase activation by AKH occurs. A concentration of ≥ 1.5 mmol Ca^{2+} l^{-1} is required for maximal activation by each of the three hormones, which is consistent with the physiological levels of Ca^{2+} normally present in the extracellular fluid of eukaryotes (1–2 mmol l^{-1}). In the absence of AKH, Ca^{2+} in the medium does not affect the enzyme activity. The presence of Ca^{2+} was also shown to be indispensable for the induction of both fat body phosphorylase activation and the synthesis and release of trehalose in the cockroach *Periplaneta americana* by corpus cardiacum extract or hyper-trehalosaemic peptides (Steele & Paul, 1985; Orr *et al.*, 1985), and recently the presence of extracellular Ca^{2+} was found to be necessary for a full stimulation of fat body phosphorylase by hypertrehalosaemic hormone in the cockroach *Blaberus discoidalis* (Park & Keeley, 1995, 1996). An obvious way for Ca^{2+} to exert its role in phosphorylase activation is by direct stimulation of phosphorylase kinase, which converts inactive phosphorylase into the active form. Ca^{2+}-calmodulin-dependent phosphorylase kinase activity has been demonstrated in cell-free preparations from *Locusta* fat body (Van Marrewijk, Van den Broek & Beenakkers, 1991), and stimulation of phosphorylase kinase by Ca^{2+} has also been shown in several other insects, including the fruitfly *Drosophila melanogaster* (Dombrádi *et al.*, 1987) and *P. americana* (Pallen & Steele, 1988). However, a Ca^{2+}-independent phosphorylase kinase also occurs in insect fat body, as has been demonstrated for *Phylosamia cynthia* (Hayakawa & Chino, 1983).

A Ca^{2+} dependence similar to that demonstrated for hormone-induced carbohydrate mobilisation appears to exist for the effect of AKH on the mobilisation of lipids, as both Lum & Chino (1990) and Wang, Hayakawa & Downer (1990) have demonstrated that AKH, which is known to stimulate the production of DAG from triacylglycerol stores in the locust fat body, requires the presence of extracellular Ca^{2+} to evoke an increase in intracellular levels of DAG. Again, Ca^{2+} alone in the absence of AKH did not significantly change the fat body DAG content. These studies on the Ca^{2+} dependence of hormonally controlled lipid mobilisation were conducted exclusively with AKH-I. However, an assessment of the Ca^{2+} dependence of the DAG production induced by AKH-II and AKH-III will also be important in providing some insight into why three different AKHs occur in locusts.

AKH-induced cyclic AMP production and calcium dependence

All three AKHs of *L. migratoria* are capable of elevating intracellular levels of cyclic AMP in the fat body *in vitro* (Vroemen, Van Marrewijk & Van der Horst, 1995b). This elevation appeared to be very rapid; a significant increase was measured within 15 s of adding AKH to the medium, whereas maximal levels of cyclic AMP (200% increase compared to the control level) were reached after 1 min (Fig. 3). Dose–response studies have shown that at 40 nmol l^{-1} the three AKHs are equally potent in enhancing intracellular cyclic AMP, this massive dose probably eliciting a maximal response. However, at a more physiological level (4 nmol l^{-1}), their potencies in elevating cyclic AMP levels were different, decreasing in the order AKH-III>AKH-II>AKH-I. Correspondingly, at this concentration AKH-III was also the most potent in inducing activation of glycogen phosphorylase, although results obtained with AKH-I and AKH-II did not differ significantly.

While the hormonally induced activation of phosphorylase described above was dependent on the presence of extracellular calcium ions, the ability of the AKHs to raise levels of cyclic AMP in the fat body appeared to be equally Ca^{2+}-dependent: when Ca^{2+} (normally present at 2 mmol l^{-1}) was eliminated from the incubation medium, none of the peptides was able to elevate cyclic AMP levels (Vroemen *et al.*, 1995b). For AKH-I, the requirement of extracellular Ca^{2+} to increase fat body cyclic AMP was shown by Wang *et al.*, (1990). Altogether, the above data on DAG levels, activation of

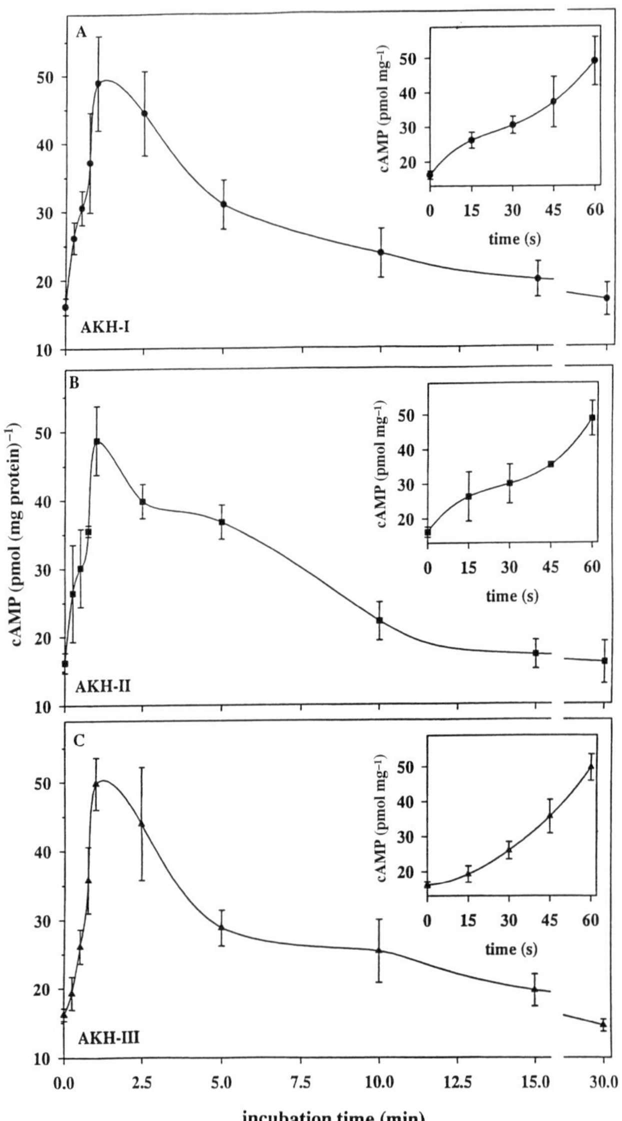

Fig. 3. Time course for the effect of a saturating dose of AKH on intracellular cyclic AMP levels in fat body of *L. migratoria*. Fat body was incubated for the times given with 40 nmol AKH-I (A), II (B) or III (C) l⁻¹, and subsequently cyclic AMP and protein contents were determined in homogenates. For each time point, $n=6$ (after Vroemen *et al.*, 1995b).

glycogen phosphorylase and increase in cyclic AMP levels provide convincing evidence that extracellular calcium ions are necessary for the AKHs to exert their effects on the fat body.

Regulation of the influx and efflux of calcium

The activating effect of AKH on phosphorylase can be mimicked by the calcium ionophore A23187 (Van Marrewijk *et al.*, 1991). Like AKH, this compound also increases the levels of cyclic AMP and DAG in locust fat body tissue incubated *in vitro* (Wang *et al.*, 1990). Although responses to the ionophore were smaller than those elicited by AKH, these results suggest that at least part of the action of AKH consists of enabling extracellular calcium ions to enter fat body cells, leading to enhanced concentrations of intracellular Ca^{2+}. Support for this suggestion was obtained by Wang *et al.* (1990), who measured a 2.5-fold increase in the uptake of $^{45}Ca^{2+}$ from the medium into the fat body upon incubation in the presence of 100 nmol AKH-I l^{-1} for 1 h. However, results from this study are not conclusive, because the hormone may also stimulate Ca^{2+} efflux (see below). The effects of 40 nmol AKH-I, II and III l^{-1} on the uptake of $^{45}Ca^{2+}$ from the medium were examined during the first 5 min after peptide addition because, in the case of a role for extracellular calcium in signal transduction, effects on Ca^{2+} fluxes may be expected to occur shortly after applying the hormone (Van Marrewijk, Van den Broek & Van der Horst, 1993; Vroemen *et al.*, 1995a). All three AKHs significantly increased the uptake of Ca^{2+} from the medium, the effect being evident within 1 min. No significant differences between the effects of the three AKHs were observed, even when AKH concentrations were reduced to 4 nmol l^{-1}.

In addition to stimulating the influx of extracellular Ca^{2+} into the fat body, all three hormones enhance the efflux of Ca^{2+} from the fat body, possibly as a result of the increased levels of intracellular Ca^{2+} (Van Marrewijk *et al.*, 1993; Vroemen *et al.*, 1995a). Fat body tissue was pre-loaded to equilibrium with $^{45}Ca^{2+}$ and the radiolabel released into the medium in the absence or presence of AKH-I, II or III was measured. At a saturating dose of 40 nmol l^{-1}, and assuming that during the first 15 s of incubation 're-influx' of $^{45}Ca^{2+}$ released into the medium was negligible, 50% of the pre-loaded $^{45}Ca^{2+}$ was estimated to have been extruded from the fat body within about 1 min of incubation in the presence of an AKH. This indicates an efflux rate that seems to be appropriate for a rapid second-messenger system. The maximal Ca^{2+} efflux evoked by saturating doses of AKH-I, II and III did not differ. However, at a more physiological level (4 nmol l^{-1}), a small but interesting difference in potency was observed; the AKH-induced increase in Ca^{2+} efflux decreased in the order AKH-III>AKH-II>AKH-I (Fig. 4), as

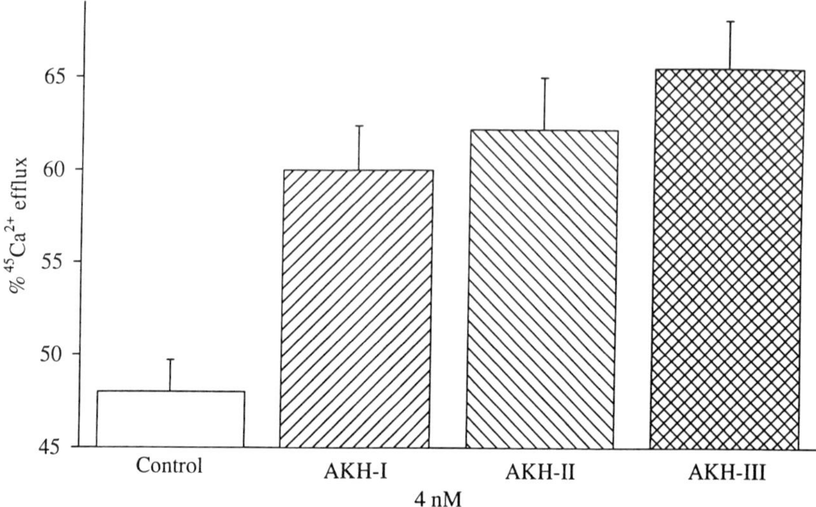

Fig. 4. Effect of a physiological dose of AKH on the efflux of radioactive calcium from *L. migratoria* fat body *in vitro*. Fat body pre-loaded with ^{45}Ca^{2+} was incubated with 4 nmol AKH-I, II or III l^{-1}, and radioactivity released into the medium was determined at several time intervals. The figure shows the percentage of pre-loaded radioactivity that has been released after 5 min of incubation. For each bar $n=12$. The effect induced by AKH-III was significantly higher than that by AKH-I ($P<0.05$) (after Vroemen *et al.*, 1995a).

was the case for stimulation of cyclic AMP production and phosphorylase activation. From the time courses of both the influx and the efflux of Ca^{2+} it could be calculated that the influx exceeds the efflux, resulting in an increase in the intracellular concentration of Ca^{2+}. This fits the above observations that the AKH-induced formation of cyclic AMP and DAG in the fat body, and the activation of glycogen phosphorylase are dependent on extracellular Ca^{2+}. Furthermore, recently Lee & Goldsworthy (1995) described a method for the preparation of metabolically active and AKH-sensitive dispersed cells from *Locusta* fat body, and preliminary data from fluorimetric measurements in these cells indicate that the AKHs cause rapid increases (<10 s) in intracellular Ca^{2+} (Lee, Hyde & Goldsworthy, 1995). Enhancement of intracellular Ca^{2+} levels is not necessarily the sole function of extracellular Ca^{2+}, as they may also be required for the binding of the AKHs to their receptor(s) on fat body membranes as recently demonstrated for *Manduca sexta* AKH (Ziegler, Jasensky & Morimoto, 1995). However, results from recent binding studies with *Locusta* AKHs have not yet revealed evidence of Ca^{2+}-dependent binding.

Mobilisation of calcium from intracellular stores: involvement of inositol phosphates

In addition to the influx of extracellular Ca^{2+}, the mobilisation of Ca^{2+} from intracellular stores appears to be important for the effects of AKH on the fat body. For instance, the addition of TMB-8, an inhibitor of intracellular Ca^{2+} release, to a Ca^{2+}-containing medium led to a significant dose-dependent reduction of the phosphorylase activating action of AKH-I (Van Marrewijk *et al.*, 1993). From this it can be concluded that for a maximal effect of AKH on phosphorylase activity in locust fat body, release of intracellular Ca^{2+} is required in addition to the availability of extracellular Ca^{2+}. Intracellular Ca^{2+} release is also a significant step in the hypertrehalosaemic hormone signal transduction cascade for trehalose biosynthesis in the fat body of *B. discoidalis* (Keeley & Hesson, 1995). Inclusion in the incubation medium of dantrolene, which inhibits the release of intracellular Ca^{2+}, partially suppressed the stimulating effect of the hypertrehalosaemic hormone on trehalose synthesis in the fat body *in vitro*.

In the regulation of Ca^{2+} mobilisation from intracellular stores, inositol phosphates have been shown to play a pivotal role in a variety of cell types (Berridge, 1989, 1994; Berridge & Irvine, 1989). In the hormone-stimulated inositol lipid cycle, binding of the agonist to its cell surface receptor leads to the activation of G-protein-coupled phospholipase C and, as a result, the conversion of the membrane phospholipid phosphatidylinositol 4,5-bisphosphate into two potential second messengers: *sn*-1,2-DAG, an activator of protein kinase C, and *myo*-inositol 1,4,5-trisphosphate (IP_3), whose role in the mobilisation of Ca^{2+} from intracellular stores has been well established (Berridge & Irvine, 1989). IP_3 might be expected to have a role in the AKH signal transduction mechanism because, in locusts, AKH has been shown to enhance intracellular Ca^{2+} levels in the fat body. In *Locusta* fat body *in vitro*, the formation of [^{3}H]inositol phosphates from phosphoinositides prelabelled with *myo*-[^{3}H]inositol indicates the functioning of the inositol lipid cycle. Phosphoinositide turnover was stimulated by AKH-I, as evidenced by the increased production of [^{3}H]inositol phosphates within 1 min of adding the hormone to the incubation medium (Van Marrewijk *et al.*, 1992). Preliminary data have demonstrated that, in *Locusta* fat body, IP_3 is among the inositol phosphates that respond to AKH-I with a strong increase in concentration. Possible effects of AKH-II and III on fat body IP_3 levels remain to be established, and may provide clues as to whether the three AKHs share the same signalling pathways and/or receptor(s). Furthermore, in *Schistocerca gregaria*, both AKH-I and II were shown to significantly stimulate IP_3 formation in the fat body (Pancholi *et al.*, 1991; Stagg & Candy, 1995). These data indicate that the action of AKH on the insect fat body may involve mobilisation of intracellular Ca^{2+} by IP_3. This role for IP_3 in hormone-induced mobilisation of Ca^{2+} from intracellular stores may be common among insects, as in *B. discoidalis* fat body hypertrehalosaemic hormone has been demonstrated to induce intracellular Ca^{2+} release as well as to greatly increase IP_3 levels (Park & Keeley, 1996).

Capacitative calcium entry

Agonist-induced Ca^{2+} influx into cells has been proposed to depend on a direct activation of plasma membrane Ca^{2+} channels by IP_3 or inositol 1,3,4,5-phosphate (IP_4) (Kuno & Gardner, 1987; Irvine, 1991). However, several recent studies (Takemura *et al.*, 1989; Bird *et al.*, 1992; Bird & Putney, 1993) using the sesquiterpene lactone tumour promoter thapsigargin, which depletes Ca^{2+} from IP_3-sensitive pools by inhibition of microsomal Ca^{2+}-ATPase activity without affecting inositol phosphate levels, have lent considerable support in favour of a mechanism termed capacitative calcium entry. According to this concept, the state of filling of the intracellular Ca^{2+} stores is considered as the decisive factor that regulates the influx of extracellular Ca^{2+}. Thus, depletion of the intracellular Ca^{2+} stores would, in some way, activate the mechanism of Ca^{2+} influx across the plasma membrane (for a recent review, see Berridge, 1995). To test whether this mechanism applies also to insects, thapsigargin was added to fat body incubated *in vitro*, and its effect was assessed by measuring the influx of $^{45}Ca^{2+}$ (Van Marrewijk *et al.*, 1993). The presence of thapsigargin led to a significant increase in $^{45}Ca^{2+}$ uptake, implying that in the locust fat body depletion of intracellular Ca^{2+} stores leads to an enhanced influx of extracellular Ca^{2+}. Furthermore, thapsigargin induced a marked activation of glycogen phosphorylase only in the presence of extracellular Ca^{2+}, which is consistent with the view that depletion of intracellular Ca^{2+} stores leads to a stimulation of the Ca^{2+}-entry mechanism across the plasma membrane of insect fat body cells. However, capacitative calcium entry does not exclude the possibility that other mechanisms are involved in regulating the influx of extracellular Ca^{2+}, such as a direct activation of Ca^{2+} channels by AKH. Moreover, the relative importance of extracellular versus intracellular Ca^{2+} for the signal transduction process may vary among insects. In studies on trehalose synthesis by the fat body of *B. discoidalis*, both thimerosal, which sensitises IP_3 receptors of the endoplasmic reticulum for Ca^{2+} release (Bootman, Taylor & Berridge, 1992), and thapsigargin were shown to produce significant hypertrehalosaemic responses, whereas the ionophores A23187 and ionomycin did not significantly increase trehalose synthesis (Keeley & Hesson, 1995). Omission of Ca^{2+} from the incubation medium did not lower the stimulation of trehalose synthesis by thimerosal, nor did

A23187-induced Ca^{2+} influx increase the thimerosal response. For the hypertrehalosaemic response, the presence of extracellular Ca^{2+} was important only if intracellular Ca^{2+} was depleted. It was suggested that for transducing the message of the hypertrehalosaemic hormone in *Blaberus* fat body, intracellular Ca^{2+} release is more important than extracellular Ca^{2+} entry. This relatively low effectiveness of extracellular Ca^{2+} influx in *B. discoidalis* contrasts with the situation in locust fat body where, as described above, extracellular Ca^{2+} is essential for the AKH-induced mobilisation of both carbohydrate and lipid.

Interactions between the second-messenger systems

At least a major part of the action of AKH may involve depletion of intracellular Ca^{2+} stores, because in locust fat body the three AKHs produce an increased uptake of radioactive Ca^{2+} from the medium that is similar to the Ca^{2+} uptake caused by thapsigargin. A likely mechanism would be that AKH increases the concentration of IP_3, which induces the emptying of intracellular Ca^{2+} stores and this, in turn, in some as yet unknown way, stimulates the influx of extracellular Ca^{2+}. This extracellular Ca^{2+} influx is apparently required for the AKH-induced production of cyclic AMP, possibly by activating a Ca^{2+}-dependent adenylyl cyclase; the cyclic nucleotide subsequently causes activation of glycogen phosphorylase. However, experiments in which cyclic AMP levels in the fat body were increased either by activation of the catalytic subunit of the adenylyl cyclase complex with forskolin or by inhibition of cyclic nucleotide phosphodiesterase with theophylline, both in the presence and in the absence of Ca^{2+} in the incubation medium, have demonstrated that the activating effect on phosphorylase by cyclic AMP itself is independent of extracellular Ca^{2+} (Van Marrewijk *et al.*, 1993). Compared with the phosphorylase activation induced by AKH, the effects caused by the Ca^{2+} ionophore A23187, as well as by enhanced intracellular levels of cyclic AMP produced by forskolin or theophylline are only partial. Therefore it still remains to be established whether the AKH-induced activation of phosphorylase occurs exclusively via the serial effects of Ca^{2+} and cyclic AMP, respectively, or whether there is additionally a parallel effect of Ca^{2+}, leading directly to phosphorylase activation independent of cyclic AMP.

Orr *et al.* (1985) showed that cyclic AMP is not always necessarily involved in Ca^{2+}-dependent hormonal activation of insect fat body phosphorylase. In fat body of *P. americana* incubated in a Ca^{2+}-containing medium, two potent hypertrehalosaemic factors from the corpus cardiacum stimulated activation of phosphorylase and increased production of trehalose without any associated elevation of cyclic AMP levels, indicating that the effects of these factors were not mediated by the cyclic nucleotide. In the absence of external Ca^{2+}, neither phosphorylase activation nor stimulation of trehalose production by the hypertrehalosaemic factors occurred.

Involvement of G-protein-coupled receptors in AKH signalling

The AKH-induced increases in the levels of cyclic AMP and IP_3 in the locust fat body described above point to a hormonal activation of both adenylyl cyclase and phospholipase C. The activation of these enzymes by many hormones that bind to cell surface receptors has been shown to occur via G-protein-coupled receptors, which may also mediate the action of AKHs. Within the group of G-proteins, G_s and G_i have been the most extensively studied; the G_s α-subunit stimulates adenylyl cyclase (and activates Ca^{2+} channels), whereas the G_i α-subunit inhibits adenylyl cyclase (and activates phospholipase C, phospholipase A_2 and K^+ channels) (for reviews, see Hepler & Gilman, 1992; Gilman, 1995). Several bacterial toxins are generally used to study the role of these G-proteins in signal transduction: cholera toxin (CTX) which irreversibly activates G_s by preventing GTPase activity of the α-subunit via ADP-ribosylation of an arginine residue, and pertussis toxin (PTX), which irreversibly inhibits G_i by ADP-ribosylation of the α-subunit at a cysteine residue, thus stabilising the GDP form of the G-protein.

To examine a possible involvement of G_s- and G_i-proteins in the AKH signal transduction mechanism, the effects of CTX and PTX on the production of cyclic AMP and the activation of glycogen phosphorylase were determined in fat body *in vitro* upon addition of the toxins to the incubation medium (Vroemen *et al.*, 1995b). PTX had no effect on either process, whereas CTX caused a significant elevation of intracellular cyclic AMP levels and also stimulated the activation of phosphorylase. These results suggest that for AKH, transduction of the signal in the fat body occurs through G_s-protein-coupled receptor(s). CTX, when applied together with AKH-I, II or III, increased the stimulatory effects of all three AKHs on both cyclic AMP levels and phosphorylase activation, but the individual effects of CTX and AKH were not additive. This suggests that CTX and AKH act via a common pathway involving the G_s-protein. Again, PTX had no effect on AKH activity. Additional evidence for the involvement of G-proteins in the signal transduction of AHK-I, II and III was obtained from experiments with GDPβS, which is a competitive inhibitor of G-protein activation by GTP. When fat body cells, permeabilised with streptolysin O and pre-incubated in the presence of GDPβS, were incubated with AKH-I, II or III, phosphorylase activation by all three hormones appeared to be strongly reduced. The involvement of G-proteins (such as G_q-α or G_s-$\beta\gamma$) in the

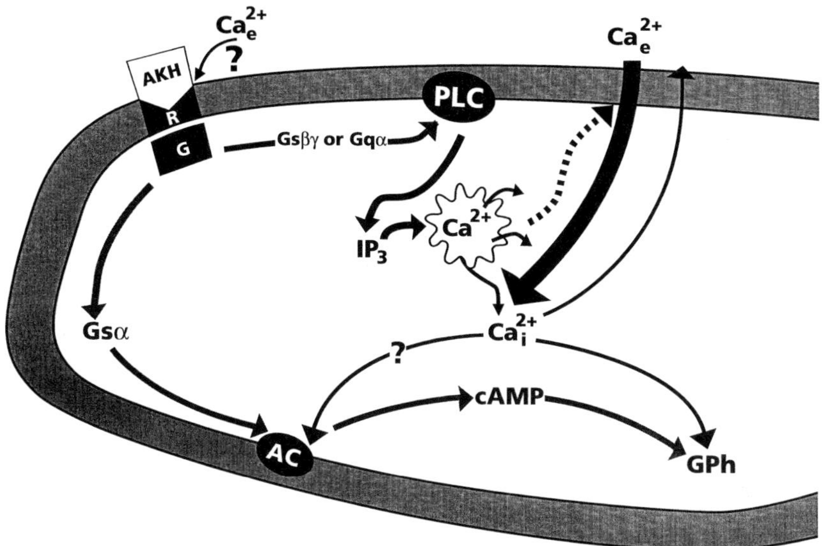

Fig. 5. Proposed model for the coupling of AKH signalling pathways in a locust fat body cell. R, receptor; G, G protein; PLC, phospholipase C; IP$_3$, inositol 1,4,5-trisphosphate; AC, adenylyl cyclase; Gph, glycogen phosphorylase.

AKH-induced activation of phospholipase C in locust fat body is currently under investigation. Figure 5 shows a proposed model for the AKH signal transduction pathways in locust fat body, based on the various data discussed in this chapter.

References

Beenakkers, A. M. Th., Van der Horst, D. J. & Van Marrewijk, W. J. A. (1985). Biochemical processes directed to flight muscle metabolism. In *Comprehensive Insect Physiology, Biochemistry and Pharmacology*, ed. G. A. Kerkut & L. J. Gilbert, Vol. 10, pp. 451–486. Oxford: Pergamon Press.

Berridge, M. J. (1989). Inositol trisphosphate and diacylglycerol: two interacting second messengers. *Annual Review of Biochemistry* **56**, 159–193.

Berridge, M. J. (1994). The biology and medicine of calcium signalling. *Molecular and Cellular Endocrinology* **98**, 119–124.

Berridge, M. J. (1995). Capacitative calcium entry. *Biochemical Journal* **312**, 1–11.

Berridge, M. J. & Irvine, R. F. (1989). Inositol phosphates and cell signalling. *Nature* **341**, 197–205.

Bird, G. St J. & Putney, J. W. Jr (1993). Inhibition of thapsigargin-induced calcium entry by microinjected guanine nucleotide analogues. Evidence for the involvement of a small G-protein in capacitative calcium entry. *Journal of Biological Chemistry* **268**, 21486–21488.

Bird, G. St J., Takemura, H., Thastrup, O., Putney, J. W. Jr & Menniti, F. S. (1992). Mechanisms of activated Ca^{2+} entry in the rat pancreatoma cell line, AR4-2J. *Cell Calcium* **13**, 49–58.

Bogerd, J., Kooiman, F. P., Pijnenburg, M. A. P., Hekking, L. H. P., Oudejans, R. C. H. M. & Van der Horst, D. J. (1995). Molecular cloning of three distinct cDNAs, each encoding a different adipokinetic hormone precursor, of the migratory locust, *Locusta migratoria. Journal of Biological Chemistry* **270**, 23038–23043.

Bootman, M. D., Taylor, C. W. & Berridge, M. J. (1992). The thiol reagent, thimerosal, evokes Ca^{2+} spikes in HeLa cells by sensitizing the inositol 1,4,5-trisphosphate receptor. *General and Comparative Endocrinology* **99**, 373–381.

Dombrádi, V., Risnik, V., Erdödi, F., Bot, G. & Friedrich, P. (1987). Regulation of phosphorylase kinase in *Drosophila melanogaster. Insect Biochemistry* **17**, 579–585.

Gilman, A. G. (1995). G proteins and regulation of adenylyl cyclase. *Bioscience Reports* **15**, 65–97.

Hayakawa, Y. & Chino, H. (1983). Insect fat body phosphorylase kinase is Ca^{2+}-independent and acts even at 0°C. *Biochimica et Biophysica Acta* **746**, 14–17.

Hepler, J. R. & Gilman, A. G. (1992). G proteins. *Trends in Biochemical Sciences* **17**, 383–387.

Irvine, R. F. (1991). Inositol tetrakisphosphate as a second messenger: confusions, contradictions, and a potential resolution. *BioEssays* **13**, 419–427.

Johnson, L. N. (1992). Glycogen phosphorylase: control by phosphorylation and allosteric effectors. *FASEB Journal* **6**, 2274–82.

Keeley, L. L. & Hesson, A. S. (1995). Calcium-dependent signal transduction by the hypertrehalosemic hormone in the cockroach fat body. *General and Comparative Endocrinology* **99**, 373–381.

Kodrík, D. & Goldsworthy, G. J. (1995). Inhibition of RNA synthesis by adipokinetic hormones and brain factor(s) in adult fat body of *Locusta migratoria. Journal of Insect Physiology* **41**, 127–133.

Kuno, M. & Gardner, P. (1987). Ion channels activated by inositol 1,4,5-trisphosphate in plasma membrane of human T-lymphocytes. *Nature* **326**, 301–304.

Lee, M. J. & Goldsworthy, G. J. (1995). The preparation and use of dispersed cells from fat body of *Locusta migratoria* in a filtration plate assay for adipokinetic peptides. *Analytical Biochemistry* **228**, 155–161.

Lee, M. J., Hyde, D. & Goldsworthy, G. J. (1995). Inhibition of acetate uptake and changes in intracellular calcium in dispersed cells from fat body of *Locusta migratoria* in response to adipokinetic peptides. *Physiological Zoology* **68**, 129.

Lum, P. Y. & Chino, H. (1990). Primary role of adipokinetic hormone in the formation of low density lipophorin in insects. *Journal of Lipid Research* **31**, 2039–2044.

Orr, G. L., Gole, J. W. D., Jahagirdar, A. P., Downer, R. G. H. & Steele, J. E. (1985). Cyclic AMP does not mediate the action of synthetic hyper-trehalosemic peptides from the corpus cardiacum of *Periplaneta americana. Insect Biochemistry* **15**, 703–709.

Oudejans, R. C. H. M., Dijkhuizen, R. M., Kooiman, F. P. & Beenakkers, A. M. Th. (1992). Dose–response relationships of adipokinetic hormones (Lom-AKH-I, II and III) from the migratory locust, *Locusta migratoria. Proceedings of the Section Experimental and Applied Entomology of the Netherlands Entomological Society* **3**, 165–166.

Oudejans, R. C. H. M., Mes, T. H. M., Kooiman, F. P. & Van der Horst, D. J. (1993). Adipokinetic peptide hormone content and biosynthesis during locust development. *Peptides* **14**, 877–881.

Oudejans, R. C. H. M., Vroemen, S. F., Jansen, R. F. R. & Van der Horst, D. J. (1996). Locust adipokinetic hormones: carrier-independent transport and differential inactivation at physiological concentrations during rest and flight. *Proceedings of the National Academy of Sciences USA* **93**, 8654–8659.

Pallen, C. J. & Steele, J. E. (1988). A putative role for calmodulin in corpus cardiacum stimulated trehalose synthesis in fat body of the American cockroach (*Periplaneta americana*). *Insect Biochemistry* **18**, 577–584.

Pancholi, S., Barker, C. J., Candy, D. J., Gokuldas, M. & Kirk, C. J. (1991). Effects of adipokinetic hormones on inositol phosphate metabolism in locust fat body. *Biochemical Society Transactions* **19**, 104S.

Park, J. E. & Keeley, L. L. (1995). *In vitro* hormonal regulation of glycogen phosphorylase activity in fat body of the tropical cockroach, *Blaberus discoidalis. General and Comparative Endocrinology* **98**, 234–243.

Park, J. H. & Keeley, L. L. (1996). Calcium-dependent action of hyper-trehalosemic hormone on activation of glycogen phosphorylase in cockroach fat body. *Molecular and Cellular Endocrinology* **116**, 199–205.

Stagg, L. E. & Candy, D. J. (1995). Specificity of adipokinetic hormone action on locust fat body. *Physiological Zoology* **68**, 131.

Steele, J. E. & Paul, T. (1985). Corpus cardiacum stimulated trehalose efflux from cockroach (*Periplaneta americana*) fat body: control by calcium. *Canadian Journal of Zoology* **63**, 63–66.

Takemura, H., Hughes, A. R., Thastrup, O. & Putney, J. W. Jr (1989). Activation of calcium entry by the tumor promoter thapsigargin in parotid acinar cells. Evidence that an intracellular calcium pool, and not an inositol phosphate, regulates calcium fluxes at the plasma membrane. *Journal of Biological Chemistry* **264**, 12266–12271.

Van Marrewijk, W. J. A., Van den Broek, A. Th. M. & Beenakkers, A. M. Th. (1991). Adipokinetic hormone is dependent on extracellular Ca^{2+} for its stimulatory action on the glycogenolytic pathway in locust fat body *in vitro. Insect Biochemistry* **21**, 375–380.

Van Marrewijk, W. J. A., Van den Broek, A. Th. M. & Beenakkers, A. M. Th. (1992). AKH and the mechanism of signal transduction in locust fat body. *Proceedings of the Section Experimental & Applied Entomology of the Netherlands Entomological Society* 3, 171–172.

Van Marrewijk, W. J. A., Van den Broek, A. Th. M. & Van der Horst, D. J. (1993). Adipokinetic hormone-induced influx of extracellular calcium into insect fat body cells is mediated through depletion of intracellular calcium stores. *Cellular Signaling* 5, 753–761.

Vroemen, S. F., Van Marrewijk, W. J. A., Schepers, C. C. J. & Van der Horst, D. J. (1995a). Signal transduction of adipokinetic hormones involves Ca^{2+} fluxes and depends on extracellular Ca^{2+} to potentiate cAMP-induced activation of glycogen phosphorylase. *Cell Calcium* 17, 459–467.

Vroemen, S. F., Van Marrewijk, W. J. A. & Van der Horst, D. J. (1995b). Stimulation of glycogenolysis by three locust adipokinetic hormones involves G_s and cAMP. *Molecular and Cellular Endocrinology* 107, 165–171.

Wang, Z., Hayakawa, Y. & Downer, R. G. H. (1990). Factors influencing cyclic AMP and diacylglycerol levels in fat body of *Locusta migratoria*. *Insect Biochemistry* 20, 325–330.

Ziegler, R., Jasensky, R. D. & Morimoto, H. (1995). Characterization of the adipokinetic hormone receptor from the fat body of *Manduca sexta*. *Regulatory Peptides* 57, 329–338.

GEOFFREY M. COAST

The regulation of primary urine production in insects

Introduction

The large surface area to volume ratio of terrestrial insects imposes considerable risk of desiccation, and various devices have evolved to minimise water loss. However, insects must sometimes rid the body of excess water, usually by increasing fluid loss from the excretory system. In the excretory process, primary urine secreted by the Malpighian tubules is modified by reabsorptive and secretory mechanisms in the hindgut (see Phillips *et al.*, this volume). Diuretic hormones accelerate Malpighian tubule secretion and if this exceeds fluid uptake in the hindgut faecal water loss is increased. In the mosquito *Aedes aegypti*, a volume of urine equivalent to 40% of the plasma component of a blood meal is voided within 20 min of feeding (Williams, Hagedorn & Beyenbach, 1983), and tubule secretion is increased >100-fold during this post-feeding diuresis. In adult butterflies, a diuretic hormone is released at the time of emergence (eclosion), and urine production is increased dramatically (post-eclosion diuresis), contributing to a reduction in body weight prior to the first flight (Nicolson, 1976). Foraging male bumble bees have been seen to void drops of urine when flying (Bertsch, 1984), and a diuretic hormone may here be used to stimulate excretion of excess metabolic water. Surprisingly, a potent diuretic hormone is found also in tenebrionid beetles living in the Namib desert (Nicolson & Hanrahan, 1986). The hormone does not increase faecal water loss, but accelerates fluid recycling between haemolymph, Malpighian tubules and hindgut, thereby increasing the efficiency with which waste products are cleared from the circulation (Knowles, 1976).

Both neuropeptides and biogenic amines stimulate Malpighian tubule secretion, and have been implicated in the control of post-eclosion, post-feeding and other forms of diuresis. This chapter describes the nature and actions of these factors, and considers their status as diuretic hormones. Antidiuretic hormones, which increase fluid uptake from the hindgut, are described by Phillips *et al.*, (this volume). For references to early work on diuretic hormones, the reader should consult reviews by Gee (1977) and Phillips (1983).

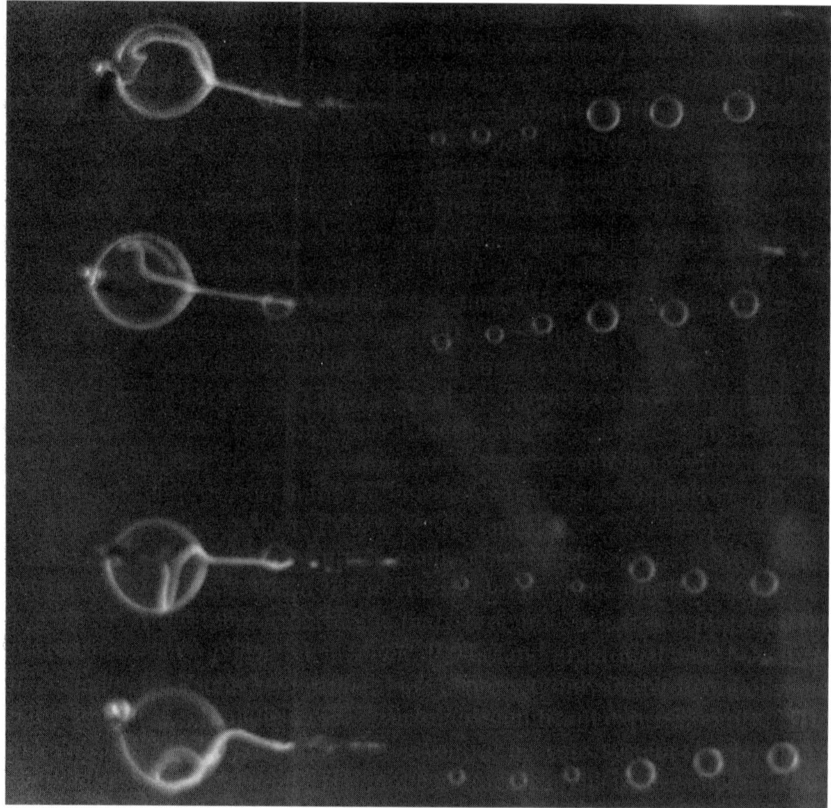

Fig. 1. The 'Ramsay assay'. Dissected tubules are held in small (5 μl) drops of saline (left of figure) under liquid paraffin. Urine escapes from a cut made close to the proximal end of the tubule and collects as a discrete droplet in the paraffin. Urine samples are collected at intervals (right of figure) and their volume determined from measurements of droplet diameter. Urine samples were collected at 20 min intervals from the cricket tubules used in this assay, and at 60 min the tubules were stimulated with 20 nmol achetakinin I l^{-1}. Diuretic activity is calculated as the difference between rates of secretion measured before and after peptide addition, with each tubule serving as its own control.

Methods used in the study of diuretic hormone activity

Diuretic activity is most conveniently measured using isolated Malpighian tubules in what has become known as the 'Ramsay assay' (Ramsay, 1954; Fig. 1). Alternative *in vitro* assays include the measurement of transepithelial potentials or cyclic AMP production in isolated tubules (Wheeler & Coast, 1990). In intact insects, weight loss has been used to follow post-eclosion diuresis in lepidopterans (Blackburn & Ma, 1994), whereas Wheelock *et al.* (1988) counted the urine drops voided per minute to quantify post-feeding diuresis in mosquitoes. However, diuretic hormones have only a modest effect on water loss in the majority of terrestrial insects, and an indirect measure of primary urine production is required. Inulin, which is used to measure glomerular filtration rate in vertebrates, is cleared too slowly to be of use in insects. On the other hand, dyes such as amaranth and indigo carmine are cleared rapidly from the haemolymph, and dye clearance has been used as a measure of diuretic activity (Mordue, 1969), although it is linked only indirectly to primary urine production (Maddrell *et al.*, 1974).

Stimulants of tubule secretion

Diuretic factors are widespread in tissues of the central nervous system (CNS), and insects may have more than one diuretic hormone (Coast, Kay & Wheeler, 1993). Several biogenic amines stimulate Malpighian tubule secretion, but the labile nature of factors from the CNS, and their susceptibility to protease digestion, suggest most diuretic hormones are neuropeptides. However, 5-hydroxytryptamine (5-HT, serotonin) functions as a diuretic hormone in the blood-sucking bug *Rhodnius prolixus* (Lange, Orchard & Barrett, 1989; Maddrell *et al.*, 1991; see also below).

Corticotropin-releasing factor-related diuretic peptides

Given the great diversity of insects, few species have been examined for the presence of diuretic peptides. Kataoka *et al.* (1989) identified a diuretic peptide from the tobacco hornworm *Manduca sexta* (*Manduca* diuretic hormone; *Manduca*-DH), which shares 29–35% sequence identity with mammalian corticotropin-releasing factor (CRF), and belongs to the same peptide superfamily (Coast *et al.*, 1993). *Manduca*-DH stimulates post-eclosion diuresis in butterflies (*Pieris rapae*), increases fluid loss via the gut in *M. sexta* larvae (Kataoka *et al.*, 1989), and accelerates tubule secretion in both insects (Coast *et al.*, 1992; Audsley, Coast & Schooley, 1993). CRF-related peptides have since been identified from a number of species (Fig. 2), including a second peptide from *M. sexta* (*Manduca*-DPII; Blackburn *et al.*, 1991), and may be present in all insects.

The primary structures of insect CRF-related peptides are well conserved in species that diverged >300 million years ago. This is particularly evident in the N-terminal half of the molecule, and results obtained with truncated analogues of *Manduca*-DH and *Acheta*-DP suggest that a conserved region encompassing residues 6–12 of *Manduca*-DH is critical for receptor activation (Coast *et al.*, 1994; Reagan, 1995a). This is followed by a

```
                              1
¹Periplaneta      T G S G P S L S I V    N P L D V L R Q R L    L L E I A
²Locusta          M G M G P S L S I V    N P M D V L R Q R L    L L E I A
³Acheta           T G A - Q S L S I V    A P L D V L R Q R L    M N E L N
⁴Stomoxys/Musca   - - N K P S L S I V    N P L D V L R Q R L    L L E I A
⁵Culex            - - - - P S L S I V    N P L D V L R Q R I    I L E M A
⁶Manduca-DH       - - R M P S L S I D    L P M S V L R Q K L    S L E - K
⁷Manduca-DPII     - - - - - S F S - V    N P A V D I L Q H -    - R Y M E
⁸Tenebrio             S P T I S I T    A P I D V L R K T W    E Q E R A

⁹Ovine CRF        - S Q E P P I S L D    L T F H L L R E - -    V L E M T

                              26
Periplaneta       R R R M R Q S Q - D    Q I Q A N R E I L Q    T  I-NH₂
Locusta           R R R L R D A E - E    Q I K A N K D F L Q    Q  I-NH₂
Acheta            R R R M R E L Q G S    R I Q Q N R Q L L T    S  I-NH₂
Stomoxys/Musca    R R Q M K E N T - R    Q V E L N R A I L K    N  V-NH₂
Culex             R R Q M R E N T - R    Q V E R N K A I L R    E  I-NH₂
Manduca-DH        E R K V H A L - - -    R A A A N R N F L N    D  I-NH₂
Manduca-DPII      K V - - - - - - - -    - A Q N R N F L N      R  V-NH₂
Tenebrio          R K Q M - - - - - -    V A Q N N R E F L N    S  L  N-OH

Ovine CRF         K A D Q L A - - - Q    Q A H S N R K L L D    I  A-NH₂
```

Results with analogues of *Acheta*-DP and *Manduca*-DH suggest the underlined region near the N-terminus is critical for activity. Note also conserved residues close to the C-terminus, [N39], [R/K40] and [L43] (the numbering refers to *Periplaneta*-DP).

References: 1, Kay *et al.* (1992); 2, Kay *et al.* (1991b); 3, Kay *et al.* (1991a); 4, Clottens *et al.* (1994); 5, T. K. Hayes & G. M. Holman (*per. comm.*); 6, Kataoka *et al.* (1989); 7, Blackburn *et al.* (1991); 8, Furuya *et al.* (1995); 9, Vale *et al.* (1981).

Neobellieria-cyclic **AMP Generating Peptide** (Neb-cGP; Spittaels *et al.*, 1996)

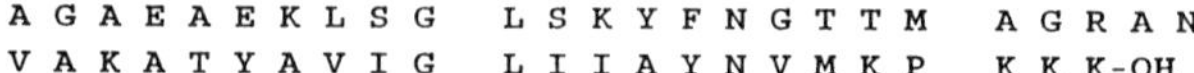

```
A G A E A E K L S G    L S K Y F N G T T M    A G R A N
V A K A T Y A V I G    L I I A Y N V M K P    K K K-OH
```

Fig. 2. The primary structures of insect CRF-related diuretic peptides. Single-letter coding is used for amino acids, and gaps are inserted to optimise the alignment. The figure includes the structures of ovine CRF and *Neobellieria*-cGP (Neb-cGP) for comparison.

receptor-binding domain (Reagan *et al.*, 1993). The C-terminal amide group is also important for receptor binding, as evidenced by the very low potency of *Manduca*-DH-COOH (Kataoka *et al.*, 1989; Audsley *et al.*, 1995). *Tenebrio*-DH is unique amongst peptides of the CRF superfamily in being non-amidated. Nevertheless, its potency on mealworm tubules ($EC_{50} \approx 3$ nmol l^{-1}; Furuya *et al.*, 1995) is comparable to that of other CRF-related peptides in conspecific assays (Audsley *et al.*, 1995). Interestingly, *Tenebrio*-DH has very low potency on *M. sexta* tubules (which respond well to other CRF-related peptides; Audsley *et al.*, 1995), whereas *Manduca*-DH is only 17 times less potent than the native peptide on mealworm tubules (Furuya *et al.*, 1995). A comparison of the structures of these two receptors might identify the ligand binding site(s).

Receptors for *Manduca*-DH and *Acheta*-DP have been cloned (Reagan, 1994, 1996). They belong to the 'gut-brain' neuropeptide family of G-protein-coupled receptors, all of which have seven putative transmembrane (TM) domains and are coupled positively to adenylyl cyclase (Chalmers *et al.*, 1996). The insect receptors share 53% sequence identity, and have about 35% sequence identity with human CRF-R_1, the most conserved regions being within or close to the seven TM domains, along with four cysteine residues in the extracellular N-terminus. Peptides acting at these receptors have a high α-helical component in a lipophilic environment, and Reagan (1995a) suggested this region is located in the membrane surrounded by the TM domains, which presumably brings the N-terminus into a position for receptor activation. Surprisingly, *Manduca*-DPII, which is just 30 residues long and shares only 37% sequence identity with *Manduca*-DH, is able to bind with high affinity and activate *Manduca*-DH receptors (Reagan, 1994; Audsley *et al.*, 1995).

The insect kinin family of diuretic peptides

Kinins are small peptides (6–14 residues) that were first described as myotropins (Holman, Nachman & Wright, 1990), but were later shown to have diuretic activity in cross-species (Hayes *et al.*, 1989) and conspecific assays (Coast *et al.*, 1990). They have been found in dictyopteran, orthopteran, dipteran and lepidopteran insects, and are characterised by the C-terminal sequence Phe-Xaa^1-Xaa^2-Trp-Gly-NH$_2$ (Xaa^1 is Asn, Ser, His or Tyr; Xaa^2 is Ser, Pro or Ala), which is all that is needed for activity. This 'active core' probably adopts a β-turn when interacting with receptors, bringing together Phe^1 and Trp^4 which are critical for activity (Nachman *et al.*, 1995; Coast, 1996; see Nachman, Holman & Coast, this volume).

Specific kinin-binding sites are present on cricket tubules, and protein bands of 53 000–57 000 M_r have been labelled using a photoaffinity probe (Chung *et al.*, 1995; J. S. Chung, unpublished observations). The size is appropriate for a G-protein-coupled receptor, and the affinity of the probe is reduced 300-fold by GppNHp, a non-hydrolysable analogue of guanosine triphosphate, which is consistent with effects at other G-protein-coupled receptors (Graeser & Neubig, 1992). A G-protein-coupled kinin receptor that shares 31% sequence identity with the neuropeptide Y_1 receptor has recently been cloned from the pond snail *Lymnaea stagnalis* (Cox *et al.*, 1996). The C-terminal pentapeptide of the native ligand (*Lymnaea* kinin) is Phe-His-Ser-Gly-NH$_2$, which is identical with that of kinins from the cockroach

Leucophaea maderae (leucokinins IV and VI), the cricket *Acheta domesticus* (achetakinin V) and the mosquito *Culex salinarius* (culekinin I).

Other peptides having diuretic activity

In addition to CRF-related peptides and kinins, several other candidate peptide diuretic hormones have been identified. The first, an arginine vasopressin-like peptide from *Locusta migratoria*, was reported to stimulate fluid secretion in a locust Malpighian tubule/midgut preparation (Proux *et al.*, 1987). However, the synthetic peptide has no effect on isolated tubules (Coast *et al.*, 1993), and the activity detected by Proux *et al.* may have been caused by release of *Locusta*-DH from endocrine cells in the posterior midgut (Montuenga *et al.*, 1996; see later). A cardioacceleratory peptide (CAP_{2b}) identified from *M. sexta* (Huesmann *et al.*, 1995) has been shown to stimulate fluid secretion by tubules of the fruitfly *Drosophila, melanogaster* (Davies *et al.*, 1995), and the octapeptide appears to be present in extracts of decapitated flies, but this has not been confirmed. CAP_{2b} has not been tested for activity on *M. sexta* tubules. Spittaels *et al.* (1996) identified a novel peptide (*Neobellieria* cyclic AMP generating peptide; Neb-cGP; Fig. 2) from the fleshfly, *Neobellieria bullata*, using as a bioassay the stimulation of cyclic AMP production by *M. sexta* tubules. This bioassay was employed in the isolation of CRF-related peptides from houseflies and stable flies (Clottens *et al.*, 1994), and it seems likely that *Neobellieria*-cGP will have diuretic activity, because there is a well-established correlation between the rate of tubule secretion and intracellular levels of cyclic AMP (see below). A single fly contains 0.6–6 pmol of *Neobellieria*-cGP, but its tissue distribution has yet to be determined as a first step towards establishing a function.

The actions of diuretic factors on Malpighian tubules

Fluid secretion is coupled to the transport of KCl and/or NaCl (for reviews, see Nicolson, 1993; Beyenbach, 1995), and diuretic activity appears to result from stimulation of these processes rather than any increase in osmotic permeability. Cation movement across the luminal membrane occurs via H^+/K^+ and/or H^+/Na^+ antiports, the driving force being a proton gradient established by a V-type ATPase. Cations enter the cells through K^+ and Na^+ channels, and via a bumetanide-sensitive $Na^+/K^+/2Cl^-$ cotransporter. The latter brings Cl^- into the cells, but a basal Cl^- conductance may also be present. Chloride movement across the luminal membrane is favoured by the lumen positive potential, and apical Cl^- channels are present in mosquito tubules (Wright & Beyenbach, 1987). Chloride may also enter the lumen via a paracellular shunt (Pannabecker, Hayes & Beyenbach, 1993).

CRF-related peptides

CRF-related peptides and kinins act via different second messengers to regulate cation and anion movements, respectively. There is overwhelming evidence of a second-messenger role for cyclic AMP in the action of CRF-related peptides. In crickets, intracellular levels of cyclic AMP increase within seconds of adding *Acheta*-DP to Malpighian tubules, and this precedes any change in fluid secretion or transepithelial voltage (Coast & Kay, 1994). The near identical dose–response curves for the stimulation of tubule secretion and cyclic AMP production by *Acheta*-DP is further evidence of the close link between them, and pharmacological agents that increase intracellular levels of the cyclic nucleotide have diuretic activity.

Cyclic AMP opens Na^+ channels on the basal membrane of *A. aegypti* tubules, producing a diuretic response accompanied by a pronounced natriuresis (Beyenbach, 1995). This is important for a blood-sucking insect, which must rapidly excrete excess fluid and salt derived from the imbibed plasma. However, cyclic AMP has a similar effect in tubules from male mosquitoes (Plawner *et al.*, 1991) and, in herbivorous insects, CRF-related peptides (or cyclic AMP) preferentially stimulate Na^+ transport, although the diet is generally low in Na^+ (Coast, 1995). These findings suggest a common mode of action unrelated to the excretion of dietary Na^+. There is evidence for CRF-related diuretic peptides stimulating $Na^+/K^+/2Cl^-$ co-transport (Audsley *et al.*, 1993, Table 1), and a putative $Na^+/K^+/2Cl^-$ co-transporter has recently been cloned using mRNA from *M. sexta* tubules (Reagan, 1995b). The encoded protein shares 42–46% sequence identity with vertebrate cotransporters and has a number of probable phosphorylation sites, including one for protein kinase A.

Insect kinins

Kinins have no effect on cyclic AMP production by Malpighian tubules, and appear to act via a Ca^{2+}-dependent mechanism. Thus pharmacological agents that mobilise cell Ca^{2+} mimic kinin activity, whereas the intracellular Ca^{2+} chelator BAPTA/AM reduces the diuretic response (Coast *et al.*, 1990, 1993; O'Donnell *et al.*, 1996; Fig. 3). Moreover, *Lymnaea* kinin increases intracellular Ca^{2+} levels in Chinese hamster ovary cells expressing a kinin receptor (Cox *et al.*, 1996). Calmodulin may be implicated in the response, because trifluoperazine blocks the activity of achetakinin I, whereas the protein kinase C inhibitor chelerythrin is without effect (G. M. Coast, unpublished observations).

In mosquito tubules, kinins open a shunt pathway for Cl^- (Pannabecker *et al.*, 1993) which, in contrast to the effect of cyclic AMP, produces a non-selective increase in NaCl and KCl transport. This shunt pathway may be

Table 1 *The activity of Acheta-DP in salines from which Na^+, K^+ or Cl^- were omitted, or to which 1 mmol bumetanide l^{-1} was added*

Saline	Δ pl mm^{-1} min^{-1}
Control	355±39.2 (12)
Na$^+$-free	25±7.1 (12)
K$^+$-free	71±16.3 (10)
Cl$^-$-free	10±12.4 (10)
+bumetanide	178±39.2 (12)

Note:
Diuretic activity is expressed as the change in rate of secretion (Δpl mm^{-1} min^{-1}) following addition of 10 nmol *Acheta*-DP l^{-1}, and results are given as the mean±1 SEM for the number of determinations shown in parentheses. Diuretic activity is greatly reduced if any one of the three ions is omitted, and is partially blocked by bumetanide. The data are consistent with *Acheta*-DP stimulating a bumetanide-sensitive $Na^+/K^+/2CL^-$ co-transporter.

paracellular or transcellular, through stellate cells located at the junctions of the principal cells (Beyenbach, 1995; O'Donnell *et al.*, 1996). The response of locust and cricket tubules to kinin stimulation is also a non-selective increase in Na^+ and K^+ transport (Coast, 1995, and unpublished observations), which suggests a similar mode of action. However, stellate cells are not present in cricket tubules (Hazelton, Parker & Spring, 1988), and any shunt pathway for Cl^- would have to be through or between the principal cells.

Serotonin

Serotonin stimulates Malpighian tubule secretion in many insects, acting via cyclic AMP in *R. prolixus* (Montoreano *et al.*, 1990), and by a cyclic AMP-independent mechanism in locusts and crickets (Morgan & Mordue, 1984; G. M. Coast, unpublished observation). In *R. prolixus*, serotonin reportedly

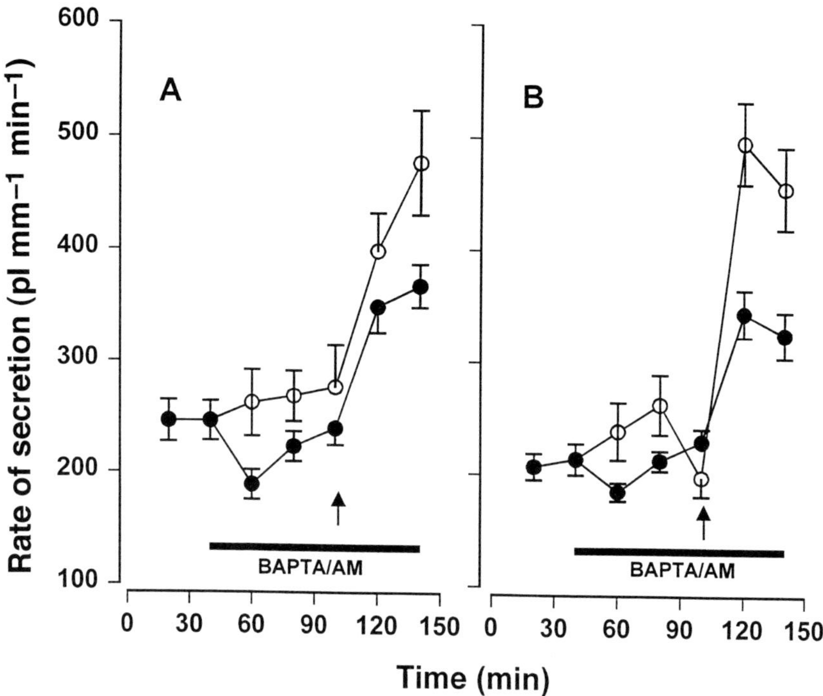

Fig. 3. The effect of the intracellular Ca^{2+} chelator BAPTA/AM on the response of cricket tubules to (A) 10 μmol thapsigargin l^{-1} or (B) 20 nmol achetakinin I l^{-1}. Data points are the means and vertical lines ± 1 SEM for 8–15 determinations. The arrows show when thapsigargin or achetakinin I were added. Control tubules (open symbols) were not treated with BAPTA/AM. Thapsigargin, which mobilises Ca^{2+} from intracellular stores, and achetakinin I have similar effects on tubule secretion, and both give a truncated response in tubules treated with 100 μmol BAPTA/AM l^{-1} (solid symbols).

stimulates basal Cl^- entry via a $Na^+/K^+/2Cl^-$ co-transporter, and increases cation transport across the luminal membrane (O'Donnell & Maddrell, 1984). There is also evidence that serotonin causes a progressive increase in luminal membrane Cl^- conductance.

Cardioacceleratory peptide (CAP_{2b})

CAP_{2b} stimulates cyclic GMP production in *Drosophila* tubules, and exogenous cyclic GMP increases tubule secretion (Davies *et al.*, 1995). This is of interest, because nitric oxide (NO) also stimulates fluid secretion (Dow *et al.*,

1994) and, in vertebrates, NO is known to activate a soluble guanylyl cyclase. As the tubules contain nitric oxide synthase, it is possible that CAP_{2b} activity is mediated by NO. O'Donnell *et al.* (1996) suggest CAP_{2b} stimulates the apical membrane V-ATPase, which provides the driving force for cation transport and hence fluid secretion.

Evidence of a hormonal functional for diuretic factors

For a diuretic peptide or biogenic amine to be ascribed a hormonal function it must: (a) be present in neurosecretory cells that contain diuretic hormone; (b) be released into the circulation in amounts sufficient to elicit a physiological response; (c) be able to mimic diuretic hormone activity; (d) be sensitive to diuretic hormone antagonists. To date, only *Locusta*-DH and serotonin satisfy all of these criteria.

The case for *Locusta*-DH

Mordue (1969) showed that a diuretic hormone is synthesised by neurosecretory cells in the pars intercerebralis of locust brain, and is transported to storage and release sites in the nervous (storage) lobe of the corpus cardiacum (Mordue, 1969). *Locusta*-DH is present in both regions (Patel *et al.*, 1994), and is released in a Ca^{2+}-dependent manner from corpora cardiaca depolarised in high K^+ saline (N. Audsley, unpublished observation). Haemolymph from starved insects contains 0.22 nmol *Locusta*-DH 1^{-1} (Audsley, Goldworthy & Coast, 1996), which is below the threshold for significant diuretic activity (Fig. 4). However, feeding triggers release of the peptide from corpora cardiaca, and circulating levels reach 2–3 nmol 1^{-1} with meals of 16–18 min duration, which is sufficient to promote 40–50% maximal tubule secretion (Coast, 1995; Audsley *et al.*, 1996; Fig. 4).

Amaranth is cleared slowly in starved locusts, but the rate increases after feeding in response to the release of diuretic hormone (Mordue, 1969). The rate of amaranth clearance in starved locusts is stimulated by an injection of *Locusta*-DH, mimicking that observed during a post-feeding diuresis, and the potency of the peptide in this assay is comparable to that in the *in vitro* fluid secretion assay (Patel, Hayes & Coast, 1995). Significantly, an increase in the rate of amaranth clearance, whether in response to injected *Locusta*-DH or to feeding, can be prevented by immunising locusts with anti-*Locusta*-DH antibodies (Patel *et al.*, 1995), thus providing unequivocal evidence for the role of this peptide in the control of post-feeding diuresis.

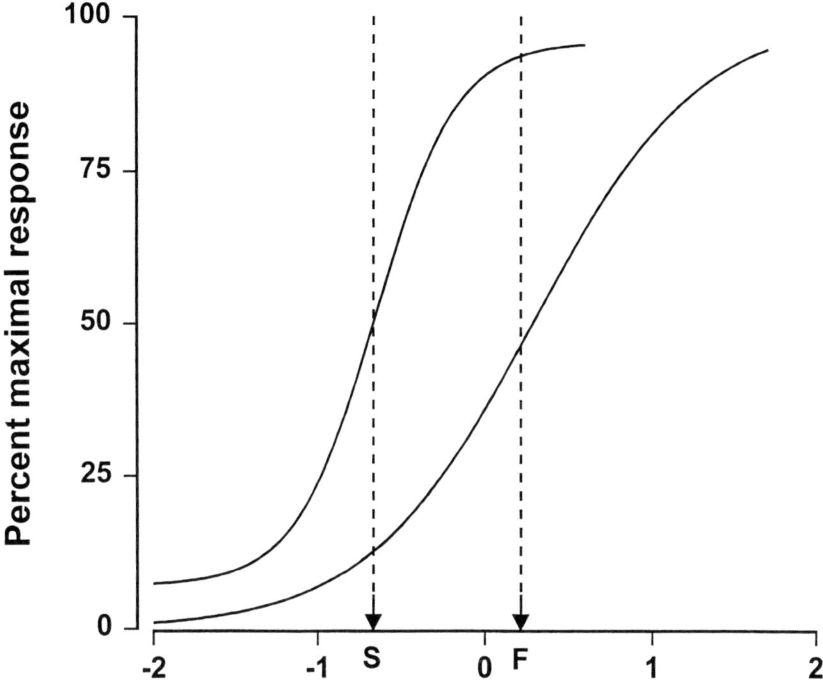

Fig. 4. Synergism between *Locusta*-DH and locustakinin. Data points have been omitted for clarity, and diuretic activity is expressed as a percentage of the response to a supramaximal dose (50 nmol l^{-1}) of *Locusta*-DH. The curves were fitted to the generalised logistic equation using the computer program Fig P (Biosoft, Cambridge). In the presence of 50 pmol locustakinin l^{-1}, the dose–response curve for *Locusta*-DH is shifted to the left and made more steep (redrawn from Coast, 1995). Arrows on the *x*-axis show circulating levels of *Locusta*-DH in starved (S) and fed (F) locusts (from Audsley *et al.*, 1996).

The case for serotonin

Serotonergic neurohaemal areas are present on abdominal nerves arising from the fused mesothoracic ganglion of *R. prolixus* (Lange *et al.*, 1989; Orchard, 1989), which Maddrell (1966) identified as the site of release of a diuretic hormone. Serotonin is released from these nerves in a Ca^{2+}-dependent manner by high K$^+$ saline, and circulating levels of serotonin increase dramatically within 1–2 min of the onset of feeding, reaching >100 nM after 5 min (Lange *et al.*, 1989), which is sufficient to partially stimulate tubule secretion. Furthermore, post-feeding diuresis is delayed by injection of 5,7-dihydroxytryptamine, which depletes serotonin stores and may act as a receptor antagonist (Cook & Orchard, 1993; Maddrell *et al.*, 1993).

The case for insect kinins

Kinin-like immunoreactive material is widespread in the CNS and associated neurohaemal areas, and kinins are released in a Ca^{2+}-dependent manner from corpora cardiaca held in high K^+ saline (Chung, Goldsworthy & Coast, 1994). However, although kinins have been detected in haemolymph (Muren, Lundquist & Nässel, 1993; Chung *et al.*, 1994), circulating levels in *A. domesticus* increase with starvation/dehydration, and fall slightly after feeding (Chung *et al.*, 1994). Such changes are not consistent with kinins having a role in the regulation of post-feeding diuresis, but suggest they are implicated in the mobilisation of food reserves during starvation, with tubule secretion being increased to facilitate the clearance of toxic wastes.

Synergism between diuretic hormones

The post-feeding increase in circulating levels of *Locusta*-DH and serotonin is not sufficient to stimulate maximal tubule secretion, and there is evidence that both *L. migratoria* and *R. prolixus* use two (or more) hormones acting synergistically to control diuresis. In locusts, a low (threshold) concentration (50 pmol 1^{-1}) of locustakinin markedly potentiates the response of isolated tubules to *Locusta*-DH, so that fluid secretion is stimulated maximally at hormone concentrations found in haemolymph from fed insects (Coast, 1995; Fig. 4). Abdominal ganglia co-localise both peptides in posterior lateral neurosecretory cells, and they may be released together from neurohaemal areas on the perivisceral organs (Thompson *et al.*, 1995). In this scenario, diuresis is initiated by an increase in the haemolymph titre of *Locusta*-DH, but the extent to which tubule secretion is stimulated may depend upon the circulating level of locustakinin. The use of a CRF-related peptide and a kinin to control tubule secretion may not be unique to locusts, because both classes of peptide are present in a number of other insects (Coast, 1995).

In *R. prolixus*, serotonin acts synergistically with a peptide hormone(s) from the mesothoracic ganglion mass (Maddrell *et al.*, 1993). The second diuretic hormone has not been identified, but is most likely a CRF-related peptide because it acts via cyclic AMP. Indeed, *Locusta*-DH-like immunoreactive material is present in posterior lateral groups of neurosecretory cells within the mesothoracic ganglion mass (I. Orchard, personal communication) that were shown to be the source of the diuretic hormone (Maddrell, 1963), and *Locusta*-DH is active on *R. prolixus* tubules (Coast, 1996).

An important advantage to be gained from having two hormones act synergistically to control tubule secretion is that a small change in the concentration of either will switch diuresis on or off. For example, when *Locusta*-DH is acting alone, its concentration must be raised >100-fold for tubule secretion to be maximally stimulated, whereas in the presence of 50 pmol 1^{-1} locustakinin a five-fold increase produces the same effect (Fig. 4). This means that primary urine production can be accelerated maximally by the release of just 100 fmol *Locusta*-DH, which is about 2% of that stored in corpora cardiaca from 14-day-old adult locusts (N. Audsley, unpublished observation).

The termination of diuresis

Rapid inactivation or removal of diuretic hormones from the circulation will terminate diuresis once the stimulus for hormone release is removed. In *R. prolixus*, 5-HT levels fall rapidly if feeding is interrupted (Lange *et al.*, 1989), and in locusts amaranth clearance returns to basal levels <60 min after feeding or injecting *Locusta*-DH (Mordue, 1969; Patel *et al.*, 1995). Serotonin is probably taken back into serotinergic terminals on the abdominal nerves, where a high-affinity uptake mechanism has been demonstrated (Orchard, 1989). On the other hand, peptide hormones are most likely inactivated by enzymatic degradation. *Manduca*-DH is cleaved rapidly on the N- and C-terminal sides of Arg^{30} when incubated with Malpighian tubules from the adult moth, and there is also evidence of methionine residues being oxidised to produce the bis-sulphoxide (D. A. Schooley, personal communication). Cleavage at Arg^{30} will produce two inactivate fragments, and methionine oxidation has been associated with a loss of activity in *Locusta*-DH (I. Kay, personal communication). In this respect, oxidation of Met^{13} Met^{11} in *Manduca*-DH may be particularly important, because this is in a region that is critical for activity (see above).

With effective means for the removal or degradation of diuretic hormones there would seem to be no need for a hormone that inhibits tubule secretion. However, Spring, Morgan & Hazelton (1988) found that a factor released from corpora cardiaca of dehydrated crickets reduced fluid secretion in isolated Malpighian tubules. This factor has no effect on fluid uptake from the rectum, and is therefore distinct from antidiuretic peptides acting at that site. Unfortunately, it has not been characterised further.

In *M. sexta* larvae, distal segments of the Malpighian tubules are closely associated with the rectum in what is referred to as a cryptonephric complex. *Manduca*-DH stimulates ion transport by these cryptonephric tubules, thereby creating a favourable gradient for osmotic withdrawal of water from the rectal lumen, an antidiuretic effect (Audsley *et al.*, 1993). However, it is the juxtaposition of the distal tubule segments and the rectum that transforms what would be a diuretic into an antidiuretic response, and there is no evidence for diuretic peptides having a direct effect on fluid or electrolyte transport by the hindgut (J. Meredith, personal communication).

A local regulation of Malpighian tubule function?

Dispersed endocrine cells in the posterior midgut and Malpighian tubule ampullae of locusts contain *Locusta*-DH (Montuenga *et al.*, 1996). These cells are of the 'open type', with a narrow apical process that extends to the lumen. This process could be used to detect changes in the composition of urine flowing through the ampullae, which may modulate release of *Locusta*-DH from secretory granules clustered at the basal surface of these cells (Montuenga *et al.*, 1996). A tachykinin-like peptide is also present in these secretory granules, and locustatachykinins are potent stimulants of tubule writhing (G. M. Coast, unpublished observation), which may reduce unstirred layers and assist circulation in the abdomen. The release of both peptides in the immediate vicinity of the tubules might therefore increase the efficiency with which waste products are cleared from the haemolymph.

Conclusions

A number of neuropeptides and biogenic amines stimulate Malpighian tubule secretion, but it is unclear whether they all have a physiological role in the control of diuresis. CRF-related peptides initiate post-feeding diuresis in locusts and stimulate post-eclosion diuresis in butterflies, and they are likely to be important factors controlling primary urine production in most insects. In this, they may act synergistically with other neurohormones such as the insect kinins or 5-HT.

Future directions

A major impetus for the study of diuretic hormones is the development of novel insecticides aimed at disrupting water balance. It is assumed that high urine flow rates will increase faecal water loss, thereby jeopardising the survival of a terrestrial insect that needs to conserve water. However, in general, faecal water loss is determined by fluid uptake in the hindgut (see Phillips *et al.*, this volume), and may not increase greatly even when tubule secretion is stimulated maximally. Indeed, the toxicity of a recombinant baculovirus which induced over expression of *Manduca*-DH in silkworm (*Bombyx mori*) larvae was only marginally increased over the wild type virus (Maeda, 1989), possibly because this peptide has both diuretic and antidiuretic activity in the larval insect (see above). Antagonists of diuretic hormones might prove more valuable, as they could prevent excretion of excess water and, more importantly, reduce the rate at which toxic wastes (including conventional insecticides) are cleared from the haemolymph. Encouragingly, non-peptide antagonists at mammalian CRF receptors have been identified in random screening strategies (Chalmers *et al.*, 1996). The cloning of receptors for insect diuretic peptides should facilitate the search for suitable lead compounds.

Acknowledgements

Original work reported in this paper was supported by grants from the UK Biotechnology and Biological Sciences Research Council, The Royal Society and NATO. I thank my colleagues who have allowed me to make reference to unpublished findings.

References

Audsley, N., Coast, G. M. & Schooley, D. A. (1993). The effects of *Manduca sexta* diuretic hormone on fluid transport by the Malpighian tubules and cryptonephric complex of *Manduca sexta*. *Journal of Experimental Biology* **178**, 231–243.

Audsley, N., Goldsworthy, G. J. & Coast, G. M. (1997). Circulating levels of *Locusta* diuretic hormone: the effect of feeding. *Peptides* **18**, 59–65.

Audsley, N., Kay, I., Hayes, T. K. & Coast, G. M. (1995). Cross reactivity studies of CRF-related diuretic peptides on insect Malpighian tubules. *Comparative Biochemistry and Physiology* **110A**, 87–93.

Bertsch, A. (1984). Foraging in male bumblebees (*Bombus lucorum* L.): maximizing energy or minimizing water load? *Oecologia* **62**, 325–336.

Beyenbach, K. W. (1995). Mechanism and regulation of electrolyte transport in Malpighian tubules. *Journal of Insect Physiology* **41**, 197–207.

Blackburn, M. B., Kingan, T. G., Bodnar, W., Shabanowitz, J., Hunt, D. F., Kempe, T., Wagner, R. M., Raina, A. K., Schnee, M. E. & Ma, M. C. (1991). Isolation and identification of a new diuretic peptide from the tobacco hornworm, *Manduca sexta*. *Biochemical and Biophysical Research Communications* **181**, 927–932.

Blackburn, M. B. & Ma, M. C. (1994). Diuretic activity of Mas-DPII, an identified neuropeptide from *Manduca sexta*; an *in vivo* and *in vitro* examination in the adult moth. *Archives of Insect Biochemistry and Physiology* **27**, 3–10.

Chalmers, D. T., Lovenberg, T. W., Grigoriadis, D. E., Behan, D. P. & De Souza, E. B. (1996). Corticotrophin-releasing factor receptors: from molecular biology to drug design. *Trends in Pharmacological Sciences* **17**, 166–172.

Chung, J. S., Goldsworthy, G. J. & Coast, G. M. (1994). Haemolymph and tissue titres of achetakinins in the house cricket *Acheta domesticus*: effect of starvation and dehydration. *Journal of Experimental Biology* **193**, 307–319.

Chung, J. S., Wheeler, C. H., Goldsworthy, G. J. & Coast, G. M. (1995). Properties of achetakinin binding sites on Malpighian tubule membranes from the house cricket, *Acheta domesticus*. *Peptides* **16**, 375–382.

Clottens, F. L., Holman, G. M., Coast, G. M., Totty, N. F., Hayes, T. K., Kay, I., Mallet, A. I., Wright, M. S., Chung, J.-S., Truong, O. & Bull, D. L. (1994). Isolation and characterization of a diuretic peptide common to the house fly and stable fly. *Peptides* **15**, 971–979.

Coast, G. M. (1995). Synergism between diuretic peptides controlling ion and fluid transport in insect Malpighian tubules. *Regulatory Peptides* **57**, 283–296.

Coast, G. M. (1996). Neuropeptides implicated in the control of diuresis in insects. *Peptides* **17**, 327–336.

Coast, G. M., Chung, J.-S., Goldsworthy, G. J., Patel, M., Hayes, T. K. & Kay, I. (1994). Corticotropin releasing factor related diuretic peptides in insects. In *Perspectives in Comparative Endocrinology*, ed. K. G. Davey, R. E. Peter & S. S. Tobe, pp. 67–73. Ottawa: National Research Council of Canada.

Coast, G. M., Hayes, T. K., Kay, I. & Chung, J.-S. (1992). Effect of *Manduca sexta* diuretic hormone and related peptides on isolated Malpighian tubules of the house cricket, *Acheta domesticus* (L.). *Journal of Experimental Biology* **162**, 331–338.

Coast, G. M., Holman, G. M. & Nachman, R. J. (1990). The diuretic activity of a series of cephalomyotropic neuropeptides, the achetakinins, on isolated Malpighian tubules of the house cricket, *Acheta domesticus*. *Journal of Insect Physiology* **36**, 481–488.

Coast, G. M. & Kay, I. (1994). The effects of *Acheta*-diuretic peptide on isolated Malpighian tubules from the house cricket, *Acheta domesticus*. *Journal of Experimental Biology* **187**, 225–243.

Coast, G. M., Kay, I. & Wheeler, C. H. (1993). Diuretic peptides in the house cricket, *Acheta domesticus* (L.): a possible dual control of Malpighian tubules. In *Molecular and Comparative Endocrinology*, Vol. 12, ed. K. W. Beyenbach, pp. 38–66. Basel: Karger.

Coast, G. M., Rayne, R. C., Hayes, T. K., Mallet, A. I., Thompson, K. S. J. & Bacon, J. P. (1993). A comparison of the effects of two putative diuretic hormones from *Locusta migratoria* on isolated locust Malpighian tubules. *Journal of Experimental Biology* **175**, 1–14.

Cook, H. & Orchard, I. (1993). The short term effects of 5,7-dihydroxytryptamine on peripheral serotonin stores in *Rhodnius prolixus* and their long-term recovery. *Insect Biochemistry and Molecular Biology* **23**, 895–904.

Cox, K. J. A., Tensen, C. P., van der Schors, R., van Heerikhuizen, H., Vreugdenhil, E., Geraerts, W. P. M. and Burke, J. F. (1996). Cloning, characterization and expression of a G-protein coupled receptor from *Lymnaea stagnalis* and identification of a leucokinin-like peptide, PSFHSWGamide, as its endogenous ligand. *Journal of Neuroscience*, **17**, 1197–1205.

Davies, S. A., Huesmann, G. R., Maddrell, S. H. P., O'Donnell, M. J., Skaer, N. J. V., Dow, J. A. T. & Tublitz, N. J. (1995). CAP_{2b}, a cardio-acceleratory peptide, is present in *Drosophila* and stimulates fluid secretion via cGMP. *American Journal of Physiology* **269**, R1321–R1326.

Dow, J. A. T., Maddrell, S. H. P., Davies, S. A., Skaer, N. J. V. & Kaiser, K. (1994). A novel role for the nitric oxide/cyclic GMP signaling pathway: the control of fluid secretion in *Drosophila*. *American Journal of Physiology* **266**, R1716–R1719.

Furuya, K., Schegg, K. M., Wang, H., King, D. S. & Schooley, D. A. (1995). Isolation and identification of a diuretic hormone from the mealworm *Tenebrio molitor*. *Proceedings of the National Academy of Sciences USA*, **92**, 12323–12327.

Gee, J. D. (1977). The hormonal control of excretion. In *Transport of Ions and Water in Animals*, ed. B. L. Gupta, R. B. Moreton, J. L. Oschman & B. J. Wall, pp. 265–281. London: Academic Press.

Graeser, D. & Neubig, R. R. (1992). Methods for the study of receptor/G-protein interactions. In *Signal Transduction: A Practical Approach*, ed. G. Milligan, pp. 1–30. Oxford: Oxford University Press.

Hayes, T. K., Pannabecker, T. L., Hinkley, D. J., Holman, G. M., Nachman, R. J., Petzel, D. H. & Beyenbach, K. W. (1989). Leucokinins, a new family of ion transport stimulators and inhibitors in insect malpighian tubules. *Life Science* **44**, 1259–1266.

Hazelton, S. R., Parker, S. W. & Spring, J. H. (1988). Excretion in the house cricket (*Acheta domesticus*): fine structure of the Malpighian tubules. *Tissue and Cell* **20**, 443–460.

Holman, G. M., Nachman, R. J. & Wright, M. S. (1990). Comparative aspects of insect myotropic peptides. In *Progress in Comparative Endocrinology*, ed. A. Epple, C. G. Scanes & M. H. Stetson, pp. 35–39. New York: Wiley-Liss.

Huesmann, G. R., Cheung, C. C., Loi, P. K., Lee, T. D., Swiderek, K. & Tublitz, N. J. (1995). Amino acid sequence of CAP_{2b}, an insect cardio-acceleratory peptide from the tobacco hornworm *Manduca sexta*. *FEBS Letters* **371**, 311–314.

Kataoka, H., Troetschler, R. G., Li, J. P., Kramer, S. J., Carney, R. L. & Schooley, D. A. (1989). Isolation and identification of a diuretic hormone from the tobacco hornworm, *Manduca sexta*. *Proceedings of the National Academy of Sciences USA* **86**, 2976–2980.

Kay, I., Coast, G. M., Cusinato, O., Wheeler, C. H., Totty, N. F. & Goldsworthy, G. J. (1991a). Isolation and characterization of a diuretic peptide from *Acheta domesticus*: evidence for a family of insect diuretic peptides. *Biological Chemistry Hoppe-Seyler* **372**, 505–512.

Kay, I., Patel, M., Coast, G. M., Totty, N. F., Mallet, A. I. & Goldsworthy, G. J. (1992). Isolation, characterization and biological activity of a CRF-related diuretic peptide from *Periplaneta americana* L. *Regulatory Peptides* **42**, 111–122.

Kay, I., Wheeler, C. H., Coast, G. M., Totty, N. F., Patel, M. & Goldsworthy, G. J. (1991b). Characterization of a diuretic peptide from *Locusta migratoria*. *Biological Chemistry Hoppe-Seyler* **372**, 929–934.

Knowles, G. (1976). The action of the excretory apparatus of *Calliphora vomitoria* in handling injected sugar solutions. *Journal of Experimental Biology* **64**, 131–140.

Lange, A. B., Orchard, I. & Barrett, F. M. (1989). Changes in haemolymph serotonin levels associated with feeding in the blood-sucking bug, *Rhodnius prolixus*. *Journal of Insect Physiology* **35**, 393–399.

Maddrell, S. H. P. (1963). Excretion in the blood-sucking bug, *Rhodnius prolixus* Stål. I The control of diuresis. *Journal of Experimental Biology* **40**, 247–256.

Maddrell, S. H. P. (1966). The site of release of the diuretic hormone in *Rhodnius*. A novel neurohaemal system in insects. *Journal of Experimental Biology* **45**, 499–508.

Maddrell, S. H. P., Gardiner, B. O. C., Pilcher, D. E. M. & Reynolds, S. E. (1974). Active transport by Malpighian tubules of acidic dyes and of acylamides. *Journal of Experimental Biology* **61**, 357–377.

Maddrell, S. H. P., Herman, W. S., Farndale, R. W. & Riegel, J. A. (1993). Synergism of hormones controlling epithelial fluid transport in an insect. *Journal of Experimental Biology* **174**, 65–80.

Maddrell, S. H. P., Herman, W. S., Mooney, R. L. & Overton, J. A. (1991). 5-Hydroxytryptamine: a second diuretic hormone in *Rhodnius. Journal of Experimental Biology* **156**, 557–566.

Maeda, S. (1989). Increased insecticidal effect of a recombinant baculovirus carrying a synthetic diuretic hormone gene. *Biochemical and Biophysical Research Communications* **165**, 1177–1183.

Montoreano, R., Triana, F., Abate, T. & Rangel-Aldao, R. (1990). Cyclic AMP in the Malpighian tubule fluid and in the urine of *Rhodnius prolixus. General and Comparative Endocrinology* **77**, 136–142.

Montuenga, L. M., Zudaire, E., Prado, M. A., Audsley, N., Burrell, M. A. & Coast, G. M. (1996). Presence of *Locusta* diuretic hormone in endocrine cells of the ampullae of locust Malpighian tubules. *Cell and Tissue Research* **285**, 331–339.

Mordue, W. (1969). Hormonal control of Malpighian tube and rectal function in the desert locust, *Schistocerca gregaria. Journal of Insect Physiology* **15**, 273–285.

Morgan, P. J. & Mordue, W. (1984). 5-Hydroxytryptamine stimulates fluid secretion in locust Malpighian tubules independently of cAMP. *Comparative Biochemistry and Physiology* **79C**, 305–310.

Muren, E. J., Lundquist, T. C. & Nässel, D. R. (1993). Quantitative determination of myotropic neuropeptide in the nervous system of the cockroach *Leucophaea maderae*: distribution and release of leucokinins. *Journal of Experimental Biology* **179**, 289–300.

Nachman, R. J., Coast, G. M., Roberts, V. A. & Holman, G. M. (1995). Incorporation of chemical/conformational components into mimetic analogs of insect neuropeptides. In *Insects: Chemical, Physiological and Environmental Aspects 1994*, ed. D. Konopinska, pp. 51–60. Poland: University of Wroclaw.

Nicolson, S. W. (1976). Diuresis in the cabbage white butterfly, *Pieris brassicae*: fluid secretion by the Malpighian tubules. *Journal of Insect Physiology* **22**, 1347–1356.

Nicolson, S. W. (1993). The ionic basis of fluid secretion in insect Malpighian tubules: advances in the last 10 years. *Journal of Insect Physiology* **39**, 451–458.

Nicolson, S. W. & Hanrahan, S. A. (1986). Diuresis in a desert beetle? Hormonal control of the malpighian tubules of *Onymacris plana* (Coleoptera: Tenebrionidae). *Journal of Comparative Physiology B* **156**, 407–413.

O'Donnell, M. J., Dow, J. A. T., Huesmann, G. R., Tublitz, N. J. & Maddrell, S. H. P. (1996). Separate control of anion and cation transport in Malpighian tubules of *Drosophila melanogaster*. *Journal of Experimental Biology* **199**, 1163–1175.

O'Donnell, M. J. & Maddrell, S. H. P. (1984). Secretion by the Malpighian tubules of *Rhodnius prolixus* Stål: electrical events. *Journal of Experimental Biology* **110**, 275–290.

Orchard, I. (1989). Serotonergic neurohaemal tissue in *Rhodnius prolixus*: synthesis, release and uptake of serotonin. *Journal of Insect Physiology* **35**, 943–947.

Pannabecker, T. L., Hayes, T. K. & Beyenbach, K. W. (1993). Regulation of epithelial shunt conductance by the peptide leucokinin. *Journal of Membrane Biology* **132**, 63–76.

Patel, M., Chung, J.-S., Kay, I., Mallet, A. I., Gibbon, C. R., Thompson, K. S. J., Bacon, J. P. & Coast, G. M. (1994). Localization of *Locusta*-DP in locust central nervous system and hemolymph satisfies initial hormonal criteria. *Peptides* **15**, 591–602.

Patel, M., Hayes, T. K. & Coast, G. M. (1995). Evidence for the hormonal function of a CRF-related diuretic peptide (*Locusta*-DP) in *Locusta migratoria*. *Journal of Experimental Biology* **198**, 793–804.

Phillips, J. E. (1983). Endocrine control of salt and water balance: excretion. In *Endocrinology of Insects*, ed. R. G. H. Downer & H. Laufer H., pp. 411–425. New York: Alan R. Liss., Inc.

Plawner, L., Pannabecker, T. L., Laufer, S., Baustian, M. D. & Beyenbach, K. W. (1991). Control of diuresis in the yellow fever mosquito *Aedes aegypti*: evidence for similar mechanisms in the male and female. *Journal of Insect Physiology* **37**, 119–128.

Proux, J. P., Miller, C. A., Li, J. P., Carney, R. L., Girardie, A., DeLaage, M. & Schooley, D. A. (1987). Identification of an arginine vasopressin-like diuretic hormone from *Locusta migratoria*. *Biochemical and Biophysical Research Communications* **149**, 180–186.

Ramsay, J. A. (1954). Active transport of water by the Malpighian tubules of the stick insect, *Dixippus morosus* (Orthoptera, Phasmidae). *Journal of Experimental Biology* **31**, 104–113.

Reagan, J. D. (1994). Expression cloning of an insect diuretic hormone receptor: a member of the calcitonin/secretin receptor family. *Journal of Biological Chemistry* **269**, 9–12.

Reagan, J. D. (1995a). Functional expression of a diuretic hormone receptor in baculovirus-infected cells: evidence suggesting that the N-terminal region of diuretic hormone is associated with receptor activation. *Insect Biochemistry and Molecular Biology* **25**, 535–539.

Reagan, J. D. (1995b). Molecular cloning of a putative Na^+-K^+-$2Cl^-$ cotransporter from the Malpighian tubules of the tobacco hornworm, *Manduca sexta*. *Insect Biochemistry and Molecular Biology* **25**, 875–880.

Regan, J. D. (1996). Molecular cloning and functional expression of a diuretic hormone receptor from the cricket, *Acheta domesticus*. *Insect Biochemistry and Molecular Biology* **26**, 1–6.

Reagan, J. D., Li, J. P., Carney, R. L. & Kramer, S. J. (1993). Characterization of a diuretic hormone receptor from the tobacco hornworm, *Manduca sexta*. *Archives of Insect Biochemistry and Physiology* **23**, 135–145.

Spittaels, K., Devreese, B., Schoofs, L., Neven, H., Janssen, I., Grauwels, L., Van Beeumen, J. & De Loof, A. (1996). Isolation and identification of a cAMP generating peptide from the flesh fly, *Neobellieria bullata* (Diptera: Sarcophagidae). *Archives of Insect Biochemistry and Physiology* **31**, 135–147.

Spring, J. H., Morgan, A. M. & Hazelton, S. R. (1988). A novel target for antidiuretic hormone in insects. *Science* **241**, 1096–1098.

Thompson, K. S. J., Rayne, R. C., Gibbon, C. R., May, S. T., Patel, M., Coast, G. M. & Bacon, J. P. (1995). Cellular co-localization of diuretic peptides in locusts: a potent control mechanism. *Peptides* **16**, 95–104.

Vale, W., Spiess, J., Rivier, C. & Rivier, J. (1981). Characterization of a 41 residue ovine hypothalamic peptide that stimulates secretion of corticotropin and β-endorphin. *Science* **213**, 1394–1397.

Wheeler, C. H. & Coast, G. M. (1990). Assay and characterization of diuretic factors in insects. *Journal of Insect Physiology* **36**, 23–34.

Wheelock, G. D., Petzel, D. H., Gillett, J. D., Beyenbach, K. W. & Hagedorn, H. H. (1988). Evidence for hormonal control of diuresis after a blood meal in the mosquito *Aedes aegypti*. *Archives of Insect Biochemistry and Physiology* **7**, 75–89.

Williams, J. C., Hagedorn, H. H. & Beyenbach, K. W. (1983). Dynamic changes in flow rate and composition of urine during the post-bloodmeal diuresis in *Aedes aegypti* (L.). *Journal of Comparative Physiology B* **153**, 257–265.

Wright, J. M. & Beyenbach, K. W. (1987). Chloride channels in apical membranes of mosquito Malpighian tubules, *Federation Proceedings* **46**, 1270.

JOHN E. PHILLIPS, JOAN MEREDITH, NEIL
AUDSLEY, MARK RING, ANDRIS MACINS,
HUGH BROCK, DAVID THEILMANN
and DWIGHT LITTLEFORD

Locust ion transport peptide (ITP): function, structure, cDNA and expression

Introduction

General background on insect excretion and regulation

The extraordinary success of insects in terrestrial environments is associated with the evolution of highly effective excretory epithelia which regulate body water content, haemolymph ionic composition and pH, and eliminate nitrogenous waste products. Early physiological observations on whole insects revealed very large changes in the volume and solute composition of the final excreta in response to severe fluctuations in external conditions, thus suggesting neuronal or endocrine control of the excretory system. For example, the desert locust, *Schistocerca gregaria*, when dehydrated and starved produces faecal pellets only infrequently and these are very dry. In contrast, when feeding on succulent plants, they daily eliminate their own body weight of fluid containing very low NaCl (for a review, see Phillips, 1981). Accordingly, the neuroendocrine control of insect excretory function received early attention (for reviews, see Phillips, 1981, 1983). The recent successful isolation and sequencing of diuretic hormones that control secretion of isosmotic KCl-rich primary urine by Malpighian tubules of several insects is the subject of the chapter by Coast, this volume. However, ionic and osmotic regulation in most insects studied to date depends ultimately on selective, active, and controlled reabsorption of solutes and water from this primary urine in anterior (ileum) and posterior (rectum) hindgut segments.

Locust hindgut mechanisms

The hindgut epithelial mechanisms responsible for selective reabsorption and their cellular location (epithelial models) have been well characterised only in the desert locust (for reviews, see Phillips *et al.*, 1986, 1988; Phillips &

This review is dedicated to the memory of Dwight Littleford, an exceptionally promising Ph.D. student, whose work on cloning ileal receptors was tragically ended after two years by his premature death in April 1996.

Audsley, 1995), including more recently the role of hindgut segments in acid–base balance, ammoniagenesis, and nitrogen excretion (Harrison & Kennedy, 1994; Phillips *et al.*, 1994).

In both hindgut segments of the desert locust, the dominant transepithelial active transport mechanism is an unusual (compared to vertebrates) electrogenic Cl^- pump located in the apical membrane (for reviews, see Phillips *et al.*, 1996). When the transepithelial potential difference is clamped at zero millivolts (mV), the change in applied current (the short-circuit current I_{sc}) required to maintain the clamp is a direct continuous measure of the increase in Cl^- transport rate after stimulation by cyclic AMP or neuroendocrine extracts. Chloride exits the rectal cells passively via a basolateral conductance (putative channel). Potassum, the major cation absorbed, follows Cl^- passively by electrical coupling via cation channels with different properties in apical and basolateral membranes. The levels of Na^+ in the primary urine and hence in fluid entering the hindgut lumen are quite low (20 mmol 1^{-1}), and active reabsorption of this cation is therefore quantitatively less important. Na^+ enters rectal cells passively by several mechanisms (a conductance pathway; in exchange for NH_4^+ and H^+; and cotransport with glycine; Black *et al.*, 1987) down a very large electrochemical potential gradient at the apical border, and is actively removed from cells basolaterally by a Na/K-ATPase pump. Stimulation of fluid transport in the absence of, or against, osmotic concentration differences across both hindgut segments is dependent on salt transport, expecially Cl^- (Proux, Proux & Phillips, 1984; Lechleitner, Audsley & Phillips, 1989a,b). An electrogenic H^+ pump (probably an ATPase) in the apical membrane causes equal rates of acid secretion into the lumen in both hindgut segments, and this is associated with passive exit of base equivalents (OH^- and HCO_3^-) to the haemocoel side (Thomson & Phillips, 1992; see also Phillips *et al.*, 1994).

While locust ilea and recta share many common epithelial transfer mechanisms, some important differences have been observed. Fluid transport (J_v) is always close to isosmotic in the ileum and can be increased four-fold by stimulants (see below), as compared to a less than two-fold increase for rectal J_v. The rectum can extract an hypo-osmotic absorbate to concentrate the lumen content to final osmotic concentrations several times that of the haemolymph; that is, this segment can create a strongly hyperosmotic urine. This is thought to be achieved by solute recycling at elaborate lateral intercellular channels that are not observed in the ileum (for reviews, see Phillips *et al.*, 1986; Irvine *et al.*, 1988). Moreover, a high capacity proline pump (Meredith & Phillips, 1988) in the apical membrane of the rectum (but not the ileum) can drive additional water extraction even in the absence of luminal Na^+, K^+ and Cl^- (Lechleitner & Phillips, 1989). Proline recovered from the rectal lumen also provides the principal substrate for cellular respiration and ammoniagenesis leading to apical NH_4^+ secretion in exchange for luminal Na^+ (Chamberlin & Phillips, 1983; Thomson, Thomson & Phillips, 1988). Water and CO_2 from cellular respiration provide the H^+ and HCO_3^- for acid–base transport and pH regulation, which is aided by ammonia production which traps H^+ to form NH_4^+. In contrast, several neutral amino acids recovered apically from the primary urine serve the same functions as substrates for cellular respiration, ammoniagenesis and acid–base equivalents (except enhanced J_v) in the ileum.

The ileum is the predominant site of Na^+ reabsoption and NH_4^+ secretion in the locust excretory system, and the only site where there is some evidence for endocrine control of these processes (Irvine *et al.*, 1988; Phillips *et al.*, 1994).

Control of locust hindgut transport

Stimulation of locust ileal and rectal Cl^- (measured as I_{sc}) and fluid (J_v) transport by exogenous cyclic AMP, cyclic GMP (which has a lesser effect), and crude homogenates of brain, nervous (storage) lobe of the corpus cardiacum (NCC) and ventral ganglia (VG) have long suggested hormonal control of these hindgut excretory processes (for a review, see Phillips & Audsley, 1995).

Hanrahan & Phillips (1984: for a review, see Phillips *et al.*, 1986) used double-barrelled ion-sensitive glass microelectrodes and electrophysiology to show that elevated intracellular cyclic AMP acts directly on the apical Cl^- pump and on both apical K^+ and basolateral Cl^- conductances (ion channels) to stimulate KCl absorption across locust rectum, while apical H^+ secretion was inhibited (for a review, see Phillips *et al.*, 1994) and Na^+ reabsorption was unaffected. Chamberlin & Phillips (1988) reported elevation of rectal tissue cyclic AMP levels in parallel with increases in I_{sc} caused by extracts of NCC: all treatments and conditions associated with high I_{sc} were accompanied by increased tissue titres of cyclic AMP.

More recently, Richardson (1993) used similar electrophysiological approaches, but with single-barrelled electrodes, to study electropotentials, resistances and specific ion conductances (by external ion substitutions) across apical and basolateral membranes of locust ilea stimulated with exogenous cyclic AMP (Fig. 1). The results indicate a large K^+ and a small Na^+ conductance increase in the apical membrane without any drop in apical electropotential difference. The latter result could be explained if cyclic AMP also directly stimulated an apical Cl^- pump to balance enhanced cation entry into the ileal epithelial cells. In contrast to the situation in the rectum, the basolateral membrane resistance of the ileum is very low relative to the apical membrane. The ileal basolateral membrane had

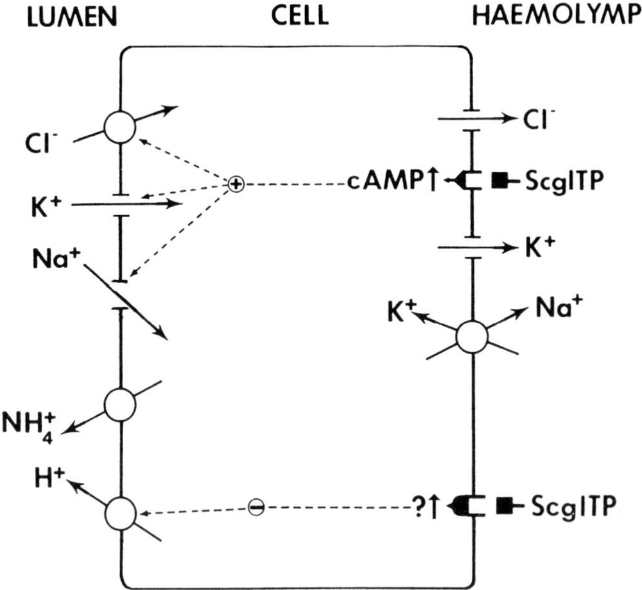

Fig. 1. Proposed model for control of ion transport across desert locust ileum. Ion transport peptide (ITP) is believed to act via cyclic AMP at the apical membrane to increase both K^+ and Na^+ conductances and stimulate the electrogenic Cl^- pump directly. ITP must act via some other second messenger to inhibit apical acid secretion. Arrows through circles indicate carrier-mediated mechanisms, and arrows through gaps indicate membrane ion channels. Based on electrophysiological studies by Richardson (1993). Scg, *Schistocerca gregaria*.

high conductances for both K^+ and Cl^- even in the unstimulated state. Not surprisingly therefore, no increase in basolateral Cl^- conductance was measured after cyclic AMP stimulation. The ileum response to cAMP therefore differs from that of the rectum in three respects: (1) an apical Na^+ conductance appears, but (b) there is no change in the large basolateral Cl^- conductance, and (c) cyclic AMP has no effect on ileal acid secretion (J_H) whereas strong inhibition is observed for the rectum. However, crude extracts of corpora cardiaca at high doses inhibit J_H nearly completely in both hindgut segments, indicating a second messenger other than cyclic AMP must be involved in the case of the ileum.

In summary, a great deal is known about locust hindgut functions, their underlying cellular transport mechanisms, and the specific points of control by cyclic AMP. These epithelial transport studies prompted the search for the natural stimulants in the neuroendocrine system.

Purification and chemical structure of neuropeptides that stimulate locust rectum

Early reports of putative diuretic (DH) and antidiuretic (ADH) factors influencing fluid transport by rectal sacs *in vitro* are reviewed by Phillips (1983). These studies identified the retrocerebral complex, consisting of the median neurosecretory cells of the pars intercerebralis (PI) and their axons to storage and release sites in the NCC, as the major source of stimulants. However, initiation of stimulant release from NCC using high external K^+ in the presence of Ca^{2+} has not been successful to date (Phillips *et al.*, unpublished observations). Stimulatory activity is also present in VG4 to VG7 of locusts (Lechleitner & Phillips, 1989; Audsley & Phillips, 1990) and in terminal abdominal ganglia of cockroaches (Goldbord *et al.*, 1970). The locust ventral ganglial stimulant has different properties from that of the corpora cardiaca with regard to the time course for changes in hindgut I_{sc}, solubility, acid liability and heat stability. Bilgen (1994) has recently partially purified the acid-labile ventral ganglia factor of desert locusts, which had an approximate molecular mass by size-exclusion chromatography of 37 000 Da, several times that of factors from the NCC discussed below.

Three neuropeptides have been purified fully or partially from the NCC of locusts (for a review, see Phillips & Audsley, 1995) Neuroparsin B (NpB) is a homodimer of a 78 amino acid residue peptide isolated and sequenced from *Locusta migratoria* (Girardie *et al.*, 1989, 1990), although an alternative monomeric structure has more recently been proposed (Hietter, Van Dorsselaer & Luu, 1991) NpB is claimed to be the only antidiuretic factor in the NCC of *L. migratoria*, because an antibody to this neuropeptide abolished all of the antidiuretic activity of crude gland extracts on fluid transport (J_v) by everted rectal sacs (Fournier & Girardie, 1988). NpB acts on rectal J_v by stimulating the phosphoinositol-Ca^{2+} cascade (for a review, see Fournier, 1991). However, Jeffs & Phillips (1996; see also Jeffs, 1993) observed no effect of neuroparsins over several hours and at high doses on either rectal J_v, or on rectal and ileal I_{sc} of the desert locust (Table 1). One explanation might be the presence of different major stimulants in the corpora cardiaca of these two locust species, which would be surprising. Reciprocal bioassays of crude gland extracts from *L. migratoria* and *S. gregaria* on rectal and ileal I_{sc} and J_v in these two locust species have now been conducted (J. Meredith, unpublished observations). Corpora cardiaca extracts from either species were equally effective on hindgut bioassays of both locusts, suggesting similar stimulants are present. *Locusta migratoria* NCC must therefore contain a rectal stimulant of *Schistocerca* I_{sc} and J_v other than NpB, because NpB had no action on *Schistocerca* hindgut (Jeffs,

Table 1 *Effect of neuroparsins (Nps) and crude nervous corpus cardiacum (NCC) extracts on hindgut absorption* (Schistocerca gregaria)

	Nps	NCC	n
	(μequiv. cm^{-2} h^{-1})		
Ileal ΔI_{sc}[a]	0.45±0.56	12.5±3.7	4
Rectal ΔI_{sc}[b]	−0.20±0.17	5.6±1.1	8
	(μl h^{-1} (everted sacs)$^{-1}$)		
Rectal ΔJ_{v}[b]	−0.32±0.35	4.4±0.51	8
Dosages:	Nps=5 CC-equiv. ml^{-1}		
	NCC=0.5 to 2.0 CC-equiv. ml^{-1}		

Notes:

CC-equiv., corpora cardiaca equivalent.

[a] L. B. Jeffs and J. Meredith, unpublished data.

[b] Jeffs & Phillips (1996).

1993; Jeffs & Phillips, 1996). An alternative explanation is that the bioassays used by the two research groups are very different. Phillips & Audsley (1995) have pointed out that rectal J_v is Cl$^-$-dependent. However, the *L. migratoria* bioassay used by Fournier & Girardie (1988) involves pre-incubating everted rectal sacs in Cl-free saline, a treatment known to quickly deplete rectal tissue levels of Cl$^-$ (Williams *et al.*, 1977). Following this preincubation, the increase in J_v is measured in the presence of stimulants (such as NpB) in Cl$^-$-containing saline, and any increase in J_v above that observed in controls is deemed to be due to hormonal activity. Phillips & Audsley (1995) believe that this *L. migratoria* assay may measure short-term cell volume changes associated with restoring Cl$^-$, rather than longer-term steady-state rates of salt and water transport as used by Jeffs & Phillips (1996) with *S. gregaria*. NpB has not been tested directly by Fournier & Girardie for effects on rectal ion transport processes.

Spring & Phillips (1980a–c) used the increase in I_{sc} across *S. gregaria* recta to detect and partially purify (Phillips *et al.*, 1980, 1982) a factor named chloride transport stimulating hormone (CTSH) of about 8000 Da, which was estimated to be maximally active at less than 7 nmol l^{-1}. The acid-lability of CTSH has prevented further purification and sequencing of this neuropeptide. Proux *et al.* (1985) provided some evidence that CTSH is produced in the pars intercerebralis region of the brain and is transported down the nervi corpori cardiaci I for storage in the NCC. Haemolymph from recently fed locusts contains more CTSH-like activity than that from starved locusts

(Phillips *et al.*, 1982), and cardiatectomy is also reported to reduce blood levels (Spring & Phillips, 1980c). CTSH appears to act on rectal I_{sc} of desert locusts (*Schistocerca gregaria*) via cyclic AMP (for a review, see Phillips & Audsley, 1995), unlike NpB. Other observations suggest that CTSH (not sequenced) is different from NpB (fully sequenced). For instance, acetic acid (0.1 mol 1^{-1}), used to extract NpB from corpora cardiaca, completely destroys the biological activity of CTSH on *S. gregaria* rectal I_{sc}. In summary, the relative importance of NpB and CTSH in controlling locust rectal reabsorption requires further clarification. Their possible actions on the ileum are completely unknown, except for the demonstration that NpB has no effect on *S. gregaria* ileal I_{sc} (Table 1).

HPLC purification and partial sequencing of locust ion transport peptide, an ileal stimulant

Using ileal I_{sc} as the bioassay, Audsley & Phillips (1990) surveyed the whole central nervous system of *S. gregaria* for stimulatory activity. Proteinaceous factors were detected in the brain and NCC, and also in VG4 to VG7, that stimulated ileal I_{sc} in a dose-dependent manner. Similar results were observed using ileal J_v as the bioassay (Lechleitner *et al.*, 1989a,b). Both corpora cardiaca and ventral ganglia stimulated ileal J_v four-fold, but only if external Cl^- was present. As for the rectum, the NCC and VG factors produce different time courses for the I_{sc} increase in the ileum. Moreover, the VG factor is much more heat labile, and its biological activity is destroyed by extraction with 0.2 mol 1^{-1} acetic acid, unlike that in the NCC.

Audsley *et al.* (1992a) used reversed-phase high-performance liquid chromatography (HPLC) and the ileal I_{sc} bioassay to isolate the predominant stimulant in *Schistocerca* NCC, which was named ion transport peptide (ITP). A second more hydrophobic factor has not yet been isolated, and accounts for 30% of the total stimulation by crude NCC extracts. A third fraction had little effect on ileal I_{sc}, but stimulated J_v, presumably by acting on a solute transport process other than Cl^- (Audsley, 1991). Purified ITP has an unblocked N-terminus and a molecular mass estimated by mass spectroscopy to be 8652 ± 3 Da. A partial amino acid sequence (33 residues) was published (Audsley, McIntosh & Phillips, 1992a; Audsley *et al.*, 1994), and further unpublished sequence data to residue 55 was withheld for confirmation (Audsley, 1991). Several attempts to complete the ITP sequence by enzymatic digestion and HPLC separation of the products were unsuccessful. This partial ITP sequence was 44–59% identical to hyperglycaemic (CHH), moult-inhibiting (MIH) and vitellogenesis-inhibiting (VIH) hormones of crustaceans, indicating that locust ITP was the first member of this peptide family to be detected in other animal groups.

Actions of HPLC-purified locust ITP

Purified ITP at a dosage of 5 pmol added to 2 ml of bathing saline (2.5 nmol 1^{-1}) had the same range of actions as crude locust NCC extracts on the ileum: namely it caused large increases in I_{sc}, Cl^- transport (10-fold), Na^+ transport (2-fold), K^+ permeability (3-fold) and isosmotic fluid absorption (4-fold), and inhibited active acid secretion almost completely at high doses (Audsley *et al.*, 1992b, 1994; Phillips & Audsley, 1995). ITP had no effect on ileal ammonia secretion. Exogenous cyclic AMP had the same range of actions on ileal transport processes as NCC extracts and purified ITP, except that this second messenger did not inhibit acid secretion, but did stimulate ammonia secretion. Thus a single neuropeptide, ITP, can account for all the actions of crude NCC extracts on locust ileum. The results are consistent with cyclic AMP being the second messenger for ITP actions except for the control of acid secretion (Fig. 1). ITP causes an elevation of cyclic AMP levels in ileal tissue after 1 h, but the time course for this change and effects on other potential second messengers have yet to be investigated.

ITP-like activity that co-eluted with ITP on the first HPLC purification step was also detected in locust haemolymph. This is the only evidence to date that ITP might normally be released into the haemocoel to influence ileal transport activities *in vivo*.

Given the similar molecule size and source of locust ITP and CTSH, which has not been sequenced, are they actually different neuropeptides? Audsley *et al.* (1992b, 1994) tested HPLC-purified ITP on locust rectal transport activities. At maximum doses, ITP caused only partial (41%) stimulation of rectal I_{sc} as compared to crude NCC extracts. Moreover, ITP did not cause significant increases in rectal fluid transport (J_v) or K^+ permeability, as do crude NCC extracts and cyclic AMP. Audsley *et al.* (1992b) concluded that different neuropeptides in locust NCC must stimulate the rectum (CTSH) and ileum (ITP), especially given that the biological activity of CTSH, but not ITP, was destroyed by extraction with acetic acid.

Complete ITP amino acid sequence deduced from its cDNA

Meredith *et al.* (1995, 1996) used the partial amino acid sequence of ITP obtained by Audsley *et al.* (1992a) to produce degenerate oligonucleotide primers for residues 2–8 (sense) and 24–30 (antisense). These were used with a locust brain cDNA library and the polymerase chain reaction (PCR) to clone cDNA that exactly encoded the known partial amino acid sequence of ITP. This nucleotide sequence was extended by anchored PCR to the start and end of the ITP message using the 5' and 3' RACE (rapid amplification of complementary ends) system. The resulting partial cDNA of 517 bp encoded a complete open reading frame for an ITP prepropeptide of 130 amino acid residues (Meredith *et al.*, 1996). This consisted of a 55 residue leader sequence with a methionine start signal and a dibasic cleavage site preceding the start of the known ITP sequence from Audsley *et al.* (1992a). A final sequence (Gly-Lys-Lys-stop) suggested terminal amidation at a second dibasic cleavage site to give ITP comprising 72 amino acid residues. The deduced prepropeptide sequence is shown below, numbered relative to the start of ITP at number 1, with dibasic cleavage sites indicated in bold text.

$^{-55}$Met-His-His-Gln-Lys-$^{-5}$-Gln–Gln–Gln–Gln–Gln–Lys–Gln–Gln–Gly–Glu-$^{-40}$Ala–Pro–Cys–Arg–His–Leu–Gln–Trp–Arg–*Leu-$^{-30}$Ser–Gly–Val–Val–Leu–Cys–Val–Leu–Val–Val-$^{-20}$Ala–Ser–Leu–Val–Ser–Thr–Ala–Ala–Ser–Ser-$^{-10}$Pro–Leu–Asp–Pro–His–His–Leu–Ala–**Lys**–**Arg**-^{1}Ser–Phe–Phe–Asp–Ile–Gln–Cys–Lys–Gly-^{10}Val–Tyr–Asp–Lys–Ser–Ile–Phe–Ala–Arg–Leu-^{20}Asp–Arg–Ile–Cys–Glu–Asp–Cys–Tyr–Asn–Leu-^{30}Phe–Arg–Glu–Pro–Gln–Leu–His–Ser–Leu–Cys-^{40}Arg–Ser–Asp–Cys–Phe–Lys–Ser–Pro–Tyr–Phe-^{50}Lys–Gly–Cys–Leu–Gln–Ala–Leu–Leu–Leu–Ile-^{60}Asp–Glu–Glu–Glu–Lys–Phe–Asn–Gln–Met–Val-^{70}Glu–Ile-^{72}Leu–Gly–**Lys**–**Lys**–STOP

(Note: the nucleotide sequence indicates Leu at position –31 of the leader sequence rather than Lys as incorrectly reported by Meredith *et al.*, 1996.)

This deduced ITP sequence agrees completely with the known partial sequence of the HPLC-purified ITP for residues 1–34, and also with additional unpublished sequence data (Audsley, 1991) for residues 35–45, 47, 49, 53 and 54. Two PCR products coding for amino acids –55 to +23 and 1 to 75 of ITP were used to screen a locust brain cDNA library. Six positive ITP clones were identified and sequenced. All of these clones confirmed the above ITP sequence derived using PCR and degenerate primers (Meredith *et al.*, 1996).

The complete deduced ITP sequence shares 39–42% sequence identity with CHH of the crab *Carcinus maenas*, the lobster *Homarus americanus* and the woodlouse *Armadillidium vulgare*, and 29 and 30% with lobster VIH and crab MIH, respectively. Including neutral amino acid substitutions, the percentage similarity of ITP to CHH is 71–67, and to MIH and VIH it is 53–54. All six cysteines at positions 7, 23, 26, 39, 43, 52 are conserved in all members of this peptide family. Based on disulphide bridge positions determined for crab CHH by Kegel *et al.* (1989), Meredith *et al.* (1996) predicted identical bridges at 7–43, 23–39 and 26–52 in ITP (see below).

The molecular mass of ITP (oxidised form with three disulphide bridges) predicted from the cDNA is 8558 Da, which is 94 Da less than that determined by mass spectroscopy for HPLC-purified ITP (Audsley *et al.*, 1994).

The difference may be due to phosphorylation or sulphation of native ITP, as suggested originally by Audsley *et al.* (1994). Alternatively, a potassium salt of ITP may have been analysed by mass spectroscopy (suggested by D. A. Schooley; see Meredith *et al.*, 1996).

Distribution of ITP among insects and specificity of the locust ITP receptor

Meredith *et al.* (1996) reported that only brain-NCC and ventral ganglia of insects closely related to *S. gregaria* stimulated locust ileal I_{sc}. These include other locusts (*L. migratoria*, J. E. Phillips *et al.*, unpublished observation) and grasshoppers (*Melanoplus sanguinipes*), crickets and cockroaches. A *Locusta* cDNA that encodes an ITP prepropeptide very similar to that of *S. gregaria* has recently been obtained (A. Macins, unpublished observations). While there are several nucleotide substitutions, the predicted amino acid sequence for the biologically active 72 amino acid residue peptide is unchanged; however, there are nine amino acid changes and two deletions in the leader sequence. The much slower time course for changes in locust ileal I_{sc} caused by cricket NCC at maximum doses suggests that there must be some amino acid sequence differences in the ITP from this species.

In contrast, brain-NCC of dipteran, lepidopteran, hymenopteran and hemipteran species that have been tested had no stimulatory action on locust ileal I_{sc}. This most likely indicates significant sequence divergence of ITP-related neuropeptides in these groups rather than their absence. Southern blots conducted at high stringency indicate ITP-related molecules in the dipteran and lepidopteran species that were tested, as well as in a nematode (*C. elegans*); however, sequences have not been determined (D. Sinclair & C. Wiens, unpublished observations).

Several identified insect neuropeptides and neurotransmitters have been tested at high doses on locust ileal I_{sc} without any effect. These include representatives of several myotropin classes, *Locusta* CRF-related diuretic hormone (J. Meredith & G. M. Coast, unpublished observations), and three crustacean hormones with high sequence similarity to locust ITP (crab CHH, Meredith *et al.*, 1996; lobster CHH and MIH, J. E. Phillips *et al.*, unpublished observations). Moreover, MIH at high dose (30 nmol 1^{-1}) did not inhibit ITP stimulation of locust ileal I_{sc}. These observations are indicative of the very high specificity of the locust ITP bioassay and the ITP receptor.

Expression of ITP cDNA using a recombinant baculovirus

To confirm that the ITP cDNA obtained does encode an active stimulant, Meredith *et al.* (1996) cloned the complete open reading frame of the ITP prepropeptide, including the natural leader sequence, in the baculovirus transfer vector pVL 1393. Insect SF9 cell cultures transfected with the recombinant virus secreted about 60 maximum doses per 3 ml culture medium of ITP-like stimulant as assayed by increase in ileal I_{sc}, whereas control cultures treated with the wild-type virus produced no such secretion. Given the high selectivity of the ITP bioassay, this experiment provides evidence that the ITP amino acid sequence deduced from the cDNA is correct. However, the extent of processing and post-translational modification of expressed ITP remains unknown. M. Ring & D. Theilmann (unpublished observations) have explored optimal conditions for ITP expression. Maximum production by Sf9 cells cultures in various media occurs rather late at 72 h, but other cell expression systems appear more promising. A trial is in progress to determine whether a recombinant baculovirus containing the ITP message enhances the rate of kill of cabbage loopers (*Trichoplusia ni*).

Synthetic ITP and its biological actions

D. King *et al.* (unpublished observations) synthesised ITP exactly in the chemical form proposed by Meredith *et al.* (1996), with amidation of residue 72 and with three disulphide bridges based on crab CHH. This is the first member of this large family of arthropod neuropeptides to be synthesised. Of particular note was that during the oxidation step of the synthesis only one combination of disulphide bridges was formed without the assistance of any cellular constituents. The biological activities of this synthetic ITP were consistently similar to those of ITP purified from locust NCC by Audsley *et al.* (1992a, 1994). Dose–response curves for natural and synthetic ITP were not significantly different given experimental uncertainties, and indicated that the concentration for half-maximum stimulation of ileal I_{sc} (EC_{50}) is 1.1 to 2.3 nmol 1^{-1}. The pattern of ileal I_{sc} time courses at different doses were similar, as were the increase in open-circuit transepithelial electropotential difference (three-fold maximum) and the 50% decrease in transepithelial resistance. Using a different bioassay, synthetic ITP at maximum doses caused a four-fold increase in fluid transport (J_v) across everted ileal sacs, identical to that caused by native ITP.

Synthetic ITP also caused maximum stimulation of rectal I_{sc} in a few preparations, but more extensive studies indicated that on average rectal I_{sc} was only partially stimulated (40% of maximum) and there was no significant effect on rectal J_v. Thus, the effects of synthetic ITP on locust rectum are similar to those of the native peptide. Moreover, western blots with an antibody specific to the C-terminus of ITP indicated co-migration of synthetic and native ITP from NCC (J. Meredith, unpublished observation). This

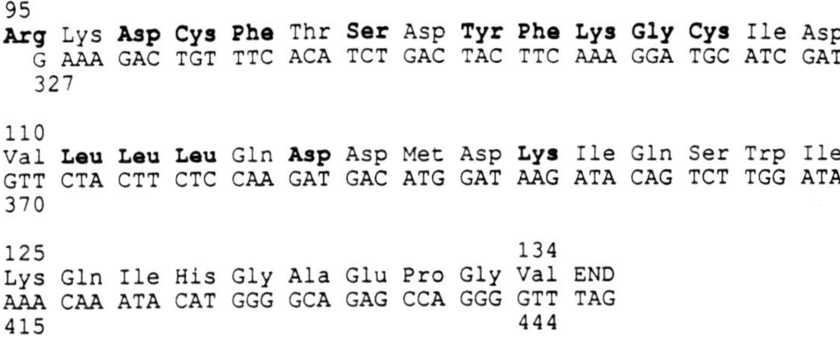

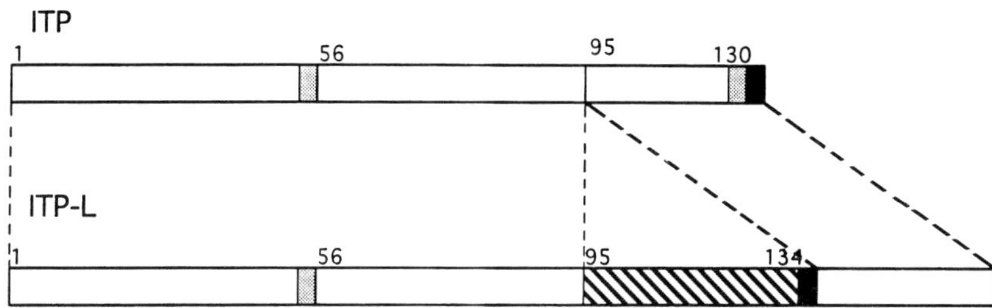

Fig. 2. Nucleotide and deduced amino acid sequences of the unique ITP-L insert. Sequence identity of amino acid residues in ITP and ITP-L are indicated by bold type. The position of the ITP-L insert is indicated in the lower diagram as the hatched area in the open reading frame. Shaded boxes represent dibasic cleavage sites and black boxes represent termination codons. Open boxes indicate identical sequences in ITP and ITP-L, with alignment indicated by broken lines. (From Meredith *et al.*, 1996.)

extensive series of bioassay and antibody studies with synthetic and natural ITP provides strong support for the chemical structure of ITP proposed by Meredith *et al.* (1996).

An ITP-like (ITP-L) cDNA in locusts

Meredith *et al.* (1996) probed a locust ileal mRNA library with the ITP cDNA sequence primers (5′ and 3′ extensions). An ITP-L clone was sequenced that was identical to the brain cDNA for ITP except for an additional 121 bp insert at amino acid position 40 of ITP, suggesting alternative splicing of genomic DNA. ITP-L has an open reading frame (134 residues) that is four residues longer than that of ITP (Fig. 2). All six cysteines are conserved. The unique C-terminal of ITP-L has only 14 of the last 36 amino acid residues in common with ITP, most of the difference being over the last 20 residues. Using reverse transcription PCR, ITP-L mRNA was detected in many tissues (such as flight muscle, hindgut and Malpighian tubules) that have no stimulatory effect in the locust ileal I_{sc} bioassay. In contrast, ITP mRNA was restricted to the brain and NCC which do stimulate ileal I_{sc}. J. Meredith & M. Ring (unpublished observation) have expressed ITP-L in SF9 cells and detected a secreted product using an antibody specific to the unique C-terminal of ITP-L. In preliminary studies, this expressed form of ITP-L had no stimulatory action on ileal I_{sc} at high (estimated) doses.

The function of ITP-L is unknown, but the identical initial sequences of ITP and ITP-L suggest that ITP-L might either act as an antagonist at the ITP receptor to shut off hindgut fluid reabsorption, or act to reduce synthesis and release of ITP at the brain-NCC level. Thus, excess reabsorption of dilute fluid in the hindgut due to stimulation by ITP might cause tissue swelling leading to widespread release of ITP-L from various tissues, which by antagonising ITP or reducing ITP synthesis/release would have a diuretic effect. In this sense, ITP-L might be analogous to vertebrate atrial natriuretic hormone, which is found in many tissues and which inhibits renal salt and hence fluid reabsorption. This working hypothesis is currently under examination using expressed ITP-L.

Summary

Ion transport peptide (ITP) from desert locusts is the first fully characterised insect neuropeptide shown to stimulate reabsorption of the major ions and to inhibit acid secretion (a pH regulatory role) in the anterior hindgut (ileum), which is functionally analogous to the proximal tubules of vertebrate kidneys. ITP has a lesser effect on the posterior hindgut (rectum). ITP has antidiuretic activity because enhanced salt transport drives isosmotic fluid absorption in the ileum. Thus, ITP is a putative insect hormone with physiological roles analogous to those of vasopressin (ADH) and aldosterone on mammalian kidneys. Insect ITP has a high degree of sequence identity with a crustacean family of hormones with hyperglycaemic (CHH), moult-inhibiting (MIH) and vitellogenesis-inhibiting (VIH) activities, and is the first member reported in insects. It is also the first member of this peptide family to be expressed and synthesised.

Future directions

It remains to be established whether ITP is naturally released into the haemolymph of locusts in response to a stimulus such as feeding or dehydration. Although the NCC is thought to be the site of ITP release, there has been no direct experimental demonstration of this. Indeed, attempts to cause release of the ileal stimulant with high K^+ in the presence of Ca^{2+} have been unsuccessful to date; therefore ITP neurohaemal sites could be elsewhere. It is anticipated that ITP is released after feeding to act in concert with diuretic hormones that stimulate Malpighian tubules. The resulting rapid recycling of fluid through the excretory system could serve to remove toxic waste products ingested with the meal or produced by metabolism of the meal (see Coast, this volume). The recent production of specific antibodies for ITP and ITP-L should help to address these issues.

All of the expression and assay methods are now in place to undertake structure–function studies on ITP using site-directed mutagenesis. Identification of specific ITP residues that interact with the receptor should be aided by the sequencing of ITP-like proteins from other insects that have no activity in the locust ileal I_{sc} bioassay. The results to date suggest there is a family of ITP-related insect neuropeptides that have yet to be isolated and sequenced. The availability of synthetic ITP should make possible the isolation of the ITP receptor cDNA by expression cloning and binding of a radioactive ligand.

Lastly, the use of baculovirus containing the ITP message to enhance the rate of kill of specific insects by causing dehydration is promising. The specificity both of baculoviruses and of ITP for different insect groups may offer the possibility of selective and environmentally safe methods for insect pest control.

Acknowledgements

This work was supported by research grants from NSERC Canada to J. E. Phillips.

References

Audsley, N. (1991). Purification of a neuropeptide from the corpus cardiacum of the desert locust which influences ileal transport. Ph.D. thesis. University of British Columbia, Vancouver.

Audsley, N., McIntosh, C. & Phillips, J. E. (1992a). Isolation of a neuropeptide from locust corpus cardiacum which influences ileal transport. *Journal of Experimental Biology* **173**, 261–274.

Audsley, N., McIntosh, C. & Phillips, J. E. (1992b). Actions of ion-transport peptide from locust corpus cardiacum on several hindgut transport processes. *Journal of Experimental Biology* **173**, 275–288.

Audsley, N., McIntosh, C., Phillips, J. E., Schooley, D. A. & Coast, G. M. (1994). Neuropeptide regulation of ion and fluid reabsorption in the insect excretory system. In *Perspectives of Comparative Endocrinology*, ed. K. G. Davey, R. E. Peter & S. S. Tobe, pp. 74–80. Ottawa: National Research Council of Canada.

Audsley, N. & Phillips, J. E. (1990). Stimulants of ileal salt transport in neuroendocrine system of the desert locust. *General and Comparative Endocrinology* **80**, 127–137.

Bilgen, T. (1994). Investigation of an ion transport peptide in desert locust ventral ganglia. M.Sc. thesis, University of British Columbia.

Black, K. T., Meredith, J., Thomson, B., Phillips, J. & Dietz, T. (1987). Mechanisms and properties of sodium transport in locust rectum. *Canadian Journal of Zoology* **65**, 3084–3092.

Chamberlin, M. E. & Phillips, J. E. (1983). Oxidative metabolism in the locust rectum. *Journal of Comparative Physiology* **151**, 191–198.

Chamberlin, M. E. & Phillips, J. E. (1988). Effects of stimulants of electrogenic ion transport on cyclic AMP and cGMP levels in locust rectum. *Journal of Experimental Zoology* **245**, 9–16.

Fournier, B. (1991). Neuroparsins stimulate inositol phosphate formation in locust rectal cells. *Comparative Biochemistry and Physiology* **99B**, 57–64.

Fournier, B. & Girardie, J. (1988). A new function for the locust neuroparsins: stimulation of water reabsorption. *Journal of Insect Physiology* **34**, 309–313.

Girardie, J., Girardie, A., Huet, J.-C. & Pernollet, J.-C. (1989). Amino acid sequence of locust neuroparsins. *FEBS Letters* **248**, 4–8.

Girardie, J., Huet, J.-C. & Pernollet, J.-C. (1990). The locust neuroparsin A: sequence and similarities with vertebrate and insect polypeptide hormones. *Insect Biochemistry* **20**, 659–666.

Goldbard, G. A., Sauer, J. R. & Mills, R. R. (1970). Hormonal control of excretion in the American cockroach. II. Preliminary purification of a diuretic and antidiuretic hormone. *Comparative and General Pharmacology* **1**, 82–86.

Hanrahan, J. W. & Phillips, J. E. (1984). KCl transport across an insect epithelium. II. Electrochemical potentials and electrophysiology. *Journal of Membrane Biology* **80**, 27–47.

Harrison, J. F. & Kennedy, M. J. (1994). *In vivo* studies of the acid–base physiology of grasshoppers: the effect of feeding state on acid–base and nitrogen excretion. *Physiological Zoology* **67**, 120–141.

Hietter, H., Van Dorsselaer, A. & Luu, B. (1991). Characterization of three structurally-related 8–9 kDa monomeric peptides present in the corpora cardiaca of *Locusta*: a revised structure for the neuroparsins. *Insect Biochemistry* **21**, 259–264.

Irvine, B., Audsley, N., Lechleitner, R., Meredith, J., Thomson, B. & Phillips, J. E. (1988). Transport properties of locust ileum *in vitro*: effects of cyclic AMP. *Journal of Experimental Biology* **137**, 361–385.

Jeffs, L. (1993). A pharmacological study of signal transduction mechanisms controlling fluid reabsorption and ion transport in the locust rectum. M.Sc. thesis. University of British Columbia.

Jeffs, L. B. & Phillips, J. E. (1996). A pharmacological study of the second messengers that control rectal ion and fluid transport in the desert locust

(*Schistocerca gregaria*). *Archives of Insect Biochemistry and Physiology* **31**, 169–184.

Kegel, G., Reichwein, B., Weese, S., Gaus, G., Peter-Katalinic, J. & Keller, R. (1989). Amino acid sequence of the crustacean hyperglycemic hormone (CCH) from the shore crab, *Carcinus maenas. FEBS Letters* **255**, 10–14.

Lechleitner, R. A., Audsley, N. & Phillips, J. E. (1989a). Antidiuretic action of cyclic AMP, corpus cardiacum and ventral ganglia on fluid absorption across locust ileum *in vitro. Canadian Journal of Zoology* **67**, 2655–2661.

Lechleitner, R. A., Audsley, N. & Phillips, J. E. (1989b). Composition of fluid transported by locust ileum: influence of natural stimulants and luminal ion ratios. *Canadian Journal of Zoology* **67**, 2662–2668.

Lechleitner, R. A. & Phillips, J. E. (1989). Effects of corpus cardiacum, ventral ganglia, and proline on absorbate composition and fluid transport by locust hindgut. *Canadian Journal of Zoology* **67**, 2669–2675.

Meredith, J., Cheng, N., Littleford, D., Ring, M., Audsley, N., Brock, H. & Phillips, J. E. (1995). Amino acid sequence of locust ion transport peptide (ITP). *FASEB Journal* **9**, A355.

Meredith, J. & Phillips, J. E. (1988). Sodium-independent proline transport in the locust rectum. *Journal of Experimental Biology* **137**, 341–360.

Meredith, J., Ring, M., Macins, A., Marschall, J., Cheng, N. N., Theilmann, D., Brock, H. W. & Phillips, J. E. (1996). Locust ion transport peptide (ITP): primary structure, cDNA and expression in a baculovirus system. *Journal of Experimental Biology* **199**, 1053–1061.

Phillips, J. E. (1981). Comparative physiology of insect renal function. *American Journal of Physiology* **241**, R241–R257.

Phillips, J. E. (1983). Endocrine control of salt and water balance: excretion. In *Endocrinology of Insects*, ed. H. Laufer & R. Downer, pp. 411–425. New York: Alan R. Liss, Inc.

Phillips, J. E. & Audsley, N. (1995). Neuropeptide control of ion and fluid transport across locust hindgut. *American Zoologist* **35**, 503–514.

Phillips, J. E., Audsley, N., Lechleitner, R., Thomson, B., Meredith, J. & Chamberlain, M. (1988). Some major transport mechanisms of insect absorptive epithelia. *Comparative Biochemistry and Physiology* **90A**, 643–650.

Phillips, J. E., Mordue, W., Meredith, J. & Spring, J. (1980). Purification and characteristics of the chloride transport stimulating factor from locust corpora cardiaca: a new peptide. *Canadian Journal of Zoology* **58**, 1851–1860.

Phillips, J. E., Hanrahan, J., Chamberlin, M. & Thomson, B. (1986). Mechanisms and control of reabsorption in insect hindgut. *Advances in Insect Physiology* **19**, 330–422.

Phillips, J., Spring, J., Hanrahan, J., Mordue, W. & Meredith, J. (1982). Hormonal control of salt reabsorption by the excretory system: Isolation of a new protein. In *Neurosecretion: Molecules, Cells, Systems*, ed. D. S. Farner & K. Lederis, pp. 371–380. New York: Plenum Press.

Phillips, J. E., Thomson, R. B., Peach, J. L., Audsley, N. & Stagg, A. P. (1994). Mechanisms of acid–base transport and control in locust excretory system. *Physiological Zoology* **67**, 95–119.

Phillips, J. E., Wiens, C., Audsley, N., Jeffs, L., Bilgen, T. & Meredith, J. (1996). Nature and control of chloride transport in insect absorptive epithelia. *Journal of Experimental Zoology* **275**, 292–299.

Proux, B. H., Proux, J. P. & Phillips, J. E. (1984). Antidiuretic actions of corpus cardiacum (CTSH) on long-term fluid absorption across locust recta in vitro. *Journal of Experimental Biology* **113**, 409–421.

Proux, J., Proux, B. & Phillips, J. E. (1985). Source and distribution of factors in locust nervous system which stimulate rectal Cl transport. *Canadian Journal of Zoology* **63**, 37–41.

Richardson, N. (1993). Ion transport in the ileum of the desert locust, *Schistocerca gregaria* Förskal. M.Sc. thesis, University of British Columbia, Vancouver, Canada.

Spring, J. H. & Phillips, J. E. (1980a). Studies on locust rectum. I. Stimulants of electrogenic ion transport. *Journal of Experimental Biology* **86**, 211–223.

Spring, J. H. & Phillips, J. E. (1980b). Studies on locust rectum. II. Identification of specific ion transport processes regulated by corpora cardiaca and cyclic AMP. *Journal of Experimental Biology* **86**, 225–236.

Spring, J. H. & Phillips, J. E. (1980c). Studies on locust rectum. III. Stimulation of electrogenic chloride transport by hemolymph. *Canadian Journal of Zoology* **58**, 1933–1939.

Thomson, R. B. & Phillips, J. E. (1992). Electrogenic proton secretion in the hindgut of the desert locust. *Journal of Membrane Biology* **125**, 133–154.

Thomson, R. B., Thomson, J. M. & Phillips, J. E. (1988). NH_4^+ transport in acid-secreting insect epithelium. *American Journal of Physiology* **254**, R348–R356.

Part III

'Myotropic and myoinhibitory' arthropod neuropeptides: structures and functions?

HANNE DUVE, ALAN THORPE, ANDERS H.
JOHNSEN, JOSÉ-LUIS MAESTRO,
ALAN G. SCOTT and PETER D. EAST

The dipteran Leu-callatostatins: structural and functional diversity in an insect neuroendocrine peptide family

Introduction

The allatostatins are a family of neuroendocrine peptides, the first members of which were identified in the cockroach *Diploptera punctata* (Pratt *et al.*, 1989; Woodhead *et al.*, 1989) (see Weaver *et al.*, this volume). They were given the name allatostatins on account of their allatostatic activity in cockroaches, since *in vitro* they induce a reduction in the biosynthesis of juvenile hormone. They vary in length from six to 18 amino acid residues and most are characterised by the C-terminal pentapeptide sequence Tyr-Xaa-Phe-Gly-Leu-NH$_2$, where Xaa in the peptides so far known, is Ala, Asn, Asp, Gly or Ser (Stay, Tobe & Bendena, 1994).

The callatostatins are dipteran homologues of this family that were identified in the blowfly, *Calliphora vomitoria* (Duve *et al.*, 1993). They were designated callatostatins to indicate their structural similarity to the cockroach allatostatins and the species of origin. They have not been shown to have an allatostatic effect in flies, although in heterologous assays on cockroach tissues they are as effective as the endogenous cockroach peptides themselves (Duve *et al.*, 1993). The terms Leu- and Met-callatostatin were later used (Duve & Thorpe, 1994) to distinguish between the group of peptides ending C-terminally with Leu-NH$_2$ and another sub-group of the same family, not found in cockroaches, that end C-terminally with Met-NH$_2$ (Duve *et al.*, 1994b, 1995a).

This account provides details of advances in our knowledge of the dipteran Leu-callatostatins by: (a) describing the Leu-callatostatin prohormone genes of *C. vomitoria* and the Australian sheep blowfly, *Lucilia cuprina*; (b) analysing post-translational precursor processing by comparing the structures of the peptides isolated from *C. vomitoria* with the putative peptide structures deducible from the prohormone DNA sequence; (c) assessing possible peptide functions by tracing neuronal and endocrine pathways immunocytochemically using antisera raised against the endogenous peptides of *C. vomitoria*; and (d) examining the roles of the peptides as regulators of gut motility.

The Leu-callatostatin prohormone gene

Gene cloning

Genomic clones for the Leu-callatostatin precursor gene were isolated from two species of calliphorid blowflies, *C. vomitoria* and *L. cuprina*, and a cDNA clone was obtained from a head cDNA library for the latter species (East *et al.*, 1996). A genomic clone containing the *C. vomitoria* gene was isolated using an oligonucleotide probe based on the coding region of the previously isolated hexadecapeptide, Leu-callatostatin 1 (Asp-Pro-Leu-Asn-Glu-Glu-Arg-Arg-Ala-Asn-Arg-Tyr-Gly-Phe-Gly-Leu-NH$_2$). Initial attempts at library screening with a long oligonucleotide based on this peptide were unsuccessful, presumably due to the high degree of degeneracy required to cover codon usage at the multiple Leu and Arg residues. To circumvent this problem, short minimally degenerate sense and antisense primers were designed from peptide sequences at the N- and C-termini, respectively, of Leu-callatostatin 1. These oligonucleotides were used to prime a polymerase chain reaction (PCR) with *C. vomitoria* genomic DNA as template. Several independent amplicons of the expected size (44 bp) were cloned and sequenced, allowing the design of a long, unique sequence oligonucleotide which was radiolabelled and used to screen a *C. vomitoria* genomic DNA library. A single clone was identified that contained two extended open reading frames, each of which encoded Leu-callatostatin peptide sequences. This *C. vomitoria* genomic DNA fragment was used as a hybridisation probe to isolate the homologue from a *L. cuprina* genomic DNA library. Sequence analysis again revealed two extended open reading frames encoding Leu-callatostatin peptides. To confirm that these two open reading frames represented the complete Leu-callatostatin precursor polypeptide sequence, a cDNA clone was isolated from a *L. cuprina* cDNA library prepared from mRNA extracted from whole adult heads as an enriched source of neural tissue. Sequence analysis of this cDNA allowed the structure of the complete prohormone open reading frame and the intron–exon organisation of the gene to be deduced (East *et al.*, 1996).

Organisation of the gene and prohormone

The dipteran Leu-callatostatin gene consists of at least three exons, the first of which is non-coding (Fig. 1A). Although the genomic organisation of the gene outside of the coding region has not been determined in detail, the first and second exons are separated by a large intron (<2 kb). Exons two and three, which contain the putative Leu-callatostatin prohormone, are separated by a small intron of only 63 bp. The complete open reading frame is 537 bp in *L. cuprina*, encoding a deduced precursor of 179 amino acid residues.

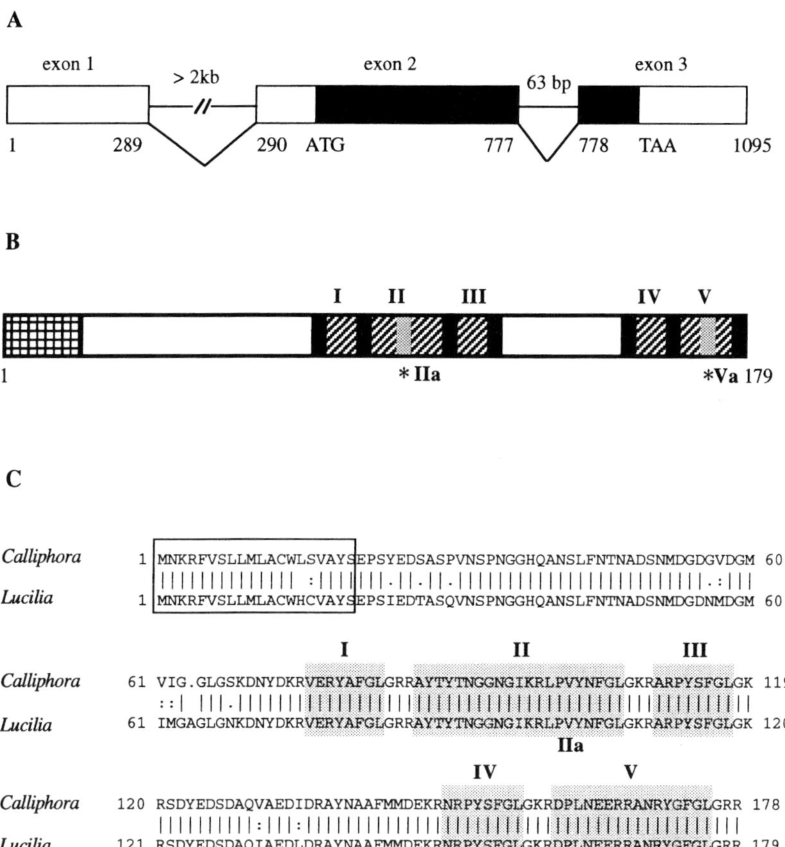

Fig. 1. Structure and organization of the dipteran Leu-callatostatin gene and prohormone. (A) Prohormone gene organization. Exons are numbered and shown as boxes and introns are indicated by horizontal lines. Protein coding regions are represented by filled boxes, and open boxes are the 5′ and 3′ transcribed but untranslated flanking sequences. The sites of translation initiation (ATG) and termination (TAA) are marked. (B) Structure of the deduced prohormone. The putative signal peptide is marked by cross-hatching. The regions encoding Leu-callatostatin peptides are shaded by diagonal lines and numbered with roman numerals. Dibasic proteolytic processing sites are indicated by vertical bars. Processing at the black bars would yield Peptides I, II, III, IV and V, and processing at the stippled bars marked with asterisks would generate Peptides IIa and Va. (C) Alignment of the Leu-callatostatin prohormone sequences of *Calliphora vomitoria* and *Lucilia cuprina*. The putative Leu-callatostatin peptide sequences are shaded and indicated by roman numerals. The putative signal peptide is boxed.

Table 1 *Leu-callatostatin (Leu-cast) putative peptide sequences deduced from the* C. vomitoria *prohormone gene compared with the peptides isolated from tissues extracts (single-letter code)*

	Precursor	Isolated peptides		
No.	Sequence	Sequence	Trivial name	M_r^a
I	VERYAFGL	VERYAFGL-NH$_2$	Leu-cast 7[b]	953.1
II	AYTYTNGGNGIKRLPVYNFGL			
IIa	LPVYNFGL	LPVYNFGL-NH$_2$	Leu-cast 8[b]	921.1
III	ARPYSFGL	ARPYSFGL-NH$_2$	Leu-cast 5[b]	909.0
IV	NRPYSFGL	NRPYSFGL-NH$_2$	Leu-cast 4[b,c]	952.1
		RPYSFGL-NH$_2$	Leu-cast 6[b]	838.0
V	DPLNEERRANRYGFGL	DPLNEERRANRYGFGL-NH$_2$	Leu-cast 1[c]	1906.1
		LNEERRANRYGFGL-NH$_2$	Leu-cast 2[c]	1693.9
Va	ANRYGFGL	ANRYGFGL-NH$_2$	Leu-cast 3[c]	896.0

Notes:

[a] M_r: Calculated relative molecular mass.
[b] Peptides described in Duve *et al.*, 1996.
[c] Peptides described in Duve *et al.*, 1993.

Both the 5' and 3' regions of the cDNA flanking this open reading frame have multiple in-frame translation stop codons. The homologous region of the *C. vomitoria* gene has a deduced open reading frame of 534 bp, encoding a precursor 178 amino acid residues in length. The putative pre-prohormone of both species begins with a short hydrophobic domain of 19 residues that has the properties expected of a secretion signal sequence according to the algorithm of von Heijne (1986). The precursor contains five copies of the sequence -Tyr-Xaa-Phe-Gly-Leu that is characteristic of the Leu-callato-statin peptide family. Each of these sequences is followed by a single Gly residue at the C-terminus that would provide a potential substrate for pep-tidyl-glycine α-amidating mono-oxygenase, thus allowing for carboxyamida-tion during processing. The putative Leu-callatostatin peptides are present in two small blocks separated by a short spacer region. The structure of the prohormone precursor is depicted in Fig. 1B.

The prohormone contains nine dibasic amino acid pairs that constitute potential endoproteolytic processing sites. If all of these sites were cleaved to completion during prohormone maturation, five different Leu-callatostatin octapeptides would be produced (Fig. 1B, Table 1, putative peptides I, IIa, III, IV, Va). Examination of the precursor sequence reveals that two addi-tional Leu-callatostatins, a heneicosapeptide (Peptide II) and a hexadeca-peptide (Peptide V) could be produced if processing at two of the dibasic sites (Fig. 1B) did not proceed to completion. Since one of these, the hexa-decapeptide, has been identified from *C. vomitoria* tissue extracts (Duve *et al.*, 1993), and the other is structurally similar to the octadecapeptide ASB2 of the cockroach *D. punctata* (Pratt *et al.*, 1991), the production of seven Leu-callatostatin peptides by differential endoproteolytic processing in *C. vomitoria* is a possibility. However, as described below, peptide isolation studies have demonstrated that the situation *in vivo* is more complex than this (Duve *et al.*, 1993, 1996).

Comparative aspects

Comparison of the deduced prohormones from the two blowflies *C. vomito-ria* and *L. cuprina* indicates a high degree of sequence conservation (Fig. 1C). The two proteins are 93% identical over 178 residues, and the sequences of the putative Leu-callatostatin peptides are the same in both species. The extent of the similarity between the two blowflies suggests that there has been some constraint on peptide sequence evolution, presumably imposed by a functional requirement to conserve the sequences of the biologically active peptides. This is in marked contrast to the calliFMRFamide family of pep-tides (FMRF=Phe-Met-Arg-Phe), where only five of the 13 different pep-tides encoded on the precursors are identical in these same two blowfly species (Duve *et al.*, 1994a). However, the absolute conservation of Leu-callatostatin peptide sequences does not extend to other insect orders. For example, gene cloning has established that cockroach species potentially express a much larger repertoire of peptides than do flies. Although the peptide precursors of the cockroaches *D. punctata* (Donly *et al.*, 1993) and *P. americana* (Stay *et al.*, 1994) encode 13 and 14 putative allatostatins respectively, only one peptide is identical between these and the two blowfly species. The prohormones of cockroaches and flies have the same basic organisation, with the biologically active peptides arranged in tandem blocks separated by acidic spacer regions (Donly *et al.*, 1993; East *et al.*, 1996), but there have obviously been substantial changes in peptide number between the insect orders during evolution. From the evidence currently available, it is not possible to say whether dipterans have lost peptides or cockroaches have gained them. The two blocks of Leu-callatostatin peptides in the blowflies are contained on separate exons and it is conceivable that acquisition or loss of peptides is achieved through the gain or loss of exons over time. In this context, it would be of considerable interest to obtain information on the genomic organisation of the cockroach allatostatin gene, and both genomic and cDNA sequences for the homologous gene from other insect orders.

Isolation and identification of the Leu-callatostatin peptides from *C. vomitoria*

Four Leu-callatostatin peptides (designated Leu-callatostatins 1–4, see Table 1) had been identified from extracts of *C. vomitoria* (Duve *et al.*, 1993) prior to the isolation of the Leu-callatostatin gene (East *et al.*, 1996). Of these, the hexadecapeptide Leu-callatostatin 1 was identified serendipitously during the purification from thoracic ganglia of a series of structurally unrelated peptides, the calliFMRFamides (Duve *et al.*, 1992). Recognition that this peptide had a C-terminal pentapeptide sequence closely similar to that of the cockroach allatostatins (Tyr-Gly-Phe-Gly-Leu-NH$_2$) initiated the production of antisera and the development of a Leu-callatostatin radio-immunoassay (RIA) based on the C-terminal octapeptide (Ala-Asn-Arg-Tyr-Gly-Phe-Gly-Leu-NH$_2$) sequence of Leu-callatostatin 1. This RIA was then used to monitor peptide purification from other extracts in studies that confirmed the identification of Leu-callatostatin 1, and led to the discovery of two N-terminally truncated forms of Leu-callatostatin 1, the tetra-decapeptide, Leu-callatostatin 2 and the octapeptide itself, designated Leu-callatostatin 3 (Table 1). In addition, an octapeptide with the sequence (Asn or Asp)-Arg-Pro-Tyr-(Ser)-Phe-Gly-Leu-NH$_2$ was also identified and designated Leu-callatostatin 4. Quantification of this peptide based on RIA data compared with that based on amino acid sequencing showed that it exhibited low cross-reactivity in the assay. The reason for this was assumed to be due primarily to the presence of Ser instead of Gly in the key, post-tyrosyl position of the molecule and the poor recognition of this in the RIA.

With the identification of the Leu-callatostatin gene, it became apparent that two of the peptides that had been identified in extracts, that is Leu-callatostatins 1 and 4, were present as putative peptides on exon 3. Leu-callatostatins 2 and 3 did not exist as separate entities on the gene, and must therefore have been present in extracts either as a result of precursor processing or by post-translational modification of Leu-callatostatin 1. In addition, examination of the gene revealed that exon 2 contained a minimum of three other putative peptides of the Leu-callatostatin type, in which the post-tyrosyl residues were Ala, Asn and Ser, respectively (Fig. 1C, Table 1). The actual number of peptides depends on whether or not the internal processing cleavage site within the potentially longer putative peptide (Fig. 1B, Peptide II) is used. It is of interest that these unidentified peptides are all present together on exon 2, whereas the ones previously isolated are on exon 3 (exon 1 being non-coding). There are several possible explanations to account for this including differential splicing, differential processing or, that an RIA based on -Tyr-Gly-Phe-Gly-Leu-NH$_2$ fails to recognise adequately those Leu-callatostatins in which the post-tyrosyl residue is Ala, Ser or Asn. In order to resolve some of these problems, to elucidate the way in which the precursor is processed in *C. vomitoria*, and to ascertain levels of expression of the putative peptides, an extract of 40 000 blowfly heads was prepared and peptides were purified by reversed-phase high-performance liquid chromatography (HPLC) and monitored with a series of newly developed Leu-callatostatin RIAs.

Antisera were generated against the three putative octapeptides deduced from the gene sequence that had not previously been isolated, i.e. Val-Glu-Arg-Tyr-Ala-Phe-Gly-Leu-NH$_2$, Ala-Arg-Pro-Tyr-Ser-Phe-Gly-Leu-NH$_2$ and Leu-Pro-Val-Tyr-Asn-Phe-Gly-Leu-NH$_2$ (Duve *et al.*, 1996). These antisera were then used to develop three separate RIAs, each of which proved to be highly specific for its respective peptide (less than 0.1% recognition at the 50 nmol 1^{-1} level for any other peptide family member with a different post-tyrosyl residue). The reversed-phase HPLC procedures used to isolate the peptides have been described (Duve *et al.*, 1996). From a practical point of view, the simple expedient of changing between acetonitrile gradients containing 0.1% trifluoroacetic acid (TFA) as a counter ion and identical gradients containing 10 mmol ammonium acetate 1^{-1} proved extremely useful in resolving closely related compounds. After a six- or seven-step purification procedure, four Leu-callatostatin peptides additional to the four previously identified (Duve *et al.*, 1993) were isolated and identified by mass spectrometry and amino acid sequencing (Duve *et al.*, 1996). Table 1 provides details of the complete family of Leu-callatostatin peptides of *C. vomitoria* resulting from the isolation studies, and compares them with the putative peptides deducible from the gene sequence.

Processing of the prohormone

Cleavage at internal basic sites of putative peptides

It can be seen that each of the putative Leu-callatostatin octapeptides is processed. Three of these (Val-Glu-Arg-Tyr-Ala-Phe-Gly-Leu-NH$_2$, Ala-Arg-Pro-Tyr-Ser-Phe-Gly-Leu-NH$_2$ and Asn-Arg-Pro-Tyr-Ser-Phe-Gly-Leu-NH$_2$) are encoded incontrovertibly as octapeptides on the gene sequence and their appearance in tissue extracts is predictable. Each is preceded by a Lys-Arg cleavage site and each is terminated by either Gly-Arg-Arg or Gly-Lys-Arg. The fact that the post-translational process of carboxyamidation, suggested by the presence of the Gly residue preceding the dibasic pair, actually occurs, has been confirmed for all isolated peptides by the process of methylation and re-measurement of the M_r (Talbo *et al.*, 1991). For these three peptides, examination of the prohormone precursor shows that there is no possibility of a longer, N-terminally extended form being generated. However, for the two other octapeptides present in extracts, (Leu-Pro-Val-Tyr-Asn-Phe-Gly-Leu-NH$_2$ and Ala-Asn-Arg-Tyr-Gly-Phe-Gly-Leu-NH$_2$), a dibasic cleavage site is present a short distance (11 and 6 residues, respectively) upstream from the one that gives rise to the octapeptides. Cleavage at these sites, and a lack of cleavage at the other internal site, provides the potential for generating 21 residue and 16 residue peptides, respectively. It is of interest that only one of these peptides occurs in extracts, namely the hexadecapeptide Asp-Pro-Leu-Asn-Glu-Glu-Arg-Arg-Ala-Asn-Arg-Tyr-Gly-Phe-Gly-Leu-NH$_2$ reported previously (Duve *et al.*, 1993). In contrast, there is no evidence of the heneicosapeptide Ala-Tyr-Thr-Tyr-Thr-Asn-Gly-Gly-Asn-Gly-Ile-Lys-Arg-Leu-Pro-Val-Tyr-Asn-Phe-Gly-Leu-NH$_2$ (Duve *et al.*, 1996). It is worth noting that the potential internal cleavage site in these two peptides is different (Arg-Arg compared with Lys-Arg), and it may be that the endoproteolytic cleavage enzymes in *C. vomitoria* have a preference for Lys-Arg rather than Arg-Arg sites. This phenomenon, which is well known from vertebrate studies (Steiner *et al.*, 1992), could explain the apparent complete cleavage of the heneicosapeptide resulting in 100% of the octapeptide Leu-Pro-Val-Tyr-Asn-Phe-Gly-Leu-NH$_2$, and the apparently partial cleavage of the hexadecapeptide resulting in only a small amount (<10%) of the octapeptide Ala-Asn-Arg-Tyr-Gly-Phe-Gly-Leu-NH$_2$. As will be seen in the section dealing with physiology, these events of post-translational processing appear to be extremely significant in determining potency. It is of interest that a peptide similar to the putative dipteran heneicosapeptide has been recovered from extracts of the cockroach *D. punctata* (Pratt *et al.*, 1991). Significantly, it has been shown that this octadecapeptide, Ala-Tyr-Ser-Tyr-Val-Ser-Glu-Tyr-Lys-Arg-Leu-Pro-Val-Tyr-Asn-Phe-Gly-Leu-NH$_2$, requires the integrity of the dibasic cleavage site Lys^9-Arg^{10} for high potency as an allatostatic peptide (Pratt *et al.*, 1991). In complete contrast to *C. vomitoria*, the octapeptide that would result from cleavage at this site has not been recovered from cockroach extracts, and when tested as a synthetic peptide, showed an allatostatic effect at a concentration two orders of magnitude lower than the naturally occurring, full-length peptide (Pratt *et al.*, 1991).

N-terminal trimming

Evidence of probable N-terminal dipeptide trimming of the hexadecapeptide Asp-Pro-Leu-Asn-Glu-Glu-Arg-Arg-Ala-Asn-Arg-Tyr-Gly-Phe-Gly-Leu-NH$_2$ was provided earlier by the appearance in extracts of a small amount (<10%) of the tetradecapeptide Leu-Asn-Glu-Glu-Arg-Arg-Ala-Asn-Arg-Tyr-Gly-Phe-Gly-Leu-NH$_2$ (Duve *et al.*, 1993). This was confirmed from the prohormone gene sequence, since only the hexadecapeptide is coded for (Fig. 1B,C). It appears that this truncated peptide must result from the activity of a dipeptidyl peptidase, an enzyme that cleaves peptides with Pro at position 2 from the N-terminus (Mentlein, 1988).

However, another unexpected form of single-residue N-terminal trimming was detected by the presence in the extract of the heptapeptide Arg-Pro-Tyr-Ser-Phe-Gly-Leu-NH$_2$. Study of the prohormone (Fig. 1B,C) reveals that this peptide must be derived by trimming of the Ala and/or Asn residues of the parent octapeptides Ala-Arg-Pro-Tyr-Ser-Phe-Gly-Leu-NH$_2$ or Asn-Arg-Pro-Tyr-Ser-Phe-Gly-Leu-NH$_2$. Since all three peptides were present in approximately the same amounts it seems likely that both octapeptides contributed equally to the production of the heptapeptide. This process of N-terminal trimming is possibly a mechanism for generating structurally diverse peptides of functional significance. If it reflected a general aminopeptidase degradative mechanism, all available octapeptides should have been affected in like manner, and this was not the case. Furthermore, degradation proceeding beyond the first residue, with resultant truncated hexa- or even pentapeptide Leu-callatostatins, would have been detected by the RIAs employed, and this also was not the case. A similar example of single-residue N-terminal trimming of invertebrate neuropeptides was demonstrated by Li *et al.* (1994) working with *Lymnaea stagnalis*. They showed that although the cDNA encoding the precursor of the light yellow cell peptides predicted a straightforward processing into three peptides at conventional dibasic sites, peptide isolation studies revealed N-terminally trimmed derivatives that were shorter by a single amino acid residue and much more abundant than their parents. The indication was that the full-length peptides represented intermediates in a peptide-processing sequence. The functional significance of the processes in both and in *C. vomitoria* and *L. stagnalis* remains to be established.

Tissue expression of the Leu-callatostatins

Immunocytochemical (ICC) studies with a range of Leu-callatostatin antisera (Duve & Thorpe, 1994; H. Duve & A. Thorpe, unpublished observations) and *in situ* hybridisation studies with a Leu-callatostatin gene probe (East, Thorpe & Duve, 1995) have provided comprehensive details of tissue expression of the Leu-callatostatin peptides and gene in *C. vomitoria*. *In situ* hybridisation has confirmed the results of ICC studies on a one-to-one basis. Selected examples are illustrated in Fig. 2. It has been possible therefore to map the neuronal and neuroendocrine Leu-callatostatin pathways with certainty. The antisera that provided a high degree of post-tyrosyl residue specificity when used in RIA, gave virtually identical patterns of immunoreactivity when used in the non-competitive technique of ICC. The reasons for this could be that all cells and axonal pathways contain the complete variety of Leu-callatostatin peptides, or it could be that minor populations of non-specific antibodies with minimal epitopes such as -Phe-Gly-Leu-NH$_2$ are present. The effect of these would be insignificant in RIA, but their presence could be sufficient to produce immunostaining in ICC. The specificity of the antisera used for ICC has been assessed by carrying out liquid-absorption control experiments. In cross-absorption experiments, other members of the peptide family were used in place of the homologous peptide. While the results of these experiments are not entirely conclusive, the immunostaining with some antisera is undoubtedly abolished when preabsorbed with some heterologous peptides. On the basis of these results, it is not possible to state that cross-reactivity does not occur and, therefore, reference to Leu-callatostatin immunoreactivity assumes a participation of all the peptides found in tissue extracts.

Leu-callatostatin neurones are found throughout the central nervous system, the peripheral nervous system and in gut endocrine cells. The brain and suboesophageal ganglion contain symmetrical pairs of neurones (exemplified in Fig. 2A,B). Some of these have extremely long and complex ipsi- and contralateral axonal trajectories with terminal arborisations in extensive regions of proto-, deutero- and tritocerebral neuropil. Some have axons that extend from the brain to the thoracico-abdominal ganglion (arrowed cells in

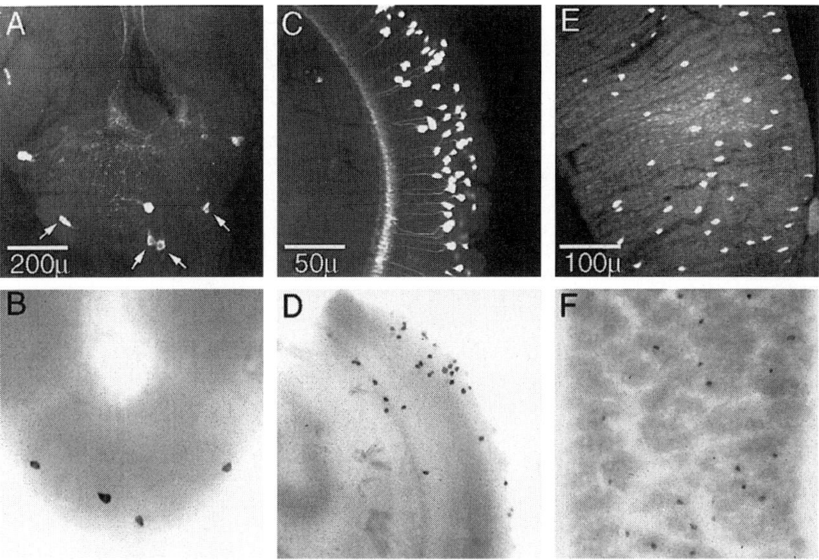

Fig. 2. Comparison of immunocytochemistry and *in situ* hybridization in whole-mounts of the brain and of the midgut of adult *C. vomitoria*. (A, C and E) Confocal laser scanning images of Leu-callatostatin immuno-reactive cells using the immunofluorescence technique and an antiserum raised against the octapeptide Leu-callatostatin 3. (B, D and F) *In situ* hybridization using digoxigenin-labelled Leu-callatostatin gene probe. (A) Throughfocus image of the suboesophageal ganglion region. Cells indicated with arrows are located on the posterior surface. The other perikarya are positioned at different levels. (B) *In situ* hybridization preparation viewed from the posterior surface of the suboesophageal ganglion showing the four cells arrowed in (A). These cells project axons to the thoracico-abdominal ganglion (Duve & Thorpe, 1994). (C) Confocal image of an optic lobe demonstrating numerous interneurones of the medulla. (D) Similar view to (C) showing *in situ* hybridization. (E) Confocal image showing Leu-callatostatin immunoreactivity in endocrine cells of the poste-rior part of the midgut. (F) *In situ* hybridization preparation showing expression of the Leu-callatostatin gene in the posterior part of the midgut. μ, μm.

Fig. 2A). The functions of the Leu-callatostatins of these brain and sub-oesophageal ganglion neurones are unknown, although they most likely have transmitter or neuromodulatory roles. There are also groups of inter-neurones situated at various positions within the brain and thoracico-abdominal ganglion. Interesting examples of this cell type are the visual interneurones present in the medulla of the optic lobe (Fig. 2C,D). Certain of

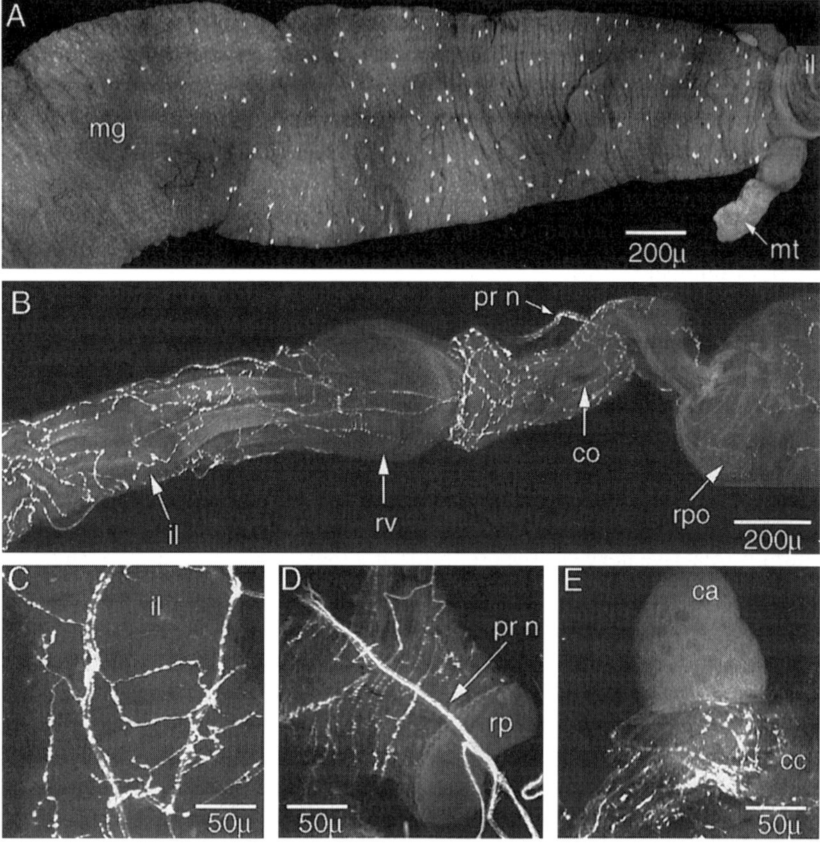

Fig. 3. (A) Montage of a series of digitised confocal images, assembled using Adobe Photoshop 3.0.1®, showing the posterior tenth of the midgut. Malphigian tubules arise at the junction of the midgut and the ileum. Presence of Leu-callatostatin peptides in the endocrine cells visualised by immunocytochemistry using a Leu-callatostatin-3 antiserum. (B, C and D). Confocal images of the hindgut of the adult blowfly showing Leu-callatostatin innervation. (B) Montage showing the pattern of innervation in the ileum and rectum. The proctodeal nerve branches out at the level of the colon. Some of the axons divide further over the colon surface, innervating the longitudinal and circular muscle fibres, in particular those close to the rectal valve. Axons cross the surface of the rectal valve, to innervate the longitudinal and circular muscles of the ileum. Other Leu-callatostatin immunoreactive fibres project posteriorly to innervate the surface muscles of the rectal pouch. (C) Detail of part of the ileum showing immunoreactive axons on the surface of longitudinal and circular muscle fibres. (D) Leu-callatostatin innervation of the rectal pouch. The proctodeal nerve

the median neurosecretory cells of the pars intercerebralis also express the Leu-callatostatin gene and, although their pathways have not been traced with certainty, it is probable that the Leu-callatostatin immunoreactivity of axons and varicosities present in the corpus cardiacum (Fig. 3E) originates in these cells. In contrast to the corpus cardiacum, the corpus allatum shows no Leu-callatostatin immunoreactivity (Fig. 3E), consistent with the lack of effect of these peptides as true allatostatins in flies (Duve *et al.*, 1993). In larvae too, there is no evidence of Leu-callatostatins in the ring gland (P. East, unpublished observations).

The thoracico-abdominal ganglion contains six prominent Leu-callatostatin neurones. They are located in the last abdominal ganglion and innervate the hindgut via axons that exit contralaterally in the median abdominal nerve. These fibres form the proctodeal nerve (Fig. 3B) which subsequently branches to provide an extensive innervation, posteriorly over the rectal pouch (Fig. 3D) and the distal rectum just before the anus, and anteriorly over the surface of the colon (also referred to as the proximal rectum), the rectal valve and the ileum (Fig. 3B, see also Fig. 4C). Leu-callatostatin immunoreactive nerve fibres are found in close association with both longitudinal and circular muscle fibres throughout the hindgut (Fig. 3B–D; see also East *et al.*, 1995), strongly suggesting that one function of this family of peptides is in the control of hindgut motility. The foregut and almost the entire length of the midgut of *C. vomitoria* contain no sign of Leu-callatostatin immunoreactive nerve fibres. In contrast, in the cockroach *Leucophaea maderae*, Leu-callatostatin immunoreactive nerve fibres are distributed extensively throughout the length of the gut (Duve, Wren & Thorpe, 1995b). Neurosecretory cells of the peripheral nervous system are present at intervals along the length of certain nerve and muscle fibres, notably on the surface of the major nerve trunks to the legs and in association with integumental muscle (Duve & Thorpe, 1994; East *et al.*, 1995). The functions of the Leu-callatostatins at these sites are unknown. In addition to the neuronal distribution of Leu-callatostatin, endocrine cells of the midgut express the Leu-callatostatin gene (Fig. 2E,F). These cells have a

---

Fig. 3. (*cont.*)

(arrowed) is shown passing over the surface of a rectal papilla. (E) Confocal image of whole-mount of the retrocerebral complex showing Leu-callatostatin immunoreactivity in varicosities and axons within the cardiac-recurrent nerve bundle in the corpus cardiacum. Note the absence of immunoreactivity in the corpus allatum. Abbreviations: ca, corpus allatum; cc, corpus cardiacum; co, colon; il, ileum; mg, midgut; mt, Malpighian tubules; pr n, proctodeal nerve; rp, rectal papilla; rpo, rectal pouch, rv, rectal valve; μ, μm.

restricted distribution, occurring only in the last tenth of this gut segment, immediately anterior to the evagination of the Malpighian tubules (Fig. 3A).

Leu-callatostatins as regulators of hindgut motility

The anatomy of the hindgut of *Calliphora* has been described in detail by Graham-Smith (1934) and the main features are illustrated in Fig. 4C. The patterns of muscle contraction of the hindgut are complex and there are distinct differences between the ileal and rectal regions. It is common to see peristaltic waves of contraction in newly dissected, meat-fed flies passing down the ileum (approximately 10 min^{-1}) and ceasing in the neighbourhood of the rectal valve. The rectal region has a more complex structure than the ileum, and consists of three anatomically distinct segments, the colon (proximal rectum), the rectal pouch and the distal rectum leading to the anus. Contractions of the proximal and distal rectal segments are primarily peristaltic, whereas movements of the rectal pouch originate from contractions of circular, longitudinal and oblique muscle fibres that result in a complex writhing action.

The *in vitro* preparation used to test the peptides has been described in detail (Duve & Thorpe, 1994), and allows for simultaneous observation of the ileum and rectum as well as the midgut, foregut, heart and oviducts. The entire preparation is superfused with *Calliphora* Ringer's solution plus peptide, alternating with Ringer's solution alone. The effects of peptides at different concentrations are observed for 2 min intervals, interrupted with washes, over periods of 2–3 h for a single preparation. Thus far, four members of the Leu-callatostatin peptide family have been synthesised and tested for their activity on the hindgut. These represent three octapeptides varying at the post-tyrosyl residue (Asn, Gly, Ser) and a fourth representing an N-terminally extended form of one of these. The peptides tested, Leu-callatostatins 1, 3, 5 and 8 (Table 1), had an effect only on the hindgut, consistent with the intense innervation of this region described above.

Myoinhibition of the ileum

All four peptides inhibited spontaneous ileal contraction, but differed both qualitatively and quantitatively in their effects. Leu-callatostatins 1 and 3 gave monophasic responses, with an IC_{50} value <1 fmol l^{-1} for the octapeptide, compared with approximately 10 nmol l^{-1} for the hexadecapeptide (Fig. 4A). The effects of both these peptides were immediate and reversible. The effects of Leu-callatostatins 5 and 8 were quite different. Thus, although inhibitory effects on this region could be observed, they only occurred with

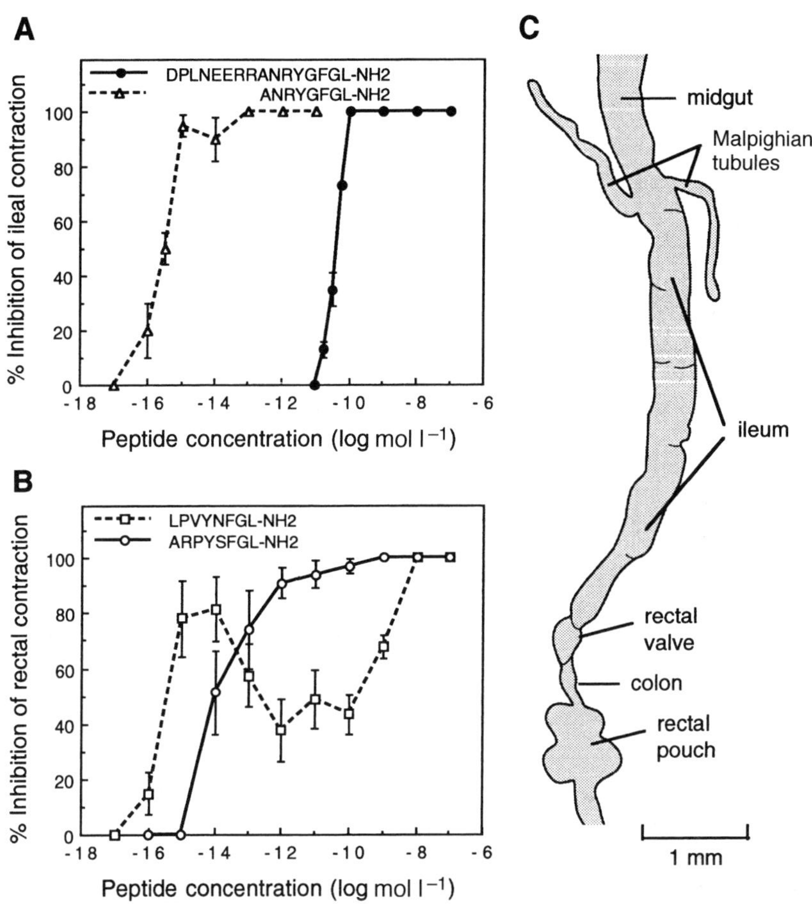

Fig. 4. Effect of Leu-callatostatin peptides on the hindgut of *C. vomitoria*. (A) Dose–response curves for inhibition of peristalsis of the ileum of the adult female blowfly by Leu-callatostatins 1 and 3. (B) Dose–response curves for inhibition of contraction of the rectum by Leu-callatostatins 5 and 8. Percentage inhibition is given as peptide-exposed guts relative to Ringer's solution-exposed guts. Each point for Leu-callatostatin treatment is the mean of four to six experiments, the vertical error bars show standard error of the mean. (C) Drawing of the posterior part of the midgut and hindgut of an adult blowfly illustrating the main features referred to in the immunocytochemical and physiological studies. Single-letter code used for amino acids.

higher peptide concentrations of 10 nmol 1^{-1} and above and the normal pattern of peristalsis contraction was not restored with washing.

Myoinhibition of the rectum

Leu-callatostatins 1 and 3 had no detectable effects on rectal contractions when tested over the same range of concentrations (1 μmol 1^{-1} to 1 fmol 1^{-1}) that gave such potent effects on the ileum. In contrast, both Leu-callatostatins 5 and 8 showed a potent, reversible inhibitory effect on the rectum (Fig. 4B). The octapeptide Ala-Arg-Pro-Tyr-Ser-Phe-Gly-Leu-NH$_2$ (Leu-callatostatin 5) had a monophasic effect, with an IC$_{50}$ value of 10 fmol 1^{-1}, and Leu-Pro-Val-Tyr-Asn-Phe-Gly-Leu-NH$_2$ (Leu-callatostatin 8) had an even more potent inhibitory effect, showing a biphasic dose–response curve with IC$_{50}$ values of approximately 1 fmol 1^{-1} and 1 nmol 1^{-1}.

These results are significant in showing that different members of a single peptide family have potent, immediate and reversible myoinhibitory effects on different regions of the gut.

Concluding comments

The Leu-callatostatins/allatostatins provide an intriguing case study of a large and complex family of insect neuroendocrine peptides. They represent the first example of brain–gut peptides to be identified in insects. The combination of immunocytochemistry and *in situ* hybridisation has established unequivocally that the Leu-callatostatins are expressed in the central nervous system, in efferent nerves extensively associated with the hindgut, in peripheral neurosecretory cells, and in intrinsic endocrine cells of the midgut of the flies *C. vomitoria* and *L. cuprina*. Similarly, in the cockroach *D. punctata*, immunocytochemistry and *in situ* hybridisation have shown expression of the allatostatin gene and peptides in brain (Stay, Chan & Woodhead, 1992; Donly *et al.*, 1993) and midgut endocrine cells (Reichwald *et al.*, 1994). Immunocytochemical studies on representatives of several other orders including Lepidoptera and Hymenoptera (H. Duve, A. Thorpe & P. East, unpublished observations) suggest that this brain–gut association is likely to be phylogenetically widespread in insects.

The Leu-callatostatins have provided a rare opportunity to fully characterise an insect peptide family at both the genetic and biochemical levels, and the results demonstrate the value of adopting a multi-disciplinary approach. For the first time, it has been possible to show that all the putative peptides encoded on a prohormone are actually expressed in the insect. More significantly, it has established that diversity of peptide structures can be achieved in insects by post-translational processes including differential proteolytic processing and

N-terminal trimming (Duve *et al.*, 1996). A major challenge for the future will be to ascertain whether the complete family of peptides seen at the level of the whole animal is manifest also at the level of individual cell types, or whether there is some degree of differential expression. At a gross level, it has been shown that cells of the midgut and the brain express cDNAs capable of encoding the full complement of Leu-callatostatin peptides (East *et al.*, 1996). However, differential processing or vesicle trafficking, as possible mechanisms for generating diversity in peptide expression at the level of individual cell types, are presently uninvestigated.

Our knowledge of Leu-callatostatin function presently lags well behind our understanding of the genetics and biochemistry of these peptides. Nevertheless, the little information that is available points to an impressive functional diversity. The immunocytochemical data clearly implicate members of this family in functions as neurotransmitters or neuromodulators within the central nervous system. Their expression in midgut endocrine cells and neuro-haemal organs such as the corpus cardiacum is indicative of hormonal functions, and as described above, these peptides are clearly involved in complex regulation of gut motility. The Leu-callatostatins provide the first example in insects of functional integration of a single organ, the hindgut, achieved through differential effects of individual members of a multi-peptide family. It seems likely that other activities on the gut will be identified when the function of the Leu-callatostatin endocrine cells of the posterior midgut is understood.

Acknowledgements

We thank Allan Kastrup and Karen Tregenza for valuable technical assistance. This work was supported by the Agricultural and Food Research Council of Great Britain (AG68/022), the Leverhulme Trust (F/476/Q), the Royal Society, the British Council and the Central Research fund of the University of London (A.T.), the Danish Biotechnology Program for Signal Peptides, the Danish Medical Research Council, and the Lundbeck, Velux, John and Birthe Meyer, Brdr. Hartman and Novo Nordisk Foundations (A.H.J.) and the Australian Wool Research and Development Corporation (CEC017) (P.D.E.). J.L.M. is in receipt of a post-doctoral research grant from CIRIT, Generalitat de Catalunya, Spain.

References

Donly, B. C., Ding, Q., Tobe, S. S. & Bendena, W. G. (1993). Molecular cloning of the gene for the allatostatin family of neuropeptides from the cockroach *Diploptera punctata. Proceedings of the National Academy of Sciences USA* **90**, 8807–8811.

Duve, H., Johnsen, A. H., Scott, A. G., Yu, C. G., Yagi, K. J., Tobe, S. S. & Thorpe, A. (1993). Callatostatins: neuropeptides from the blowfly *Calliphora vomitoria* with sequence homology to cockroach allatostatins. *Proceedings of the National Academy of Sciences USA* **90**, 2456–2460.

Duve, H., Johnsen, A. H., Sewell, J. C., Scott, A. G., Orchard, I., Rehfeld, J. F. & Thorpe, A. (1992). Isolation, structure, and activity of -Phe-Met-Arg-Phe-NH$_2$ neuropeptides (designated calliFMRFamides) from the blowfly *Calliphora vomitoria. Proceedings of the National Academy of Sciences USA* **89**, 2326–2330.

Duve, H., Johnsen, A. H., East, P. & Thorpe, A. (1994a). Comparative aspects of the FMRFamides of blowflies: isolation of the peptides, genes, and functions. In *Perspectives in Comparative Endocrinology*, ed. K. G. Davey, R. E. Peter & S. S. Tobe, pp. 91–96. Ottawa: National Research Council of Canada.

Duve, H., Johnsen, A. H., Maestro, J.-L., Scott, A. G., East, P. D. & Thorpe, A. (1996). Identification of the dipteran Leu-callatostatin peptide family: the pattern of precursor processing revealed by isolation studies in *Calliphora vomitoria. Regulatory Peptides* **67**, 11–19.

Duve, H., Johnsen, A. H., Scott, A. G., East, P. & Thorpe, A. (1994b). [Hyp^3]Met-callatostatin: identification and biological properties of a novel neuropeptide from the blowfly *Calliphora vomitoria. Journal of Biological Chemistry* **269**, 21 059–21 066.

Duve, H., Johnsen, A. H., Scott, A. G. & Thorpe, A. (1995a). Isolation, identification and functional significance of [Hyp^2]Met-callatostatin and desGly-Pro Met-callatostatin, two further post-translational modifications of the blowfly neuropeptide Met-callatostatin. *Regulatory Peptides* **57**, 237–245.

Duve, H. & Thorpe, A. (1994). Distribution and functional significance of Leu-callatostatins in the blowfly *Calliphora vomitoria. Cell and Tissue Research* **276**, 367–379.

Duve, H., Wren, P. & Thorpe, A. (1995b). Innervation of the foregut of the cockroach *Leucophaea maderae* and inhibition of spontaneous contractile activity by callatostatin neuropeptides. *Physiological Entomology* **20**, 33–44.

East, P. D., Thorpe, A. & Duve, H. (1995). Leu-callatostatin gene expression in the blowflies *Caliphora vomitoria* and *Lucilia cuprina* studied by in situ hybridisation: comparison with Leu-callatostatin confocal laser scanning immunocytochemistry. *Cell and Tissue Research* **280**, 355–364.

East, P. D., Tregenza, K., Duve, H. & Thorpe, A. (1996). Identification of the dipteran Leu-callatostatin peptide family: characterisation of the prohormone gene from *Calliphora vomitoria* and *Lucilia cuprina. Regulatory Peptides* **67**, 1–9.

Graham-Smith, G. S. (1934). The alimentary canal of *Calliphora erythrocephala* L., with special reference to its musculature and to the proventriculus, rectal valve and rectal papillae. *Parasitology* **26**, 176–248.

Li, K. W., Hoek, R. M., Smith, F., Jiménez, C. R., van der Schors, R. C., van Veelen, P. A., Chen, S., van der Greef, J., Parish, D. C., Benjamin, P. R. & Geraerts, W. P. M. (1994). Direct peptide profiling by mass spectrometry of single identified neurons reveals complex neuropeptide-processing pattern. *Journal of Biological Chemistry* **269**, 30288–30292.

Mentlein, R. (1988). Proline residues in the maturation and degradation of peptide hormones and neuropeptides. *FEBS Letters* **234**, 251–256.

Pratt, G. E., Farnsworth, D. E., Fok, K. F., Siegel, N. R., McCormack, A. L., Shabanowitz, J., Hunt, D. F. & Feyereisen, R. (1991). Identity of a second type of allatostatin from cockroach brains: an octadecapeptide amide with a tyrosine-rich address sequence. *Proceedings of the National Academy of Sciences USA* **88**, 2412–2416.

Pratt, G. E., Farnsworth, D. E., Siegel, N. R., Fok, K. F. & Feyereisen, R. (1989). Identification of an allatostatin from adult *Diploptera punctata*. *Biochemical and Biophysical Research Communications* **163**, 1243–1247.

Reichwald, K., Unnithan, G. C., Davis, N. T., Agricola, H. & Feyereisen, R. (1994). Expression of the allatostatin gene in endocrine cells of the cockroach midgut. *Proceedings of the National Academy of Sciences USA* **91**, 11894–11898.

Stay, B., Chan, K. K. & Woodhead, A. P. (1992). Allatostatin-immunoreactive neurons projecting to the corpora allata of adult *Diploptera punctata*. *Cell and Tissue Research* **270**, 15–23.

Stay, B., Tobe, S. S., & Bendena, W. G. (1994). Allatostatins: identification, primary structures, functions and distribution. *Advances in Insect Physiology* **25**, 269–337.

Steiner, D. F., Smeekens, S. P., Ohagi, S. & Chan, S. J. (1992). The new enzymology of precursor processing endoproteases. *Journal of Biological Chemistry* **267**, 23435–23438.

Talbo, G., Højrup, P., Rahbek-Neilsen, H., Andersen, S. O. & Roepstorff, P. (1991). Determination of the covalent structure of an N- and C-terminally blocked glycoprotein from endocuticle of *Locusta migratoria*. Combined use of plasma desorption mass spectrometry and Edman degradation to study post-translationally modified proteins. *European Journal of Biochemistry* **195**, 495–504.

von Heijne, G. (1986). A new method for predicting signal cleavage sites. *Nucleic Acids Research* **14**, 4683–4690.

Woodhead, A. P., Stay, B., Seidel, S. L., Khan, M. A. & Tobe, S. S. (1989). Primary structure of four allatostatins: neuropeptide inhibitors of juvenile hormone synthesis. *Proceedings of the National Academy of Sciences USA* **86**, 5997–6001.

DICK R. NÄSSEL, C. TOMAS LUNDQUIST,
J. ERIC MUREN and ÅSA M. E. WINTHER

An insect peptide family in search of functions: the tachykinin-related peptides

Introduction

One of the very successful bioassays used for monitoring high-performance liquid chromatography (HPLC) fractions during insect peptide purification is the sensitive cockroach hindgut contraction assay (Holman *et al.*, 1991; Schoofs, Vanden Broek & De Loof, 1993). This assay, which utilises the hindgut of the cockroach *Leucophaea maderae*, has been instrumental in the isolation of a large number of myostimulatory or myoinhibitory peptides both from this and from other insect species. However, one major problem that has accompanied the successful isolation of myotropic peptides is the large number of chemically characterised insect neuropeptides for which the physiological functions are unknown. The determination of physiological roles is acute also for peptides isolated in other *in vitro* assays. This problem is further complicated by the possibility that each neuropeptide may have a range of physiological functions, as indicated by the demonstration that certain insect neuropeptides may have multiple actions *in vitro* (Goldsworthy *et al.*, 1992; Schoofs *et al.*, 1993; Nässel, 1994; Lange, Bendena & Tobe, 1995). For instance, the leucokinin-related peptides of *L. maderae* have been shown *in vitro* to have myotropic and diuretic actions (Coast, Holman & Nachman, 1989; Holman, Nachman & Wright, 1990; Coast, 1995), and to act as stimulators of lipid release and inhibitors of protein synthesis in fat body (Goldsworthy *et al.*, 1992). Whereas all these actions may be mediated by peptides released into the circulation, there are indications from immunocytochemistry that leucokinins are present in interneurones and may therefore have modulatory roles in the central and stomatogastric nervous systems the nature of which have yet to be revealed (Nässel, Cantera & Karlsson, 1992a; Muren, Lundquist & Nässel, 1993). Another example is furnished by *Locusta* diuretic hormone (Lom-DH), isolated as a diuretic factor and known to act as a neurohormone to induce secretion in Malpighian tubules. This peptide has been shown to be present not only in neurosecretory cells and their release sites, but also in neurones in different parts of the nervous system, indicating further functional roles (Patel *et al.*,

1994; Patel, Hayes & Coast, 1995). These are only two examples of a dilemma that probably encompasses most of the known insect neuropeptides: little is known about their diverse physiological roles.

A further complication in revealing neuropeptide functions is that many neuropeptides are known to exist in multiple related forms in a given insect species. For example, there are 13 allatostatins in the cockroach *Diploptera punctata* (Donly *et al.*, 1993), 8 leucokinins in the cockroach *Leucophaea maderae* (Holman *et al.*, 1990) and 13 extended FMRFamides (FMRF=Phe-Met-Arg-Phe) in the blowfly *Calliphora vomitoria* (Duve *et al.*, 1992). Do these isoforms represent functionally distinct peptides with different distributions and different affinities for receptors? Why do multiple isoforms exist for some peptide families and apparently not for others, such as proctolin and crustacean cardioactive peptide (Orchard, Belanger & Lange, 1989; Dircksen, 1994)? The situation is the reverse to that faced by classical endocrinologists (Lafont, 1991): factors (in this case neuropeptides) have been identified but not their functions.

How does one search for the functions of a recently identified neuropeptide, with the only clue being that it stimulates contractions of visceral muscle (possibly by cross-reactivity on a population of promiscuous receptors)? Before screening for further actions, it is helpful first to localise the sites of synthesis of the neuropeptide in question, and then to determine its putative release sites. The most common method to pursue cellular localisation is by immunocytochemistry using an antiserum raised to the neuropeptide in question, or a specific part thereof. In this chapter, the search for sites of actions and physiological functions of neuropeptides is exemplified with work on a family of insect peptides designated tachykinin-related peptides, TRPs (Schoofs *et al.*, 1990a,b, 1993). The insect TRPs can be used to highlight several interesting aspects of neuropeptides, such as expression in a variety of neurone types and tissues, and the existence of multiple isoforms within a species. They can also be used to illustrate the frustrating lack of knowledge about physiological function(s). This review describes work on the chemical characterisation of insect TRPs, and studies of their cellular distributions and *in vitro* actions. Special emphasis is given to studies on locust, cockroach and blowfly tissues.

Tachykinin-related peptides (TRPs) of invertebrates

True tachykinins have primarily been studied in vertebrates, and a brief outline of this peptide family is given before turning to the TRPs of invertebrates. Tachykinins constitute a large and widespread family of neuropeptides with a vast range of actions in the nervous system and other tissues of vertebrates (Erspamer, 1981; Pernow, 1983; Maggio, 1988; Helke *et al.*, 1990;

Otsuka & Yoshioka, 1993). The name of the family was given by Erspamer & Melchiorri (1973) as a description of their relatively fast contractile action on smooth muscle. The most well-known tachykinin, substance P, had already been isolated in fairly pure form from horse brain and intestine in 1931 (von Euler & Gaddum, 1931), and was used extensively for experimental purposes (see Pernow, 1983; Helke *et al.*, 1990; Otsuka & Yoshioka, 1993). However, it was not until 40 years later that the sequence of substance P was determined (Chang, Leeman & Niall, 1971). In fact, the first tachykinin to be isolated in pure form and sequenced was derived from the salivary glands of an invertebrate, the cephalopod *Eledone moschata*, and designated eledoisin (Erspamer & Anastasi, 1962). Since then many more members of the tachykinin family have been isolated from a variety of vertebrates (see Erspamer, 1981; Maggio, 1988; Helke *et al.*, 1990). The true tachykinins are characterised by the C-terminal pentapeptide Phe-Xaa-Gly-Leu-Met-NH$_2$ (where Xaa is variable), and the sequence of substance P is Arg-Pro-Lys-Pro-Gln-Gln-Phe-Phe-Gly-Leu-Met-NH$_2$.

Apart from eledoisin, no invertebrate tachykinins or TRPs were known before 1990. Antisera to substance P had been used for immunocytochemical investigations of insects to reveal a small number of immunoreactive neurones in the central nervous system (CNS), but the chemical structure of the native peptide(s) was not known (Verhaert & De Loof, 1985; Lundquist & Nässel, 1990; Nässel *et al.*, 1990). In 1990, the first reports on the isolation and sequencing of tachykinin-related peptides in insects were published (Schoofs *et al.*, 1990a,b). These peptides, which were isolated from dissected brains and corpora cardiaca–corpora allata of the locust *Locusta migratoria*, did not have the complete C-terminal pentapeptide characteristic of the vertebrate tachykinins and eledoisin, but had identical amino acids in a few strategic positions (about 40% sequence identity). Thus it was suggested that these so called locustatachykinins I–IV (LomTK I–IV) are ancestrally related to the vertebrate type tachykinins, but should be placed in a separate subfamily characterised by the C-terminal pentapeptide Phe-Xaa-Gly-Val-Arg-NH$_2$ where Xaa can be His or Tyr (Nachman *et al.*, 1990; Schoofs *et al.*, 1990a,b, 1993). However, invertebrate TRPs have a slightly more variable C-terminal pentapeptide, Phe-Xaa$_1$-Gly-Xaa$_2$-Arg-NH$_2$ amide (where Xaa$_1$ and Xaa$_2$ are variable).

The first insect TRPs were isolated with the aid of the cockroach hindgut assay. Additional to LomTK I–IV, there are two known forms of TRPs in the mosquito *Culex salinarius*, CusTK I and II (see Clottens *et al.*, 1993), and two in the blowfly *Calliphora vomitoria*, CavTK I and II (Lundquist *et al.*, 1994a). More recently, five TRPs were isolated from the midgut of *L. maderae*, using a combination of the hindgut contraction assay and a radioimmunoassay for LomTK-like peptides (Muren & Nässel, 1996b). The cock-

Table 1   *Amino acid sequences (single-letter code) of invertebrate tachykinin-related peptides*

Isolated from insects[a]	
LemTRP 1	APSGFLGVRa
LemTRP 2	APEESP**KR**APSGFLGVRa
LemTRP 3	NGERAPGS**KK**APSGFLGTRa
LemTRP 4*	APSGFMGMRa
LemTRP 5	APAMGFQGVRa
LomTK I	GPSGFYGVRa
LomTK II	APLSGFYGVRa
LomTK III	APQAGFYGVRa
LomTK IV	APSLGFHGVRa
LomTK V	XPSWFYGVRa
CavTK I	APTAFYGVRa
CavTK II	GLGNNAFVGVRa
CTK I*	APSGFMGMRa
CTK II	APYGFTGMRa
Isolated from other invertebrates	
UruTK I	LAQSQFVGSRa
UruTK II	AAGMGFFGARa
AncTK	pQYGFHAVRa

[a] LemTRP 6–10 were isolated from *L. maderae* brain (Muren & Nässel, 1997). Underlined amino acid residues are shared by LemTRP 1 and other tachykinin-related peptides. Amino acids shown in bold indicate putative endoproteolytic cleavage sites. LemTRPs (*Leucophaea* tachykinin-related peptides, cockroach) from Muren & Nässel (1996b), LomTKs (locustatachykinins, locust) from Schoofs *et al.* (1990a,b, 1993), CTKs (cultetachykinins; mosquito) from Clottens *et al.* (1993), CavTKs (callitachykinins, blowfly) from Lundquist *et al.* (1994b), UruTKs (urechistachykinins; echiuroid worm) from Ikeda *et al.* (1993), AncTK (anodontatachykinin, bivalve mollusc) from Fujisawa *et al.* (1994). Peptides marked with asterisk are identical.

roach peptides, designated *L. maderae* tachykinin-related peptides 1–5 (LemTRP 1–5), displayed somewhat larger variabilities in their amino acid sequences (including the C-terminus) than the previously isolated insect TRPs (Table 1). It was also found that two of the LemTRPs are extended forms, with putative endoproteolytic cleavage sites indicated by the presence of the adjacent basic residues Lys-Arg and Lys-Lys, respectively (Andrews,

Brayton & Dixon, 1987). Eight LemTRPs have now also been isolated from *L. maderae* brain (Muren & Nässel, 1997). Three of these are identical to LemTRP 1, 2 and 5, the others (LemTRP 6–10) are novel forms. The possible tissue–specific expression of TRPs is dealt with in a later section.

A few other invertebrate TRPs have been isolated: urechistachykinins I and II (UruTK I and II) from the echiuroid worm *Urechis unicinctus* (Ikeda *et al.*, 1993), and anodontatachykinin (AncTK) from the bivalve mollusc *Anodonta cygnea* (Fujisawa *et al.*, 1994). These display minor similarities to the insect TRPs. The presence of TRPs in several groups of invertebrates indicates an early evolutionary origin of the extended tachykinin family.

After the isolation of native insect TRPs it was important to determine whether these were the peptides accounting for the substance P immunoreactivity shown earlier in the insect CNS (Verhaert & De Loof, 1985; Lundquist & Nässel, 1990; Nässel *et al.*, 1990). Antisera were raised against LomTK I and used for immunocytochemistry (Nässel, 1993b). It became clear for several insect species that LomTK-like immunoreactivity was located in a set of neurones distinct from those that reacted to polyclonal antisera raised against substance P or the tachykinin neurokinin A (Nässel, 1993a,b; Lundquist *et al.*, 1994b; Muren, Lundquist & Nässel, 1995). However, a monoclonal antibody to substance, P, and an antiserum to the frog tachykinin kassinin, reacted with the same neurones as the LomTK antisera (Nässel, 1993a; Lundquist *et al.*, 1994b). The conclusion drawn was that in the insects studied there are neurones containing: (1) LomTK-like peptide(s) recognised by antisera to LomTK I, and cross-reacting with antiserum to kassinin and a monoclonal antibody to substance P; and (2) additional neurones containing a peptide(s) recognised by several polyclonal antisera to substance P or neurokinin A (Nässel, Lundquist & Brodin, 1992b; Nässel, 1993a; Lundquist *et al.*, 1994b; see also Würden & Homberg, 1995). These findings were puzzling until recently, when Champagne & Ribeiro (1994) isolated two peptides with a great similarity to substance P, the sialokinins I and II, from salivary glands of the mosquito *Aedes aegypti*. The existence of the sialokinins, and the demonstration of LomTK-like immunoreactivity in the midgut of the same mosquito species (Veenstra *et al.*, 1995), is further evidence that insects may produce both insect TRPs and vertebrate type tachykinins. This possibility has yet to be fully explored, and it needs to be shown whether peptides resembling the vertebrate tachykinins and insect TRPs can both be synthesised in nervous tissue. In the present account, only the insect TRPs are considered, and the emphasis is on their distribution and possible functions. As relatively little experimental work has been performed on the insect TRPs, much of the speculation about possible actions is based on what has been revealed about the distribution of the peptides in different tissues and cell types.

Distribution and action of tachykinin-related peptides in insects

LomTKs in Locusta migratoria

When the first antisera were raised to LomTK I it was natural to test these on nervous tissue of *Locusta migratoria* (Nässel, 1993b), although the major aim was to use the LomTK antisera to screen for related peptides in other insect species. Thus, this first investigation emphasised the specificity of the two LomTK antisera, and presented a rather superficial mapping of LomTK-like immunoreactive (LomTK-LI) neurones in the brain of the locust. Here the distribution LomTK-LI neurones in the brain of *L. migratoria* is reviewed, and some additional information is provided on the distribution in the ganglia of the ventral nerve cord and the stomatogastric system. Detailed descriptions of the morphology of some of the identifiable LomTK-LI interneurones are also presented.

The locust brain, including the optic lobes, was found to contain about 800 LomTK-LI neuronal cell bodies. The anterior cell bodies are shown in Fig. 1. Most of the LomTK-LI cell bodies are small in size, and belong to interneurones whose individual morphologies could not be resolved. However, it could be concluded that many of the so-called glomerular neuropils are supplied by LomTK-LI processes from these small sized neurones: the different compartments of the central complex (Fig. 2a); layers in all the neuropil regions of the optic lobes; the peripheral ocellar neuropil (Fig. 2b); parts of the mushroom body calyces (Fig. 2a), peduncles and β-lobes; and most, if not all, of the glomeruli of the antennal lobes. An abundance of thin varicose LomTK-LI fibres arborise in non-glomerular neuropil in proto-, deuto- and tritocerebrum, including the pars intercerebralis and neuropil surrounding the mushroom bodies (Fig. 2a).

Some of the smaller neurones could be untangled to yield at least part of their projections and arborisations: neurones connecting the protocerebral bridge and fan-shaped body have cell bodies dorsal to the protocerebral bridge, and some neurones innervating the antennal glomeruli have their cell bodies anterior to the antennal lobe. Many of the LomTK-LI neurones have processes in sensory neuropils, including 'first order sensory neuropils' where the synapses between sensory cells and first order interneurones are located. Such first-order sensory neuropils are the lamina of the optic lobe, the peripheral ocellar neuropils, and the antennal glomeruli.

A few of the larger neurones could be resolved in some detail. In each of the optic lobes, there is one large LomTK-LI cell body that gives rise to wide field processes in the basal and distal medulla (and accessory medulla), and in a layer proximal to the lamina (Fig. 1b). This neurone was also recently described in *Schistocerca gregaria* and suggested to be part of the optic lobe

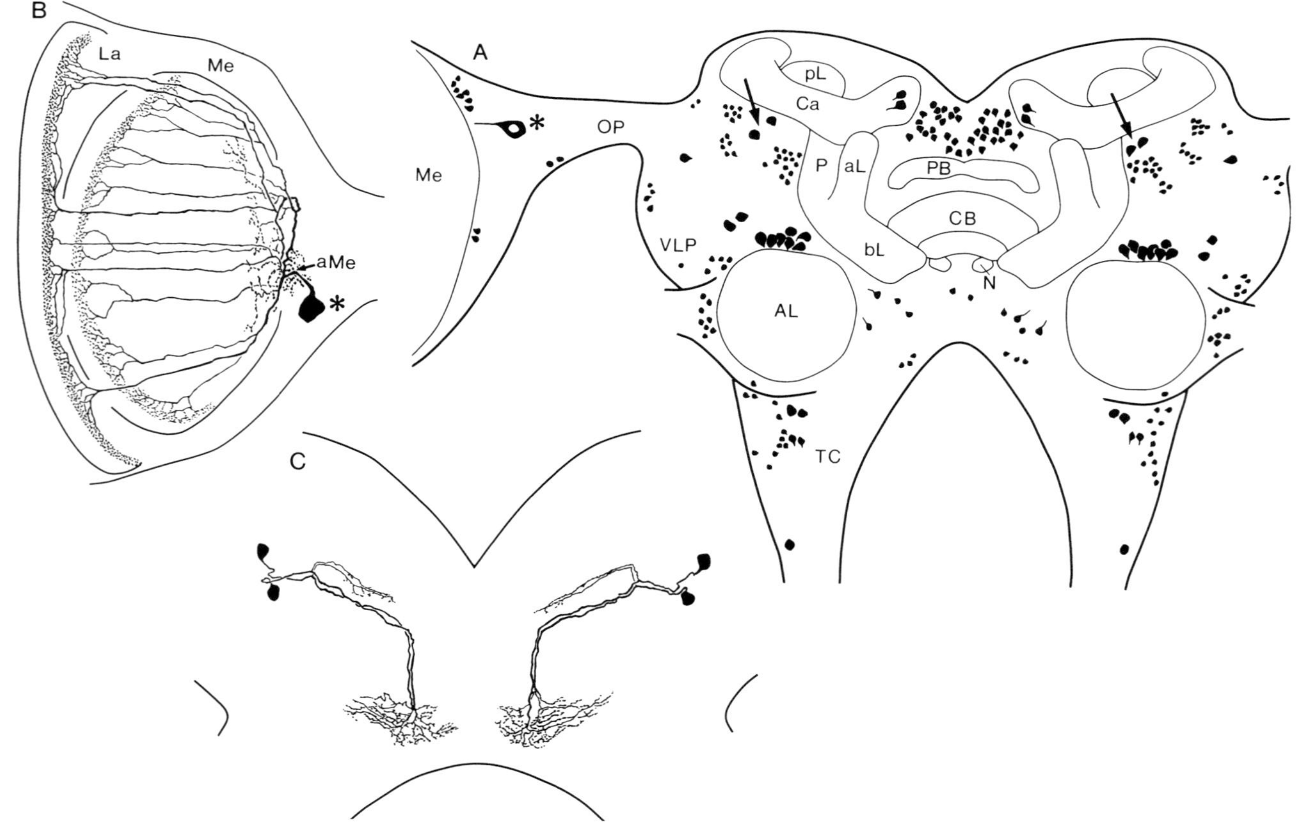

A
B
C
La
Me
aMe
*
OP
pL
Ca
P
aL
PB
CB
N
bL
VLP
AL
TC

circadian pacemaker circuit (Würden & Homberg, 1995). Another set of neurones in the protocerebrum that could be traced in detail is shown in Fig. 1c. These constitute two pairs of neurones (one pair in each hemisphere) with dorsal cell bodies in the anterior protocerebrum, and anterior processes that arborise dorsally (in the pars intercerebralis) and ventrally below the β-lobes of the mushroom bodies. Sets of about four or five neurones in each of the lateral neurosecretory cell groups have axons projecting to the glandular lobe of corpora cardiaca where they form connections with the adipokinetic hormone (AKH) producing glandular cells (see Fig. 5c) (Nässel *et al.*, 1995c; P. Passier *et al.*, unpublished observations). These neurones, which were identified by a combination of Lucifer yellow back filling from corpora cardiaca and LomTK immunocytochemistry (P. Passier *et al.*, unpublished observations), are likely to serve in regulating the release of adipokinetic hormone from the glandular cells (Nässel *et al.*, 1995c) and would thus not qualify as neurosecretory cells. Release of AKH from corpora cardiaca could be induced *in vitro* by LomTK I in a range of 10–200 μmol l^{-1}, with an ED$_{50}$ of about 50 μmol l^{-1} (Nässel *et al.*, 1995c).

In the ganglia of the ventral nerve cord, there is also a relatively large number of LomTK-LI cell bodies (Fig. 3b) and an abundance of neuropil processes (Fig. 4a,b). The metathoracic ganglion (T3) with the fused abdominal ganglia A1–3 contains about 60 LomTK-LI cell bodies, about 30 of which are in the T3 proper. An unfused abdominal ganglion has about 24 LomTK-LI cell bodies. Even in the ventral ganglia most of the neurones appear to be interneurones. Some are local interneurones, with branches confined to one hemiganglion, others interconnect the segmental ganglia either ipsilaterally or bilaterally. Only in the suboesophageal ganglion and the terminal abdominal ganglion could LomTK-LI neurones be found with axons that exit through segmental nerve roots (in maxillary nerves and the last set of ventral abdominal nerves, respectively). The ventral neuropils of suboesophageal, thoracic and abdominal ganglia are especially densely supplied by LomTK-LI processes.

Fig. 1. Tracing of LomTK-LI neurones in the brain of the locust *Locusta migratoria*. Abbreviations: AL, antennal lobe; CB, central body, Ca, mushroom body calyx; La, lamina; aL, bL and pL, lobes of of mushroom body; Me, medulla; aMe, accessory medulla; N, nodulus; OP, optic peduncle; PB, protocerebral bridge; TC, tritocerebrum; VLP, ventrolateral protocerebrum. (a) Cell bodies in the anterior portion of the brain. (b) Detail of the large identifiable optic lobe neurone marked with an asterisk in A. This neurone connects accessory medulla, medulla and lamina with wide field branches. (c) Two pairs of neurones in the anterior protocerebrum (cell bodies indicated by large arrows in a). (a) is redrawn from Nässel (1993b).

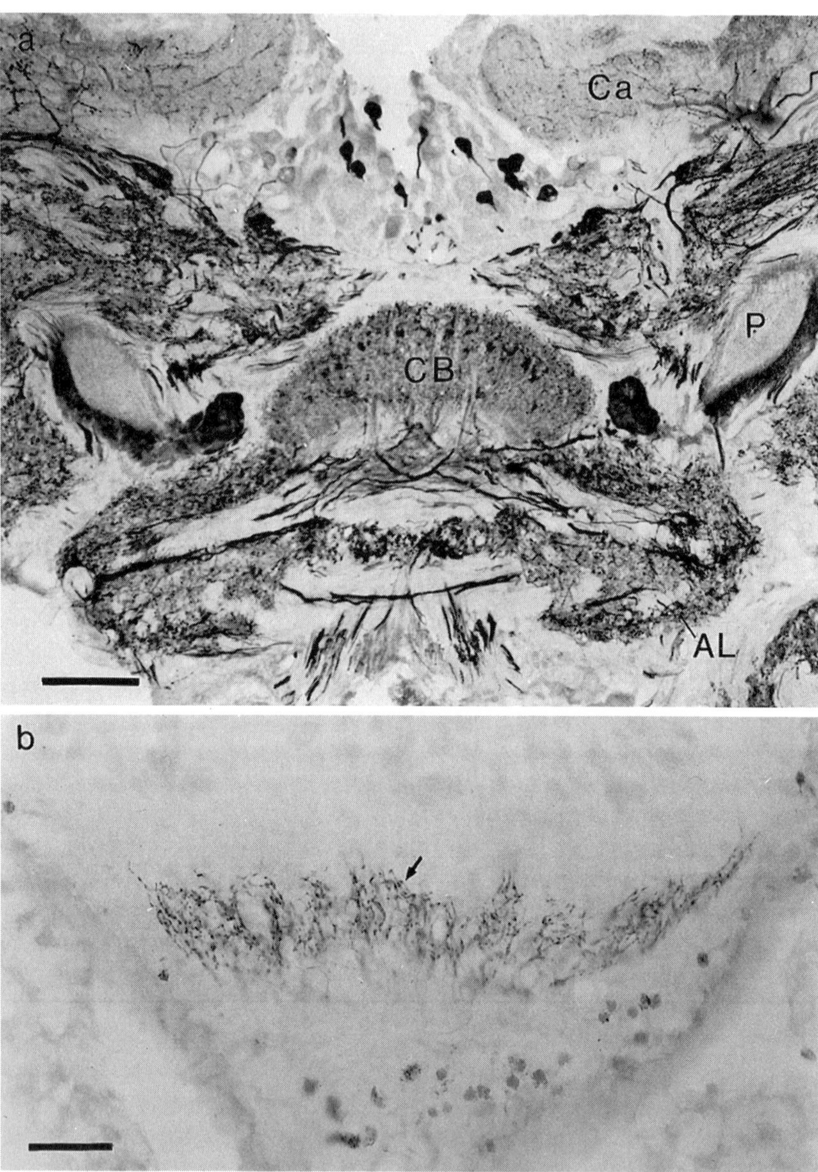

Fig. 2. Micrographs of LomTK-LI neurones in the locust brain. (a) Cryostat section through the centre of the brain with intense labelling in the central body (CB) and in associated accessory lobes (AL). Labelling is also seen in the calyces (Ca) and part of the peduncle (P) of the mushroom bodies, as well as in non-glomerular neuropil dorsal and lateral to the central body. (b) Peripheral ocellar neuropil with LomTK-LI fibers (arrow). Scales: a, b=100 μm.

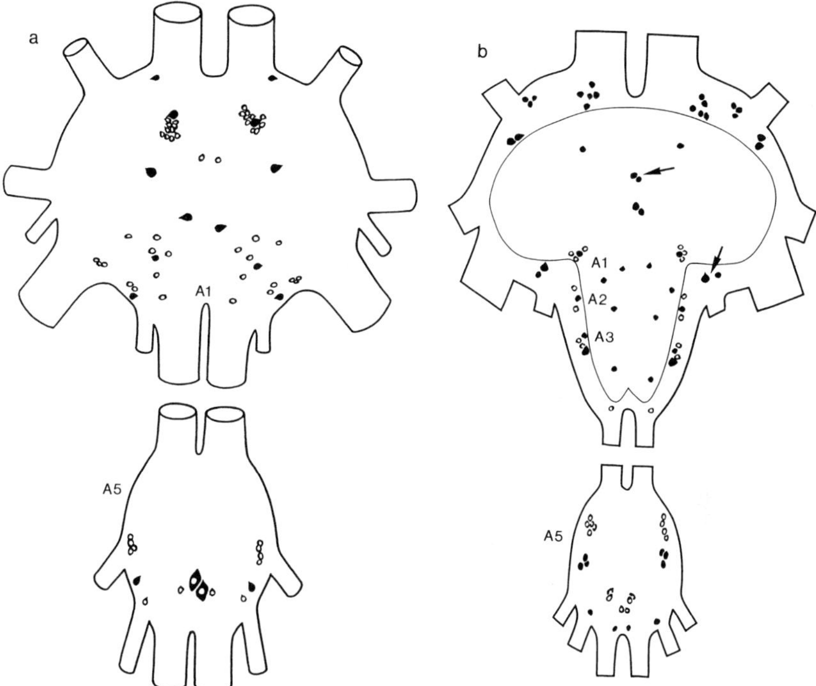

Fig. 3. Tracing of LomTK-LI cell bodies in metathoracic and abdominal ganglia. Filled cell bodies are intensely immunolabelled, whereas cell bodies with the nucleus shown are weakly labelled. (a) Ganglia of the cockroach *L. maderae*. The first abdominal ganglion, A1, is fused with the metathoracic ganglion (T3). In A5, the large median cell bodies are dorsal. (b) Ganglia of *L. migratoria*. The first three abdominal ganglia (A1–3) are fused with T3. The cell bodies indicated by arrows are the only ones with a dorsal location.

The stomatogastric nervous system consists of the frontal ganglion, the hypocerebral ganglion and the ingluvial ganglion. The frontal ganglion is connected to the hypocerebral ganglion by the recurrent nerve and to the tritocerebrum by the frontal connectives. There are about 20 large and small LomTK-LI cell bodies in the frontal ganglion of *L. migratoria* (Fig. 5b). These have arborisations in the frontal ganglion and processes that branch in the hypocerebral ganglion (Fig. 5b). No immunoreactive cell bodies were found in the latter ganglion. It was not possible to detect any LomTK-LI fibres in the muscle layer of the foregut, indicating that any effect of LomTKs on foregut contractility (see Schoofs *et al.*, 1993) would be by action on the neurones of the stomatogastric system. It should be noted that

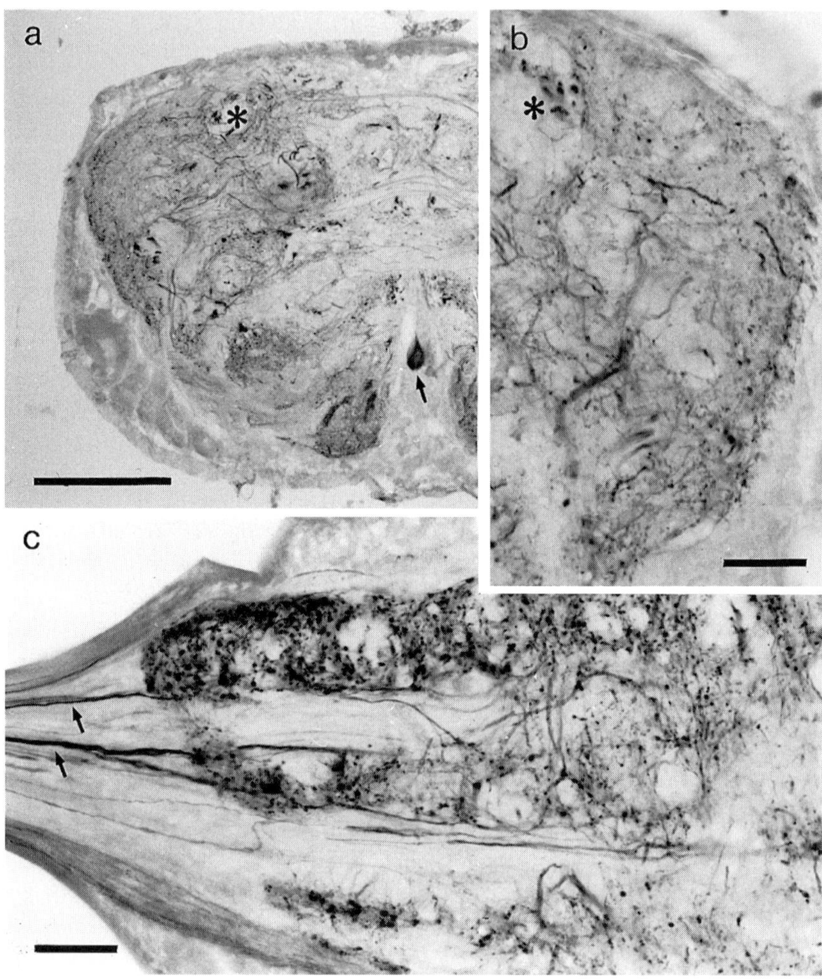

Fig. 4. Micrographs of LomTK-LI neurones in the ventral nerve cord of *L. migratoria*. (a) Transverse section through the metathoracic ganglion. Immunoreactive fibers are seen in most neuropil regions. One of the ventral midline neurones is seen at the arrow. The asterisk indicates a fibre tract that is seen also in the right hemisphere in b. (b) Detail of part of the right hemisphere at higher magnification. (c) Sagittal section through the posterior portion of the metathoracic ganglion with the neuropil of the three fused abdominal ganglia A1–3. Note interganglionic axons in the connective (arrows). Scales: a=200 μm; b, c=50 μm.

in our hands the deganglionated foreguts of *L. migratoria* and *S. gregaria* did not respond to LomTK I in a range of concentrations (Å. M. E. Winther, R. H. Osborne & D. R. Nässel, unpublished observations).

The entire intestine of *L. migratoria* was screened for LomTK-LI neurones. The only immunoreactive cells seen were endocrine cells of the midgut. These were also described by Agricola & Bräunig (1994). No other tissues in the thorax and abdomen of *L. migratoria* were investigated with the LomTK antiserum. Since the oviducts have been shown to respond to the different LomTKs (Schoofs *et al.*, 1993) it is possible that they are innervated by LomTK-LI fibres.

On the basis of immunocytochemistry, LomTKs in *L. migratoria* may have a role in modulating neurotransmission in the CNS, especially in sensory centres, but also in integrative centres that may be involved in motor control and learning. Peripherally, these peptides appear to be important in regulating foregut motility and in the control of enzyme release and/or motility of the midgut, as well as in relase of AKH. It seems that all of the LomTK-LI neurones in the locust brain are interneurones, and only a few examples of immunoreactive efferent neurones were detected in the ventral nerve cord. Another feature of the LomTK-LI neurones is that they are relatively numerous and their arborisations invade most of the neuropils of the CNS. Many other neuropeptides have much more restricted distributions in the brain (Nässel, 1993a; Homberg, 1994) and ventral nerve cord (Nässel, 1996a).

LemTRPs in *Leucophaea maderae*

The distribution of the LomTK-LI neurones in the Madeira cockroach, *L. maderae*, was analysed in some detail by Muren *et al.* (1995) and are only briefly summarised here. It is important to note that all the recently identified LemTRPs were isolated with a combination of bioassay and radio-immunoassay (RIA) using the LomTK antiserum (Muren & Nässel, 1996b, 1997). All of the identified LemTRPs were immunoreactive in the RIA, and it can therefore be assumed that the immunocytochemistry identifies most, if not all, native LemTRPs.

There are about 460 LomTK-LI cell bodies in the brain including the optic lobes. Most of these are in the protocerebrum and deutocerebrum, but some were also found in the tritocerebrum. About the same distribution of immunoreactive fibres was seen as in the locust brain. The major differences were seen in the mushroom bodies and the optic lobe. In *L. maderae*, LomTK fibres were only found in the basal layer of the double calyces of the mushroom bodies, and in the optic lobe only diffuse fibres were seen in the lobula and medulla. There are no fibres in the lamina, and no large identifiable

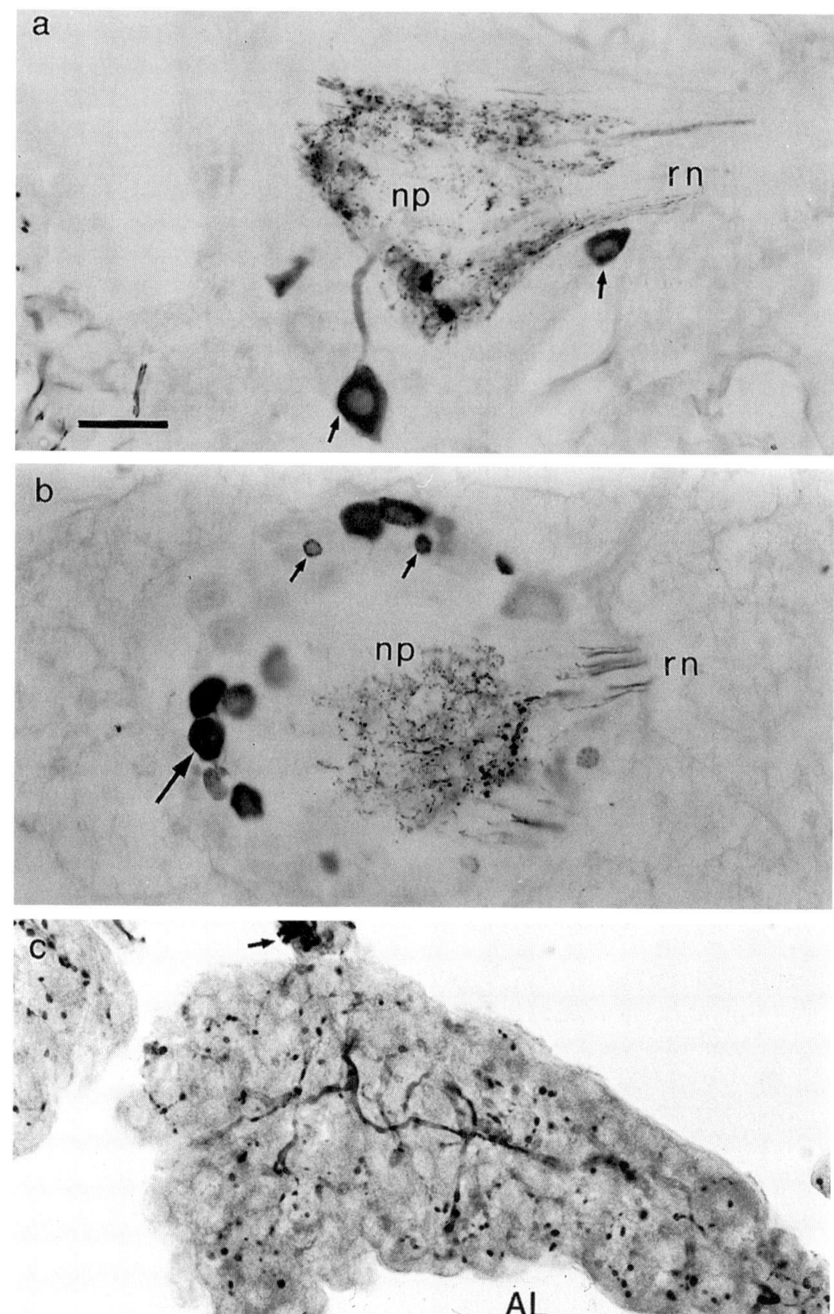

neurones in the optic lobe. No immunoreactive fibres were detected in the peripheral ocellar system.

The distribution of LomTK-LI material in the brain suggests that the actions of these peptides are similar in *L. maderae* and *L. migratoria*, with a few notable differences: in the cockroach, ocellar and laminar neuropils are devoid of immunoreactivity, suggesting that in this species the peptides are not involved in peripheral visual processing. Also the role in the mushroom bodies may differ between the species.

Cell bodies of LomTK-LI neurones were mapped in the suboesophageal, thoracic and abdominal ganglia. Of the strongly immunoreactive cell bodies, 15 were found in the suboesophageal ganglion, 6 in each of the thoracic ganglia, and 2 in each of the unfused abdominal ganglia and 16 in the terminal abdominal ganglion (Muren *et al.*, 1995). Only the abdominal neurones could be resolved in some detail, but all neurones except some in the terminal abdominal ganglion were identified as interneurones. With an improved immunocytochemical protocol, each ganglion was shown to contain additional, less strongly immunoreactive, cell bodies. Thus about 40 LomTK-LI cell bodies were identified in each of the thoracic ganglia, and about 16 in each of the unfused abdominal ganglia (Fig. 3a). It was also possible to show that about eight of the neurones in the terminal abdominal ganglion may be efferents innervating the hindgut muscle layer (Muren & Nässel, 1996a), and to identify one immunoreactive axon in each of the thoracic nerves to the legs (nerve 5). The cellular source of the thoracic efferents could not be identified.

The supply of LomTK-LI fibres in the peripheral nervous system is richer in *L. maderae* than in the locust. It was possible to resolve three LomTK-LI cell bodies and their arborisations in the frontal ganglion (Fig. 5a), branching fibres in the hypocerebral and ingluvial ganglia (Fig. 6a), as well as arborising fibres in the ingluvial (or gastric) nerves as they run on the crop (Fig. 6b). No LomTK-LI fibres innervate muscles of the oesophagus or crop. The

---

Fig. 5. LomTK-LI neurones in the retrocerebral glandular complex. (a) Frontal ganglion of *L. maderae* in horizontal section. Three LomTK-LI cell bodies are located in this ganglion, two of which are seen here (arrows). Fibres are seen arborising in the ganglionic neuropil (np) and axons enter the recurrent nerve (rn). (b) Frontal ganglion of *L. migratoria*. In this species, about 20 LomTK-LI cell bodies were found. Some smaller (small arrows) and larger (large arrow) cell bodies are present. Other abbreviations as in (a). (c) One half of the glandular lobe of the corpora cardiaca with LomTK-LI fibres arborising among glandular cells. The second corpora cardiaca nerve (NCC2) is seen at the arrow, with axons from LomTK-LI neurones that have cell bodies in the lateral neurosecretory cell group of the brain. AL, aorta lumen. Scales: a–c=50 µm.

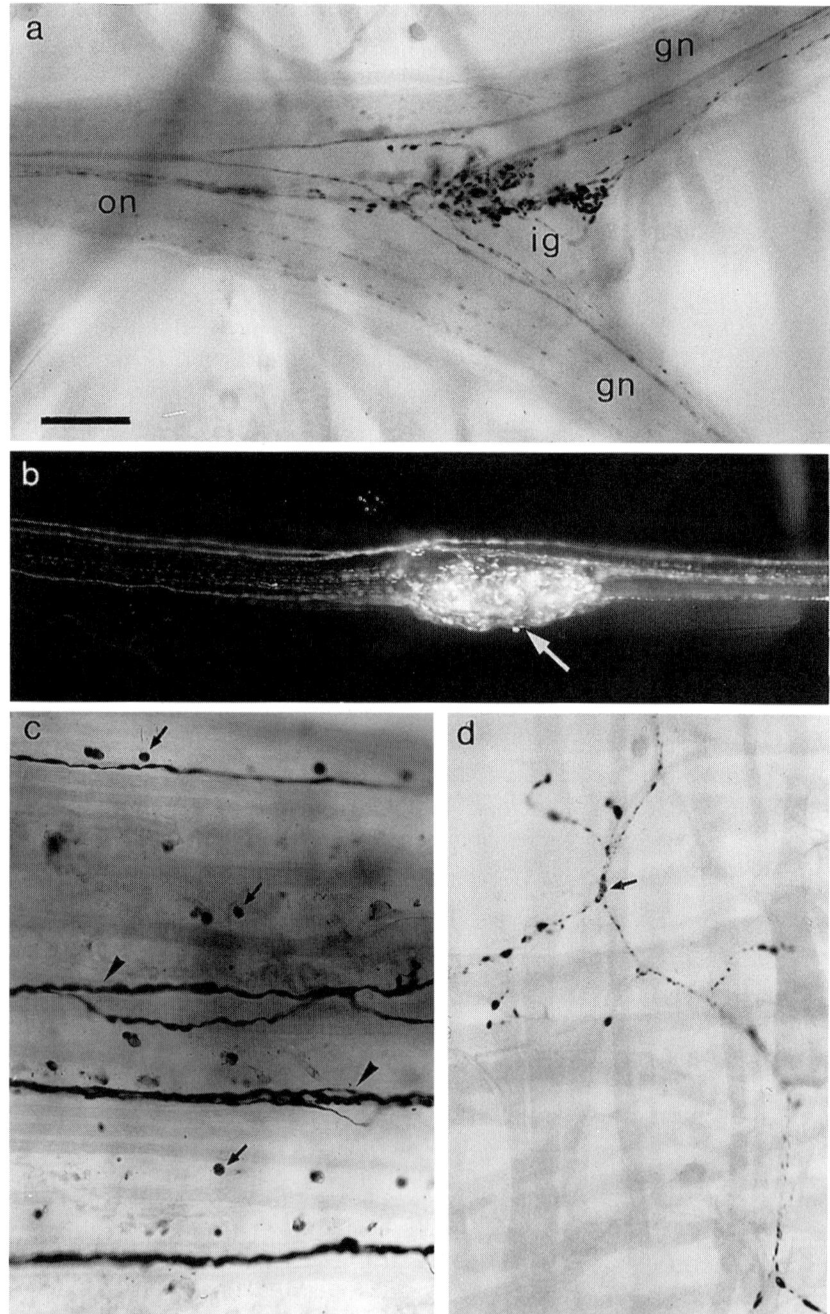

midgut contains numerous LomTK-LI endocrine cells and also a very rich supply of immunoreactive fibres over the muscle layer (Fig. 6c). In the gastric caecae, only LomTK-LI endocrine cells were detected. Finally, the muscle of the posterior part of the hindgut is also innervated by LomTK-LI fibres (Fig. 6d). Most of the pharynx dilator muscles in the head and many of the skeletal muscles attached to the posterior hindgut are supplied by LomTK-LI fibres. This supply of LomTK-LI fibres to skeletal and visceral muscle distinguishes the cockroach from the locust. It is not clear what the role of the muscle innervation is. In principle, the superficial innervation of the midgut could represent an extensive neurohaemal release site from which LemTRPs are released into the circulation. Alternatively, they may form a true innervation of the visceral muscle for the control of gut movement. The dilator muscles of the fore- and hindgut may actually be controlled by LemTRPs released from LomTK-LI fibres, which could thus have a role in gut distension.

LemTRP 1, 3, and 4–9 were tested in the isolated hindgut assay over a range of concentrations from 1 pmol 1^{-1} to 1 μmol 1^{-1} (Figs. 7 and 8). The stimulation thresholds (0.25 nmol 1^{-1}) and ED_{50} values (1 nmol 1^{-1}) of the different peptides do not differ significantly in spite of differences in their primary structures (Fig. 8). In the hindgut assay, the ED_{50} of the LemTRPs are similar to that of proctolin, but preliminary data indicate that these families of peptides act on different receptors and have different temporal dynamics (Muren, 1996). LomTK I and two of the LemTRPs so far tested also stimulate contractions of the isolated foregut of the cockroach *Periplaneta americana* (although a slightly higher concentration was required), but not that of *L. maderae* (H. Penzlin, J. E. Muren, M. Eckert & D. R. Nässel, unpublished observations).

CavTKs in *Calliphora vomitoria*

In the blowfly *C. vomitoria*, the distribution of TRP immunoreactive neurones has been investigated in the brain, thoracic abdominal ganglion and intestine of larvae and adults using an antisera to LomTK I and CavTK II (Lundquist *et al.*, 1994a; Nässel, Kim & Lundquist, 1995a). With these antisera, which recognise both CavTK I and II, it was also possible to determine

---

Fig. 6. LomTK-LI material in gut-related structures of *L. maderae*. (a) Immunoreactive fibres in the ingluvial ganglion (ig), oesophageal (on) and gastric nerves (gn). Note arborisations in ig. (b) In the gastric nerves, the LomTK-LI fibres arborise (arrow) at regular intervals in node-like structures. (c) Fibres (arrowheads) and endocrine cells (arrows) in the midgut. (d) Fibres in the muscle layer of the hindgut. Scales: a–d=50 μm.

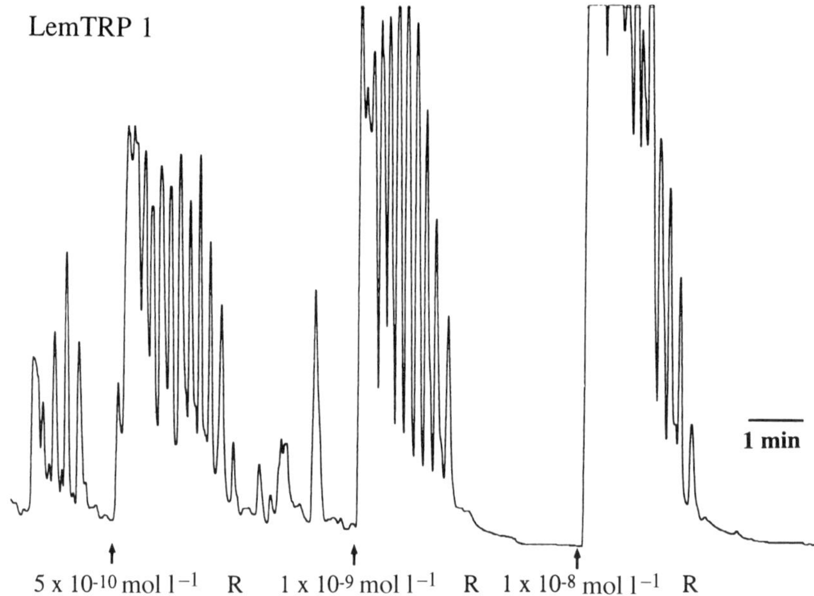

Fig. 7. Response of *L. maderae* hindgut *in vitro* to differing concentrations of LemTRP1 (applied at the arrows). At 10 nmol l⁻¹ the response amplitude is out of the range of the printer. The hindgut responds with increases in amplitude, frequency and tonus. R, rinse.

the distribution of CavTK isoforms in extracts of different portions of the CNS and in the intestine using HPLC in combination with an enzyme-linked immunosorbent assay (ELISA) and a RIA (Nässel *et al.*, 1995b; Kim *et al.*, 1997). This was essential, because CavTKs had been isolated from whole animals and information was needed on the relative distribution of the two isoforms. Both CavTK I and II were found in about the same amounts in the brain, the fused thoracic-abdominal ganglion, and in the midgut of adults and larvae. Possibly further forms are present in all tissues in slightly smaller amounts and therefore escaped detection during the isolation of CavTKI and II.

Immunocytochemistry with different antisera to the TRPs revealed the same labelling of neurones in *C. vomitoria* (Lundquist *et al.*, 1994a; Nässel *et al.*, 1995b). About 160 LomTK-LI neuronal cell bodies were found in the adult brain, optic lobes and suboesophageal ganglion (Fig. 9). These are distributed as follows: 60 in the protocerebrum, 26 in the deutocerebrum, 60 in the tritocerebrum and 8 in the suboesophageal ganglion. In the fused thoracic-abdominal ganglion of adults, about 46 LomTK-LI cell bodies were

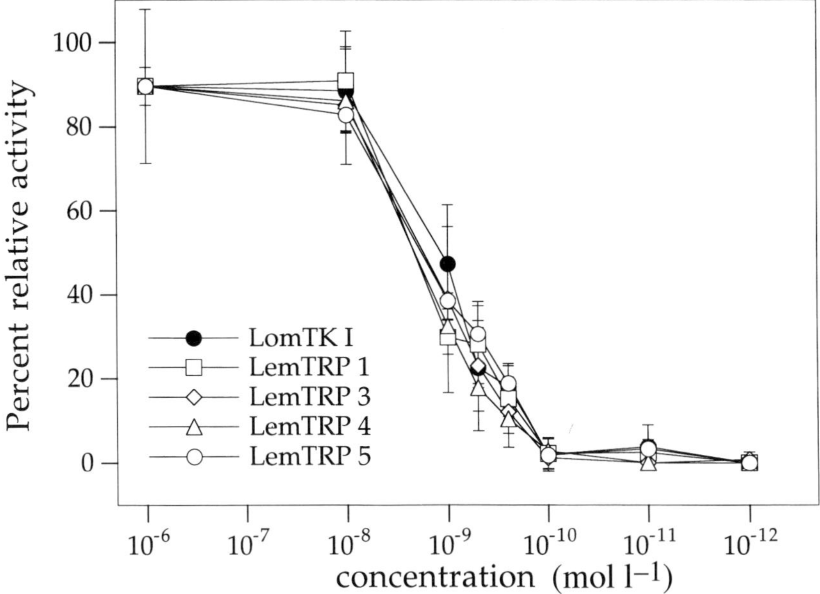

Fig. 8. Dose–response curves for four different LemTRPs and LomTK I on the *L. maderae* hindgut. The maximal response is given as 100%. Five different guts were tested with each of the peptides.

detected (Fig. 10a). Of these, 10 are located in the abdominal neuromeres. In the larval CNS, the cell bodies were distributed as follows (Lundquist *et al.*, 1994a): 12 in the brain plus suboesophageal ganglion and about 40 in the fused thoracic-abdominal ganglion (of the latter, 28 are in the abdominal neuromeres).

As seen in the schematic tracing in Fig. 9, many of the CavTK-containing neurones supply fibres to the antennal glomeruli and the fan-shaped body of the central complex. The fan-shaped body is innervated by neurones with cell bodies in at least two pairs of clusters, one pair in the protocerebrum the other in the tritocerebrum. The dorsal protocerebrum (pars intercerebralis and optic tubercles) is innervated by protocerebral and tritocerebral cell bodies, whereas each of the antennal lobes appears to be innervated by three groups of cell bodies all of which are in the deutocerebrum. Thus, most of the neurones that could be resolved in some detail appear to have long processes connecting different portions of the brain. Two pairs of sub-oesophageal neurones resolved in both adults and larvae are descending neurones that send processes throughout the ventral nerve cord (Fig. 10a). These were the only neurones that could be followed with certainty from the

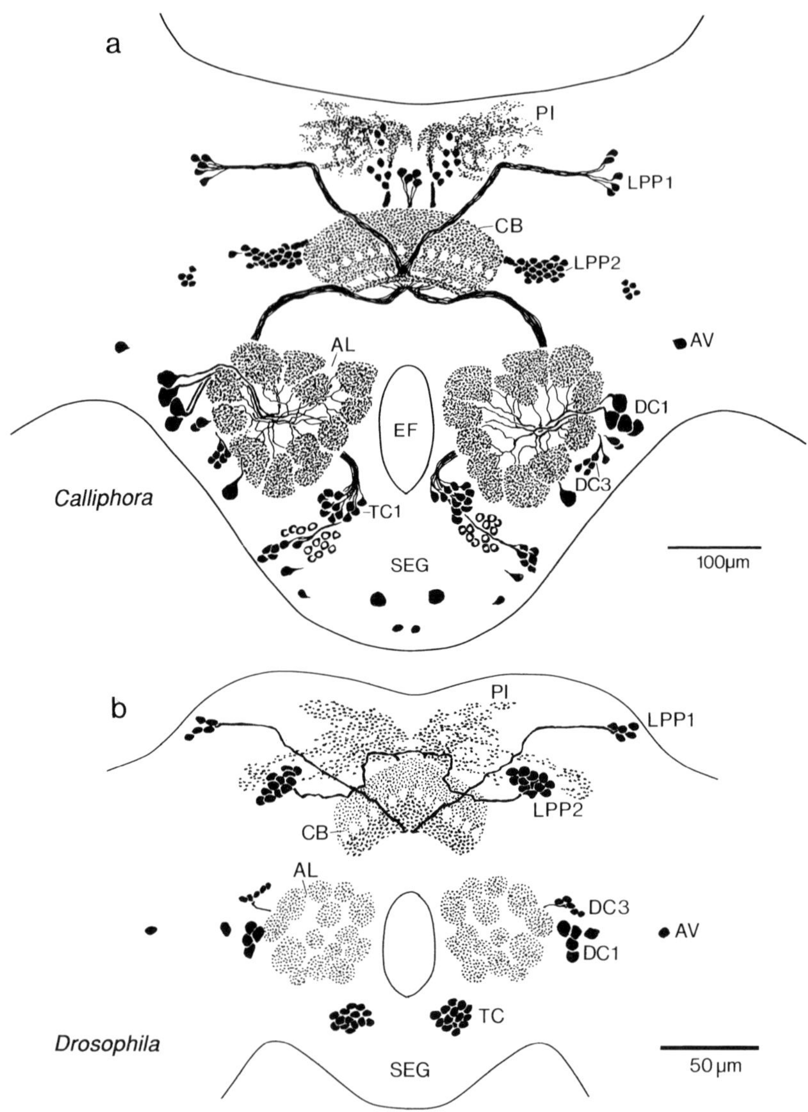

Fig. 9. Tracings of LomTK-LI neurones in the brains of *C. vomitoria* and *D. melanogaster*. Compressed frontal views are shown. Neurones that may be homologues in the two species are indicated by the same designations. Note innervation of central body (CB), antennal lobes (AL) and pars intercerebralis (PI). SEG, suboesophageal ganglion; EF, oesophageal foramen; TC, tritocerebrum; DC, deuterocerebrum; LPP, lateral posterior protocerebrum.

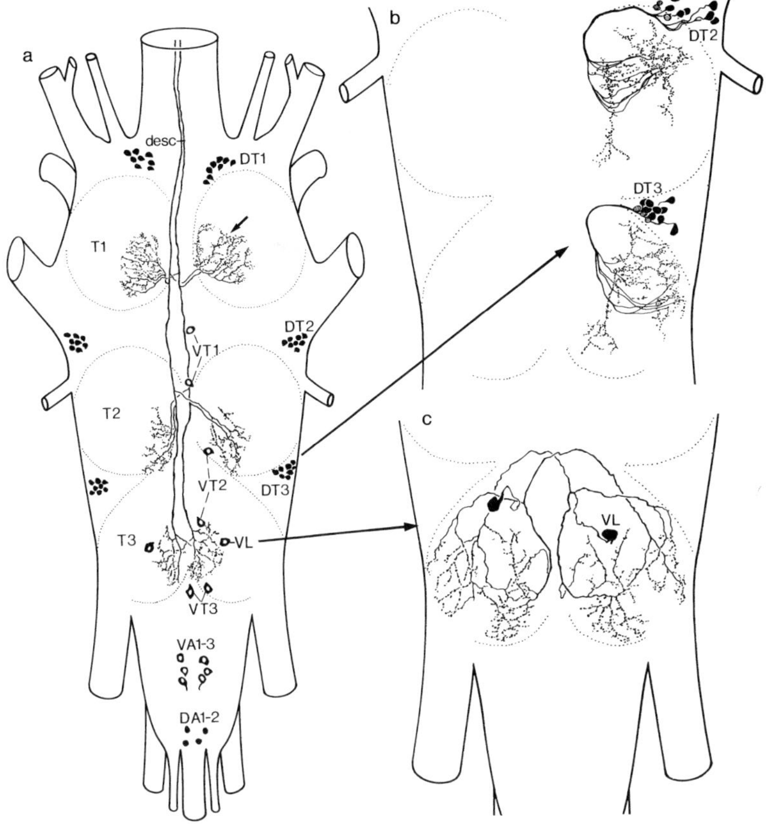

Fig. 10. LomTK-LI neurones in the fused thoracic-abdominal ganglion of *C. vomitoria* seen in horizontal view. (a) There are dorsal cell bodies (DT1–3 and DA1–2) and ventral ones (VT1–3, VL and VA1–3). A pair of axons from descending neurones with cell bodies in the suboesophageal ganglion are shown (desc). (b) Details of the neurone groups DT2 and 3 on the right side. These neurones form processes in the dorsal neuropil. (c) A pair of ventral cell bodies (VL) form bilateral processes in the ventral neuropil of the metathoracic neuromere.

larval to the adult CNS. However, it seems that several of the 28 abdominal neurones loose their CavTK phenotype (or die) during metamorphosis, because only 4–10 cells are seen in the abdominal neuromeres of the adult fly (Lundquist *et al.*, 1994a). In the thoracic neuromeres, a few sets of CavTK-LI neurones could be resolved in detail (Fig. 10). These are groups of local interneurones with dorsal cell bodies and processes in the leg motor neuropil of all three thoracic neuromeres (Fig. 10b), and a pair of neurones found only in the metathoracic neuromere, which has bilateral processes in the ventral sensory neuropil of the leg (Fig. 10c).

The only other structure in which TRPs were detected was the larval and adult midgut. A large number of LomTK- and CavTK-LI endocrine cells are distributed uniformly in the adult midgut, whereas in the larval midgut most of these cells are located in the posterior region adjacent to the Malpighian tubules (Lundquist *et al.*, 1994a; Kim *et al.*, 1997).

It is apparent that the CavTKs are distributed in a much smaller set of neurones in the blowfly than in locusts and cockroaches (see Table 2). As shown in Fig. 9 and Table 2, similar sets of LomTK-LI neurones were seen in the brain of *Drosophila melanogaster* (Nässel, 1994), and immunoreactive endocrine cells were also detected in the midgut.

No actions are known for the CavTKs in *C. vomitoria*, but these peptides stimulate contractions of the *L. maderae* hindgut *in vitro* (Lundquist *et al.*, 1994b). Preliminary experiments indicate that CavTK I and II can increase cyclic AMP levels in brain homogenates (Nässel *et al.*, 1995a). similar to the finding that LomTK I elevates cyclic AMP in locust corpora cardiaca (P. Passier *et al.*, unpublished observations).

Distribution and action of tachykinin-related peptides in other invertebrates

Aside from insects, LomTK-LI material has been detected in the stomatogastric nervous system of a crab (Blitz *et al.*, 1995) and throughout the nervous system of several molluscs (Elekes & Nässel, 1994; Elekes *et al.*, 1995). The stomatogastric nervous system of decapod crustaceans is well suited for studying the influence of neuropeptides on neural network activity (Harris-Warrick, Nagy & Nusbaum, 1992; Marder, Skiebe & Christie, 1994). Within the stomatogastric nervous system, neuronal networks generate motor patterns that control the rhythmic chewing and filtering activity of the crustacean foregut. LomTK-LI material has been detected in neurones of the stomatogastric nervous system in the crab *Cancer borealis* (Blitz *et al.*, 1995): cell bodies in the paired commissural ganglia (five larger neurones and several smaller neurones in each ganglion); neuropilar varicosities and fibers forming a club-shaped structure in the commissural ganglia; one cell body in the oesophageal ganglion; two fibres in the stomatogastric nerve; and in neuropil of the stomatogastric ganglion. Superfusion with LomTK I or II (1 μmol 1^{-1}) excited the pyloric rhythm, which is controlled by the stomatogastric nervous system, by increasing the frequency of the weak (<0.6 Hz) rhythm pattern (50–300% increase). However, rhythms faster than 1 Hz remained almost unchanged (10% increase). LomTK also elicited a pyloric

Table 2 *Distribution of LomTK-like immunoreactivity in cerebral structures of insects*

Location of fibres	Locusta	Leucophaea	Calliphora	Drosophila
Mushroom body				
Calyx	x[a]	x[b]	–	–
β-lobe	x	–	–	–
Central complex				
Fan-shaped body	x	x	x	x
Protocerebral bridge	x	x	–	–
Lateral accessory lobe	x	x	x	x
Pars intercerebralis	x	x	x	x
Optic lobe				
Lamina	x	–	–	–
Medulla	x	x	–	–
Lobula	x	x	x	x
Anterior optic tubercle	x	x	x	x
Ocellar neuropil	x	–	–	–
Antennal lobe	x	x	x	x
Tritocerebrum	x	x[c]	x	x
Subesophageal ganglion	x	x	x	x
Frontal ganglion	x	x	?	–
Hypocerebral ganglion	x	x	–	–
Corpora cardiaca	x[d]	x[e]	–	–
Pharynx dilator muscles	–	x	–	–

[a] In a special posterior lobe of the calyx.
[b] In basal portion of median and lateral calyces.
[c] Very dense distribution of LomTK-LI fibers in glomerular structures.
[d] Mainly in glandular lobe.
[e] Diffuse and weak immunolabelling.
Compiled from Nässel (1993b, 1994), Lundquist *et al.* (1994) and Muren *et al.* (1995).

rhythm in preparations where no spontaneous rhythm was present. In addition to these effects, LomTKs produced changes in impulse activity in several pyloric neurones during ongoing pyloric rhythms, either increasing or decreasing the number of spikes per burst (Blitz *et al.*, 1995). However, no effect was observed on the gastric mill rhythm.

The distribution of LomTK-LI material has also been mapped in three species of mollusc: *Helix pomatia*, *Lymnaea stagnalis* and *Anodonta cygnea*

(Elekes & Nässel, 1994; Elekes *et al.*, 1995). In each of these species, a number of smaller neurones were detected with LomTK antiserum; about 900 in the CNS of *H. pomatia*, 180 in *L. stagnalis*, and a small number of cell bodies in *A. cygnea*. In the two former species, LomTK-LI fibres were also seen in the intestinal muscle layer. These were both of intrinsic origin and derived from cell bodies in the CNS. In *A. cygnea*, the reactivity of the LomTK antiserum could be blocked with synthetic anodontatachykinin (AncTK), indicating that the antiserum binds to a native TRP.

The effects of applying AncTK and LomTK I at 10 μmol 1^{-1} to identified neurones of *H. pomatia* was studied by Elekes *et al.* (1995). These peptides induced a 5–10 mV hyperpolarization and blocked spiking activity in neurones of the D-cluster and the Rpa2 neurone. Another set of *Helix* neurones, V2, Ra1 and LPLa1, gave a depolarising response to AncTK and LomTK I. The hyperpolarisation seems to be caused by a decrease in delayed K^+ currents, and the depolarisation by an increase in Ca^{2+} currents (Elekes *et al.*, 1995).

Concluding remarks

The existence of several forms of TRPs in the locust and the cockroach is similar to other insect peptide families, such as allatostatins, FMRFamide-related peptides, myotropins and leucokinins (Holman *et al.*, 1990; Duve *et al.*, 1992; Donly *et al.*, 1993 Schoofs *et al.*, 1993). It is not known at present whether the different TRP forms have different functions, or if one form is sufficient and the others redundant. There are indications that some insect neuropeptide isoforms can have different potencies in certain *in vitro* bioassays. The best evidence is provided by the allatostatins (see Weaver, this volume), which vary in their effectiveness in inhibiting juvenile hormone biosynthesis and proctolin induced contractions of the hindgut (Woodhead *et al.*, 1993; Lange *et al.*, 1995). Some of the *Diploptera* allatostatins that are potent inhibitors of juvenile hormone biosynthesis are poor inhibitors of hindgut contractions and *vice versa*. It has been suggested that the isoforms serve different functions in different tissues, although they may be released in the same ratios at these targets (Lange *et al.*, 1995).

A different result was obtained in the mollusc *Aplysia* for another family of neuropeptides, the myomodulins. In *Aplysia*, nine myomodulins (MM_{A-I}) are encoded on the same precursor; myomodulin A is encoded in 10 copies, the others in one copy each (Lopez *et al.*, 1993; Miller *et al.*, 1993). All isoforms (except MM_B and MM_C) have the same action on contractions of one of their targets, the accessory radula closer muscle *in vitro* (Brezina *et al.*, 1995). However, the actions of the myomodulins *in vivo* may be determined by the stoichiometric ratio in which they are released. In identified motor neurones, the stoichiometric ratios between myomodulin A and B are about 5–6:1 (Cropper *et al.*, 1991). When the nine isoforms were applied to the target muscle in a cocktail with the molar ratio suggested from the transcript, the response was not different from when MM_A and MM_{D-I}, were applied individually, and the actions of MM_B and MM_C were overridden (Brezina *et al.*, 1995). Thus it was considered likely that some of these peptides are funcationally redundant at the accessory radula closer neuromuscular junction (Brezina *et al.*, 1995).

The four different intestinal LemTRPs tested so far (Fig. 8) have equal potencies in stimulating contractions of the *L. maderae* hindgut (Muren & Nässel, 1996b), whereas some of the brain-specific LemTRPs have slightly lower potencies (Muren & Nässel, 1997). Effects at other target organs have not yet been investigated. The four LomTKs isolated from locust brain and corpora cardiaca were found to be equally potent in the locust visceral muscle assays employed so far (Schoofs *et al.*, 1990a,b). The variation in the primary structure of the LemTRPs is much greater than that of the other insect TRPs (Muren & Nässel, 1996b, 1997), and it would not be surprising if some have different actions or potencies on different targets as shown for the allatostatins (Lange *et al.*, 1995). One reason for suggesting tissue-specific actions of the different LemTRPs is the finding that five of the isoforms appear to be expressed exclusively in the brain, two in the intestine only and three forms (LemTRP1, 2 and 5) are found in both tissues (Muren & Nässel, 1996b, 1997). The possibility of tissue-specific LemTRPs is interesting, because it indicates that either there are two different genes encoding the TRPs, or the same gene is differentially spliced or post-translationally processed in the brain and gut. For a more extensive discussion of the differential distribution and action of neuropeptide isoforms the reader is referred to the recent review by Nässel (1996b).

There are as yet no data to suggest whether the different brain LemTRP isoforms are co-localised in all neurones of the brain, and whether the gut isoforms coexist in endocrine cells and fibres of the midgut. This question is especially interesting in the gut, because LomTK immunoreactive nerve fibres are derived from the stomatogastric system while the endocrine cells have a different embryonic origin. The question of co-localisation of LomTKs in the locust has been addressed in a simple system where only one cell type provides LomTK-LI fibres. This is the glandular lobe of the corpora cardiaca, which is supplied by a small set of LomTK-LI lateral neurosecretory cells (LNCs) HPLC analysis of an extract of corpora cardiaca, combined with a LomTK RIA, revealed that all isoforms are likely to be present, and must therefore be co-localised in the few LNCs (P. Passier *et al.*, unpublished observations).

Although present knowledge about the actions of insect TRPs is limited to stimulation of intestinal muscle in *L. maderae*, and the stimulation of oviduct contractions and possibly the release of AKH in *L. migratoria*, the distribution of these peptides suggests they have roles in many different physiological processes. The simplest explanation for their abundant distribution in a variety of neurone types in the CNS would be that the TRPs are co-transmitters at all the sites where they are released and that they have distributed functions. Rather than having global actions in many regions of the CNS, they may act within local circuits to modulate the response to other co-released transmitters or to modulate local network properties. It is not yet clear which other substances are co-localised with insect TRPs, except that certain γ-aminobutyric acid immunoreactive neurones in the protocerebrum and antennal lobes of both blowflies and *L. maderae* contain LomTK-LI material (D. R. Nässel *et al.*, unpublished observation). A systematic screen for co-localised neurotransmitters and modulators needs to be performed.

In the periphery, it may be that the TRPs have their own primary actions, for example on visceral muscle, and that their release is episodic. For instance, it is possible that LemTRPs are released from neurones innervating the pharynx dilator muscles when there is need for extra distension of the foregut, such as when the old cuticle is shed at moults (see Muren *et al.*, 1995). The LemTRPs in gut endocrine cells may regulate synthesis or release of digestive enzymes, or regulate local reflexes controlling peristalsis. It is important to investigate whether LemTRPs can be released into the circulation to function as neurohormones in *L. maderae* and, if so, what their targets are. Certainly, much work is required to untangle the complex set of roles the insect TRPs may play even in one species. Moreover, phylogenetic variations in the distribution of TRPs indicate that these roles may vary in different species.

Acknowledgements

The original research was supported by a grant from the Swedish Natural Science Research Council (NFR) to D.R.N. We thank Anne Karlsson for technical assistance.

References

Agricola, H. & Bräunig, P. (1994). Comparative aspects of peptidergic signalling pathways in the nervous system of arthropods. In *The Nervous Systems of Invertebrates: An Evolutionary and Comparative Approach*, ed. O. Breidbach & W. Kutsch, pp. 303–328. Basel: Birkäuser Verlag.

Andrews, P. C., Brayton, K. & Dixon, J. E. (1987). Precursors to regulatory peptides: their proteolytic processing. *Experientia* **43**, 784–790.

Blitz, D. M., Christie, A. E., Marder, E. & Nusbaum, M. P. (1995). Distribution and effects of tachykinin-like peptides in the stomatogastric nervous system of the crab, *Cancer borealis. Journal of Comparative Neurology* **354**, 282–294.

Brezina, V., Bank, B., Cropper, E. C., Rosen, S., Vilim, F. S., Kupfermann, I. & Weiss, K. R. (1995). Nine members of the myomodulin family of peptide cotransmitters at the B16-ARC neuromuscular junction of *Aplysia. Journal of Neurophysiology* **74**, 54–72.

Champagne, D. E. & Ribeiro, J. (1994). Sialokinin I and II: vasodilatory tachykinins from the yellow fever mosquito *Aedes aegypti. Proceedings of the National Academy of Sciences USA* **91**, 138–142.

Chang, M. M., Leeman, S. E. & Niall, H. D. (1971). Amino acid sequence of substance P. *Nature* **232**, 86–87.

Clottens, F. L., Meola, S. M., Coast, G. M., Hayes, T. K., Wright, M. S., Nachman, R. J. & Holman, G. M. (1993). Characterization of an antiserum against an achetakinin I-analog and its use for the localization of culekinin depolarizing peptide II in the mosquito, *Culex salinarius. Regulatory Peptides* **49**, 145–157.

Coast, G. M. (1995). Synergism between diuretic peptides controlling ion and fluid transport in insect Malpighian tubules. *Regulatory Peptides* **57**, 283–296.

Coast, G. M., Holman, G. M. & Nachman, R. J. (1990). The diuretic activity of a series of cephalomyotropic neuropeptides, the achetakinins, on isolated Malpighian tubules of the house cricket, *Acheta domesticus. Journal of Insect Physiology* **36**, 481–488.

Cropper, E. C., Vilim, F. S., Alevizos, A., Tenenbaum, R., Kolks, M., Rosen, S., Kupfermann, I. & Weiss, K. R. (1991). Structure, bioactivity, and cellular localization of myomodulin B. *Peptides* **12**, 683–690.

Dircksen, H. (1994). Distribution and physiology of crustacean cardioactive peptide in arthropods. In *Perspectives in Comparative Endocrinology*, ed. K. G. Davey, R. E. Peter & S. S. Tobe, pp. 139–148. Ottawa: National Research Council of Canada.

Donly, B. C., Ding, Q., Tobe, S. S. & Bendena, W. G. (1993). Molecular cloning of the gene for the allatostatin family of neuropeptides from the cockroach *Diploptera punctata. Proceedings of the National Academy of Sciences USA* **90**, 8807–8811.

Duve, H., Johnsen, A. H., Sewell, J. C., Scott, A. G., Orchard, I., Rehfeld, J. F. & Thorpe, A. (1992). Isolation, structure, and activity of -Phe-Met-Arg-Phe-NH$_2$ neuropeptides (designated calliFMRFamides) from the blowfly *Calliphora vomitoria. Proceedings of the National Academy of Sciences USA* **89**, 2326–2330.

Elekes, K. & Nässel, D. R. (1994). Tachykinin-related neuropeptides in the central nervous system of the snail *Helix pomatia*: an immunocytochemical study. *Brain Research* **661**, 223–236.

Elekes, K., Herná di L., Kiss, T., Muneoka, Y. & Nässel, D. R. (1995). Tachykinin- and leucokinin-related peptides in the molluscan nervous system. *Acta Biologica Hungarica* **46**, 281–294.

Erspamer, V. (1981). The tachykinin peptide family. *Trends in Neuroscience* **4**, 267–273.

Erspamer, V. & Anastasi, A. (1962). Structure and pharmacological actions of eledoism, the active endecapeptide of *Eledone. Experientia* **18**, 58–61.

Erspamer, V. & Melchiorri, P. (1973). Active polypeptides of the amphibian skin and their synthetic analogues. *Pure and Applied Chemistry* **35**, 463–494.

Fujisawa, J., Muneoka, Y., Takashashi, T., Takao, T., Shimonishi, Y., Kubota, I., Ikeda, T., Minakata, H., Nomoto, K., Kiss, T. & Hiripi, L. (1994). An invertebrate-type tachykinin isolated from the freshwater bivalve mollusc, *Anodonta cygnea.* In *Peptide Chemistry 1993*, ed. Y. Okoda, pp. 161–164. Osaka: Protein Research Foundation.

Goldsworthy, G., Coast, G., Wheeler, C., Cusinato, O., Kay, I. & Khambay, B. (1992). The structure and functional activity of neuropeptides. In *Insect Molecular Science*, ed. J. M. Crampton & P. Eggleston, pp. 205–225. London: Academic Press.

Harris-Warrick, R. M., Nagy, F. & Nusbaum, M. P. (1992). Neuromodulation of stomatogastric networks by identified neurons and transmitters. In *Dynamic Biological Networks: The Stomatogastric Nervous System*, ed. R. M. Harris-Warrick, E. Marder, A. I. Selverston & M. Moulins, pp. 139–160. Cambridge, MA: MIT Press.

Helke, C. J., Krause, J. E., Mantyh, P. W., Couture, R. & Bannon, M. J. (1990). Diversity in mammalian tachykinin peptidergic neurons: multiple peptides, receptors and regulatory mechanisms. *FASEB Journal* **4**, 1606–1615.

Holman, G. M., Nachman, R. J., Schoofs, L., Hayes, T. K., Wright, M. S. & De Loof, A. (1991). The *Leucophaea maderae* hindgut preparation: a rapid and sensitive bioassay tool for the isolation of insect myotropins of other insect species. *Insect Biochemistry* **21**, 107–112.

Holman, G. M., Nachman, R. J. & Wright, M. S. (1990). Insect neuropeptides. *Annual Reviews of Entomology* **35**, 201–217.

Homberg, U. (1994). Distribution of neurotransmitters in the insect brain. *Progress in Zoology* **40**, 1–88.

Ikeda, T., Minakata, H., Nomoto, K., Kubota, I. & Muneoka, Y. (1993). Two novel tachykinin-related neuropeptides in the echiuroid worm, *Urechis unicinctus. Biochemical and Biophysical Research Communications* **192**, 1–6.

Kim, M. Y., Muren, J. E., Lundquist, C. T. & Nässel, D. R. (1997). Insect tachykinin-related neuropeptides: developmental changes in expression of callitachykinin isoforms in the central nervous system and intestine of the blowfly *Calliphora vomitoria. Archives of Insect Biochemistry and Physiology* **34**, 475–491.

Lafont, R. (1991). Reverse endocrinology, or 'hormones' seeking functions. *Insect Biochemistry* **21**, 697–721.

Lange, A. B., Bendena, W. G. & Tobe, S. S. (1995). The effect of thirteen Dip-allatostatins on myogenic and induced contractions of the cockroach (*Diploptera punctata*) hindgut. *Journal of Insect Physiology* **41**, 581–588.

Lopez, V., Wickham, L. & Des Groseillers, L. (1993). Molecular cloning of myomodulin cDNA, a neuropeptide precursor gene expressed in neuron L10 of *Aplysia californica*. *DNA Cell Biology* **12**, 53–61.

Lundquist, C. T., Clottens, F. L., Holman, G. M., Nichols, R., Nachman, R. J. & Nässel, D. R. (1994b). Callitachykinin I and II, two novel myotropic peptides isolated from the blowfly, *Calliphora vomitoria*, that have resemblances to tachykinins. *Peptides* **15**, 761–768.

Lundquist, C. T., Clottens, F. L., Holman, G. M., Riehm, J. P., Bonkale, W. & Nässel, D. R. (1994a). Locustatachykinin immunoreactivity in the blowfly central nervous system and intestine. *Journal of Comparative Neurology* **341**, 225–240.

Lundquist, C. T. & Nässel, D. R. (1990). Substance P-, FMRFamide-, and gastrin/cholecystokinin-like immunoreactive neurons in the thoraco-abdominal ganglia of the flies *Drosophila* and *Calliphora*. *Journal of Comparative Neurology* **294**, 161–178.

Marder, E., Skiebe, P. & Christie, A. (1994). Multiple modes of network modulation. *Verhandlungen des Deutschen Zoologischen Gesellschaft* **87**, 177–184.

Maggio, J. E. (1988). Tachykinins. *Annual Reviews of Neuroscience* **11**, 13–28.

Miller, M. W., Beushausen, S., Vitek, A., Stamm, S., Kupfermann, I., Brosius, J. & Weiss, K. R. (1993). The myomodulin-related neuropeptides: Characterization of a gene encoding a family of peptide cotransmitters in *Aplysia*. *Journal of Neuroscience* **13**, 3358–3367.

Muren, J. E. (1996). Tachykinin-related peptides in the Madeira cockroach: structures, distributions and actions. Ph.D. thesis, University of Stockholm.

Muren, J. E., Lundquist, C. T. & Nässel, D. R. (1993). Quantitative determination of myotropic neuropeptides in the nervous system of the cockroach *Leucophaea maderae*: distribution and release of leucokinins. *Journal of Experimental Biology* **179**, 289–300.

Muren, J. E., Lundquist, C. T. & Nässel, D. R. (1995). Abundant distribution of locustatachykinin-like peptide in the nervous system and intestine of the cockroach *Leucophaea maderae, Philosophical Transactions of the Royal Society London B* **348**, 423–444.

Muren, J. E., & Nässel, D. R. (1996a). Radioimmunoassay determination of tachykinin-related peptide in different portions of the central nervous system and intestine of the cockroach *Leucophaea maderae*. *Brain Research* **739**, 314–321.

Muren, J. E. & Nässel, D. R. (1996b). Isolation of five tachykinin-related peptides from the midgut of the cockroach *Leucophaea maderae*: existence of N-terminally extended isoforms. *Regulatory Peptides* **65**, 185–196.

Muren, J. E. & Nässel, D. R. (1997). Seven tachykinin-related peptides isolated from the brain of the Madeira cockroach: evidence for tissue specific expression of isoforms. *Peptides* **18**, 7–15.

Nachman, R. J., Roberts, V. A., Holman, G. M. & Trainer, J. A. (1990). Consensus chemistry and confirmation of an insect neuropeptide family analogous to tachykinins. In *Progress in Comparative Endocrinology*, ed. A. Epple, C. G. Scanes & M. H. Stetson, pp. 60–66. New York: Wiley-Liss.

Nässel, D. R. (1993a). Neuropeptides in the insect brain: a review. *Cell and Tissue Research* **273**, 1–29.

Nässel, D. R. (1993b). Insect myotropic peptides: differential distribution of locustatachykinin- and leucokinin-like immunoreactive neurons in the locust brain. *Cell and Tissue Research* **274**, 27–40.

Nässel, D. R. (1994). Neuropeptides, multifunctional messengers in the nervous system of insects. *Verhandlungen des Deutschen Zoologischen Gesellschaft* **87**, 59–81.

Nässel, D. R. (1996a). Neuropeptides, amines and amino acids in an elementrary insect ganglion: functional and chemical anatomy of the unfused abdominal ganglion. *Progress in Neurobiology* **48**, 325–420.

Nässel, D. R. (1996b). Peptidergic neurohormonal control in invertebrates. *Current Opinion in Neurobiology* **6**, 842–850.

Nässel, D. R., Cantera, R. & Karlsson, A. (1992a). Neurons in the cockroach nervous system reacting with antisera to the neuropeptide leucokinin I. *Journal of Comparative Neurology* **322**, 45–67.

Nässel, D. R., Karlsson, A., Kim, M. Y., Lundquist, C. T., Muren, J. E. & Winther, Å. (1995a). Tachykinin-related neuropeptides in the insect nervous system: structure, distribution and putative functions. In *Insects. Chemical, Physiological and Environmental Aspects*, ed. D. Konopinska, pp. 242–247. Wroclaw: Wroclaw University Press.

Nässel, D. R., Kim, M.-Y. & Lundquist, C. T. (1995b). Several forms of callitachykinins are distributed in the central nervous system and intestine of the blowfly *Calliphora vomitoria*. *Journal of Experimental Biology* **198**, 2527–2536.

Nässel, D. R., Lundquist, C. T. & Brodin, E. (1992b). Diversity in tachykinin-like peptides in the insect brain. In *Neurobiology of invertebrates*, ed. J. Salanki, K. S. Rozsa & K. Elekes. *Acta Biologica Hungarica* **43**, 175–188.

Nässel, D. R., Lundquist, T., Höög, A. & Grimelius, L. (1990). Substance P-like immunoreactive neurons in the nervous system of *Drosophila*. *Brain Research* **507**, 225–233.

Nässel, D. R., Passier, P. C. C. M., Elekes, K., Dircksen, H., Vullings, H. G. B. & Cantera, R. (1995c). Evidence that locustatachykinin I is involved in release of adipokinetic hormone from locust corpora cardiaca. *Regulatory Peptides* **57**, 297–310.

Orchard, I., Belanger, J. H. & Lange, A. B. (1989). Proctolin: a review with emphasis on insects. *Journal of Neurobiology* **20**, 470–496.

Otsuka, M. & Yoshioka, K. (1993). Neurotransmitter functions of mammalian tachykinins. *Physiological Reviews* **73**, 229–308.

Patel, M., Chung, J. S., Kay, I., Mallet, A. I., Gibbon, C. R., Thompson, K., Bacon, J. P. & Coast, G. M. (1994). Localization of *Locusta*-DP in locust CNS and hemolymph satisfies initial hormonal criteria. *Peptides* **15**, 591–602.

Patel, M., Hayes, T. K. & Coast, G. M. (1995). Evidence for the hormonal function of a CRF-related diuretic peptide (*Locusta*-DP) in *Locusta migratoria*. *Journal of Experimental Biology* **198**, 793–804.

Pernow, B. (1983). Substance P. *Pharmacological Reviews* **35**, 85–141.

Schoofs, L., Holman, G. M., Hayes, T. K., Nachman, R. J. & De Loof, A. (1990a). Locustatachykinin I and II, two novel insect neuropeptides with homology to peptides of the vertebrate tachykinin family. *FEBS Letters* **261**, 397–401.

Schoofs, L., Holman, G. M., Hayes, T. K., Kochansky, J. P., Nachman, R. J. & De Loof, A. (1990b). Locustatachykinin III and IV: two additional insect neuropeptides with homology to peptides of the vertebrate tachykinin family. *Regulatory Peptides* **31**, 199–212.

Schoofs, L., Vanden Broek, J. & De Loof, A. (1993). The myotropic peptides of *Locusta migratoria*: structures, distribution, functions and receptors. *Insect Biochemistry and Molecular Biology* **23**, 859–881.

Veenstra, J. A., Lau, G. W., Agricola, H. J. & Petzel, D. H. (1995). Immunohistological localization of regulatory peptides in the midgut of the female mosquito *Aedes aegypti*. *Histochemistry Cell Biology* **104**, 337–347.

Verhaert, P. & De Loof, A. (1985). Substance P-like immunoreactivity in the central nervous system of the blattarian insect *Periplaneta americana* L. revealed by a monoclonal antibody. *Histochemistry* **83**, 501–507.

Von Euler, U. S. & Gaddum, J. H. (1931). An unidentified depressor substance in certain tissue extracts. *Journal of Physiology London* **72**, 74–86.

Woodhead, A. P., Asano, W. Y. & Stay, B. (1993). Allatostatins in the haemolymph of *Diploptera punctata* and their effect in vivo. *Journal of Insect Physiology* **39**, 1001–1005.

Würden, S. & Homberg, U. (1995). Immunocytochemical mapping of serotonin and neuropeptides in the accessory medulla of the locust *Schistocerca gregaria*. *Journal of Comparative Neurology* **362**, 305–319.

Note added in proof: Since this paper was submitted a number of relevant findings have been made:
Christie, A. C. *et al.* (1997). *J. Exp. Biol.* **200**, 2279–2294; Lundquist, C. T. & Nässel, D. R. (1997). *J. Neurobiol.*, in press; Winther, Å. E. M. *et al.* (1997). *Peptides*, in press.

IAN ORCHARD and ANGELA B. LANGE

The distribution, biological activity, and pharmacology of SchistoFLRFamide and related peptides in insects

Introduction

The tetrapeptide FMRFamide (Phe-Met-Arg-Phe-NH$_2$), a cardioacceleratory peptide originally sequenced in the clam *Macrocallista nimbosa* (Price & Greenberg, 1977), is now regarded as the primary member of an extended family of peptides found throughout the Metazoa. These N-terminally extended RFamide peptides (referred to as FMRFamide-related peptides or FaRPs) have diverse biological activities, including actions upon visceral and skeletal muscles, glands, and neurones (Painter & Greenberg, 1982; Hooper & Marder, 1984; Walther, Schiebe & Voigt, 1984; Weiss *et al.*, 1984; Holman, Cook & Nachman, 1986; Li & Calabrese, 1987; Bulloch *et al.*, 1988; Spencer, 1988; Baines, Tyrer & Mason, 1989; Lange, Orchard & Te Brugge, 1991; Robb & Evans, 1994). Although the evolutionary relatedness of members of the FaRPs is not clearly understood, it is apparent that subfamilies exist and constitute N-terminally extended FMRFamides, FLRFamides and HMRFamides (e.g. Holman *et al.*, 1986; Nachman *et al.*, 1986; Robb, Packman & Evans 1989; Duve *et al.*, 1992). One subfamily of decapeptide FaRPs (also described as myosuppressins because of their ability to inhibit visceral muscle contraction) has been described in several insect species and shown to share the common sequence Xaa-Asp-Val-Xaa-His-Xaa-Phe-Leu-Arg-Phe-NH$_2$ (see Table 1). One of these myosuppressins is referred to as SchistoFLRFamide (Robb *et al.*, 1989) and there exists a body of literature on its presence, distribution, biological activity and pharmacology in locusts. This review describes the information known about this particular member of the subfamily of myosuppressins, making reference to the other myosuppressins where appropriate.

Isolation and sequences

The first N-terminally extended RFamide to be sequenced in insects was leucomyosuppressin from the cockroach *Leucophaea maderae* which was extracted from nervous tissue, separated on reversed-phase high-performance

Table 1 *Isolation and sequence of FMRFamide-related myosuppressins in insects*

Peptide sequence[a]	Trivial name	Tissue	Organism	Reference
pQDVDHVFLRFamide	Leucomyosuppressin	Head extract	*Leucophaea maderae*	Holman *et al.*, 1986
PDVDHVFLRFamide	SchistoFLRFamide	Thoracic nervous system	*Schistocerca gregaria*	Robb *et al.*, 1989
		Brain+CC+CA+SOG	*Locusta migratoria*	Schoofs *et al.*, 1993
		Brain	*Locusta migratoria*	Peef *et al.*, 1994
		VNC	*Locusta migratoria*	Lange *et al.*, 1994
ADVGHVFLRFamide		Brain	*Locusta migratoria*	Peeff *et al.*, 1994
		VNC	*Locusta migratoria*	Lange *et al.*, 1994
TDVDHVFLRFamide	Neomyosuppressin	Head extract	*Neobellieria bullata*	Fónagy *et al.*, 1992
	Dromyosuppressin	Whole body	*Drosophila melanogaster*	Nichols, 1992
pQDVVHSFLRFamide	ManducaFLRFamide	Brain+SOG	*Manduca sexta*	Kingan *et al.*, 1990

Notes:

CC, corpus cardiacum; CA, corpora allata; SOG, suboesophageal ganglion; VNC, ventral nerve cord.

[a] Single-letter code.

liquid chromatography (HPLC) and isolated through its ability to inhibit spontaneous contractions of *L. maderae* hindgut (Holman *et al.*, 1986). Subsequently, a number of similar peptides were extracted from a diverse range of insects including locusts, fleshfly, fruitfly and tobacco hornworm using HPLC coupled to bioassay, enzyme-linked immunosorbent assay (ELISA) or radioimmunoassay (RIA), the latter two detecting extended RFamide peptides. The sequences and source of these peptides is shown in Table 1 and illustrates that these peptides share the sequence Xaa^1-Asp-Val-Xaa^4-His-Xaa^6-Phe-Leu-Arg-Phe-NH_2 (where Xaa^1 is pGlu, Pro, Thr or Ala; Xaa^4 is Asp, Gly or Val; and Xaa^6 is Val or Ser).

Gene characterisation

The gene for SchistoFLRFamide has not been cloned, but that of leucomyosuppressin from the cockroach *Diploptera punctata* has recently been characterised (Donly *et al.*, 1996). Initially these authors used HPLC separation, RIA and subsequent bioassay to demonstrate the presence of an active peptide in the brain of *D. punctata* which co-eluted with authentic leucomyosuppressin. Oligonucleotides coding for the amino acid sequence of the predicted non-modified form of leucomyosuppressin (Gln-Asp-Val-Asp-His-Val-Phe-Leu-Arg-Phe-Gly) were synthesised; the amino acid sequence beginning with glutamine, the unmodified form of pyroglutamate, and also including a C-terminal glycine required for processing into the mature, amidated peptide structure. A polymerase chain reaction (PCR)-based approach was used to isolate a fragment of the *D. punctata* leucomyosuppressin gene. A 400 bp DNA fragment representing a 3′ portion of the *D. punctata* leucomyosuppressin gene was used to screen portions of an unamplified cockroach brain cDNA library for homologous clones. From approximately 500 000 plaques screened, 68 positively hybridising clones were isolated, all but one of which were confirmed as positive leucomyosuppressin clones by PCR amplification. The cDNA sequence contained an open reading frame that upon translation would result in a prepropolypeptide of 96 amino acid residues. Proteolytic cleavage and N- and C-terminal modification of the predicted precursor could result in three peptides, one of which is leucomyosuppressin. No other RFamide products are predicted to be processed from the precursor, although of the other two peptides one does show the characteristic C-terminal RF, but without amidation. The precursor for leucomyosuppressin in *D. punctata* encodes only a single extended FLRFamide, leucomyosuppressin, and is therefore unlike the multipeptide FaRP precursors which have been deduced in other organisms (see Bendena *et al.*, 1997). Southern blot analysis indicated that the gene is present in the *D. punctata* genome in a single copy and northern blot analysis showed that the gene is predominantly expressed as a 3.8 kb mRNA in cockroach brain.

Distribution

Immunohistochemistry

In insects, immunohistochemical evidence has indicated the presence of FaRPs throughout the central and peripheral nervous system. Most of these studies used an antiserum raised against the C-terminus RFamide and therefore detected the multiple endogenous FaRPs known to be present in insects. More recently, studies have been performed using antisera directed against the N-terminal sequence of SchistoFLRFamide and dromyosuppressin, two members of the subfamily. Thus, in *Schistocerca gregaria* (Swales & Evans, 1995) immunohistochemistry using antisera raised against Pro-Asp-Val-Asp-His-Val coupled to a purified protein derivative of tuberculin, revealed immunoreactivity within a sub group of neurones within the ventral nerve cord that were also immunoreactive to an antiserum raised against bovine pancreatic polypeptide. Within the suboesophageal ganglion three groups of cells stained positively against the N-terminally specific antiserum. These included one pair of large posterior ventral cells that have been previously shown to innervate the heart and retro-cerebral glandular complex. In the thoracic and abdominal ganglia, two and three sets of neurones respectively stained. In the abdominal ganglia the immunoreactive neurones projected via the median nerves to the highly immunoreactive perivisceral neurohaemal organs. As with most immunohistochemical studies, the antiserum probably did not stain all processes since immunoreactive processes from a number of cells were too weak to trace, and no staining was observed in prothoracic nerve VI or the peripheral nerves leaving the abdominal neuromeres of the metathoracic ganglion, nerves known to stain with bovine pancreatic polypeptide antisera. However, the study does indicate that SchistoFLRFamide is widely distributed in *S. gregaria*, and may have a neurohormonal role, being present in neurohaemal organs, as well as a possible direct role, influencing the heart.

In the fruitfly *Drosophila melanogaster*, antisera raised against Thr-Asp-Val-Asp-His-Val-Cys conjugated to thyroglobulin have been used to examine the cellular expression pattern of dromyosuppressin immunoreactive material during all stages of *D. melanogaster* development (McCormack & Nichols, 1993). Immunoreactivity was first seen in two cells of the medial protocerebrum of embryos, with an increase in the number of immunoreactive cells in the brain and the first appearance of positive neurones in the ventral ganglion during the larval stage. In addition, immunoreactive processes extended from the medial protocerebrum cells into the ventral ganglion. The pupal and adult stages were characterised by an increase in the number of immunoreactive cells in the central nervous system and an increase in their arborisations. Along with immunoreactivity within the central nervous system, two immunoreactive cells were also observed in larvae, in the gut near the anus. In the adult gut, immunoreactivity was seen in two cells of the rectum and immunoreactive processes extended over the crop, apparently arising from the central nervous system. No immunoreactivity was observed in the other regions of adult or larval gut.

In situ hybridisation

Expression of the *D. punctata* leucomyosuppressin gene has recently been examined *in situ* using immunological procedures (Donly *et al.*, 1996). Production of mRNA was studied using a digoxygenin-labelled fragment of the cloned cDNA as a hybridisation probe. *In situ* hybridisation revealed expression in numerous cells of the brain of *D. punctata*, the majority of which were located in the pars intercerebralis of the protocerebrum, in a region containing the medial neurosecretory cells. In addition, apparent endocrine cells of the midgut showed intense expression of the leucomyosuppressin mRNA (Orchard *et al.*, 1997; M. Fusé, personal communication). This is an interesting observation since leucomyosuppressin has recently been shown to stimulate the secretion of amylase in weevil midgut (Nachman *et al.*, 1997), and myosuppressins have been shown to inhibit contractions of midgut (see later).

Biological activity

Visceral muscle

Oviduct muscle

Attention has been drawn more recently towards the presence of a SchistoFLRFamide-like peptide associated with the oviducts of the migratory locust *Locusta migratoria*. With the use of reversed-phase HPLC coupled with RIA for RFamide-like peptides and a bioassay, Lange, Orchard & Te Brugge (1991) isolated and quantified a SchistoFLRFamide-like peptide within the areas of the oviduct which receive extensive innervation, and quantified RFamide-like material in the oviducal nerve and VIIth abdominal ganglion. RFamide-like immunoreactive material is associated with terminals on locust oviducts (Wang & Orchard, 1995b), and immuno-gold-labelling revealed this material to be contained within electron-dense round granules which were present alongside electron-lucent vesicles in the nerve endings. SchistoFLRFamide is a potent modulator of contraction of this visceral muscle, possessing physiological effects similar to those of octopamine, which is present in dorsal unpaired median (DUM) neurones innervating the oviducts (Lange *et al.*, 1991). The location of neurones delivering the SchistoFLRFamide-like peptide is unknown. SchistoFLRFamide inhibits or reduces the amplitude and frequency of spontaneous contractions, relaxes basal tonus, and reduces the amplitude of neuronally evoked, proctolin-induced, glutamate-induced and high potassum-induced contractions. The effects are dose-dependent. Intracellular recordings from the muscle have shown that SchistoFLRFamide, at a concentration which totally abolishes neuronally evoked and myogenic contractions (10^{-6} M), has little effect on resting potential and reduces the amplitude of the excitatory junction potential by only 35%. The native peptide within the oviducts, which co-elutes with SchistoFLRFamide on two sequential HPLC systems, is also capable of reducing the amplitude of neuronally evoked and proctolin-induced contractions, and of inhibiting spontaneous contractions and relaxing basal tonus.

Digestive system

In addition to FMRFamide-like immunoreactive nerve processes extending over the various regions of the gut, immunoreactivity has also been shown to be present in cells of endocrine type in the midgut of a diversity of insects (see e.g. Brown, Crim & Lea, 1986; Jenkins, Brown & Crim, 1989; Crim *et al.*, 1990; Tsang & Orchard, 1991; Sivasubramanian, 1992; Zitnan, Sauman & Sehnal, 1993; Veenstra & Lambrou, 1995). The identification of the FaRPs contributing to this immunoreactivity is largely unknown although the expression of the *D. punctata* leucomyosuppressin gene in apparent endocrine cells of midgut of *D. punctata* indicates that leucomyosuppressin or its related peptides are likely to contribute to the staining pattern (Orchard *et al.*, 1997; M. Fusé, personal communication). These peptides are unlikely to be the sole contributors of the staining pattern since gene expression occurs only in a fraction of the total immunoreactive cells of *D. punctata* midgut, and recently a novel FaRP was isolated from *Periplaneta americana* midgut and found to have the sequence Ala-Asn-Arg-Ser-Pro-Ser-Leu-Arg-Leu-Arg-Phe-NH$_2$ (Veenstra & Lambrou, 1995).

With regard to biological activity, leucomyosuppressin, as well as neomyosuppressin, inhibit spontaneous contractions of the *L. maderae* hindgut, the assay used in their isolation (Holman *et al.*, 1986; Fónagy *et al.*, 1992). Interestingly, leucomyosuppressin does not inhibit the spontaneous contractions of visceral muscles of *L. maderae* uniformly as a group, but shows a selective suppression of activity of the foregut and hindgut (Cook & Wagner, 1991). The effects of leucomyosuppressin on *L. maderae* hindgut (Cook, Wagner & Pryor, 1993) are very similar to those reported for

SchistoFLRFamide on locust oviduct (Lange *et al.*, 1991). The presence of leucomyosuppressin gene expression in endocrine cells of the midgut of *D. punctata* implies a hormonal function for this peptide associated with digestion. Interestingly, Nachman *et al.* (1997) have shown that leucomyosuppressin stimulates the secretion of amylase from the midgut of the weevil, *Rynchophorus ferragineus*, and so the involvement of leucomyosuppressin with digestive enzyme secretion would seem to be a useful avenue of pursuit.

An involvement in feeding is implied by the work of Jenkins *et al.* (1989), who have shown that the amounts of FMRFamide-like immunoreactivity in the midgut and haemolymph of the corn earworm, *Heliothis zea*, is correlated to the feeding state of the larvae. Fed animals have about twice as many immunostained endocrine cells and four times the amount of immunoreactive material detected by RIA as starved individuals, who in turn have significantly more FMRFamide-like immunoreactivity in their haemolymph. In the midgut of the sphingid moth *Agrius convolvuli*, ManducaFLRFamide is a potent inhibitor of spontaneous contractions (Fujisawa *et al.*, 1993). The threshold concentration is 0.1 nmol 1^{-1}, with complete arrest in some preparations at 0.3 nmol 1^{-1}. Curiously, and unlike other preparations responding to the myosuppressin family, this midgut preparation is also inhibited by a range of FaRPs, including FLRFamide, FMRFamide, and YGGFMRFamide, although higher concentrations are necessary. SchistoFLRFamide is also capable of inhibiting proctolin-induced contractions of the circular muscles of *L. migratoria* midgut (A. B. Lange, unpublished observations). Using a ring preparation from *L. migratoria* midgut, it can be shown that proctolin induces a dose-dependent sustained contraction, which can be inhibited at all doses by SchistoFLRFamide (Fig. 1). It will be of interest to examine the effects of leucomyosuppressin on *D. punctata* midgut in a similar manner.

Heart muscle

FMRFamide-like immunoreactivity has been demonstrated in nerves projecting over the heart (Kingan *et al.*, 1990; Tsang & Orchard, 1991; Stevenson & Pflüger, 1994; Ude & Agricola, 1995), and members of the FaRP family shown to have modulatory effects upon amplitude and frequency of heart contractions (Cuthbert & Evans, 1989). In some cases the responses to the peptides are cardioexcitatory, in some cardioinhibitory, and in others biphasic. With regard to SchistoFLRFamide itself, this peptide was originally isolated from thoracic ganglia of *S. gregaria* using reversed-phase HPLC, RIA, and bioassay (Robb, Packman & Evans, 1989). SchistoFLRFamide has a potent cardioinhibitory effect on the amplitude and frequency of spontaneous contractions in the semi-isolated locust heart (Robb *et al.*, 1989; Robb & Evans, 1994) as does leucomyosuppressin

(Cuthbert & Evans, 1989). These effects of SchistoFLRFamide are long-lasting, and in many preparations are still present 2 min after peptide removal. Interestingly, as heart beat frequency begins to recover during wash out, contraction amplitude is potentiated for several minutes. This is particularly noticeable at high doses, although it does not occur until several minutes after the removal of the peptide. More recently, SchistoFLRFamide-like immunoreactivity has been suggested to be present in a pair of suboesophageal neurones which have previously been shown to innervate the heart (Bräunig, 1991). Thus, the modulatory effects of SchistoFLRFamide on the locust heart may well be due to release of SchistoFLRFamide from these heart neurones, but possibly could also be due to release from neurohaemal organs associated with the abdominal ganglia (Swales & Evans, 1995).

Skeletal muscle

The locust extensor tibiae muscle is sensitive to FaRPs, and a number of studies (Walther *et al.*, 1984; Evans & Meyers, 1986; Cuthbert & Evans, 1989; Walther *et al.*, 1991) have examined the modulatory actions of FaRPs on the tension generated by stimulation of the slow excitatory motoneurone (SETi). Once again, some members of the family of FaRPs are excitatory, increasing the amplitude, contraction rates and relaxation rates of twitch tension. However, leucomyosuppressin actually reduces both the amplitude and the rate of relaxation of twitch tension at concentrations above 1 μmol 1^{-1} (Cuthbert & Evans, 1989). At lower concentrations, there are small increases in these parameters. In addition, leucomyosuppressin inhibits the myogenic rhythm found in the extensor tibiae muscle. SchistoFLRFamide produces a complex dose-dependent pattern of potentiation and inhibition of the amplitude and relaxation rate of SETi-induced twitch tension (Robb & Evans, 1994). At low concentrations, SchistoFLRFamide produces a small but variable increase in the amplitude of twitch tension, whereas at higher concentrations (0.1–10 μmol 1^{-1}) the responses are complex and vary between preparations, with some producing potentiation, others inhibition, and others being biphasic. These data could be interpreted as indicating the presence of multiple receptor types for the varied subfamily of FaRPs.

Elsewhere, leucomyosuppressin attenuates evoked transmitter release from the presynaptic terminal of excitatory motoneurones innervating the ventral longitudinal muscles of the mealworm *Tenebrio molitor* (Yamamoto *et al.*, 1988). Thus, leucomyosuppressin decreases neurotransmitter quantal content and has no effect on glutamate-induced depolarisation. The net effect would be to reduce the amplitude of neuronally evoked contractions.

In contrast to these inhibitory effects, ManducaFLRFamide increases the force of neuronally evoked contractions in the flight muscles of the tobacco

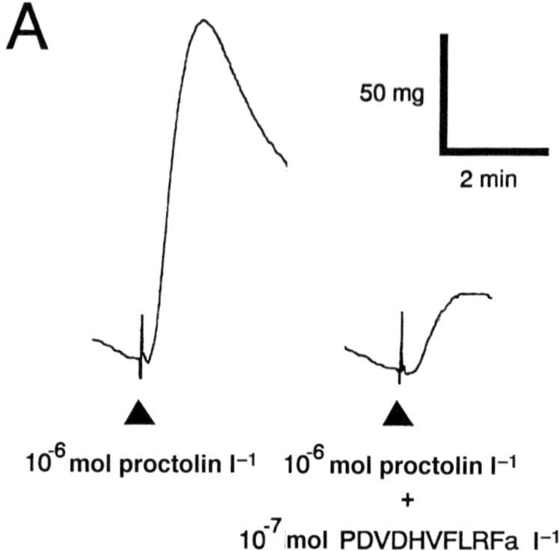

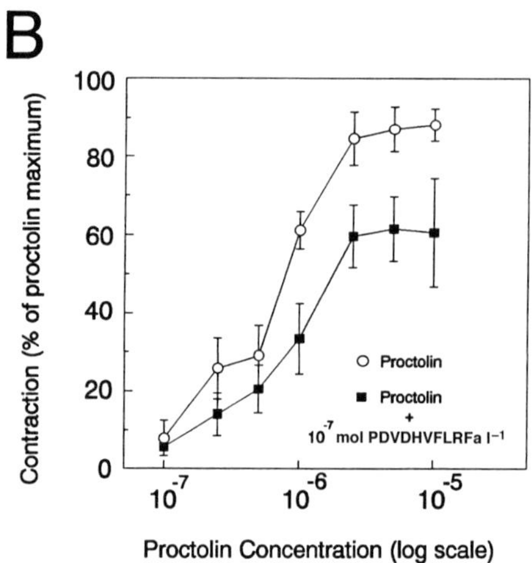

Fig. 1. Effect of SchistoFLRFamide (PDVDHVFLRFamide) on proctolin-induced contraction of the circular muscles of the midgut of the locust *Locusta migratoria*. (A) Addition of 1 μmol proctolin l⁻¹ to a ring preparation of the midgut muscle produces an increase in basal tonus. In the presence of 0.1 μmol PDVDHVFLRFamide l⁻¹, the proctolin-induced contraction is reduced. The effects of proctolin and PDVDHVFLRFamide are reversible when the preparation is washed with saline (not shown) although the preparation is slow to recover. (B) Addition of increasing concentrations of proctolin produces a dose-dependent increase in basal tonus (O) which is reduced in the presence of 0.1 μmol PDVDHVFLRFamide l⁻¹ (■). The contraction is measured as a percentage

hawkmoth, *Manduca sexta* (Kingan *et al.*, 1990), although not all preparations respond. SchistoFLRFamide has also been shown to increase the
amplitude of neuronally evoked contractions in the external protractor
muscle of *L. migratoria* (Facciponte, Miksys & Lange, 1995), an effect which
has been shown recently to be at least partially due to an enhancement of
neurotransmitter quantal content (A. B. Lange, unpublished results).

Receptors

An *in vitro* binding assay using [^{125}I-Y^1]SchistoFLRFamide (Tyr-Asp-
Val-Asp-His-Val-Phe-Leu-Arg-Phe-NH$_2$), an iodinated analogue of
SchistFLRFamide, has recently been developed to demonstrate and
characterise putative receptors for SchistoFLRFamide associated with the
locust oviduct (Wang, Orchard & Lange, 1994). Identification of the putative
receptors for these inhibitory peptides is a critical stage for the elucidation of
their mode of action and their structure–activity relationships. Preliminary
results indicated that this analogue retained physiological activity similar to
that of SchistoFLRFamide on locust oviduct and therefore would be useful
as an analogue for binding studies. Fixed amounts of membrane fraction
obtained from locust oviduct were incubated with varying concentrations of
radioligand and the bound radioactivity measured. Specific binding was
determined by deducting non-specific binding, as measured in the presence
of 50 μmol unlabelled [Y^1]SchistoFLRFamide 1^{-1}, from the total binding.
The specific binding shows a two-phase increase, the first between 0.02 nmol
1^{-1} and 8 nmol 1^{-1} and the second between 10 nmol 1^{-1} and 200 nmol 1^{-1},
suggesting the presence of two receptors. Scatchard transformation confirmed the presence of a high affinity receptor and a low affinity receptor. The
high affinity receptor has a K_d of 0.95$\pm$0.11 nmol 1^{-1} and B_{max} of 14.5$\pm$1.2
fmol (mg protein^{-1}), while the low affinity receptor has a K_d of 0.19$\pm$0.02
μmol 1^{-1} and B_{max} of 540$\pm$43 fmol (mg protein^{-1}). Binding to both receptors is saturable, specific and reversible and competitively inhibited by [Y^1]

Fig. 1. (*cont.*)
relative to the maximum contraction produced by proctolin alone. Symbols
are mean$\pm$S.E.

    Note: The midgut is dissected out of the locust under physiological saline,
two strands of thread are guided through the lumen, one of which is tied
and pinned to a Sylgard-coated dish, the other is also tied and attached to a
force transducer. This method monitors contractions of the circular
muscles of the midgut. The midgut appears as a ring shaped structure and
therefore the preparation is called a ring preparation. (Thanks to Cynthia
Thomas for technical help in this work.)

SchistoFLRFamide, [P¹]SchistoFLRFamide and [A¹]SchistoFLRFamide. Binding is dependent upon both calcium and magnesium, and is completely inhibited by 2 mmol EDTA 1^{-1}. The affinities of these two receptors are lower than the affinity of the receptor for Phe-Leu-Phe-Gln-Pro-Gln-Arg-Phe-NH$_2$ in rat spinal cord (Allard *et al.*, 1989), but similar to the affinities of the two FMRFamide receptors identified in *Helix aspersa* (Payza, 1987). The affinity of the two SchistoFLRFamide receptors on locust oviducts is consistent with the doses of SchistoFLRFamide required to produce inhibitory biological activity. Interestingly, the receptors show a regional distribution, with the majority of the high affinity receptors associated with the upper lateral oviducts, which receive little or no innervation, and most of the low affinity receptors associated with the lower lateral and common oviducts, which receive extensive innervation. This differential distribution may have some physiological relevance in locust oviduct with respect to its response to released SchistoFLRFamide. SchistoFLRFamide-like immunoreactivity is found associated mainly with the highly innervated region of the oviduct (Lange *et al.*, 1991), and is located in one class of nerve terminals (Wang & Orchard, 1995). When released from these nerve terminals, the concentration of SchistoFLRFamide would gradually decrease with diffusion towards less innervated regions. The greater density of high-affinity receptors in the less innervated areas will allow these areas to respond to the lower concentrations of released SchistoFLRFamide.

Transduction and mode of action

It seems clear that SchistoFLRFamide acts upon the locust oviduct via membrane-bound receptors. Recent studies indicate that these receptors are coupled to G-proteins. As has been mentioned earlier, there appear to be two receptors for SchistoFLRFamide in oviduct muscle membrane, a high affinity and a low affinity receptor (Wang *et al.*, 1994). The non-hydrolysable GTP analogue, GTPγS, reduces the binding of SchistoFLRFamide to both types of receptors (Wang, Lange & Orchard, 1995a). Indeed, the binding affinities (K_d) of the receptors are reduced from 0.7 nmol 1^{-1} to 1.9 nmol 1^{-1} for the high affinity site, and from 0.19 μmol 1^{-1} to 0.45 μmol 1^{-1} for the low affinity site. Maximum binding capacities are not significantly altered. Furthermore, SchistoFLRFamide increases GTPase activity in a dose-dependent manner, with 10 μmol SchistoFLRFamide 1^{-1} inducing a 75% increase in GTPase activity. Neither cholera toxin nor pertussis toxin alter specific binding to the receptors, indicating that the G-proteins that mediate the effects of SchistoFLRFamide are unlikely to be G$_s$ or G$_i$ types.

The possible mode of action of SchistoFLRFamide on locust oviduct has recently been examined (Wang *et al.*, 1995c). The ability of

SchistoFLRFamide to block such a broad range of spontaneous and induced contractions indicates a post-synaptic action beyond the level of receptor activation and somewhere in the excitation–contraction pathway. Concentrating on an influence upon calcium mobilisation, Wang *et al.* (1995c) examined the ability of SchistoFLRFamide to inhibit oviduct contractions induced by the calcium inophore A23187, caffeine or phorbol ester. A23187-induced muscle contractions include two components: one requires the influx of extracellular calcium through calcium channels that are blocked by cobalt ions; the other component is not blocked by cobalt ions. SchistoFLRFamide inhibits the cobalt-sensitive component but not the other. Caffeine induces two phases of contraction: a phasic contraction, which is probably due to the release of intracellular calcium; and a tonic contraction that requires the influx of extracellular calcium. SchistoFLRFamide inhibits the tonic contraction in a dose-dependent manner, but does not influence the phasic contraction. Phorbol 12-myristate 13-acetate induces muscle contractions which requires the presence of extracellular calcium. SchistoFLRFamide does not inhibit these induced contractions. Taken as a whole these experiments indicate that SchistoFLRFamide inhibits muscle contraction by preventing the accumulation of free intracellular calcium from the extracellular medium. SchistoFLRFamide is incapable of inhibiting contraction induced by the release of intracellular stores of calcium. It would appear that SchistoFLRFamide closes or blocks voltage-gated and some ligand-gated channels in the plasma membrane. The effects of leucomyosuppressin on both *L. maderae* hindgut and mealworm neuromuscular junction are blocked by nordihydroguiaretic acid, an inhibitor of lipoxygenase, indicating that the involvement of this second messenger pathway in these preparations (Nachman *et al.*, 1994).

Structure–activity relationships

The structure–activity relationships of leucomyosuppressin have been examined in *L. maderae* hindgut (Nachman *et al.*, 1993). In this preparation, Val-Phe-Leu-Arg-Phe-NH$_2$ (VFLRFamide) appears to be the active core for inhibitory biological activity, although its activity is only 0.2% of leucomyosuppressin. For activity of the same magnitude (in this case 40%) as the parent peptide, Asp-His-Val-Phe-Leu-Arg-Phe-NH$_2$ (DHVFLRF amide) is required. Using a series of substitution analogues, it has been found that an Ala substitution is only tolerated in positions 4 (Asp) and 6 (Val), in which 3% and 9%, respectively, of the inhibitory activity of the parent peptide is retained. Replacement of His[5], Phe[7], Arg[9] or Phe[10] leads to inactive analogues.

The development of a receptor binding assay and a bioassay for

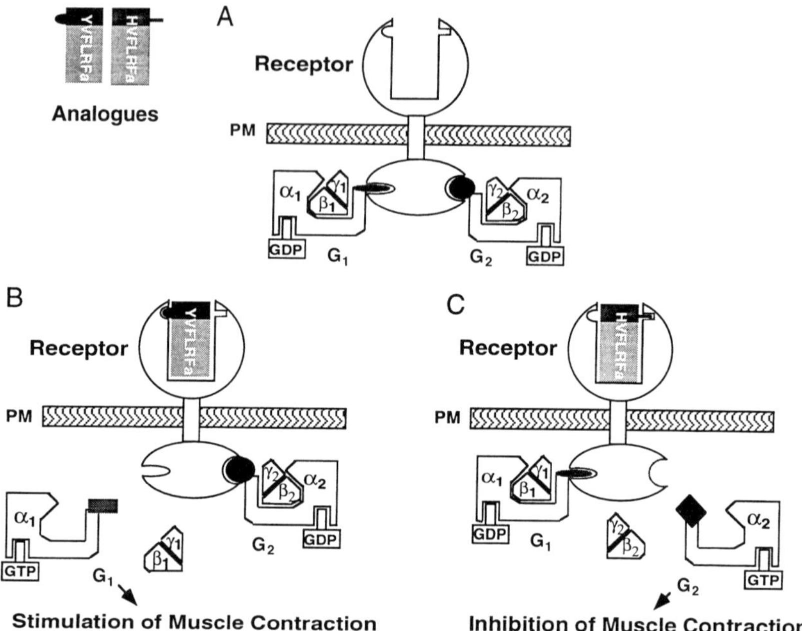

Fig. 2. Schematic model of the FaRP-receptor reaction system in the locust oviduct. Locust oviducts possess two receptor types for FaRPs, a high and a low affinity receptor (see Wang *et al.*, 1994). In this model the illustrated receptor refers to both the high and the low affinity receptor on locust oviducts. This model predicts that each receptor type can have a higher and lower affinity state. (A) In the absence of both excitatory and inhibitory FaRPs, the receptor is coupled to two G-proteins. This one receptor/two G-protein complex has high binding affinity to both excitatory and inhibitory FaRPs, but has no effect on muscle contraction. (B) Upon binding to an excitatory analogue, YVFLRFamide, the receptor activates one G-protein (G_1) through the release of GDP and the binding of GTP. Once activated the GTP-bound α_1-subunit dissociates from the β_1- and γ_1-subunits and activates an excitatory downstream signalling cascade to stimulate muscle contraction. The resulting one receptor/one G-protein (G_2) complex has low affinity to YVFLRFamide analogues. (C) Upon binding to an inhibitory analogue, HVFLRFamide, the receptor activates the other G-protein (G_2) which dissociates from the receptor and activates an inhibitory downstream signalling cascade to inhibit muscle contraction. The resulting one receptor/one G-protein (G_1) complex has low affinity to HVFLRFamide analogues. a, amide; PM, plasma membrane. (Figure adapted from Wang *et al.*, 1995a.)

SchistoFLRFamide on locust oviduct has provided an ideal model system to critically examine the relationship between the structure of the parent peptide, and its interaction with both binding and activation sites on the receptors. The amide is critical for both binding and inhibitory biological activity, because Pro-Asp-Val-Asp-His-Val-Phe-Leu-Arg-Phe (PDVD-HVFLRF)-COOH is unable to displace the radioligand in the binding assay at concentrations up to 0.1 mmol 1^{-1} (Wang *et al.*, 1995d). Using N-terminally truncated peptides, Wang *et al.* (1995d) have also shown that SchistoFLRFamide has separate binding and activation regions. VFLRFamide is the minimum sequence required for binding of comparable affinity to PDVDHVFLRFamide, whereas HVFLRFamide is the minimum sequence for comparable inhibitory biological activity (Table 2). Thus, the His residue, which does not apparently contribute to binding, is a critical amino acid for the activation of the receptor. Indeed, VFLRFamide itself shows activity reversal, displaying weak stimulatory activity on oviduct muscle contraction, a situation which differs for *L. maderae* hindgut (Nachman *et al.*, 1993). Histidine has previously been shown to be a critical residue for determining biological activity of a number of peptides, including gonadotropin-releasing hormone, angiotensin II, glucagon and luteinising hormone-releasing hormone, and analogues in which the His imidazole group have been altered have revealed details of the important features of the His residue (Rivier *et al.*, 1972; Needleman, Marshall & Rivier, 1973; Arnold *et al.*, 1974; Hsieh, Jorgensen & Lee, 1979). The properties of the His residue in determining the inhibitory biological activity of HVFLRFamide on locust oviducts have been investigated (Lange *et al.*, 1996). A number of modifications have been made in the N-terminal His residue to examine the importance of the imidazole group for biological activity, and a comparison made with the influence of these modifications on binding (Table 2). Substitution of His by the D-isomer or by Phe produces analogues with stimulatory rather than inhibitory activity confirming the importance of the His moiety, and indicating that inhibition is not simply due to the presence of an aromatic residue. In addition, inhibitory activity is retained when the His moiety is methylated at the N-3 position of the imidazole ring, but methylation of N-1 yields a peptide which stimulates contractions. Inhibitory activity is further retained when N^{α}-methyl-L-His and D,L -1',2',4'-triazole-3-Ala are substituted for His. The activity of this latter analogue is perhaps not surprising, because two of its heterocyclic ring nitrogens are at the same positions as the imidazole nitrogens of His. Interestingly, analogues which show activity reversal and result in stimulatory rather than inhibitory activity, also have weaker binding as revealed by K_i values. Thus, while the His residue in HVFLRFamide may not participate in binding directly, alterations in this residue obviously influence

Table 2 *Binding affinity and physiological activity of SchistoFLRFamide (PDVDHVFLRFamide) and some HVFLRFamide analogues on the locust oviduct*

Peptide[a]	K_i High-affinity site (nmol l^{-1})[b]	K_i Low-affinity site (μmol l^{-1})[b]	Physiological effect[c]
PDVDHVFLRFamide	0.67	0.49	I
Truncated forms			
HVFLRFamide	0.71	0.51	I
VFLRFamide	0.85	0.43	E
FLRFamide	51.2	8.55	E
LRFamide	1490	440	—
RFamide	—	—	—
Position-1 HVFLRFamide analogues			
[3-MeHis]VFLRFamide	0.75	1.3	I
[D,L-1′,2′,4′-triazoleAla]VFLRFamide	0.83	1.64	I
[D-His]VFLRFamide	12.8	17.2	E
[1-MeHis]VFLRFamide	19.9	59.1	E
FVFLRFamide	34.5	216.8	E
IVFLRFamide	0.71	0.48	E
LVFLRFamide	0.66	0.38	E
VVFLRFamiide	0.67	0.51	E
YVFLRFamide	0.7	0.34	E

Other HVFLRFamide analogues

HLFLRFamide	0.68	0.69	I
HAFLRFamide	0.7	0.9	I
HDFLRFamide	10.4	5.16	I
HVYLRFamide	0.69	0.56	I
HVALRFamide	9.23	6.54	—
HVFVRFamide	3.37	2.25	I
HVFDRFamide	15.3	9.48	—
HVFLKFamide	0.84	0.55	—
HVFLAFamide	—	—	—
HVFLRYamide	0.58	0.48	—
HVFLRAamide	48.9	23.9	—

Notes:

[a] Single-letter code.

[b] Binding affinities (K_i) were studied by competitive displacement experiments (see Wang & Orchard, 1995; Wang *et al.*, 1995a,b; Lange *et al.*, 1996).

[c] I indicates an inhibitory effect and E an excitatory effect on muscle contraction.

binding, possibly by altering the conformation of the remaining VFLRFamide sequence.

The importance of the core amino acids has been investigated through substitution (Wang *et al.*, 1995b,d). Each amino acid in the sequence VFLRFamide has been substituted with a structurally similar or dissimilar amino acid to yield a group of HVFLRFamide analogues (Table 2). Intrestingly, as long as the His residue lies in position 1, no activity reversal is achieved, the analogues are either inhibitory or possess no biological activity. The C-terminal RFamide group is critical for biological activity, but substitution of Arg^5 or Phe^6 with the structurally similar amino acids Lys^5 or Tyr^6 results in analogues with comparable binding affinity to HVFLRFamide, suggesting their potential as antagonists (see later). Elsewhere, substitution of Val^2 with Leu^2 or Ala^2 results in analogues with high affinity and strong inhibitory biological activity.

The activity reversal shown by VFLRFamide is intriguing, and implies an alternative transduction pathway linked to the receptors of locust oviducts. Changes therefore have also been made in the His residue of HVFLRFamide to form a novel group of analogues (Wang *et al.*, 1995e; see Table 2). When the His residue is substituted by Tyr, Leu, Ile, or Val, all of the analogues exert stimulatory activity on oviduct muscle contraction. Binding studies indicate that these analogues share the same binding site as the inhibitory SchistoFLRFamide or truncated HVFLRFamide. Thus the analogues can competitively displace iodinated ligand from membrane preparations of locust oviduct; the binding sites for YVFLRFamide possess identical properties to the binding sites for SchistoFLRFamide; and both unlabelled SchistoFLRFamide and HVFLRFamide competitively displace bound $[^{125}I]$Y-VFLRFamide in the same manner as unlabelled YVFLRFamide. These experiments suggest the presence of a novel ligand receptor reaction system, in which the inhibitory and stimulatory peptides share a single receptor by having the same binding sequence VFLRFamide, but are able to produce opposite muscle responses due to differences in activation sites. Subsequent experiments have indicated that the receptor is coupled to two different G-proteins (Fig. 2), the activation of one being responsible for the inhibitory effect and the activation of the other responsible for the stimulatory effect (Wang *et al.*, 1995a).

Agonists and antagonists

The development of agonists and antagonists directed against specific receptors is a critical step in the ultimate description of the physiology and mode of action of peptides. These agonists and antagonists are needed as tools for the pharmacological dissection of normal peptide–receptor interactions. In addition, there is some hope that the peptidergic system of insects may be targeted in pest control strategies. However, insect neuropeptides are susceptible to enzymatic degradation, are rapidly excreted, poorly transported in the haemolymph, and unable to pass through the insect cuticle. A fuller understanding of the active conformation of neuropeptides can aid in the development of peptidomimetics of either pseudopeptide or non-peptide structure, capable of disrupting the normal physiological processes of insects, and overcoming the inherent limitations of the natural peptides (see Nachman *et al.*, this volume).

Peptide antagonists

A detailed examination of the binding and activation regions of SchistoFLRFamide on locust oviduct (see earlier) identified possible candidates as antagonists at the SchistoFLRFamide receptor. Thus, VFLRFamide is the minimum sequence required for binding of a comparable nature to the parent peptide, and yet HVFLRFamide is the minimum sequence for inhibitory biological activity (Wang *et al.*, 1995d). One may anticipate therefore that if, as suggested by the data, VFLRFamide competes for the same receptors as SchistoFLRFamide, then VFLRFamide could be an antagonist. Indeed this has been found to be the case, and the dose–response curve for SchistoFLRFamide on inhibition of proctolin-induced contractions shows a parallel shift to the right in the presence of VFLRFamide. In the presence of 5 μmol VFLRFamide 1^{-1}, SchistoFLRFamide is approximately 30-fold less effective. Similarly, the binding data indicate that the analogues HVFLKFamide and HVFLRYamide are capable of displacing radioligand from the receptors and yet these peptides show no inhibitory biological activity. Once again, these two peptide analogues act as antagonists of the SchistoFLRFamide receptors on locust oviduct (Wang *et al.*, 1995b,d).

Bztc: a non-peptide agonist

Recently, it was observed that the non-peptide benzethonium chloride (Bztc) shares several chemical features with the sequence VFLRFamide. Both structures have zones with branched chain and basic character, and two separate zones with a phenyl ring, indicating that two or more of the structural zones of Bztc might bind with portions of the myosuppressin receptor which interact with the side chains Val^1 or Leu^3, Arg^4 and $Phe^{2,5}$ of the peptide thereby acting as a ligand for the receptor. Bztc mimics the inhibitory biological activity of leucomyosuppressin on *L. maderae* hindgut (Nachman *et al.*, 1994). Thus, Bztc reversibly inhibits spontaneous contractions of cockroach hindgut at a threshold of 60 nmol 1^{-1}. In addition, at the mealworm neuromuscular junction, Bztc also mimics the effects of leu-comyosuppressin in reversibly suppressing the amplitude of neurally evoked excitatory junction potentials (Nachman *et al.*, 1994). Using the locust oviduct as a model preparation for the effects of SchistoFLRFamide, Lange *et al.* (1995) found that Bztc reversibly inhibits proctolin-induced contractions, neurally evoked contractions and spontaneous contractions. The effects are qualitatively identical to SchistoFLRFamide although the dose–response curves are pushed to the right for Bztc. Of considerable importance is the discovery that Bztc interacts with the receptors on locust oviduct since Bztc competitively displaces [^{125}I-Y^1] SchistoFLRFamide from both high- and low-affinity receptors. As with biological activity, however, Bztc displaces the radioligand at higher concentrations than are required by the cold peptide ligand. These data (Lange *et al.*, 1995) provide clear evidence from both biological activity and displacement from binding sites that the non-peptide Bztc is recognised by the SchistoFLRFamide receptors and that Bztc is a non-peptide ligand. The discovery of this non-peptide ligand is an important prelude to studies involving the development of pest control strategies aimed at the peptidergic system of insects.

Concluding remarks

The FaRPs represent a diverse family of neuropeptides found throughout the Metazoa. Within any single insect species, there appear to be many forms, distributed in a wide range of neurones, and possibly acting upon multiple types of receptors. SchistoFLRFamide, and the FMRFamide-related myosuppressins, represent but one sub family of FaRPs that appear to have multiple actions in insects. Although classified by the structural RFamide character and ability to inhibit visceral muscle contraction, their appearance in central neurones (some of which appear to be interneurones), neuro-secretory cells and neurohaemal organs, and peripheral sites (for example heart, oviduct, and midgut endocrine cells) indicates a wide range of actions, many of which are yet to be explored. Differences between species (such as the apparent absence of dromyosuppressin in midgut endocrine cells) also indicates that members of this subfamily may play different physiological roles in different tissues and in different insect species. Multiple receptor types are also indicated by the differences in pharmacology shown in structure–activity studies.

Extensive research is needed on this subfamily of neuropeptides before a clear picture emerges. Areas to be explored include: the molecular biology of gene expression and processing; a full description of distribution using *in situ* hybridisation and specific antibodies for immunohistochemistry; biological effects on a broader range of tissues; molecular characterisation of receptors and second-messenger systems; and a more detailed description of the sequences of other endogenous FaRPs in order to assess their physiological relevance.

Acknowledgements

This work was funded by the Natural Sciences and Engineering Research Council of Canada.

References

Allard,M., Geoffre, S., Legendre, P., Vincent, J. D. & Simonnet, G. (1989). Characterization of rat spinal cord receptors to FLFQPQRFamide, a mammalian morphine modulating peptide: a binding study. *Brain Research* **500**, 169–176.

Arnold, W., Flouret, G., Morgan, R., Rippel, R. & White, W. (1974). Synthesis and biological activity of some analogs of the gonadotropin releasing hormone. *Journal of Medical Chemistry* **17**, 314–319.

Baines, R. A., Tyrer, N. M. & Mason, J. C. (1989). The innervation of locust salivary gland. II. Physiology of excitation and modulation. *Journal of Comparative Physiology A* **165**, 407–413.

Bendena, W. G., Donly, B. C., Fusé, M., Lee, E., Lange, A. B., Orchard, I. & Tobe, S. S. (1997). Molecular characterization of the inhibitory myotropic peptide leucomyosuppressin. *Peptides* **18**, 157–163.

Bräunig, P. (1991). A suboesophageal ganglion cell innervates heart and retrocerebral glandular complex in the locust. *Journal of Experimental Biology* **156**, 567–582.

Brown, M. R., Crim, J. W. & Lea, A. O. (1986). FMRFamide- and pancreatic polypeptide-like immunoreactivity in midgut endocrine cells of a mosquito. *Tissue and Cell* **18**, 419–428.

Bulloch, A. G. M., Price, D. A., Murphy, A. D., Lee, T. D. & Bowes, H. N. (1988). FMRFamide peptide in *Helisoma*: identification and physiological actions at a peripheral synapse. *Journal of Neuroscience* **8**, 3459–3469.

Cook, B. J. & Wanger, R. M. (1991). Comparative effects of leucomyosuppressin on the visceral muscle systems of the cockroach *Leucophaea maderae. Comparative Biochemistry and Physiology* **99C**, 95–99.

Cook, B. J., Wanger, R. M. & Pryor, N. W. (1993). Effects of leucomyosuppressin on the excitation–concentration coupling of insect *Leucophaea maderae* visceral muscle. *Comparative Biochemistry and Physiology* **106C**, 671–678.

Crim, J. W., Jenkins, A. C., Brown, M. R., Herzog, G. A. & Lea, A. O. (1990). FMRF-amide in the corn earworm (*H. zea*): immunoreactivity in the midgut and cerebral nervous system. In *Insect Neurochemistry and Neurophysiology*, ed. A. B. Bořkovec & E. P. Masler, pp. 401–404. Totowa, NJ: The Humana Press Inc.

Cuthbert, B. A. & Evans, P. D. (1989). A comparison of the effects of FMRFamide-like peptides on locust heart and skeletal muscle. *Journal of Experimental Biology* **144**, 395–415.

Donly, B. C., Fusé, M., Orchard, I., Tobe, S. S. & Bendena, W. G. (1996). Characterization of the gene for leucomyosuppressin and its expression in the brain of the cockroach *Diploptera punctata. Insect Biochemistry and Molecular Biology* **26**, 627–637.

Duve, H., Johnsen, A. H., Sewell, J. C., Scott, A. G., Orchard, I., Rehfeld, J. F. & Thorpe, A. (1992). Isolation, structure, and activity of -Phe-Met-Arg-Phe-NH$_2$ neuropeptides (designated calliFMRFamides) from the blowfly *Calliphora vomitoria. Proceedings of the National Academy of Sciences USA* **89**, 2326–2330.

Evans, P. D. & Meyers, C. M. (1986). The modulatory action of FMRFamide and related peptides on locust skeletal muscle. *Journal of Experimental Biology* **126**, 403–422.

Facciponte, G., Miksys, S. & Lange, A. B. (1995). The innervation of a ventral abdominal protractor muscle in *Locusta. Journal Comparative Physiology A* **177**, 645–657.

Fónagy, D., Schoofs, L., Proost, P. Van Damme, J., Bueds, H. & De Loof, A. (1992). Isolation, primary structure and synthesis of neomyosuppressin, a myoinhibiting neuropeptide from the grey fleshfly, *Neobellieria bullata. Comparative Biochemistry and Physiology* **102C**, 239–245.

Fujisawa, Y., Shimoda, M., Kiguchi, K., Ichikawa, T. & Fujita, N. (1993). The inhibitory effect of a neuropeptide, ManducaFLRFamide, on the midgut activity of the sphingid moth, *Agrius convolvuli. Zoological Science* **10**, 773–777.

Holman, G. M., Cook, B. J. & Nachman, R. J. (1986). Isolation, primary structure and synthesis of leucomyosuppressin, an insect neuropeptide that inhibits spontaneous contractions of the cockroach hindgut. *Comparative Biochemistry and Physiology* **85C**, 329–333.

Hooper, S. L. & Marder, E. (1984). Modulation of a central pattern generator by two neuropeptides, proctolin and FMRFamide. *Brain Research* **305**, 186–191.

Hsieh, K., Jorgensen, E. C. & Lee, T. C. (1979). Angiotensin II analogues. 14. Roles of the imidazole nitrogens of position-6 histidine in pressor activity. *Journal of Medical Chemistry* **22**, 1199–1206.

Jenkins, A. C., Brown, M. R. & Crim, J. W. (1989). FMRF-amide immunoreactivity and the midgut of the corn earworm (*Heliothis zea*). *Journal of Experimental Zoology* **252**, 71–78.

Kingan, T. G., Teplow, D. B., Phillips, J. M., Riehm, J. P., Rao, R., Hildebrand, J. G., Homberg, U., Kammer, A. E., Jardine, I., Griffin, P. R. & Hunt, D. F. (1990). A new peptide in the FMRFamide family isolated from the CNS of the hawkmoth, *Manduca sexta. Peptides* **11**, 849–856.

Lange, A. B., Orchard, I. & Te Brugge, V. A. (1991). Evidence for the involvement of a SchistoFLRF-amide-like peptide in the neural control of locust oviduct. *Journal of Comparative Physiology A* **168**, 383–391.

Lange, A. B., Orchard, I., Wang, Z. & Nachman, R. J. (1995). A nonpeptide agonist of the invertebrate receptor for SchistoFLRFamide (PDVDHVFLRFamide), a member of a subfamily of insect FMRFamide-related peptides. *Proceedings of the National Academy of Sciences USA* **92**, 9250–9253.

Lange, A. B., Peeff, N. M. & Orchard, I. (1994). Isolation, sequence, and bioactivity of FMRFamide-related peptides from the locust ventral nerve cord. *Peptides* **15**, 1089–1094.

Lange, A. B., Wang, Z., Orchard, I. & Starratt, A. N. (1996). Influence of methylation or substitution of the Histidine of HVFLRFamide on biological activity and binding of locust oviduct. *Peptides* **17**, 375–380.

Li, C. & Calabrese, R. L. (1987). FMRFamide-like substances in the leech. III. Biochemical characterization and physiological effects. *Journal of Neuroscience* **7**, 595–603.

McCormick, J. & Nichols, R. (1993). Spatial and temporal expression identify dromyosuppressin as a brain–gut peptide in *Drosophila melanogaster. Journal of Comparative Neurology* **338**, 279–288.

Nachman, R. J., Giard, W., Favrel, P., Sreekumar, S. & Holman, G. M. (1997). Insect myosuppressins and sulfakinins stimulate release of the digestive enzyme alpha-amylase in two invertebrates: the scallop *Pecten maximus* and insect *Rynchophorus ferrugineus. Annals of the New York Academy of Science* **814**, 335–338.

Nachman, R. J., Holman, G. M., Haddon, W. F. & Ling, N. (1986). Leucosulfakinin, a sulfated insect neuropeptide with homology to gastrin and cholecystokinin. *Science* **234**, 71–73.

Nachman, R. J., Holman, G. M., Hayes, T. K. & Beier, R. C. (1993). Structure–activity relationships for inhibitory insect myosuppressins: contrast with the stimulatory sulfakinins. *Peptides* **14**, 665–670.

Nachman, R. J., Yamamoto, D., Holman, G. M. & Beier, R. C. (1994). Pseudopeptides and a non-peptide that mimic the biological activity of the myosuppressin insect neuropeptide family. In *Insect Neurochemistry and Neurophysiology, 1993*, ed. A. B. Borkovec & M. J. Loeb, pp. 319–322. Boca Raton, FL: CRC Press.

Needleman, P., Marshall, G. R. & Rivier, J. (1973). Angiotensin II. Synthesis and biological activity of 6-(1-methylhistidine) and 6-(3-methylhistidine) analogs. *Journal of Medical Chemistry* **16**, 968–970.

Nichols, R. (1992). Isolation and structural characterization of *Drosophila* TDVDHVFLRFamide and FMRFamide-containing neural peptides. *Journal of Molecular Neuroscience* **3**, 213–218.

Orchard, I., Donly, B. C., Fusé, M., Lange, A. B., Tobe, S. S. & Bendena, W. G. (1997). FMRFamide-related peptides in insects, with emphasis on the myosuppressins. *Annals of the New York Academy of Science* **814**, 307–309.

Painter, S. D. & Greenberg, M. J. (1982). A survey of the responses of bivalve hearts to the molluscan neuropeptide FMRFamide and to 5-hydroxytryptamine. *Biological Bulletin of the Marine Biology Laboratories, Woods Hole* **162**, 311–322.

Payza, K. (1987). FMRFamide receptors in *Helix aspersa*. *Peptides*, **8**, 1065–1074.

Peeff, N. M., Orchard, I. & Lange, A. B. (1994). Isolation, sequence, and bioactivity of PDVDHVFLRFamide and ADVGHVFLRFamide peptides from the locust central nervous system. *Peptides* **15**, 387–392.

Price, D. A. & Greenberg, M. J. (1977). Structure of a molluscan cardioexcitatory neuropeptide. *Science* **197**, 670–671.

Rivier, J., Vale, W., Monahan, M., Ling, N. & Burgus, R. (1972). Synthetic thyrotropin releasing factor analogs. 3. Effect of replacement or modification of histidine residue on biological activity. *Journal of Medical Chemistry* **15**, 479–482.

Robb, S. & Evans, P. D. (1994). The modulatory effect of SchistoFLRFamide on heart and skeletal muscle in the locust *Schistocerca gregaria*. *Journal of Experimental Biology* **197**, 437–442.

Robb, S., Packman, L. C. & Evans, P. D. (1989). Isolation, primary structure and bioactivity of SchistoFLRF-amide, a FMRF-amide-like neuropeptide from the locust, *Schistocerca gregaria*. *Biochemical and Biophysical Research Communications* **160**, 850–856.

Schoofs, L., Holman, G. M., Paemen, L., Veelaert, D., Amelineky, M. & De Loof, A. (1993). Isolation, identification, and synthesis of PDVDHVFLRFamide (SchistoFLRFamide) in *Locusta migratoria* and its association with the male accessory glands, the salivary glands, the heart, and the oviduct. *Peptides* **14**, 409–421.

Sivasubramanian, P. (1992). FMRFamide-like immunoreactivity in the stomatogastric nervous system innervating the gut of the fly, *Sarcophaga bullata*. *Comparative Biochemistry and Physiology* **103C**, 333–337.

Spencer, A. N. (1988). Effects of Arg-Phe-amide peptides on identified motor neurons in the hydromedusa *Polyorchis penicillatus*. *Canadian Journal of Zoology* **66**, 639–645.

Stevenson, P. A. & Pflüger, H.-J. (1994). Colocalization of octopamine and FMRFamide related peptide in identified heart projecting (DUM) neurones in the locust revealed by immunohistochemistry. *Brain Research* **638**, 117–125.

Swales, L. S. & Evans, P. D. (1995). Distribution of SchistoFLRFamide-like immunoreactivity in the adult ventral nervous system of the locust, *Schistocerca gregaria*. *Cell and Tissue Research* **281**, 339–348.

Tsang, P. W. & Orchard, I. (1991). Distribution of FMRFamide-related peptides in the blood-feeding bug, *Rhodnius prolixus*. *Journal of Comparative Neurology* **311**, 17–32.

Ude, J. & Agricola, H. (1995). FMRFamide-like and allatostatin-like immunoreactivity in the lateral heart nerve of *Periplaneta americana*: colocalization at the electron-microscopic level. *Cell and Tissue Research* **282**, 69–80.

Veenstra, J. A. & Lambrou, G. (1995). Isolation of a novel RFamide peptide from the midgut of the American cockroach, *Periplaneta americana*. *Biochemical and Biophysical Research Communications* **213**, 519–524.

Walther, C., Schiebe, M. & Voigt, K. H. (1984). Synaptic and non-synaptic effects of molluscan cardioexcitatory neuropeptides on locust skeletal muscle. *Neuroscience Letters* **45**, 99–104.

Walther, C., Zittlau, K. E., Murck, H. & Nachman, R. J. (1991). Peptidergic modulation of synaptic transmission in locust skeletal muscle. In *Comparative Aspects of Neuropeptide Function*, ed. E. Florey & G. Stefano, pp. 175–186. Manchester: Manchester University Press.

Wang, Z., Lange, A. B. & Orchard, I. (1995a). Coupling of a single receptor to two different G proteins in the signal transduction of FMRFamide-related peptides. *Biochemical and Biophysical Research Communications* **212**, 531–538.

Wang, Z. & Orchard, I. (1995). Ultrastructural and immunocytochemical studies of neuromuscular junctions in oviduct of *Locusta migratoria*. *Cell and Tissue Research* **279**, 591–599.

Wang, Z., Orchard, I. & Lange, A. B. (1994). Identification and characterization of two receptors for SchistoFLRFamide on locust oviduct. *Peptides* **15**, 875–882.

Wang, Z., Orchard, I. & Lange, A. B. (1995b). Binding affinity and physiological activity of some HVFLRFamide analogues on the oviducts of the locust, *Locusta migratoria*. *Regulatory Peptides* **57**, 339–346.

Wang, Z., Orchard, I., Lange, A. B. & Chen, X. (1995c). Mode of action of an inhibitory neuropeptide SchistoFLRFamide on the locust oviduct visceral muscle. *Neuropeptides* **28**, 147–155.

Wang, Z., Orchard, I., Lange, A. B. & Chen, X. (1995d). Binding and activation regions of the decapeptide PDVDHVFLRFamide (SchistoFLRFamide). *Neuropeptides* **28**, 261–266.

Wang, Z., Orchard, I., Lange, A. B., Chen, X. & Starratt, A. N. (1995e). A single receptor transduces both inhibitory and stimulatory signals of FMRFamide-related peptides. *Peptides* **16**, 1181–1186.

Weiss, S., Goldberg, J. I., Chohan, K. S., Stell, W. K., Drummond, G. I. & Lukowiak, K. (1984). Evidence for FMRFamide as a neurotransmitter in the gill of *Aplysia*. *Journal of Neuroscience* **4**, 1994–2000.

Yamamoto, D., Ishikawa, S., Holman, G. M. & Nachman, R. J. (1988). Leucomyosuppressin, a novel insect neuropeptide, inhibits evoked transmitter release at the mealworm neuromuscular junction. *Neuroscience Letters* **95**, 137–142.

Zitnan, D., Sauman, I. & Sehnal, F. (1993). Peptidergic innervation and endocrine cells of insect midgut. *Archives of Insect Biochemistry and Physiology* **22**, 113–132.

HEINRICH DIRCKSEN

Conserved crustacean cardioactive peptide (CCAP) neuronal networks and functions in arthropod evolution

Introduction

Since the crustacean cardioactive peptide (CCAP) was first discovered and identified by Stangier *et al.* (1987) as a cardioacceleratory factor in the pericardial organs (POs) of the crab *Carcinus maenas*, the authentic peptide has been found in several species from different arthropod groups. It is a C-terminally amidated nonapeptide which shows an unique cyclic structure with an intramolecular disulphide bridge between the cysteines. Evidence has been provided for the existence of CCAP-related molecules in species from other invertebrate phyla, for example an echiuroid worm, *Urechis unicinctus* (UEP$_C$, *Urechis* excitatory peptide C; Ikeda *et al.*, 1991), and the Roman snail, *Helix pomatia* (CCAP-RP, CCAP-related peptide; Minakata *et al.*, 1993; Muneoka *et al.*, 1994). These molecules share obvious similarities in the position of an intramolecular disulphide bridge between cysteines, but differ with regard to the amidation of the C-terminus or in length (Fig. 1). It is not clear whether these structural similarities are indicative of an early diversification in invertebrate evolution. Furthermore, the possible existence of other forms of CCAP in arthropods, which may occur in addition to authentic CCAP, cannot be excluded, as has been shown from studies in *Limulus polyphemus* (Groome & Lehman 1995; see below). Nonetheless, it is a striking feature of CCAP that, to our present knowledge, its primary structure has obviously been highly conserved in arthropods. Authentic CCAP has now been identified in two other crustacean and five insect species. Apart from the heart, several visceral muscles, and even the crustacean retina and the stomatogastric ganglion are target organs. Much has been learned about the distribution of CCAP in the different protostomian invertebrate phyla since specific antisera against CCAP became available for immunocytochemistry (ICC) and radioimmunoassay (RIA) (Dircksen & Keller, 1988; Stangier *et al.*, 1988). Strikingly similar CCAP-immunoreactive structures and substances have hitherto been demonstrated in many species of crustaceans, insects, chelicerates, myriapods, some molluscs and annelids, which is indicative of a long evolutionary history of CCAP and perhaps similar molecules

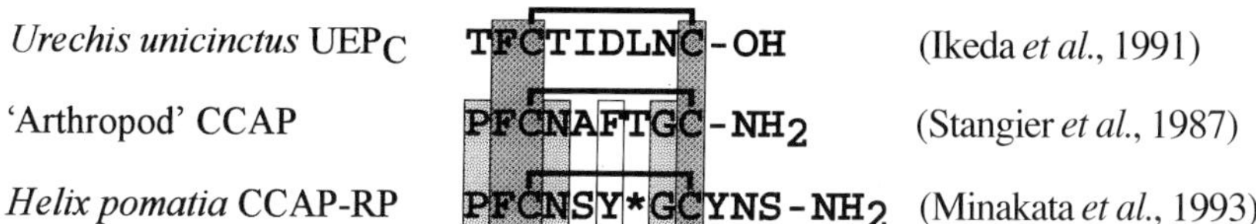

Urechis unicinctus UEP$_C$	TFCTIDLNC-OH	(Ikeda *et al.*, 1991)
'Arthropod' CCAP	PFCNAFTGC-NH$_2$	(Stangier *et al.*, 1987)
Helix pomatia CCAP-RP	PFCNSY*GCYNS-NH$_2$	(Minakata *et al.*, 1993)

Fig. 1. Relationship between authentic 'arthropod' CCAP and related molecules from other invertebrate phyla (* gap introduced to maximise identity). Note the corresponding Phe[2] and Cys residues in all three, but more overlap in the arthropod and molluscan molecules, including the amino acid exchange from Phe[5] to Tyr[5] in molluscan CCAP-RP. For abbreviations see the text.

(see Table 2). There is at present no report on the occurrence of CCAP in any deuterostomian group.

CCAP in crustaceans

Biological activity

The bioactivity of CCAP has been established in bioassays using isolated or semi-isolated heart preparations. CCAP displays strong chronotropic effects on the isolated heart of *Carcinus maenas* (Stangier *et al.*, 1987). For comparison, proctolin, another cardioactive factor isolated earlier from crab pericardial organs (Stangier, Dircksen & Keller, 1986), produced mainly inotropic effects (Stangier, 1991). Other workers, measuring heart rate, stroke volume, arterial pulse pressure and ventricular pressure upon application of CCAP determined cardiac performance in terms of cardiac output, stroke work and power in the same species. They similarly found that CCAP increased cardiac output mainly by increasing the heart rate, but found only moderate increases in stroke volume at threshold concentrations of <1 nmol l^{-1}, and a mean effective dose of 0.1 μmol l^{-1} (Wilkens & McMahon, 1992; Wilkens & Mercier, 1993). However, the cardioexcitatory effects of CCAP seem to be species-specific, because only moderate to strong positive chronotropic and inotropic effects are described for the hearts of the crayfishes *Orconectes limosus* (Stangier & Keller, 1990) and *Procambarus clarkii* (Wilkens, 1995), whereas responses to CCAP were weak or absent on hearts of the American lobster, *Homarus americanus*, and of the Dungeness crab *Cancer magister* (Stangier, 1991; McGaw, Airriess & McMahon, 1994; Wilkens, 1995). Interestingly, CCAP causes pronounced increases in scaphognathite beat frequency in *H. americanus* and in the edible crab, *Cancer pagurus* (Stangier, 1991), and *C. magister*. In the latter, this is obviously accompanied by a redirection of haemolymph flow through the anterolateral arteries and a reduction in flow through the sternal arteries (McGaw *et al.*, 1994; Wilkens *et al.*, 1996). Some workers concluded that CCAP may not be a physiologically relevant cardiac neurohormone in Cancridae (McGaw *et al.*, 1994). This might be true: threshold concentrations were at least two orders of magnitude higher than the haemolymph concentration of about 10 pmol CCAP 1^{-1} previously measured for the shore crab by radioimmunoassay (RIA; Stangier *et al.*, 1988). A release of about 1% of CCAP from the neurohaemal POs (containing 40 pmol CCAP per pair of POs or 868 pmol (mg protein)$^{-1}$) would, in a medium-sized crab with a blood volume of 5–10 ml, raise the level of circulating CCAP above threshold, as determined for semi-isolated hearts in this species (0.1 nmol CCAP 1^{-1}; Stangier *et al.*, 1988; Stangier, 1991). It is also still not clear whether heart muscle cells are the target of CCAP, or whether the peptide exerts its effect directly on certain cardiac ganglion cells, as has been suggested by Yazawa & Kuwasawa (1992) from their work on the giant hermit crabs *Aniculus aniculus* and *Dardanus crassimanus*.

Structure and distribution of CCAP immunoreactive neurones

Immunocytochemical analysis revealed that CCAP in the POs originates from projections of large neurosecretory cells in dorsal lateral positions of each neuromere of the crab thoracic ganglia complex (Dircksen & Keller, 1988; Figs. 2, 3a). On re-examination, these cells were found to have contralateral intraganglionic projections (H. Dircksen, unpublished results; see

---

Fig. 2. CCAP-immunoreactive neurones in the central nervous system of the shore crab, *C. maenas*. The segmentally homologous large type-1 and small type-2 cells in the ventral nerve cord the type-1 neurones showing projection patterns to the segmental nerves and neurohaemal terminals in the anterior ramifications and the pericardial organs. Note the contralateral pathway of type-1 neurones in a cross-section plane through the mouth part ganglia. ANP, antennal neuropils; AR, anterior ramifications; CG, cerebral ganglia; COG, connective ganglia; DAMC, dorsal anterior median cell cluster; LG, lamina ganglionaris; LNNP$_2$, leg nerve neuropil of leg 2; LT, CCAPir lateral tract; ME medulla externa; MI, medulla interna; MPG, mouth part ganglia; MT, medulla terminalis; OL, olfactory lobe; P$_1$, cross-section plane 1; P$_2$, cross-section plane 2; PO, pericardial organ; SA sternal artery; SG, sinus gland; SN, segmental nerve; STN, stomatogastric nervous system; TG, thoracic ganglia complex; VNP, ventral neuropils; XO, X-organ. (Modified from Dircksen & Keller, 1988.)

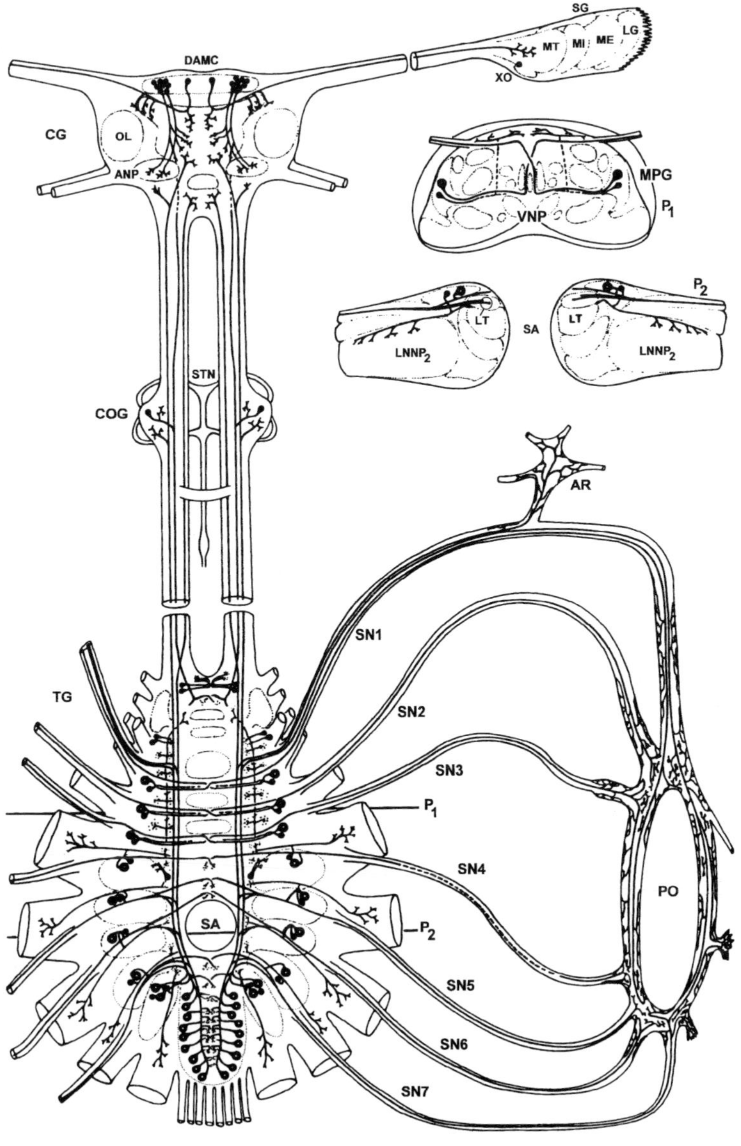
SG
LG
MT MI ME
XO
DAMC
CG
OL
ANP
MPG
VNP
P1
P2
STN
LT
SA
LT
COG
LNNP2
LNNP2
AR
SN1
SN2
SN3
TG
P1
SN4
P2
SA
PO
SN5
SN6
SN7

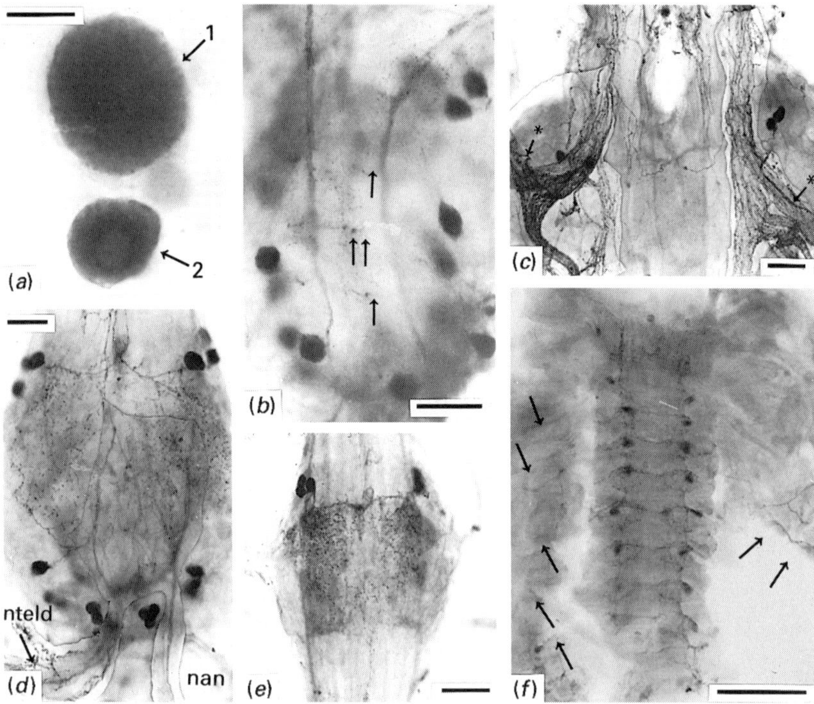

Fig. 3. CCAPir neurones in crustaceans. (a) Somata of large type-1 (1) and small type-2 (2) neurones in the fifth thoracic neuromere of the shore crab, *C. maenas*. (b) Contralateral pathways (arrows) of large type-1 neurones in the abdominal ganglia of *C. maenas* (a, b: H. Dircksen, unpublished observations). (c) Fourth thoracic ganglion of juvenile crayfish (*A. astacus*) showing central contralateral pathways of cnc- and cdn-neurones (type-1 and 2) and the neurohaemal areas associated with the third roots at the dorsal surface of the ganglion (asterisks; photomontage). (d) Terminal abdominal ganglion of a juvenile *A. astacus* showing three groups of cnc- and cdn neurones (type-1 and 2) and their projections into the nervus telsonos dorsalis (nteld, from cnc-type-1) and the nervus ani (nan, from cdn-type-2) (photomontage). (e) Fourth abdominal ganglion of *A. astacus* showing duplicate cnc-type-1 and cdn-type-2 neurones. (f) Thoracic neuromeres of an *A. astacus* embryo (about 97% of development) showing well established segmentally homologous neurones, lateral tracts and peripheral projection sites (arrows) (c–f: A. Trube, U. Audehm & H. Dircksen, unpublished observations). Bars=20 μm in (a); in all others 100 μm.

Figs. 2, 3b, 4b), with axons running through the corresponding segmental nerves (SNs; Maynard, 1961) into the POs and into the anterior ramifications complex (ARC), a neurohaemal system spreading across the respiratory muscles (Fig. 2). The cells give rise to masses of cortical terminals containing the largest identified type of electron dense neurosecretory granules both in the POs and the ARC (Dircksen & Keller, 1988; Dircksen, 1990, 1991). These peripheral distributions of CCAP-immunoreactive (CCAPir) terminals are in close proximity to the suspected targets of CCAP as established from physiological experiments. Furthermore, in each suboesophageal and thoracic neuromere, another type of contralaterally projecting CCAPir neurone with a smaller perikaryon is found next to the neurosecretory neurone, which is most likely an interneurone (Figs. 3a, 4b). Other less well-characterised neurones are found in the eyestalk, brain and connective ganglia of the crab (Dircksen & Keller, 1988; Fig. 2). Peripheral ganglia, such as the stomatogastric ganglion, do not contain CCAPir fibres or terminals. However, lateral pyloric and gastric motoneurones in the crab *Cancer borealis* respond strongly to CCAP (Weimann, Heinzel & Marder, 1992; Skiebe-Corrette *et al.*, 1993; Weimann *et al.*, 1997), by dramatically changing the bursting modes of neurones in the network and the movements of the valve between the gastric mill and the pyloric chamber of the stomach. Anatomically, this ganglion is not a surprising target for humoral activites, because it lies within the cor frontale (Baumann, 1917) within the cephalic artery (one of the major vessels from the heart). Thus, the ganglion can directly be perfused with CCAP released from the POs into the pericardial cavity.

Studies on the more plesiomorphic central nervous system of crayfish have led to a deeper understanding of the strikingly similar pathways of likely homologous CCAPir neurones in brachyurans and astacurans, and have suggested other possible targets for CCAP (Trube, 1992; Audehm, Trube & Dircksen, 1993; Trube, Audehm & Dircksen, 1994). In both crayfish and lobster (Stangier & Keller, 1990; Schwiemann, 1991), authentic CCAP elicits pronounced myotropic effects on another target, namely the hindgut (Stangier, 1991). The question therefore arose of whether this effect was brought about by CCAP released from intramural terminals in this tissue. However, Audehm *et al.* (1993) found that this was not the case, and CCAP is probably released from serially homologous CCAPir neurones in the crayfish sixth abdominal ganglion (Fig. 3d) which terminate in neurohaemal areas of the so-called peri-intestinal ring vessels (Baumann, 1918). In addition, newly discovered muscle targets such as telson flexor muscles have been found to be heavily innervated by CCAPir fibres via the nervi telsonos dorsales (Audehm *et al.*, 1993; Fig. 3d). The CCAPir neurones innervating these neurohaemal areas and muscles arise from at least three pairs of large-type neurosecretory cells in the sixth abdominal ganglion which are obvious serial homologues of

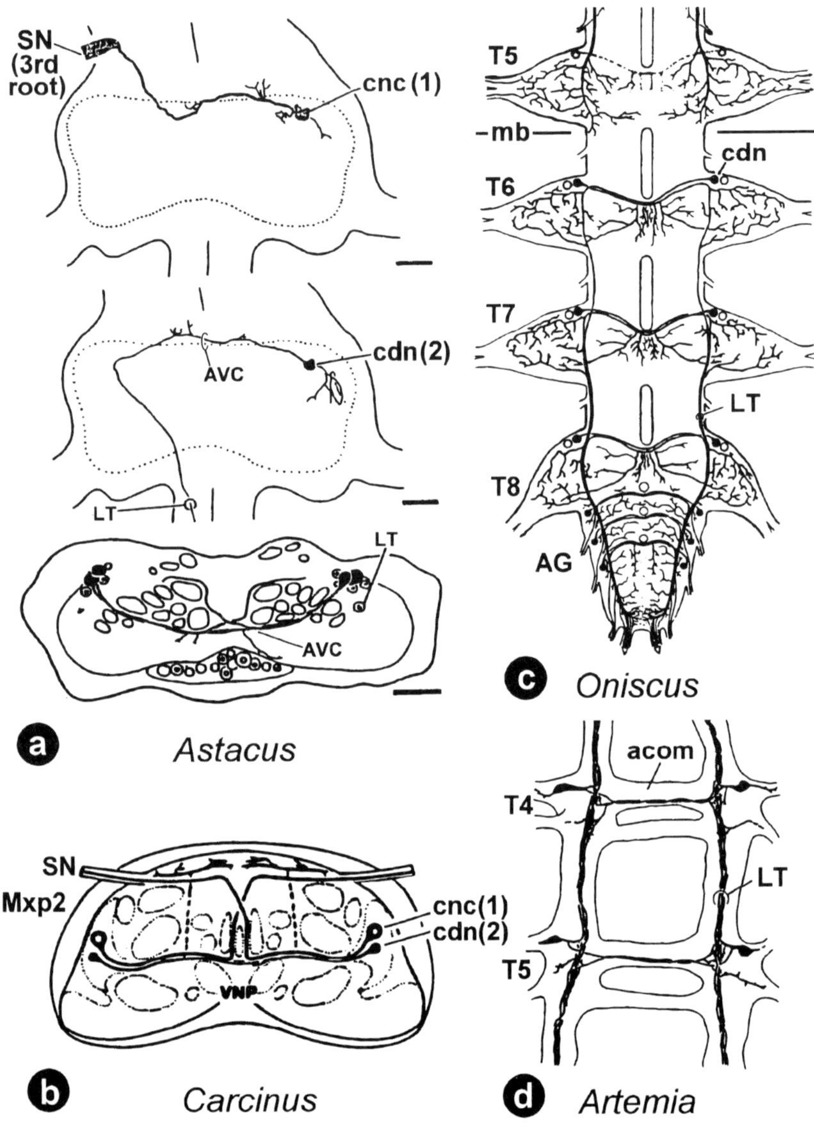

SN (3rd root)
cnc(1)
AVC
cdn(2)
LT
LT
AVC
a Astacus
T5
-mb-
cdn
T6
T7
LT
T8
AG
c Oniscus
SN
Mxp2
cnc(1)
cdn(2)
VNP
b Carcinus
acom
T4
LT
T5
d Artemia

CCAPir neurones occurring in each neuromere of crayfish ventral nerve cords (VNC; cnc-neurones; Audehm *et al.*, 1993; Trube *et al.*, 1994; Fig. 3d). In the suboesophageal (SOG), thoracic (TG) and abdominal (AG) ganglia, the cnc-neurones show a characteristic central pathway with a contralateral ascending axonal projection into the so-called third root of the next anterior neuromere (Figs. 3c, 4a). In the SOG and TG neuromeres, the axons can be followed through the segmental nerves into the POs (Trube, 1992). Fibres branching off from these axons extend close to the surfaces of all roots, across the dorsal ganglionic perineurium and in the POs, forming extensive varicosities and neurohaemal terminals, and thus comprise a newly discovered large putative neurohaemal area (Fig. 3c). These terminals closely resemble those of the crab large-type neurones in that they contain large neurosecretory granules from which CCAP is presumably released into the haemolymph (Trube *et al.*, 1994). They also closely resemble the insect type-1 neurones (see below) in terms of projection patterns. Within the AG neuromers only, the cnc-neurones occur in pairs (Fig. 3e). Their segmentally iterated occurrence proved the existence of what are obviously serially homologous cells in at least eight abdominal neuromeres (Audehm *et al.*, 1993; Trube *et al.*, 1994), a distribution which agrees well with neuromer numbers determined by other authors in studies on the expression of

---

Fig. 4. Identified CCAPir neurones in malacostracan and entomostracan Crustacea. (a) Single cnc-type-1 and cdn-type-2 neurones in a crayfish third thoracic ganglion; and a cross-section through the second thoracic ganglion on the bottom showing their pathways. Note the dorsally directed cnc-type-1 (1) axon and the cdn-type-2 (2) neuronal pathway through the anterior ventral commissure (from Trube *et al.*, 1994). (b) The cnc-type-1 and cdn-type-2 neurones in a schematized cross-section through the second maxillipedal ganglion of the crab *C. maenas* showing contralateral dorsally directed pathways of cnc-neurones into the segmental nerves (SN; H. Dircksen, unpublished observation). (c) The cdn-like neurones in the thoracic neuromeres (T6) down to the abdominal ganglia of the woodlouse *O. asellus* (schematic drawing) stain more intensely than those anterior (T1–T5) to the moulting border (mb), and cnc-neurones are barely visible (open circles). Note also interrupted lateral tract between T5 and T6 (H. Dircksen & T. Nussbaum, unpublished observation). (d) Serially homologous cdn-type-2-like neurones in the ventral nerve cord of *A. salina* showing contralateral projections through the anterior (ventral) commissure into the lateral tract. (Q. Zhang & H. Dircksen, unpublished observations.) acom, anterior commisure; AVC, anterior ventral commissure; cdn, CCAPir decending neurone; cnc, CCAPir neurosecretory cell; LT, CCAPir lateral tract; mb, moulting border; Mxp2, second maxillipedal ganglion; SN, segmental nerve; VNP, ventral neuropils; T, thoracic ganglion.

another segment marker, the product of the *engrailed* gene, in crayfish (Patel, Kornberg & Goodman, 1989a; Patel *et al.*, 1989b; Scholtz, 1995). The same distribution was found for a single small-type neurone adjacent to the cnc-neurone in each VNC neuromere (cdn-neurone; Figs. 3c,d,e, 4a). These neurones descend contralaterally through the anterior ventral commissure into the ventral lateral tract, spanning through more than one (and forming terminals in each) caudally following neuromere, finally terminating in the anal nerve (Trube *et al.*, 1994; Fig. 3d). They resemble the small-type neurones in the crab and the insect type-2 neurones (see below). The principal distribution patterns of all CCAPir crayfish VNC-neurones and the peripheral projections form a network which is already established at about 80% of embryonic development and persists into the adult stage (Trube, 1992; Fig. 3f). All other ganglia, including the eyestalks, contain less well-investigated CCAPir putative interneurones (Trube, 1992). The neurohaemal sinus gland does not contain CCAPir terminals; thus the observation by Gaus & Stieve (1992) that CCAP reduces (by about 30%) the light-flash-evoked electro-retinogram in the isolated retina of the crayfish *Orconectes limosus* suggests that any humoral effect of CCAP cannot arise from local release of CCAP in the eyestalks.

Comparable CCAPir cells occur in other malacostracans and in phylogenetically more distant orders of crustaceans, for example the entomostracan branchiopods and cirripedes. In the latter, the neurosecretory types of CCAPir neurones may be missing or their segmentally homologous distribution may be lost. In isopod malacostracans, CCAPir neurones occur throughout the brain, and the VNC contain one strongly stained serially homologous neurone per neuromer, which is similar to the decapod cdn-type-2 neurone (Fig. 4c). Nevertheless, a decapod cnc-type neurosecretory neurone exhibits only weak CCAP-immunoreactivity or may be missing completely. Interestingly, during the initiations of the biphasic moult (anterior moult) of the woodlouse *Oniscus asellus*, CCAPir cdn-like VNC-neurones anterior to the thoracic segment T5/T6-borderline stain very weakly in contrast to those in the posterior segments, but during and after the posterior moult all these neurones regain full immunoreactivity (Dircksen, 1994; Fig. 4c). These changes start in premoult, during a drop in haemolymph ecdysteroid titre, and coincide with a transient but 10-fold increase in the central nervous system (CNS) content of CCAPir material which terminates within a few hours of the posterior moult as measured by enzyme immunoassay (Johnen, von Glisynski & Dircksen, 1995). Johnen *et al.*, therefore, have suggested that CCAP plays a role in isopod moulting processes and exuviation behaviour.

One pair of CCAPir cells with contralateral projection patterns partly similar to the decapod cdn-neurones has been found in each VNC-neuromere of the brine shrimp, *Artemia salina* (Fig. 4d), but only one single pair of a similar cdn-like type of cell in the entire CNS occurs in the SOG of the waterflea, *Daphnia magna*. However, the cells in both species are bipolar and show contralateral ascending and descending branches forming a lateral tract. In *D. magna*, they terminate next to the muscles of the so-called steering spines of the tail (Zhang, Keller & Dircksen, 1993; Zhang & Dircksen, 1994; Q. Zhang & H. Dircksen, unpublished observations). More recently, in each of three neuromeres in the anterior thoracic ganglia of three adult barnacle species (*Balanus balanus, Semibalanus balanoides* and *Chirona hameri*), CCAPir cells with similarities to decapod cdn-neurons have been found to show contralateral descending projections and fibres leaving the ganglia into the cirral nerves. No CCAPir neurohaemal areas have hitherto been observed (S. G. Webster, unpublished observation).

CCAP in insects

Biological activity

Stangier, Hilbich & Keller (1989) first discovered authentic CCAP in CNS extracts of the migratory locust, *Locusta migratoria*, as a potent myoactive substance on isolated, spontaneously active locust hindguts (threshold <1 nmol 1^{-1}) and detected CCAP quantitatively throughout the entire CNS. CCAP was later found to exhibit even stronger excitatory effects on the oviduct (Burdzik, 1990, cited by Dircksen, 1994), inotropic rather than chronotropic effects on the heart (Dircksen, Müller & Keller, 1991), and excitatory effects on the metathoracic extensor tibia 'jumping' muscle of locusts (Walther *et al.*, 1991; Lehman *et al.*, 1993). Furthermore, it has recently been shown that CCAP exerts a secretagogue effect on the corpora cardiaca of *L. migratoria* causing release of adipokinetic hormones (AKH; Passier, 1996). This putative humoral effect of CCAP released from neurohaemal terminals in the corpora cardiaca (see below) occurs at much lower concentrations (1 nmol 1^{-1}) than that of locustatachykinin I (100 nmol 1^{-1}), previously described as an AKH-release stimulating factor derived from synaptoid terminals directly abutting on the AKH cells (Nässel *et al.*, 1995; see also Nässel *et al.*, this volume). More pronounced cardioacceleratory activities of CCAP were found for the heart of the tobacco hawkmoth, *Manduca sexta* (Davis *et al.*, 1990), and the blowfly, *Calliphora vicina* (Jahn, Käuser & Koolman, 1991, 1992; Jahn, 1992). The heart of *M. sexta* later served as a bioassay system for the identification of authentic CCAP in this species, and in the flour beetle, *Tenebrio molitor*, and the armyworm, *Spodoptera eridania* (Hildebrand *et al.*, 1990; Cheung *et al.*, 1992; Furuya *et al.*, 1993; Lehman *et al.*, 1993). CCAP exerts a strong chronotropic effect on the heart of *M. sexta* with a threshold concentration at about 50 pmol 1^{-1}, an effect which is still weaker than the effect of proctolin (Furuya *et al.*, 1993). Interestingly, it appeared that CCAP is identical to cardioactive peptide 2a (CAP_{2a}) one of five CAPs described previously in *M. sexta* (Tublitz & Truman 1985; Cheung *et al.*, 1992; Loi *et al.*, 1992). The excitatory activity on the hindgut of the Madeira cockroach, *Leucophaea maderae* (threshold concentration 36 pmol 1^{-1}), has also proved advantageous in a heterologous bioassay system for the identification of authentic CCAP from the stablefly *Stomoxys calcitrans* (G. M. Holman, personal communication).

Structure and distribution of CCAP immunoreactive neurones

The first detailed immunocytochemical study was made on the VNC and peripheral targets of four identified types of CCAPir neurones in *L. migratoria* (Dircksen *et al.*, 1991) which occur partially as segmental homologues. Most interesting are the highly conserved so-called type-1 and type-2 neurones, which occur as a pair of single neurones adjacent to each other in each VNC-neuromere down to the terminal abdominal ganglion, and share striking structural similarities with the crustacean cnc- and cdn-neurones, respectively (see figs. 2, 4). The type-1 neurones project ventrally from a dorsal–lateral soma to the mid-sagittal plane, where they climb to the dorsal side and descend into the intersegmental nerve of the adjacent posterior ganglion. Another distinct feature of these neurones is the presence of a further anterior contralateral branch that runs into the intersegmental nerve of the same or the anterior segment (i.e. the dorsal nerve N1) similar to the axons of the crustacean cnc-neurones. This is most prominent in the SOG and the TGs, but is sometimes also seen in the AG neuromeres. These neurones terminate at various peripheral sites: in the paramedian nerves; in neurohaemal organs (the perisympathetic or perivisceral organs and corpora cardiaca); on pharyngeal dilator muscles, ventral diaphragm muscles and alary muscles; in the lateral heart nerve; and on the heart itself. The type-2 neurones were supposed to be interneurones which show a ventral contralateral projection, and split into anterior and posterior branches running in a lateral tract up and down to terminals in one or more neuromeres. However, in the thoracic segments, one branch of this neurone type enters the leg (N5) of the corresponding segment and forms terminals on flexor muscles (Dircksen *et al.*, 1991; Dircksen, 1994; Dircksen & Homberg, 1995; Fig. 5d). Both types of neurones, but none of the others, occur in locust embryos from about 45% of development onwards (Fig. 6a), with all peripheral projections of the type-1 neurones being fully established by about 70% of development (Fig. 6d). They are also seen around the first ecdysis in digging vermiform and first-instar larvae (Truman, Ewer & Ball, 1996; H. Dircksen, unpublished observations). The type-1 neurones show the strongest CCAP-immunoreactivity around and after the imaginal moult (Dircksen & Homberg, 1995). Other serially homologous CCAPir cells are restricted to locust AG neuromeres. These are the weakly CCAPir type-3 neurones in AGl-7 which seem to be descending interneurones (Fig. 5f), and the type-4 neurones in AGl-6 which are large neurosecretory projection neurones. The latter neurones show dorsal and ventral central branches, and an axon which exits the ganglia ipsilaterally through the ventral nerve (N2). Their extensive peripheral projections contribute to the majority of CCAPir terminals within the perisympathetic organs, in the transverse and link nerve-associated branches to the spiracles, and in the lateral heart nerve, all of which contain typical neurosecretory granules (Dircksen *et al.*, 1991; Nässel, Bayraktaroglu & Dircksen, 1994). The CCAPir type-4 neurones show similar intraganglionic positions and branches as well as ispilateral peripheral projections as the locust leucokinin-I immunoreactive neurones, but they are not co-localised (Dircksen & Nässel, 1993; Chen *et al.*, 1994). Similar peripheral CCAPir terminals are formed in the glandular lobe of the corpora cardiaca by another type of neurone originating in the ventral median part of the SOG that exits to the periphery via the nervi corporis cardiaci II (Dircksen *et al.*, 1991; Dircksen & Homberg, 1995).

Type-1 and type-2 neurones with very similar morphology to those in the locust have subsequently been found to occur throughout the VNC of *T. molitor*, with the exception of the terminal ganglion. For instance, the type-1 neurones take the same exit routes to varicosities and terminals next to the surface of the perisympathetic organs and on certain dorsal longitudinal muscles. These neurones persist from early stages (45% of embryonic development; see Fig. 6b) throughout post-embryogenesis into the adults in terms of positions and principal projection patterns (Breidbach & Dircksen, 1991; Dircksen & Breidbach, 1991a,b; Dircksen, 1994). However, none of the other locust neurone types occurs in this species. CCAPir neurones and pathways similar to those in *T. molitor* have been described in the VNC of the cricket *Gryllus bimaculatus*. Neurones of apparent type-1 morphology show projections to terminals in the distal nerve roots and the perisympathetic organs associated with the transverse nerves (Spörhase-Eichmann *et al.*, 1991; Spörhase-Eichmann, 1993; Helle *et al.*, 1995). More recently, CCAPir neurones of apparent (putative) similarity to the type-1 and 2 neurones of *T. molitor* and *L. migratoria* (Figs. 5b,e and 6c) have been found in another cricket (*G. campestris*), in apterygotes (*Lepisma saccharina* and *Thermobia domestica*), and in a mosquito (Ewer & Truman 1996; H. Dircksen, unpublished observations). A more complex set of CCAPir neurones has been described in the VNC of larvae, pupae and adults of lepidopteran species

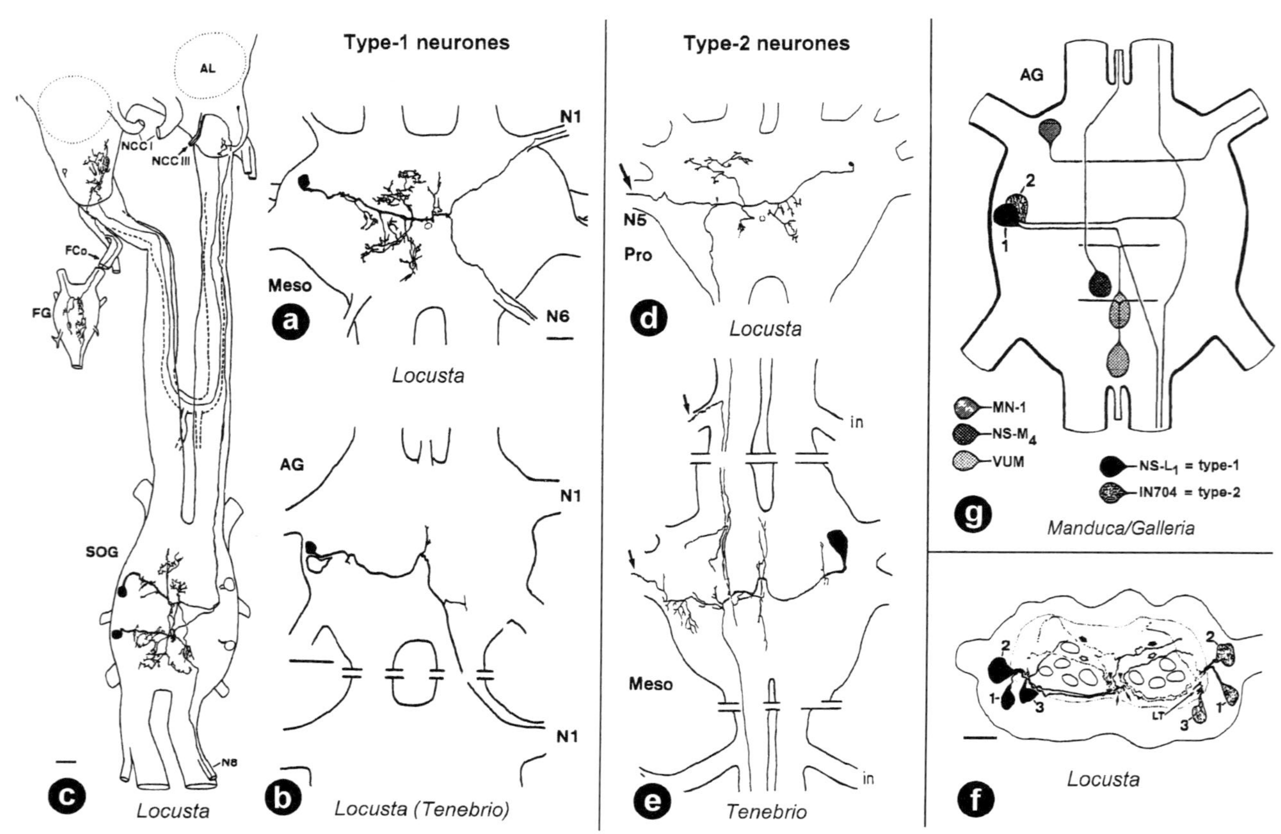

Type-1 neurones
Type-2 neurones
AL
NCC I
NCC III
FCo
FG
SOG
N8
Locusta
c
N1
Meso
N6
a
Locusta
AG
N1
N1
b
Locusta (Tenebrio)
N5
Pro
d
Locusta
in
Meso
e
Tenebrio
in
AG
2
1
MN-1
NS-M4
VUM
NS-L1 = type-1
IN704 = type-2
g
Manduca/Galleria
2
1
3
2
LT
3
f
Locusta

such as *M. sexta* (see Marshall & Reynolds, this volume) and *Galleria mellonella*, which contain additional types of CCAPir neurones not seen in any other species (Davis *et al.*, 1993; Dircksen, 1994; Fig. 5g). Many of the cells persist from the first larval into the adult stage; others become immunoreactive late, or only transiently, during post-embryonic development. There are three lateral pairs of contralaterally projecting neurones including the neurosecretory type NS-L$_1$ (identical to cell 27 of Taghert & Truman, 1982) and the interneurone IN-704 closely resembling the locust type-1 and type-2 morphology, respectively, and a pair of motoneurones (MN1). The IN-704 has recently been named 'cell 704', because these cells in the larval thoracic ganglia, which were considered homologues of those in the abdominal ganglia, have an efferent branch into the ventral nerve (Ewer & Truman, 1996) similar to the type-2 neurones in the locust thoracic ganglia. In the abdominal ganglia, the additional CCAPir cells are the posterior ventral unpaired median (VUM) neurones, a medial pair of cells M4 and a peripheral cell NS-L1 (Davis *et al.*, 1993). NS-L$_1$ (cell 27) also contains the tanning hormone, bursicon (Taghert & Truman, 1982), while M4 contains *Manduca* diuretic hormone (*Manduca*-DH) and pheromone biosynthesis-activating

---

Fig. 5. Identified type-1 and type-2 neurones and other CCAPir neurones in insects. (a) Locust mesothoracic type-1 neurone. Note anterior contralateral branch into nerve 1 (N1 and N6 are parts of the thoracic intersegmental nerve). (b) Abdominal type-1 projection neurone in locust (similar to the ones in *T. molitor*). An anterior contralateral branch as in (a) is occasionally observed (H. Dircksen, unpublished observation). (c) Maxillary and labial neurones projecting via nervus corporis cardiaci III to pharyngeal dilator muscles. Note also tritocerebral neurone projecting into the frontal ganglion (from Dircksen & Homberg, 1995). (d) Prothoracic type-2 neurones in the locust showing peripheral projections into the leg nerve (arrow, N5). (e) Mesothoracic type-2 neurone in *T. molitor* showing similar projections (arrows) into the leg nerve and intersegmental nerve. (f) Cross-section through a locust abdominal ganglion showing pathways of type-1, 2, and 3 neurones. (g) Schematic drawing of moth CCAPir neurones, note pathways of NS-L$_1$ and IN-704 identical to type-1 and 2 neurones and additional cells. AG, abdominal ganglion; AL, antennal lobe; NCC, nervus corporis cardiaci; FCo, frontal connective; FG frontal ganglion; in, intersegmental nerve; IN704, interneurone 704; LT, CCAPir lateral tract; Meso, mesothroacic ganglion; MN-1, motor neurone 1; N1 etc., first nerve of a ventral nerve cord ganglion; NS-L1, lateral neurosecretory cell 1; NS-M4, median neurosecretory cell 4; Pro, prothoracic ganglion; SOG, suboesophageal ganglion; VUM, ventral unpaired median cells. (a, d, f, slightly modified from Dircksen *et al.*, 1991; c, from Dircksen & Homberg, 1995; e, from Breidbach & Dircksen, 1991; g, from Davis *et al.*, 1993.)

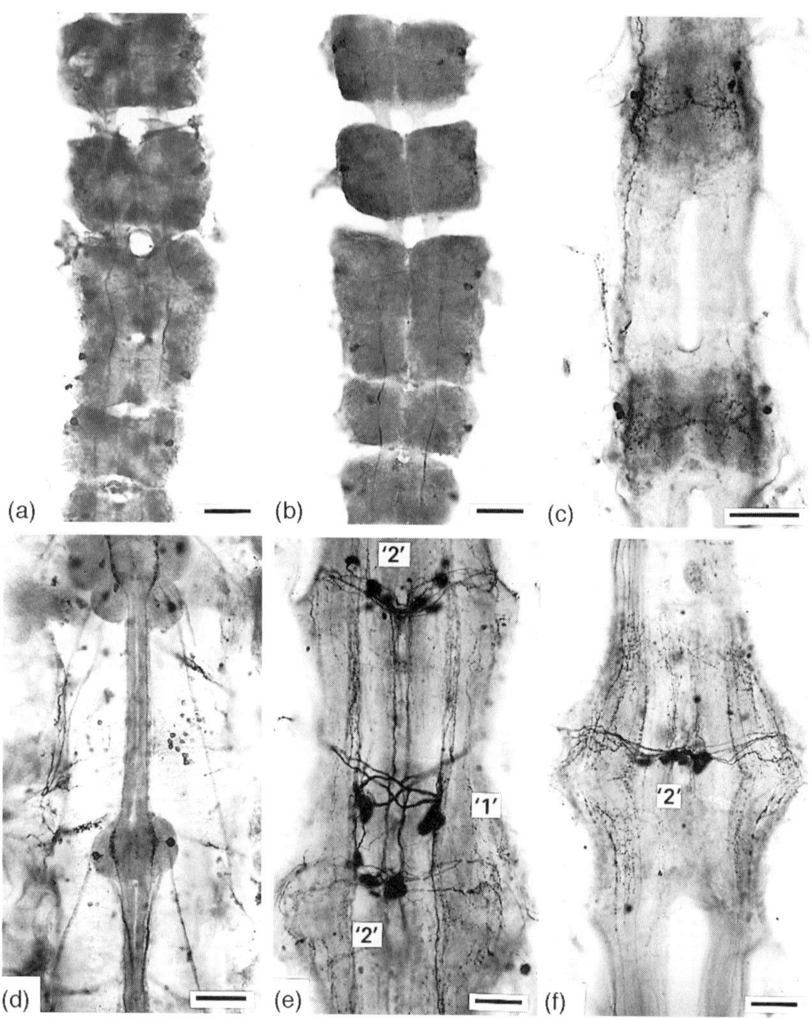

Fig. 6. CCAPir neurones in embryonic and adult insects and a centipede. (a, b) Type-1 and 2 neurones in thoracic and first three abdominal neuromeres of a *L. migratoria* (a) and a *T. molitor* (b) embryo at about 50% development. (c) Type-1 and 2 neurones in fifth and sixth abdominal ganglia of the apterygote *Lepisma saccharina*. (d) *In situ* whole mount of central CCAPir neurones and terminals in the distal perisympathetic organs (asterisks) of a locust embryo at 70% development. (e, f) Large type-1-like neurones ('1') and serially homologous descending neurones ('2') in the suboesophageal (e) and sixth abdominal ganglion (f) of the centipede *L. forficatus*. (a, c–f: H. Dircksen, unpublished observations; b, courtesy of R. Urbach). Bars=100 μm.

neuropeptide (PBAN) or a related peptide (Chen *et al.*, 1994; Veenstra, Lehman & Davis, 1994). In the flies *Drosophila melanogaster, Calliphora vicina, Calliphora vomitoria,* and *Lucilia cuprina,* the whole set of serially homologous CCAPir VNC neurones appear only in the third larval stage. In the adult fly, only four neurones with projections to the hindgut can be detected in the posterior abdominal neuromeres (Jahn *et al.*, 1991, 1992; Jahn, 1992; Dircksen, 1994; H. Dircksen, unpublished observations; Kutsch & Breidbach, 1994; Nässel *et al.*, 1994; Ewer & Truman, 1996). In lower Diptera, such as the mosquito *Aedes aegypti,* CCAPir cells exhibit the same position and morphology as those in *T. molitor* from the larval into the adult stage (Ewer & Truman, 1996).

All insects investigated to date contain CCAPir neurones in the brain, some of which show striking interspecies similarities. Detailed accounts and reconstructions of 12 groups out of the total of about 120 neurones in the brain and the 380 neurones in the optic lobes of *L. migratoria,* which carries the largest amount of CCAPir neurones hitherto described in insects, have recently been given by Dircksen & Homberg (1995). The distinct brain neurones are associated with almost all neuropiles with the exception of the mushroom bodies, suggesting important neuromodulatory roles of CCAP. Some of the locust optic lobe neurones associated with the accessory medulla (cam-neurones) seem to contain more than one peptide, and may be part of a neurone group controlling circadian rhythmicity in this insect (Würden & Homberg, 1995). In all other species examined, up to five protocerebral pairs of CCAPir neurones have been described, but none in the optic lobes. Interestingly, these cells, such as those of the first and the second subgroup in *T. molitor* (Breidbach & Dircksen, 1991), the interneurones IN-ADs and IN-MPs in *M. sexta,* and similar ones in *G. mellonella* (Davis *et al.*, 1993; H. Dircksen unpublished observations), the locust cp_6-neurones (Dircksen & Homberg, 1995), and those in flies (Kutsch & Breidbach, 1994; H. Dircksen & O. Breidbach, unpublished observations), show almost identical soma positions and projection patterns. In some species, the whole set of five neurones appears early in development, for example from 50–55% embryonic development onwards in *T. molitor* (R. Urbach, personal communication) and in *L. migratoria* (H. Dircksen, unpublished observations), and seem to persist into the adult stages (Breidbach & Dircksen, 1991). In other species, for example in moths, they become CCAPir at distinct times during post-embryogenesis (Davis *et al.*, 1993). No brain neurones have been shown to innervate the retrocerebral complex in any insect, with the possible exception of a locust tritocerebral cell that could be traced into the frontal ganglion but not further into the recurrent nerve (Fig. 5c). The CCAPir innervation of the locust and the moth retrocerebral complex, and that of the pharyngeal dilator muscles, arise exclusively from type-1 neurones (*M. sexta*: NS-L[Mx]) and anterior ventral median cells (*M. sexta*: IN-VM) in the SOG (Davis *et al.*, 1993; Ewer, De Vente & Truman, 1994; Dircksen & Homberg, 1995).

CCAP in myriapods

Preliminary data are available on the distribution of CCAPir neurones in myriapods. Although nothing is known about possible functions and the identity of CCAP in this arthropod group, immunopositive neurones have been detected in a centipede, *Lithobius forficatus*. In the VNC, at least, the CCAPir cells outnumber those found in crustaceans and insects. CCAPir neurones occur throughout the central nervous system of this species. Apart from neurones innervating several brain neuropiles, the suboesophageal ganglia contain pairs of large contralaterally projecting neurones with central descending fibres and peripheral axons exiting the VNC through an interseg- mental nerve with unknown destination. The projection patterns of these SOG neurones closely resemble the type-1 neurones (Fig. 6e; H. Dircksen, unpublished observations). Another medial group of segmentally homolo- gous neurones occurs in each VNC ganglion. It comprises six to eight contra- laterally descending putative interneurones (Kutsch & Breidbach, 1994; H. Dircksen, unpublished observations; Fig. 6f) that share similarities with the type-1 neurones in terms of their projection projection. Possible neurohaemal release sites for CCAP have not yet been found in this species.

CCAP in Chelicerata

Although authentic CCAP has not been fully identified in chelicerates, immunoreactive material, co-eluting in HPLC with authentic CCAP, as well as a rather distinct later-eluting immunoreactive peak has been isolated from ventral nerve cord extracts of a xiphosuran, the horseshoe crab *Limulus poly- phemus* (Groome & Lehman, 1995). CCAP has little effect on the frequency and amplitude of heart beat, but strongly increases the tonus and frequency of spontaneous contractions of the midgut of this species (thresholds at 3 and 0.1 nmol 1^{-1}, respectively) as does the other as yet unidentified CCAP- related substance. Immunopositive neurones occur as ganglion cells associ- ated with the so-called central body, among the medullary and the ventral medial cell groups of the brain, as clusters of six to eight cells in the circum- oesophageal ring neuromers, and as 12 cells in the anterior and six to eight cells in the posterior quadrants of apparently all abdominal neuromeres. Additional cells were found adjacent to the lateral vascular sheath or near the medial neuropil. Interestingly, fibres in the anterior haemal nerves and posterior branchial nerves are seen to innervate the midgut and the posterior gill plates. However, these fibres could not be related to any of the central neurones. Other fibres and small monopolar and bipolar cells occur in the cardiac ganglion and in the myocardium.

CCAPir neurones have been detected in the CNS of a harvestman and a spider (Dircksen, 1994). The protocerebrum of seven arachnid species (see Table 1, Chelicerata) has recently been shown to contain 14 reconstructed clusters of CCAPir neurones associated with the arachnid optic lobes and the so-called central body. Their positions, pathways and projection sites are obviously indicative of a common general pattern in the arachnid brain (Breidbach, Dircksen & Wegerhoff, 1995). This may identify them as true homologues among phylogenetically distant species in this arthropod group, according to the criteria of Kutsch & Breidbach (1994). The occurrence of CCAPir fibres in similar neuropilar structures, such as the central body, suggests that similar cells are present in both arachnid and xiphosuran brains, although this conclusion may have to await more detailed analyses of the xiphosuran types of these protocerebral CCAPir cells. Furthermore, all VNC neuromers in an opilionid, *Rilaena triangularis*, obviously contain two paired CCAPir neurones with central projection patterns to the mid-sagittal plane and in mid-dorsal direction (Fig. 7; H. Dircksen, unpublished observations), which are essentially similar to the type-1 and -2 neurones of crustacean and insect VNCs.

CCAP in non-arthropod invertebrates: molluscs and worms

Apart from the above-mentioned CCAP-related peptide (CCAP-RP) of *Helix pomatia*, which is regarded as a member of the CCAP family (see Muneoka *et al.*, 1994; Fig. 1), there is hitherto only immunocytochemical evidence for the existence of CCAP in invertebrates other than arthropods. The antiserum used in these studies is specific for the C-terminal of CCAP as determined by RIA (Stangier *et al.*, 1988), and detects authentic CCAP but not necessarily the C-terminally extended molluscan CCAP-RP. The molluscan CCAP related peptide from ganglia markedly potentiates phasic contractions of the anterior byssus retractor muscle of *H. pomatia* in response to repeated electrical stimulation. Interestingly, this peptide is also excitatory on the crop of the cricket *Gryllus bimaculatus*, but is not active (even at concentrations of 1 μmol 1^{-1}) on the oesophagus of an annelid *Perinereis vancaurica*, the body wall muscle of an echiuroid worm *Urechis unicinctus*, the radial longitudinal muscle of the sea cucumber, *Stichopus japonicus*, the duodenum of African clawed toad, *Xenopus laevis*, and the rectum of the Japanese quail *Coturnix japonica* (Minakata *et al.*, 1993).

CCAPir putative interneurones have been detected in the cerebral ganglia of the planorbid snail *Biomphalaria glabrata* and in the CNS of the leeches. *Theromyzon rude, Placobdella parasitica* and *Hirudo medicinalis* (see

Table 1 *Authentic CCAP, and CCAP-immunoreactivity (CCAPir) in protostomian invertebrate species*

	Authentic CCAP	ICC	CCAPir RIA/ ELISA	Source/Reference
Crustacea				
Artemia salina		+		Q. Z. Zhang & H. Dircksen, unpublished
Astacus astacus		+		Audehm *et al.*, 1991, 1993
		+		Trube *et al.*, 1994
Astacus leptodactylus		+		Audehm *et al.*, 1993
Balanus balanus		+		S. G. Webster, unpublished
Cancer borealis		+		Christie, Skiebe & Marder, 1995
Cancer pagurus			+	Stangier, 1991
Carcinus maenas	+			Stangier *et al.*, 1987
		+		Dircksen & Keller, 1988
			+	Stangier *et al.*, 1988
Chirona hameri		+		S. G. Webster, unpublished
Daphnia magna		+		Zhang *et al.*, 1993
Euphausia superba		+		Dircksen, 1994
Homarus americanus	+		+	Schwiemann, 1991
Hyas araneus			+	Stangier, 1991
Idothea balthica		+		Dircksen, 1994
Liocarcinus puber			+	Stangier, 1991
Oniscus asellus		+		Nussbaum, Dircksen & Keller, 1991
		+	+	Johnen *et al.*, 1994
Orconectes limosus	+		+	Stangier & Keller, 1990

		+		Audehm *et al.*, 1991, 1993
		+		Trube *et al.*, 1994
Procambarus clarkii		+		Audehm *et al.*, 1993
Semibalanus balanoides		+		S. G. Webster, unpublished
Insecta				
Aedes aegypti		+		Ewer & Truman, 1996
Calliphora vicina		+	+	Jahn *et al.*, 1992
		+		Kutsch & Breidbach, 1994
Calliphora vomitoria		+		Nässel *et al.*, 1994
Ctenolepisma longicaudata		+		Ewer & Truman, 1996
		+	+	Jahn, 1992
Drosophila melanogaster		+		Dircksen, 1994
		+		Kutsch & Breidbach, 1994
		+		Ewer & Truman, 1996
Galleria mellonella		+		Dircksen, 1994
Gryllus bimaculatus		+		Spörhase-Eichmann *et al.*, 1991
		+		Spörhase-Eichmann, 1993
		+		Helle *et al.*, 1995
Gryllus campestris		+		Ewer & Truman, 1996
Lepisma saccharina		+		H. Dircksen, unpublished
Locusta migratoria	+		+	Stangier *et al.*, 1989
		+		Dircksen *et al.*, 1991
		+		Dircksen & Homberg 1995
Lucilia cuprina		+		H. Dircksen, unpublished
Manduca sexta	+	+		Hildebrand *et al.*, 1990
		+		Davis *et al.*, 1990, 1993

Table 1 (*cont.*)

	Authentic CCAP	ICC	CCAPir RIA/ ELISA	Source/Reference
	+			Cheung *et al.*, 1992
	+		+	Lehman *et al.*, 1993
		+		Ewer & Truman, 1996
Musca domestica		+		Ewer & Truman, 1996
Periplaneta americana		+		Davis *et al.*, 1990
Schistocerca gregaria		+		Würden & Homberg, 1995
Spodoptera eridania	+			Furuya *et al.*, 1993
Stomoxys calcitrans	+			G. M. Holman, unpublished
Tenebrio molitor		+		Breidbach & Dircksen, 1991
		+		Dircksen & Breidbach, 1991a, 1991b
	+			Furuya *et al.*, 1993
		+		Ewer & Truman, 1996
Thermobia domestica		+		Ewer & Truman, 1996
Chelicerata				
Limulus polyphemus		+		Dircksen, 1994
		+	+	Groome & Lehman, 1995
Araneus diadematus		+		Breidbach *et al.*, 1995
Brachypelma albopilosa		+		Breidbach *et al.*, 1995
Cupiennius salei		+		Breidbach *et al.*, 1995
Meta segmentata		+		Breidbach *et al.*, 1995

Nephila clavipes	+	Breidbach *et al.*, 1995
Rilaena triangularis	+	Breidbach *et al.*, 1995
	+	H. Dircksen, unpublished
Tegenaria atrica	+	Breidbach *et al.*, 1995
Myriapoda		
Lithobius forficatus	+	Kutsch & Breidbach, 1994
	+	H. Dircksen, unpublished
Mollusca		
Biomphalaria glabrata	+	Dircksen, 1994
Helix pomatia	$(+)^a$	Minakata *et al.*, 1993
Annelida		
Theromyzon rude	+	Ng, 1992
Placobdella parasitica	+	Ng, 1992
Hirudo medicinalis	+	H. Dircksen & J. R. Groome, unpublished

Notes:

CCAP, crustacean cardioactive peptide; ICC, immunocytochemistry; RIA, radioimmunoassay; CCAPir, CCAP-immunoreactive; ELISA, enzyme-linked immunosorbent assay.
[a] molluscan CCAP-related peptide; compare Fig. 1.

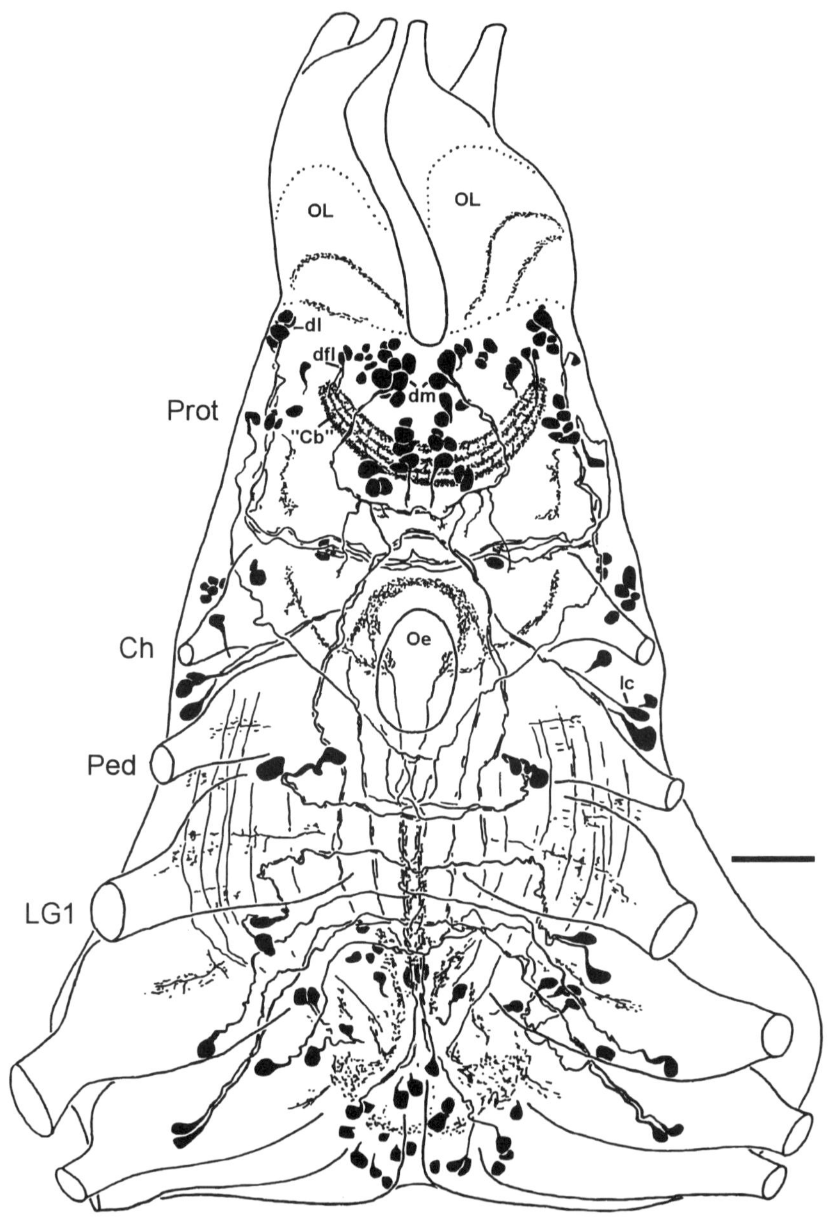
OL
OL
dl
dfl
dm
Prot
"Cb"
Ch
Oe
lc
Ped
LG1

Dircksen, 1994). CCAPir neurones in the first two species of leech are restricted to the brain in leech embryos (Ng, 1992), but are found in brain neuromers and all VNC segmental ganglia of adults of *H. medicinalis* (H. Dircksen & J. R. Groome, unpublished observations). Immunopositive cells were not detectable in adult *Lumbricus terrestris* (H. Dircksen, unpublished observation). No immunocytochemical results are available on the possible existence of CCAPir cells in the echiuroid worm *Urechis unicinctus*, where the above-mentioned CCAP-RP has been identified as the *Urechis* excitatory peptide UEP_c, which is active on the inner circular body-wall muscle (Ikeda *et al.*, 1991; Fig. 1). This peptide lacks the C-terminal amidation necessary for the successful immunodetection of CCAP (Stangier *et al.*, 1988), and would probably not be recognized by CCAP-specific antibodies.

Concluding remarks

The distribution of many CCAPir neurones, and their principal central and peripheral projection patterns are remarkably similar among the different arthropod groups, but our present knowledge is unfortunately too sparse to draw generalizations in other articulate and molluscan groups. Type-1 and 2 VNC neurones are obviously highly conserved, and can be considered as true arthropod homologous neurones (Fig. 8) according to the criteria of Kutsch & Breidbach (1994), differing only with respect to their peripheral targets among the species examined. The situation for similar CCAPir neurones in the arthropod cerebral ganglia is more complex, not only in terms of the different numbers observed, but inasmuch as the brain architectures of the Chelicerata (Merostomata, Arachnida) obviously differ considerably from those of the Mandibulata (Crustacea, Insecta) (see Arbas, Meinertzhagen & Shaw, 1991; Wegerhoff & Breidbach 1994; Breidbach *et al.*, 1995). It may be suggested that the VNC-associated neuronal networks and their conserved peptide product, CCAP, serve important myotropic and neuromodulatory functions indispensible for all arthropods. Since the type-1 neurone terminals are in most cases observed in close proximity to respiratory- and heart-associated neurohaemal areas, and pronounced moult-cycle-related

---

Fig. 7. CCAPir neurones in the central nervous system of the harvestman, *Rilaena triangularis* (Opilionidae; H. Dircksen, unpublished observation). Note neurones innervating the so-called central body (Cb), optic lobes (OL) and other neuropiles in the protocerebrum (Prot) (nomenclature according to Breidbach *et al.*, 1995). Note also the two pairs of centrally projecting neurones in each of the cheliceral (Ch), pedicellar (Ped), leg ganglia (LG) and other ventral nerve cord neuromeres (H. Dircksen, unpublished observations). Oesophageal orifice (*Oe*). Bar=100 μm.

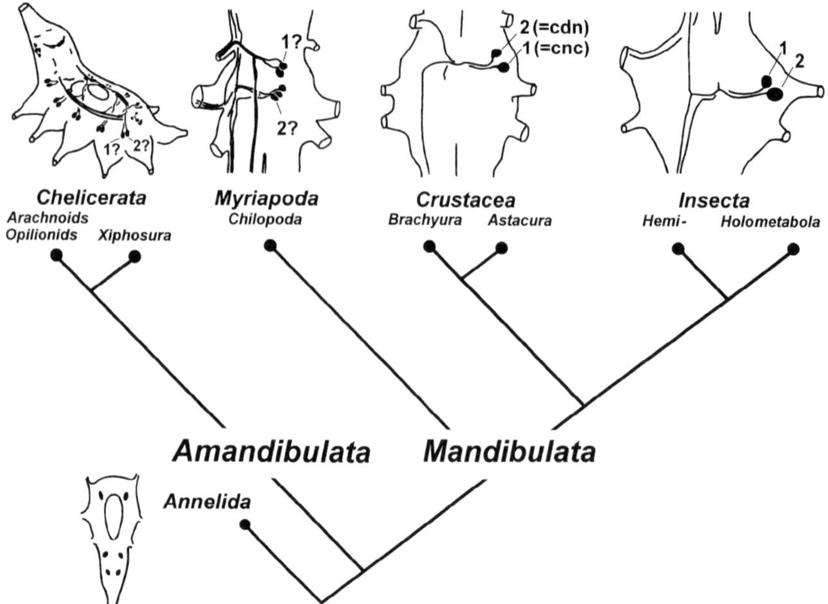

Fig. 8. Comparison of the principal positions and intraganglionic projection patterns of the highly conserved type-1 and type-2 CCAPir neurones, and their distribution in phylogenetically distant amandibulate and mandibulate arthropods.

differences are observed in terms of cdn-type-2 neurone-associated CCAP contents in isopod CNS and CCAP immunoreactivity in locusts, it has been hypothesised that CCAP might be involved in the control of moulting behaviour (Dircksen & Homberg, 1995; Johnen *et al.*, 1995), a prerequisite for the survival and the evolutionary success of all arthropods. In insects, for instance, moulting behaviour is associated with air swallowing as well as with circulatory and respiratory changes that are known to occur at ecdysis (Reynolds, 1980). Further support for this hypothesis came from recent studies on certain neurones in insects that respond to application of an eclosion hormone by transient changes in the contents of cyclic GMP immunoreactivity during moult, which have been interpreted as possibly being associated with release activity in these cells. These neurones (or at least major subsets of them) are identical to the CCAPir neurones of type-1 and type-2 morphology in *M. sexta* and several other insect species with the possible exception of some fly species (Ewer *et al.*, 1994; Ewer & Truman, 1996; Truman *et al.*, 1996). Some of the type-1 neurones described previously in the locust maxillary and labial neuromeres (Dircksen & Homberg, 1995), are obviously identical to the CN2 and CN3 neurones which were long known to lose their electrical activity after the imaginal moult (Bräunig, 1990; Bräunig, Agricola & Dircksen, 1995). These observations on members of phylogenetically distant arthropod groups with highly diversified behavioural repertoires provide compelling evidence for CCAP being of paramount importance as an arthropod neurohormone and neuromodulator which has been remarkably conserved during evolution. Moreover, the co-localisation of CCAP with the tanning hormone bursicon and diuretic factors in several neurones of the CCAPir network in orthopterans and lepidopterans (Kostron *et al.*, 1996; see also Nässel, 1996) suggests that the conserved CCAPir network may not only be involved in the dynamic control of moulting behaviour, but also in the subsequent associated events concerned with adult emergence, such as sclerotisation and diuresis.

Acknowledgements

The author thanks Dr S. G. Webster, Bangor, Wales, for encouragement and helpful discussions during this work.

References

Arbas, E. A., Meinertzhagen, I. A. & Shaw, S. R. (1991). Evolution in nervous systems. *Annual Review of Neuroscience* **14**, 9–38.

Audehm, U., Dircksen, H. & Keller, R. (1991). Projections of crustacean cardioactive peptide (CCAP)-immunoreactive neurons of the 6th abdominal ganglion of crayfishes. *General and Comparative Endocrinology* **82**, 284.

Audehm, U., Trube, A. & Dircksen, H. (1993). Patterns and projections of crustacean-cardioactive-peptide-immunoreactive neurones of the terminal ganglion of crayfish. *Cell and Tissue Research* **272**, 473–485.

Baumann, H. (1917). Das Cor frontale bei dekapoden Krebsen. *Zoologischer Anzeiger* **49**, 137–144.

Baumann, H. (1918). Das Gefäßsystem von *Astacus fluviatilis* (*Potamobius astacus* L.). *Zeitschrift für wissenschaftliche Zoologie* **118**, 246–312.

Bräunig, P. (1990). The morphology of subesophageal ganglion cells innervating the nervus corporis cardiaci III of the locust. *Cell and Tissue Research* **260**, 95–108.

Bräunig, P., Agricola, H. & Dircksen, H. (1995). Thoracic homologues of the locust suboesophageal CN cells: complex neurosecretory cells with diverse targets. In *Learning and Memory*, ed. N. Elsner & R. Menzel, p. 673. Stuttgart, New York: Thieme.

Breidbach, O. & Dircksen, H. (1991). Crustacean cardioactive peptide-immunoreactive neurons in the ventral nerve cord and the brain of the meal beetle *Tenebrio molitor* during postembryonic development. *Cell and Tissue Research* **265**, 129–144.

Breidbach, O., Dircksen, H. & Wegerhoff, R. (1995). Common general morphological pattern of peptidergic neurons in the arachnid brain: crustacean cardioactive peptide-immunoreactive neurons in the protocerebrum of seven arachnid species. *Cell and Tissue Research* **279**, 183–197.

Chen, Y., Veenstra, J. A., Davis, N. T. & Hagedorn, H. H. (1994). Leucokinin and diuretic hormone immunoreactivity in the tobacco hornworm, *Manduca sexta*, and co-localization of this immunoreactivity in lateral neurosecretory cells of the abdominal ganglia. *Cell and Tissue Research* **278**, 493–507.

Cheung, C. C., Loi, P. K., Sylwester, A. W., Lee, T. D. & Tublitz, N. J. (1992). Primary structure of a cardioactive neuropeptide from the tobacco hawkmoth, *Manduca sexta. FEBS Letters* **313**, 165–168.

Christie, A. E., Skiebe, P. & Marder, E. (1995). Matrix of neuromodulators in neurosecretory structures of the crab *Cancer borealis. Journal of Experimental Biology* **198**, 2431–2439.

Davis, N. T., Miller, T. A., Lehman, H. & Hildebrand, J. G. (1990). Crustacean cardioactive peptide: immunoreactive neurons and physiological actions in *Manduca sexta. Society of Neuroscience Abstracts* **16**, 856.

Davis, N. T., Homberg, U., Dircksen, H., Levine, R. B. & Hildebrand, J. G. (1993). Crustacean cardioactive peptide-immunoreactive neurons in the hawkmoth *Manduca sexta* and changes in their immunoreactivity during postembryonic development. *Journal of Comparative Neurology* **338**, 612–627.

Dircksen, H. (1990). Immunocytochemical identification of the neurosecretory products of the pericardial organs of *Carcinus maenas*. In *Frontiers in Crustacean Neurobiology*, ed. K. Wiese, W. D. Krenz, J. Tautz, H. Reichert, B. Mulloney, pp. 487–491. Basel, Boston, Berlin: Birkhäuser.

Dircksen, H. (1991). Neurosecretory endings in the pericardial organs of the shore crab *Carcinus maenas* L., and their identification by neuropeptide immunocytochemistry. In *Comparative Aspects of Neuropeptide Function*, ed. E. Florey & G. B. Stefano, pp. 198–200. Manchester: Manchester University Press.

Dircksen, H. (1994). Distribution and physiology of the crustacean cardioactive peptide in arthropods. In *Perspectives in Comparative Endocrinology*, ed. K. G. Davey, R. E. Peter, S. S. Tobe, pp. 139–148. Ottawa: National Research Council of Canada.

Dircksen, H. & Breidbach, O. (1991a). Crustacean cardioactive peptide (CCAP)-immunoreactive neurons in the central and peripheral nervous system of the locust *Locusta migratoria* and the metamorphosing meal beetle *Tenebrio molitor. General and Comparative Endocrinology* **82**, 286.

Dircksen, H. & Breidbach, O. (1991b). Peripheral projection sites of crustacean cardioactive peptide (CCAP)-immunoreactive neurons of the central nervous system of the locust *Locusta migratoria* and the metamorphosing meal beetle *Tenebrio molitor*. In Synapse – Transmission – Modulation, ed. N. Elsner & H. Penzlin, p. 14. Stuttgart, New York: Thieme.

Dircksen, H. & Homberg, U. (1995). Crustacean cardioactive peptide-immunoreactive neurons innervating brain neuropils, retrocerebral complex and stomatogastric nervous system of the locust, *Locusta migratoria. Cell and Tissue Research* **279**, 495–515.

Dircksen, H. & Keller, R. (1988). Immunocytochemical localization of CCAP, a novel crustacean cardioactive peptide, in the nervous system of the shore crab, *Carcinus maenas* L.. *Cell and Tissue Research* **254**, 347–360.

Dircksen, H. & Nässel, D. R. (1993). Immunoreactive leukokinin-like peptides and crustacean cardioactive peptide show similar, but distinct neuronal distributions in neurohaemal systems of the locust abdomen. In *Genes – Brain – Behaviour* ed. N. Elsner & M. Heisenberg, p. 519. Stuttgart, New York: G. Thieme.

Dircksen, H., Müller, A. & Keller, R. (1991). Crustacean cardioactive peptide in the nervous system of the locust, *Locusta migratoria*: an immunocytochemical study on the ventral nerve cord and peripheral innervation. *Cell and Tissue Research* **263**, 439–457.

Ewer, J., De Vente, J. & Truman, J. W. (1994). Neuropeptide induction of cGMP increases in the insect CNS: resolution at the level of single identifiable neurons. *Journal of Neuroscience* **14**, 7704–7712.

Ewer, J. & Truman, J. W. (1996). Increases in cyclic 3′,5′-guanosine monophosphate (cGMP) occur at ecdysis in an evolutionary conserved crustacean cardioactive peptide immunoreactive insect neuronal network. *Journal of Comparative Neurology* **370**, 330–341.

Furuya, K., Liao, S., Reynolds, S. E., Ota, R. B., Hackett, M. & Schooley, D. A. (1993). Isolation and identification of a cardioactive peptide from *Tenebrio molitor* and *Spodoptera eridania. Biological Chemistry Hoppe-Seyler* **374**, 1065–1974.

Gaus, G. & Stieve, H. (1992). The effect of neuropeptides on the ERG of the crayfish *Orconectes limosus. Zeitschrift für Naturforschung, Section C Biosciences* **47**, 300–303.

Groome, J. R. & Lehman, H. K. (1995). Characterization of crustacean cardioactive peptide-like immunoreactivity in the horseshoe crab, *Limulus polyphemus. Journal of Comparative Neurology* **357**, 36–51.

Helle, J., Dircksen, H., Eckert, M., Nässel, D. R., Spörhase-Eichmann, U. & Schürmann, F. W. (1995). Putative neurohemal areas in the peripheral nervous system of an insect, *Gryllus bimaculatus*, revealed by immunocytochemistry. *Cell and Tissue Research* **281**, 43–61.

Hildebrand, J. G., Lehman, H. K., Miller, T. A. & Davis, N. T. (1990). A peptide related to crustacean cardioactive peptide in *Manduca sexta*: characterization and distribution. *Society of Neuroscience Abstracts* **16**, 856.

Ikeda, T., Kubota, I., Kitajima, Y. & Muneoka, Y. (1991). Structures and actions of neuropeptides isoplated from an echiuroid worm *Urechis uni-cinctus*. In *Comparative Aspects of Neuropeptide Function*, ed. E. Florey & G. B. Stefano, pp. 29–41. Manchester: Manchester University Press.

Jahn, G. (1992). Cardioaktive Peptide bei Dipteren. Ph.D. thesis, University of Marburg/Lahn.

Jahn, G., Käuser, G. & Koolman, J. (1991). Localization of crustacean cardioactive peptide (CCAP) in the central nervous system of larvae of the blowfly, *Calliphora vicina*. *General and Comparative Endocrinology* **82**, 287.

Jahn, G., Käuser, G. & Koolman, J. (1992). Immunochemical characterization of a CCAP-like peptide in the blowfly *Calliphora vicina*. In *Rhythmogenesis in Neurons and Networks*, ed. N. Elsner & D. W. Richter, p. 539. Stuttgart, New York: Thieme.

Johnen, C., von Gliscynski, U. & Dircksen, H. (1995). Changes in haemolymph ecdysteroid levels and CNS contents of crustacean cardioactive peptide-immunoreactivity during the moult cycle of the isopod *Oniscus asellus*. *Netherlands Journal of Zoology* **45**, 38–40.

Kostron, B., Kaltenhauser, U., Seibel, B., Bräunig, P. & Honegger, H. W. (1996). Localization of bursicon in CCAP-immunoreactive cells in the thoracic ganglia of the cricket *Gryllus bimaculatus*. *Journal of Experimental Biology* **199**, 367–377.

Kutsch, W. & Breidbach, O. (1994). Homologous structures in the nervous system of Arthropoda. *Advances in Insect Physiology* **24**, 1–113.

Lehman, H. K., Murgiuc, C. M., Miller, T. A., Lee, T. D. & Hildebrand, J. G. (1993). Crustacean cardioactive peptide in the sphinx moth, *Manduca sexta*. *Peptides* **14**, 735–741.

Loi, P. K., Cheung, C. C., Lee, T. D. & Tublitz, N. J. (1992). Amino acid sequence and molecular analysis of insect cardioactive peptides. *Society of Neuroscience Abstracts* **18**, 472.

Maynard, D. M. (1961). Thoracic neurosecretory structures in Brachyura. II. Secretory neurons. *General and Comparative Endocrinology* **1**, 237–263.

McGaw, I. J., Airriess, C. N. & McMahon, B. R. (1994). Peptidergic modulation of cardiovascular dynamics in the Dungeness crab, *Cancer magister*. *Journal of Comparative Physiology* **164**, 103–111.

Minakata, H., Ikeda, T., Fujita, T., Kiss, T., Hiripi, L., Muneoka, Y. & Nomoto, K. (1993). Neuropeptides isolated from *Helix pomatia*. Part 2. FMRFamide-related peptides, S-Iamide peptides, FR peptides and others. In *Peptide Chemistry 1992*, ed. N. Yanahara, pp. 579–582. Leiden, The Netherlands: ESCOM Science Publishers BV.

Muneoka, Y., Takahashi, T., Kobayashi, M., Ikeda, T., Minakata, H. & Nomoto, K. (1994). Phylogenetic aspects of structure and action of molluscan neuropeptides. In *Perspectives in Comparative Endocrinology*, ed. K. G. Davey, R. E. Peter, S. S. Tobe, pp. 109–118. Ottawa: National Research Council of Canada.

Nässel, D. R. (1996). Neuropeptides, amines and amino acids in an elementary insect ganglion: functional and chemical anatomy of the unfused abdominal ganglion. *Progress in Neurobiology* **48**, 325–420.

Nässel, D. R., Bayraktaroglu, E. & Dircksen, H. (1994). Neuropeptides in neurosecretory and efferent neural systems of insect thoracic and abdominal ganglia. *Zoological Science* **11**, 15–31.

Nässel, D. R., Passier, P. C. C. M., Elekes, K., Dircksen, H., Vullings, H. G. B. & Cantera, R. (1995). Evidence that locustatachykinin I is involved in release of adipokinetic hormone from locust corpora cardiaca. *Regulatory Peptides* **57**, 297–310.

Ng, L. (1992). Distribution of crustacean cardioactive peptide immunoreactivity in the embryonic nervous system of the glossiphoniid leech, *Theromyzon rude*. B.Sc. thesis, University of California.

Nussbaum, T., Dircksen, H. & Keller, R. (1991). Immunocytochemical mapping of peptidergic neurons in the central nervous system of an isopod crustacean, *Oniscus asellus*. *General and Comparative Endocrinology* **82**, 287–288.

Passier, P. C. C. M. (1996). The multifactorial control of the release of the adipokinetic hormones from the locust corpus cardiacum. Ph.D. thesis, Rijkuniversiteit Utrecht, Utrecht.

Patel, N. H., Kornberg, T. B. & Goodman, C. S. (1989a). Expression of *engrailed* during segmentation in grasshopper and crayfish. *Development* **107**, 201–212.

Patel, N. H., Marin-Blanco, E., Coleman, K. G., Poole, S. J., Ellis, M. C., Kornberg, T. B. & Goodman, C. G. (1989b). Expression of *engrailed* proteins in arthropods, annelids, and chordates. *Cell* **58**, 955–968.

Reynolds, S. E. (1980). Integration of behaviour and physiology at ecdysis. *Advances in Insect Physiology*, **15**, 475–595.

Scholtz, G. (1995). Expression of the *engrailed* gene reveals nine putative segment-anlagen in the embryonic pleon of the freshwater crayfish *Cherax destructor* (Crustacea, Malacostraca, Decapoda). *Biological Bulletin* **188**, 157–165.

Schwiemann, K. (1991). Vorkommen, Verteilung und biochemische Charakterisierung von CCAP (Crustacean cardioactive peptide) beim Hummer. Diploma-thesis, Unversity of Bonn.

Skiebe-Corrette, P., Weimann, J. M., Heinzel, H. G. & Marder, E. (1993). Crustacean cardioactive peptide (CCAP) influences the pyloric and the gastric rhythms in the stomatogastric ganglion of the crab *Cancer borealis*. In *Genes – Brain – Behavior*, ed. N. Elsner & M. Heisenberg, p. 600. Stuttgart, New York: G. Thieme.

Spörhase-Eichmann, U., Dircksen, H., Hecht, T., Helle, J. & Schürmann, F. W. (1991). Neurohaemal-like fiber networks in an insect. In *Synapse – Transmission – Modulation*, ed. N. Elsner & H. Penzlin, p. 339. Stuttgart, New York: G. Thieme.

Spörhase-Eichmann, U. (1993). *Aminerge und peptiderge Neuronen im Zentralnervensystem der Grille*. Ph.D. thesis. Göttingen: Cuvillier.

Stangier, J. (1991). Biological effects of crustacean cardioactive peptide (CCAP), a putative neurohormone/neurotransmitter from crustacean pericardial organs. In *Comparative Aspects of Neuropeptide Function*, ed. G. B. Stefano & E. Florey, pp. 201–210. Manchester: Manchester University Press.

Stangier, J. & Keller, R. (1990). Occurrence of the crustacean cardioactive peptide (CCAP) in the nervous system of the crayfish *Orconectes limosus*. In *Frontiers in Crustacean Neurobiology*, ed. K. Wiese, W. D. Krenz, J. Tautz, H. Reichert, B. Mulloney, pp. 394–400. Basel: Birkhäuser.

Stangier, J., Dircksen, H. & Keller, R. (1986). Identification and immunocytochemical localization of proctolin in the pericardial organs of the shore crab, *Carcinus maenas. Peptides* 7, 67–72.

Stangier, J., Hilbich, C., Beyreuther, K. & Keller, R. (1987). Unusual cardioactive peptide (CCAP) from pericardial organs of the shore crab *Carcinus maenas. Proceedings of the National Academy of Sciences USA* **84**, 575–579.

Stangier, J., Hilbich, C., Dircksen, H. & Keller, R. (1988). Distribution of a novel cardioactive neuropeptide (CCAP) in the nervous system of the shore crab *Carcinus maenas. Peptides* **9**, 795–800.

Stangier, J., Hilbich, C. & Keller, R. (1989). Occurrence of crustacean cardioactive peptide (CCAP) in the nervous system of an insect, *Locusta migratoria. Journal of Comparative Physiology B* **159**, 5-11.

Taghert, P. H. & Truman, J. W. (1982). Identification of the bursicon-containing neurons in abdominal ganglia of the tobacco hornworm, *Manduca sexta. Journal of Experimental Biology* **98**, 385–401.

Trube, A. (1992). Immuncytochemische Darstellung des myotropen Neuropeptids CCAP im Nervensystem von Flußkrebsen. Ph.D. thesis, University of Bonn.

Trube, A., Audehm, U. & Dircksen, H. (1994). Crustacean cardioactive peptide-immunoreactive neurons in the ventral nervous system of crayfish. *Journal of Comparative Neurology* **348**, 80–93.

Truman, J. W., Ewer, J. & Ball, E. E. (1996). Dynamics of cyclic GMP levels in identified neurones during ecdysis behaviour in the locust *Locusta migratoria. Journal of Experimental Biology* **199**, 749–758.

Tublitz, N. J. & Truman, J. W. (1985). Insect cardioactive peptides. II. Neurohormonal control of heartbeat by two cardioacceleratory peptides in the tobacco hawkmoth, *Manduca sexta. Journal of Experimental Biology* **114**, 381–395.

Veenstra, J. A., Lehman, H. K. & Davis, N. T. (1994). Allatotropin is a cardioacceleratory peptide in *Manduca sexta. Journal of Experimental Biology* **188**, 347–354.

Walther, C., Zittlau, K. E., Murck, H. & Nachman, R. J. (1991). Peptidergic modulation of synaptic transmission in locust skeletal muscle. In *Comparative Aspects of Neuropeptide Function* ed. E. Florey & G. B. Stefano, pp. 175–186. Manchester: Manchester University Press.

Wegerhoff, R. & Breidbach, O. (1994). Comparative aspects of the Chelicerata nervous system. In *The Nervous Systems of Invertebrates – A Comparative Approach*, ed. O. Breidbach & W. Kutsch, pp. 159–179. Basel, Birkhäuser.

Weimann, J. M., Heinzel, H. G. & Marder, E. (1992). Crustacean cardioactive peptide activation of the pyloric network in the STG of the crab, *Cancer borealis. Society of Neuroscience Abstracts* **18**, 1056.

Weimann, J. M., Skiebe, P., Heinzel, H. G., Soto, C., Kopell, N., Jorge-Riviera, J. C. & Marder, E. (1997). Modulation of oscillator interactions in the crab stomatogastric ganglion by crustacean cardioactive peptide. Journal of Neuroscience **17**, 1748–1760.

Wilkens, J. L. (1995). Regulation of cardiovascular performance in crayfish. *American Zoologist* **35**, 37–48.

Wilkens, J. L. & McMahon, B. R. (1992). Intrinsic properties and extrinsic neurohormonal control of crab cardiac hemodynamics. *Experientia* **48**, 827–834.

Wilkens, J. L. & Mercier, A. J. (1993). Peptidergic modulation of cardiac performance in isolated hearts from the shore crab *Carcinus maenas. Physiological Zoology* **66**, 237–256.

Wilkens, J. L., Kuramoto, T. & McMahon, B. R. (1996). The effects of six pericardial hormones and hypoxia on the semi-isolated heart and sternal arterial valve of the lobster *Homarus americanus. Comparative Biochemistry and Physiology* **114C**, 57–65.

Würden, S. & Homberg, U. (1995). Immunocytochemical mapping of serotonin and neuropeptides in the accessory medulla of the locust *Schistocerca gregaria. Journal of Comparative Neurology* **362**, 305–319.

Yazawa, T. & Kuwasawa, K. (1992). Intrinsic and extrinsic neural and neurohumoral control of the decapod heart. *Experientia* **48**, 834–840.

Zhang, Q., Keller, R. & Dircksen, H. (1993). Immunocytochemical mapping of peptidergic neurons in the nervous system of *Daphnia magna* (Crustacea; Cladocera). In *Genes – Brain – Behavior*, ed. N. Elsner & M. Heisenberg, p. 545. Stuttgart, New York: Thieme.

Zhang, Q. & Dircksen, H. (1994). Immunoreactive crustacean hyperglycemic hormone (CHH) and crustacean cardioactive peptide (CCAP) in the nervous system of *Artemia salina* L. (Crustacea: Anostraca). In *Sensory Transduction*, ed. N. Elsner & H. Breer, p. 686. Stuttgart, New York: G. Thieme.

ANNA K. MARSHALL
and STUART E. REYNOLDS

Control of the insect oviduct: the role of the neuropeptide CCAP in the tobacco hornworm, *Manduca sexta*

Introduction

In insects, the oviducts are the route by which eggs are transported from the ovarioles to the ovipositor prior to egg laying. The ovarioles open into paired lateral oviducts which fuse to form the common oviduct (Snodgrass, 1935). The muscular tubes of the oviducts contract spontaneously to move the eggs, but their motility is additionally subject to the regulatory effects of innervation, circulating hormones and perhaps the exogenous secretions of the male insect introduced during mating. These factors act to orchestrate the activity of the whole of the reproductive system during egg laying, including fine tuning of the myogenic contractions of the oviducts.

Knowledge of these controls and how they are integrated is fragmentary. Insect oviducts are generally innervated, and neural controls on spontaneous muscular activity are probably important in most insects. It has been suggested that in the stick-insect *Carausius morosus* (Thomas & Mesnier, 1973; Thomas, 1979) and in the migratory locust *Locusta migratoria* (Lange *et al.*, 1984; Kalogianni & Pflüger, 1992) neuronally driven contractions of the common oviduct and the junctional area act to retain eggs in the lateral oviducts, and thus to prevent oviposition. Interestingly, however, the locust oviduct is also innervated by a large number (16–20) of median neurones with bilateral projections. Most of these project to the lateral oviducts, but two neurones of the DUM (dorsal unpaired median) group project to the important junctional region which shows spontaneous contractions *in vitro*. These midline neurones are probably octopaminergic (Lange & Orchard, 1986b). Orthodromic stimulation of these cells inhibits the spontaneous rhythm of oviducal contraction (Kalogianni & Pflüger, 1992). It seems likely that the details of neuronal control of the oviduct vary considerably between insects, even within the Orthoptera. Central pattern generators in the last two abdominal ganglia have also been shown to drive oviducal contractions in the catantopid orthopteran *Calliptamus* species and the tettigonid orthopteran *Decticus albifrons*. However, only in *D. albifrons* was there evidence of modulation of the oviduct's spontaneous contractions by DUM cells (Kalogianni & Theophilidis, 1995).

On the other hand, a number of authors have provided evidence that the oviducts may also be subject to neurohormonal controls (for example, in the reduviid bug *Rhodnius prolixus* (Davey, 1965); in the wax moth *Galleria mellonella* (Mesnier, 1972; Mesnier & Provansal, 1975); and in the stick-insects *C. morosus* (Girardie & Lafon-Cazal, 1972), *Sphodramantis lineola* (Mesnier, 1984) and *Clitumnus extradentatus* (Mesnier, 1985). However, most papers on the subject recount the effects of injected extracts presumed to contain neurohormones. Perhaps the strongest evidence for endocrine control of egg laying comes from Davey's work on *R. prolixus*. Here, it has been shown that the circulating titre of an oviposition-stimulating factor derived from identified brain neurosecretory cells rises and falls during the gonotrophic cycle in a pattern corresponding with the timing of oviducal motility and of ovulation/oviposition (Davey & Kriger, 1985). The release of this hormone after feeding appears to be driven by the previous release of another hormone from a site in the abdomen (Mulye & Davey, 1995). In no case is the chemical identity of an ovulation- or oviposition-stimulating hormone known.

One way to investigate the physiological regulation of oviduct activity is to test the effects of known myotropins on oviduct activity *in vitro*. A number of different insect oviduct preparations have been investigated in this way. The responses from only a limited number of species (mostly orthopterans) have been studied. Both classical neurotransmitters and neuropeptides have both been investigated. It is clear that there are considerable differences between species. For example, as far as small molecule neurotransmitters are concerned, in *L. migratoria* octopamine is inhibitory to lateral oviduct contractions at physiological doses (Lange & Orchard, 1986a), but has been reported to be both excitatory and inhibitory in the cockroach *Leucophaea maderae* (Cook *et al.*, 1984), and excitatory in the cricket *Gryllus bimaculatus* (Sefiani, 1987). L-Glutamate causes contractures in the oviducts of all these three species, but is reportedly inactive in the stable fly, *Stomoxys calcitrans* (Cook & Wagner, 1991). Acetylcholine has been reported to be stimulatory in *L. maderae* (Cook *et al.*, 1984), but was inactive in *G. bimaculatus* (Sefiani, 1987). Of these neurotransmitters, octopamine at least is likely to act *in vivo*, because the oviducts of *L. migratoria* (Orchard & Lange, 1985; Lange & Orchard, 1986a) and the cockroach *Periplaneta americana* (Orchard & Lange, 1987) are innervated by octopamine-containing nerves.

Oviducts are also sensitive to neuropeptides; some peptides seem to have effects on the oviducts of most species. The pentapeptide proctolin, for example, originally isolated for its ability to cause hindgut contractions in the cockroach *P. americana* (Brown & Starratt, 1975) occurs in a wide range of insects. It has been found to be excitatory when applied to the oviducts of most species so far tested: for example, the horsefly *Tabanus proximus* (Cook, 1981); the cockroaches *L. maderae* (Holman & Cook, 1985) and *P. americana*

(Orchard & Lange, 1987); and the locust *L. migratoria* (Lange, Orchard & Adams, 1986). Moreover, proctolin is known to be present in the innervation of the locust oviduct (Lange *et al.*, 1986). A notable exception, however, is the moth *Manduca sexta*, in which proctolin is not present (Kingan & Titmus, 1983), and in which proctolin has no effect on visceral muscles, including the heart (Platt & Reynolds, 1986) and the oviduct (see below). Responses to other peptides, however, vary greatly. Sometimes responses cross taxonomic boundaries. For example, the oviduct stimulating peptide Led-OVM from the Colorado potato beetle, *Leptinotarsa decemlineata*, is active not only on this insect's own oviduct, but also on that of *Locusta migratoria* (Spittaels *et al.*, 1991). On the other hand, sometimes the responses appear to be very specific. Thus, the peptide Lom-AG-myotropin II isolated from locust male accessory glands, is active in the oviduct of the same species (Paeman *et al.*, 1991a,b), but is inactive in the oviduct of the cockroach *L. maderae* (Cook *et al.*, 1984). The oviduct of *M. sexta* is rather specific in its responses to peptides (see below), whereas the oviducts of *L. migratoria* are responsive to an almost incredibly wide range of peptides that includes locustatachykinins, locustapyrokinins, locustamyotropins, SchistoFLRFamide, crustacean cardioactive peptide (CCAP), as well as proctolin and Lom-AG-myotropin I (for a review, see Schoofs, Vanden Broeck & De Loof, 1993). The physiological significance of multiple myotropic factors acting on the same tissue is, however, unknown.

This chapter summarises work designed to shed light on how oviduct motility is regulated in the tobacco hornworm, *M. sexta*, a large sphingid moth. It will be shown that CCAP is a strong candidate for the control of the oviducts in this insect. CCAP, first isolated as a cardioaccelerator from the shore crab *Carcinus maenas* (Stangier *et al.*, 1987), is a nonapeptide with the sequence:

$$\overline{\text{Pro-Phe-Cys-Asn-Ala-Phe-Thr-Gly-Cys}} \text{ (NH}_2\text{)}$$

Since first being identified in the crab, this peptide has been positively identified in a number of insects including *L. migratoria* (Dircksen, Müller & Keller, 1991), *Spodoptera eridania* and *Tenebrio molitor* (Furuya *et al.*, 1993), and *M. sexta* (Hildebrand *et al.*, 1990; Cheung *et al.*, 1992; Lehman *et al.*, 1993). It is now evident that this peptide is a highly conserved signalling molecule in arthropods and other invertebrates (see Dircksen, this volume).

Spontaneous muscular activity of the *Manduca* oviduct

An isolated *Manduca* oviduct preparation has been developed that displays spontaneous contractile activity *in vitro*, and responds to a range of stimulatory and inhibitory agents. This preparation includes the common oviduct and both lateral oviducts. Connections with the terminal abdominal ganglion are severed, although the nerve endings are not removed. The preparation is pinned at one end through a piece of cuticle attached to the opening of the common oviduct. The cut ends of the lateral oviducts are tied with thread which is attached to a transducer. The preparation is continuously perfused with an adult *Manduca* physiological saline solution. After an initial equilibration period, the preparation displays spontaneous contractions at a frequency of approximately $0.2–0.4$ min^{-1} (Fig. 1). These spontaneous movements originate in, and are largely confined to, the region of the junction between the lateral oviducts and the common oviduct.

Effects of known myotropic factors

Substances known to affect the activity of visceral muscle in insects were tested in the *Manduca* oviduct preparation. The spontaneous contractions were affected by a number of neurotransmitters, but only when applied at very high concentrations (Table 1). Test substances were applied in a volume of 100 μl of saline solution, without interrupting the flow of the perfusing saline. Thus, the actual concentration of the chemical at the oviduct would have been somewhat less than that applied, and the test substance would have been washed away very quickly (a few seconds). The excitatory neuromuscular transmitter L-glutamate stimulated the oviduct at a threshold of 100 μmol l^{-1}. Dopamine and serotonin both stimulated oviducal contractions only at concentrations of 1 mmol l^{-1}. Octopamine was inhibitory at concentrations of 10 μmol l^{-1} and above. These responses were all very short lived.

A number of known insect peptide myotropins were also tested. The *Manduca* oviduct responded to a surprisingly narrow range of these (Table 1). Of the eleven peptides tested, only two had any activity at all. CCAP was by far the most potent, causing an increased frequency of spontaneous contractions in the oviducts of newly emerged adults at a threshold applied concentration of about 50 pmol l^{-1} (Fig. 1). The response was dose-dependent and long lasting; at higher doses the rate of spontaneous contraction took up to 20 min to return to the pre-application level. Interestingly, the threshold for the response to CCAP was about four-fold higher when the oviduct used in the bioassay was taken from a pharate adult about 24 h before adult emergence, and at earlier stages no response at all could be detected. This implies that responsiveness to CCAP develops at a late stage of pharate adult development, perhaps only just before it is required.

The only other neuropeptide tested that had any oviduct stimulating activity was *Manduca* allatotropin Mas-AT (Kataoka *et al.*, 1989).

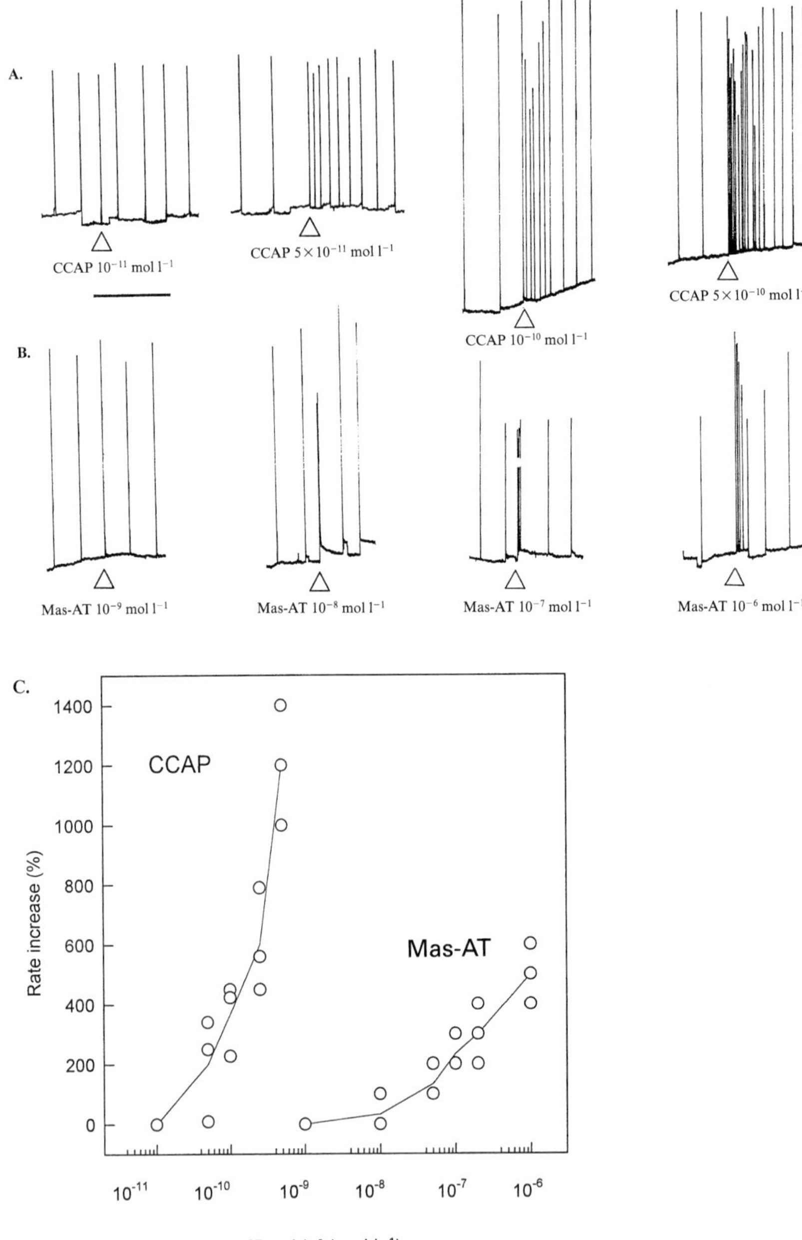
A.
CCAP 10⁻¹¹ mol l⁻¹
CCAP 5×10⁻¹¹ mol l⁻¹
CCAP 10⁻¹⁰ mol l⁻¹
CCAP 5×10⁻¹⁰ mol l⁻¹
B.
Mas-AT 10⁻⁹ mol l⁻¹
Mas-AT 10⁻⁸ mol l⁻¹
Mas-AT 10⁻⁷ mol l⁻¹
Mas-AT 10⁻⁶ mol l⁻¹
C.
Rate increase (%)
1400
1200
1000
800
600
400
200
0
CCAP
Mas-AT
10⁻¹¹   10⁻¹⁰   10⁻⁹   10⁻⁸   10⁻⁷   10⁻⁶
[Peptide] (mol l⁻¹)

Table 1 *Actions of neurotransmitters and insect neuropeptides on the* M. sexta *oviduct*

Substance	Active?	Threshold
Neurotransmitters		
Manduca FLRFamide	Excitatory	0.1 mmol 1^{-1}
Dopamine	Excitatory	1 mmol 1^{-1}
Glutamate	Excitatory	0.1 mmol 1^{-1}
Octopamine	Inhibitory	10 µmol 1^{-1}
5-Hydroxtryptamine	Excitatory	1 mmol 1^{-1}
Acetylcholine	Inactive	
Adrenaline	Inactive	
GABA	Inactive	
Noradrenaline	Inactive	
Peptides		
Manduca allatotropin	Excitatory	10 pmol 1^{-1}
CCAP	Active – excitatory	10 nmol 1^{-1}
AKH-I	Inactive	
Corazonin	Inactive	
FMRFamide	Inactive	
Lem-PK	Inactive	
Lom-TKII	Inactive	
Manduca FLRFamide	Inactive	
Hez-PBAN	Inactive	
Proctolin	Inactive	
SchistoFLRFamide	Inactive	

Fig. 1. Responses of *M. sexta* oviduct *in vitro* to (A) CCAP and (B) Mas-AT. The records show examples of oviduct contractile activity in response to the peptide. Peptides were applied as 100 µl of a solution of indicated concentration and washed off immediately. Contraction rate was measured over the 5 min period following application, and was compared to the 5 min period preceding this. The records read from left to right. Upward deflection of the record indicates contraction. The time bar indicates 10 min. (C) Dose–response curves for the actions of the two peptides on the oviduct. Each concentration was tested in three replicates, and the points show individual responses. The lines join the mean values for each concentration.

Sensitivity to this peptide was, however, some 200 times less than to CCAP (Fig. 1). The response was qualitatively different to that seen with CCAP, being short lived. Moreover, application of Mas-AT led to rapid desensitisation, and prolonged washing was required to restore the original response. Such desensitisation was not seen with CCAP. The sensitivity of the *Manduca* oviduct to Mas-AT is interesting because this peptide has strong sequence similarity (10 of 13 residues are identical) to the *Locusta* male accessory gland peptide Lom-AG-MT I (Paeman *et al.*, 1991a), which is myoactive on the locust oviduct. Moreover, the strong selectivity of the locust peptide for its action on the oviduct (it is 1000 times more active on the oviduct than on the hindgut) suggests that it plays a special role in the locust's reproductive physiology. However, there is little evidence that Mas-AT plays any role in controlling the oviduct in *M. sexta* (see below), and this peptide is not strongly selective for the oviduct, also being active on the adult heart.

Nine other insect neuropeptides known to have stimulatory activities on the oviducts of other insects, or which are known to be myotropic when tested on the larval and adult *Manduca* heart preparations (A. K. Marshall & S. E. Reynolds, unpublished observations), were all inactive on the oviduct (Table 1).

Thus the prime candidate neuroregulator of oviduct function in *M. sexta* is CCAP. The peptide is present in the abdominal nerve cord of the adult moth (Cheung *et al.*, 1992; Lehman *et al.*, 1993) so that CCAP passes two of the tests required for identification as an endogenous oviduct regulator; it is highly active, and it is present in the insect. Of course, the experiments described so far do not prove that CCAP is the regulator *in vivo*. Moreover, other unknown factors might be involved.

The effect of nerve cord extracts

Another way of investigating oviduct control is to test the effects of tissue extracts on spontaneous activity *in vitro*. Oviduct bioassays for endogenous myotropins have been reported for a limited range of insect species, mostly Orthoptera. In many insects, extracts from conspecific abdominal nerve cord have been found to stimulate oviducal contractions (an increase in frequency and/or amplitude). The usual rationale for undertaking such experiments is that because the oviducts are innervated by nerves originating in the terminal abdominal ganglion, this part of the nervous system is likely to contain substances that are used to control the oviduct *in vivo*. However, it should be remembered that myoactive substances found in this way may not be used to control the contractions of the oviducts *in vivo*.

The rationale for experiments in which extracts from one species of insect are tested for their ability to affect the oviduct of another species of insect is rather less obvious. Actually, such work has been rather successful in assisting the isolation of novel insect neuropeptides (see, for example, Schoofs *et al.*, 1993), but it is so far unclear whether these substances have any physiological role in the insects from which they were isolated.

Extracts of the abdominal nerve cord of *M. sexta* were tested for the presence of factors that affected the activity of the oviduct. Crude extracts of the abdominal nerve cord from the adult moth were prepared by homogenising the tissue in saline, boiling (3 min), and centrifuging to remove insoluble material. These extracts were strongly myoactive when applied to the *Manduca* oviduct preparation, causing an increased frequency of contractions. The stimulatory activity was present in approximately equal amounts in extracts of male and female insects. Extracts of fifth-instar larval, pupal, and adult nerve cords all stimulated contractions to a roughly equal extent. The response was dose-dependent, with larger quantities of extract causing responses that were more intense and longer lived. The conclusion must be that the abdominal nerve cord contains at least one excitatory substance. If inhibitory factors were present, then their activity was masked by the excitatory factor(s). It seemed likely that CCAP was responsible for some or all of the excitatory effects, because this peptide has been shown to be excitatory, and is known to be present in the abdominal nerve cord.

The next step was to purify the myotropic factor(s) present in the abdominal nerve cord by reversed-phase high-performance liquid chromatography (HPLC) employing a C_4 column. Bioassay of HPLC fractions after resuspending in *Manduca* saline solution revealed only a single peak of activity in the oviduct bioassay (Fig. 2A). This fraction eluted in fractions 25 and 26, with a retention time identical to that of authentic CCAP.

The same HPLC fractions were also tested for activity on the larval (Fig. 2B) and adult (Fig. 2C) heart of *M. sexta*. These assays both detect CCAP as well as additional myotropic peptides (Tublitz *et al.*, 1992). In addition to the oviduct exciting peak in fractions 25 and 26, an additional peak of myotropic activity present in fractions 28 and 29 was active in the adult heart bioassay, but not on the larval heart or the oviduct. The two peaks of cardioacceleratory activity detected in these experiments correspond to the factors previously named as CAP_2 and CAP_1, respectively (Tublitz & Truman, 1985a; Tublitz *et al.*, 1992). The failure of the second peak (fractions 28 and 29) to excite the larval heart is diagnostic of CAP_1, and it is interesting that this also fails to excite the oviduct. Both CAP_2 and CAP_1 are now known to consist of several peptides (Cheung *et al.*, 1992; N. J. Tublitz, personal communication). CAP_{2A}, the principal cardioactive component of CAP_2, is now recognised to be identical to CCAP (Cheung *et al.*, 1992).

Notably, fractions corresponding to Mas-AT did not contain significant amounts of oviduct stimulating activity. This peptide elutes from the

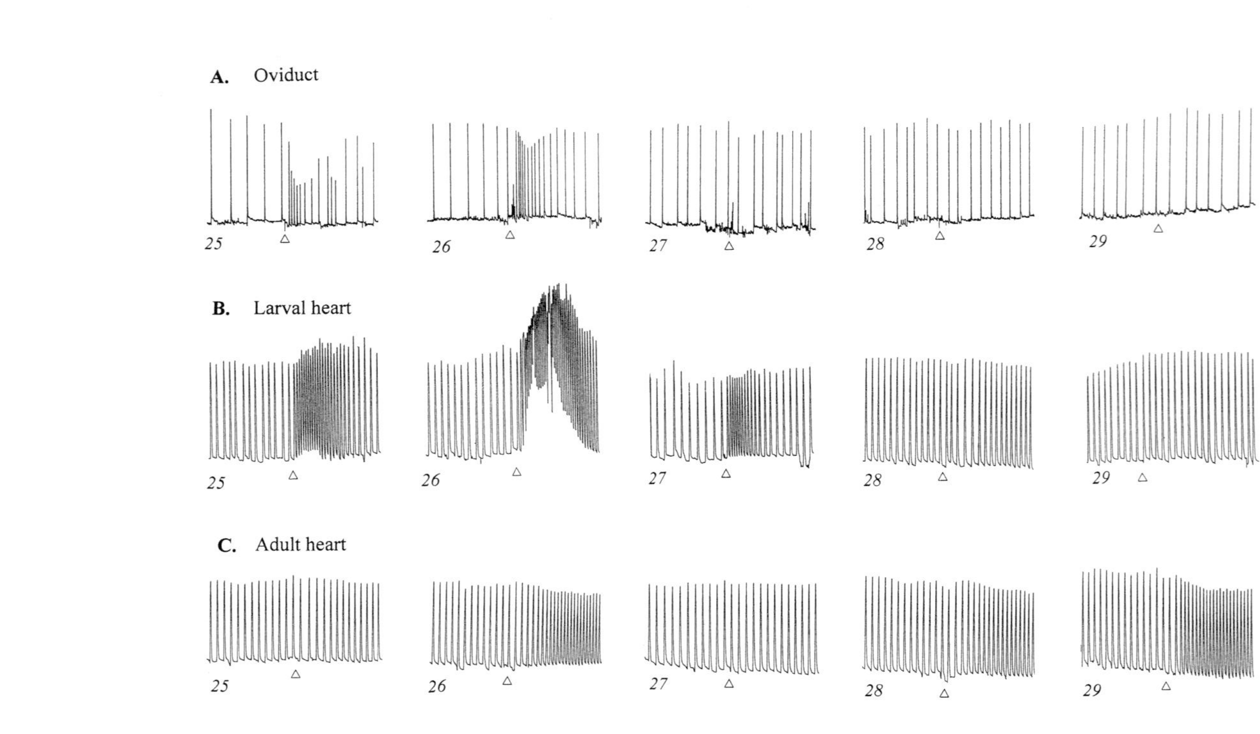
A. Oviduct
25   26   27   28   29
B. Larval heart
25   26   27   28   29
C. Adult heart
25   26   27   28   29

reversed-phase C_4 column later than CCAP, which is in accord with the suggestion of Veenstra, Lehman & Davies (1994) that Mas-AT is in fact one of the so far unidentified CAP_1 peptides. However, the very low activity of this peptide would probably have precluded detecting it unless it was present in very large amounts. Another *Manduca* cardioacceleratory peptide, CAP_{2B}, has now been sequenced (Huesman *et al.*, 1995), but this peptide is much less active on the adult heart than is CCAP. Unfortunately, it was not possible to test CAP_{2B} on the oviduct during this work, but unless it is much more active on the oviduct than on the heart it is unlikely to have been detected. Given that CCAP is active on the *Manduca* oviduct, that the oviduct-stimulating activity in nerve cord extracts co-elutes with CCAP, and that CCAP has been isolated and sequenced from extracts of the *Manduca* nerve cord, it is very likely that most if not all of the oviduct-stimulating activity in the abdominal nervous system is in fact CCAP.

Innervation of the oviducts

Anatomical studies performed in the course of this work confirm the observations of Thorn & Truman (1989) that the innervation of the female reproductive tract by nerves originating in the terminal abdominal ganglion is as shown in Fig. 3. Methylene blue staining together with forward and backward cobalt filling shows that the lateral oviducts are innervated only by ventral nerve 7 (VN7) and dorsal nerve 8 (DN8). The picture is, however,

---

Fig. 2. Reversed-phase HPLC separation of myotropic activity in extracts of abdominal nerve cord. (A) Effect on the isolated oviduct of the HPLC fractions. Each application was of 0.5 insect equivalents. Fractions other than those shown had no effect on the oviduct. The same fractions were applied to (B) the larval *M. sexta* heart bioassay and (C) the adult *M. sexta* heart bioassay. For the heart assays, each application was of 0.25 insect equivalents. Synthetic CCAP eluted in fractions 25 and 26. Extracts of abdominal nerve cords were prepared from frozen tissue (stored at $-70°C$ with a small amount of phenylthiourea) by homogenising in a medium designed to prevent proteolysis (1 Mol acetic acid l^{-1}, 20 mmol sulphuric acid l^{-1}, 1 mmol ethylenediaminetetraacetic acid l^{-1}, 0.1 mmol phenyl-methylsulphonylfluoride l^{-1}, all in methanol). The extracts were diluted to 10% methanol and loaded onto a C_{18} Sep-Pak (Waters). This was washed with 10% acetonitrile/0.1% trifluoroacetic acid (TFA). All myoactivity was eluted with 80% acetonitrile/0.1% TFA and lyophilised before further purification. HPLC was performed using a C_4 Hypersil 10 μm column and a gradient of acetonitrile. Fractions collected from the eluate were lyophilised after adding 0.1% bovine serum albumin and resuspended in *Manduca* saline before bioassay.

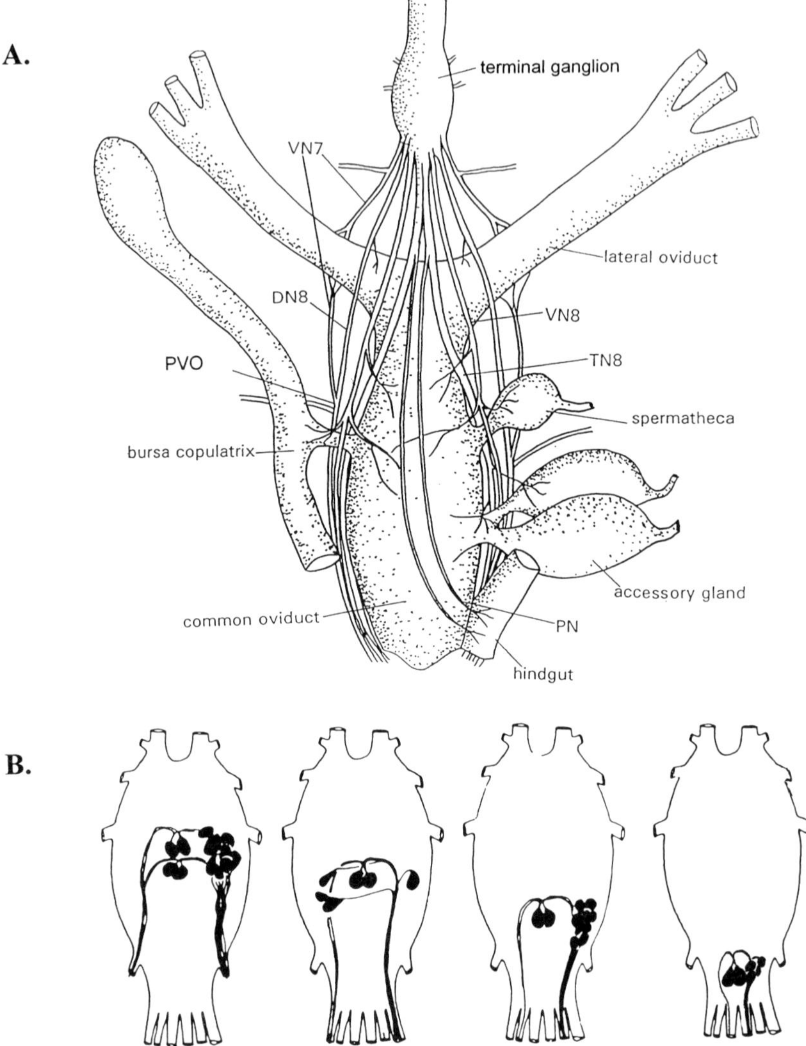

Fig. 3. (A) The oviduct and its innervation in *M. sexta*. (B) Neurones in the terminal abdominal ganglion of the adult female moth that project via nerves supplying the oviduct. Schematic of cell bodies backfilled by cobalt from the indicated nerves. Not all of these cells would necessarily project to the oviduct. See text for details. ((B) After Giebultowicz & Truman, 1984, with additional information from Thorn & Truman, 1989). VN7, ventral nerve 7; DN8, dorsal nerve 8; VN8, ventral nerve 8; PN, proctodaeal nerve; TN8, terminal nerve; PVO, perivisceral organ.

rather complicated by the fact that VN7 and DN8 meet just beyond the lateral oviducts in an anastomosis that corresponds to a swollen neurohaemal area, or perivisceral organ (PVO). Axons from both VN7 and DN8 are present in this region, which corresponds anatomically to the transverse nerves of more anterior segments. Thorn & Truman (1989) showed that the branch of VN7 joining DN8 can be backfilled to the terminal ganglion to reveal two ipsilateral neurones and two bilateral midline neurones projecting to both VN7 nerves. In contrast, the branch of VN7 innervating the lateral oviduct backfills three pairs of midline neurones in the terminal ganglion. VN7 has other branches, and Giebultowicz & Truman (1984) show additional lateral motoneurones that fill from this nerve (they found nine lateral cells and four midline cells in total), but these are unlikely to be concerned with the oviduct. These authors also showed that DN8 can be backfilled to show a single ipsilateral neurone, three contralateral neurones, and a single pair of midline cells.

The common oviduct is additionally innervated by ventral nerve 8 (VN8), which also supplies the bursa copulatrix, the spermatheca and the ovipositional musculature. Giebultowicz & Truman (1984) backfilled this nerve to reveal nine ipsilateral cells and one pair of midline cells in the terminal ganglion. Not all of these cells will be involved in controlling the oviduct. The terminal nerve (TN8) does not innervate the lateral oviduct, but does supply both the common oviduct and the accessory glands. Thorn & Truman (1989) found that the branch of this nerve supplying the oviduct could be backfilled to reveal four ipsilateral cells and two bilateral midline cells. The proctodeal nerve (actually a branch of TN8 in this insect) projects only to the hindgut and does not innervate the female reproductive apparatus at all. Summarising this information, the nerves VN7, DN8, VN8 and TN8 all contain axons supplying the lateral and/or common oviduct, and any or all of these might regulate the spontaneous contractions of the oviduct.

Knowing that the oviduct's spontaneous activity was affected by CCAP, whole-mount immunohistochemistry was used to detect the presence of immunoreactive CCAP in the terminal abdominal ganglion and its nerves. Although staining was weak in cell bodies of the terminal ganglion (this was in contrast to more anterior ganglia where strong staining of cell bodies was seen), CCAP staining was seen along the whole length of DN8 in four out of four preparations examined. Weaker staining for CCAP was also seen in some of VN7 and areas of TN8 in two of the four preparations. For all of the nerves, this positive staining took the form of blebs on the surface rather than fibres running along the nerves, indicating the presence of neurohaemal release sites (Fig. 4).

The previous observations by other workers suggest that CCAP staining would be expected at least in VN7 and DN8. Davies *et al.* (1993) found that

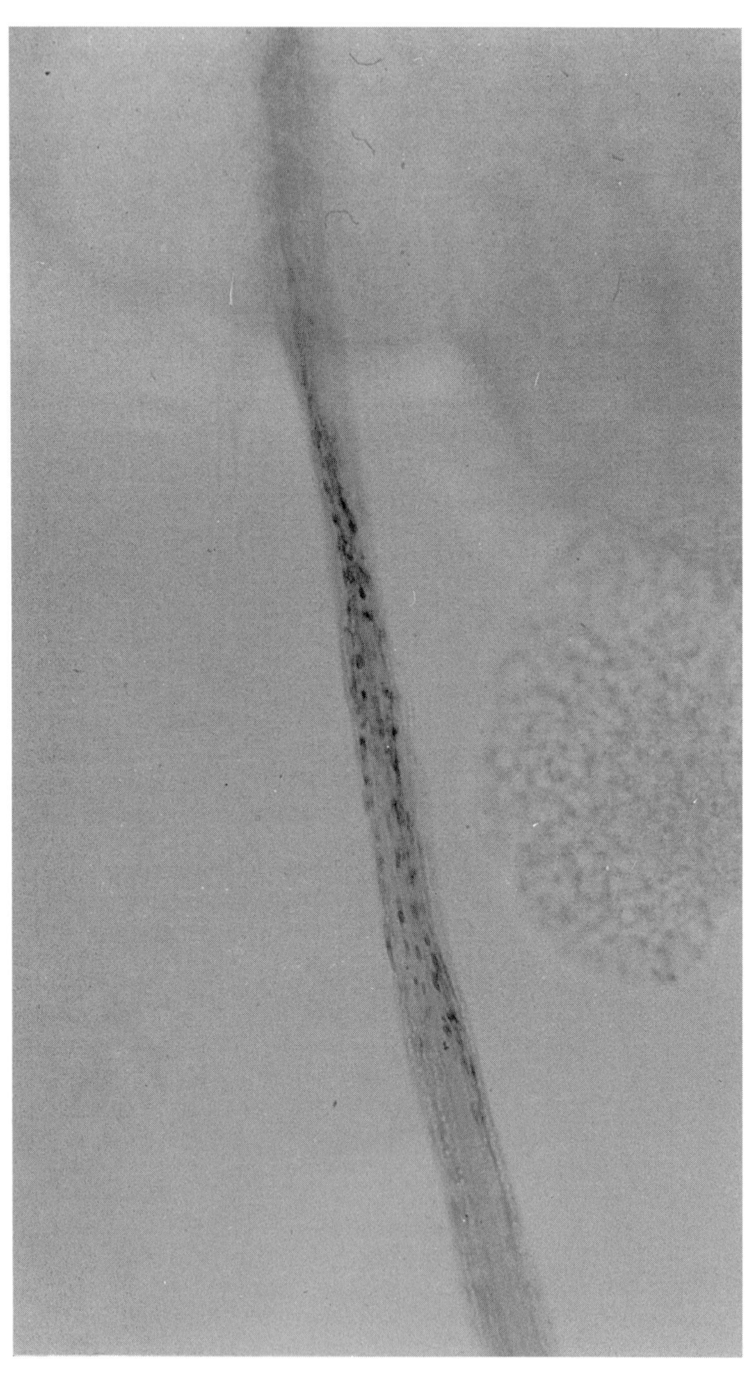

CCAP immunostaining was present in all ganglia of the *Manduca* abdominal nerve cord. The cells positive to the anti-CCAP antiserum were all neurosecretory cells that projected to the PVOs of the transverse nerves. These include four pairs of lateral cells and a pair of midline cells in each ganglion. The authors concluded that the CCAP-positive lateral cells in the abdominal ganglia are the same as those identified by Taghert & Truman (1982) as containing the tanning hormone bursicon. In each abdominal ganglion, three of these cells have axons that exit via the ipsilateral ventral nerve of the same ganglion, but the fourth cell's axon leaves the ganglion via the connective and exits the next most posterior ganglion via the dorsal nerve. Thorn & Truman (1989) also commented that the lateral cells in neuromere 7, which could be back-filled from VN7 (in the branch to the VN7–DN8 anastomosis), are probably homologues of the bursicon cells. The single lateral cell filled from DN8 by Giebultowicz & Truman (1984) may also correspond to a cell of this type.

Davies *et al.* (1993) also suggested that the pair of bilateral midline cells stained by CCAP antiserum in the abdominal ganglia was a subset of the midline neurones identified by Tublitz & Truman (1985b,c) as containing cardioactive peptides. The study of Davies *et al.* (1993) found a pair of midline neurones in abdominal neuromere 7 of the terminal abdominal ganglion that is stained by anti-CCAP antiserum. The axons of these neurones would be expected to exit from VN7.

It seems likely therefore that CCAP-containing neurones in the terminal abdominal ganglion project to both VN7 and DN8. Results from the present study (see above) suggest that DN8 is the route that contains most CCAP.

Oviduct-stimulating activity in the nerves supplying the oviducts

The observations described above indicate that CCAP is present in nerves supplying the oviduct. To confirm this, crude extracts of terminal abdominal ganglia and their associated nerves were tested in the *Manduca* oviduct bioassay. Only DN8 and the terminal abdominal ganglion contained sufficient excitatory material to stimulate contraction of the oviduct. The increase in rate of contraction with DN8 extract was 200%, roughly twice that seen with the terminal ganglion (Fig. 5). Because the CCAP antiserum also stained other nerves, the much more sensitive heart assay was also used to detect the presence of myoactive substances. In accord with the oviduct bioassay result,

---

Fig. 4. Presence of immunoreactive CCAP in DN8. CCAP is present in dark-staining varicosities along the length of the nerve. Whole mount.

A. Oviduct   **B.** Larval heart

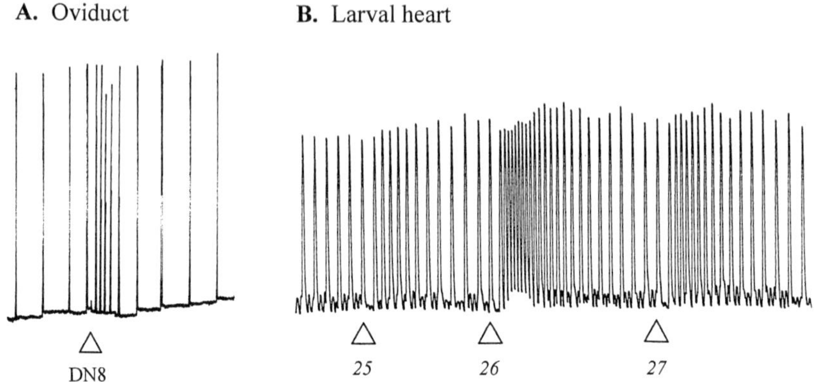

Fig. 5. Myotropic activity in the nerves supplying the oviduct. (A) Response of *M. sexta* oviduct to an extract (0.5 insect equivalents) of DN8. (B) Responses of the larval heart to HPLC fractions of a DN8 extract. Only these three fractions were stimulatory. In a separate experiment, synthetic CCAP eluted in the same fractions.

cardioacceleratory activity was found principally in DN8 and to a lesser extent in the terminal ganglion. However, smaller amounts of cardioacceleratory material were found in VN7, VN8 and (even less) in TN8. No activity was present in extracts of PN.

To some extent, the finding that DN8 has most oviduct-stimulating activity is a result of the way that the extracts were prepared (the PVOs were included with DN8 and not with VN7). However, the identification of DN8 as the nerve that contains the most material that stimulates the lateral oviducts is consistent with the CCAP immunostaining results, which show DN8 to be the principal source of CCAP-like immunoreactivity, with only rather weak staining in VN7 and in the cell bodies of the terminal ganglion.

As was the case for CCAP immunostaining, the presence of CCAP-like bioactivity in the other nerves of the terminal ganglion was much less than in DN8, and could only be detected by the use of the heart bioassay. It should be remembered, however, that the heart bioassay is much less specific than the oviduct bioassay, and that the observed cardioacceleration could have been due to the presence of other myotropins. To confirm that the oviduct-stimulating material detected in the nerve homogenates was really CCAP, the extracts were fractionated by reversed-phase HPLC. The heart bioassay was used to maximise the sensitivity of detection of myoactive substances. For DN8 (see Fig. 5), the terminal ganglion, and the other three nerves (in descending order of stimulatory effect), the only fractions that stimulated the heart were those eluting at 26 and 27 minutes. This corresponds to the retention time of synthetic CCAP. It was concluded that the presence of CCAP in these nerves accounted for their bioactivity.

CCAP as a regulator of the oviduct in insects

Observations made in the course of this study are consistent with the hypothesis that CCAP is present in nerves that innervate the lateral oviducts in *M. sexta*, and that this peptide is used to regulate the oviduct musculature's intrinsic rhythm of contraction. It is not clear, however, whether CCAP is released directly onto the oviduct surface from specialised nerve endings, or whether the peptide is released into the surrounding haemolymph from the PVO present at the peripheral anastomosis of VN7 and DN8, which is located very close to the surface of the common oviduct. To resolve this question would require separate microperfusion of the nerves and PVO, with collection and assay of the perfusate for CCAP. Quantitative immunoassays for CCAP such as those described by Stangier *et al.* (1988) and Johnen, von Glisynski & Dircksen (1995) might make this possible.

An alternative scenario is that the release of CCAP from the neurohaemal sites of DN8/VN7 is not special as far as the oviduct is concerned, and that the oviduct's contractile activity is not regulated locally, but is instead regulated by circulating CCAP (and possibly other peptide neurohormones) released generally from the PVOs of the entire abdominal nervous system. Both CAP_1s (possibly including Mas-AT) and CAP_2s (of which the major component is CCAP) are known to be released from neurohaemal sites on the segmental transverse nerves into the general circulation in adult tobacco hornworm moths, especially at the times of eclosion (Tublitz & Truman, 1985c; Tublitz & Evans, 1986) and during flight (Tublitz, 1989). It is possible that flight-related increase of CCAP is important in controlling oviduct contractile activity, because mating and in particular egg-laying are associated with episodes of flight in *M. sexta*.

The question of the generality of these findings is also of interest. As yet too few insect species have been investigated to know whether CCAP is generally involved in the regulation of reproductive function in female insects. Although Dircksen *et al.* (1991) found that CCAP stimulated contraction of the oviducts in *L. migratoria*, they also observed that the oviducts were not directly innervated by CCAP-containing neurones. Further research will doubtless provide the answer to this point.

Acknowledgements

A number of colleagues generously supplied samples of insect peptides and we are grateful to them for this. In particular, we would like to thank

Heinrich Dircksen for providing the anti-CCAP antiserum. We are indebted to Audrey Brown for ensuring an uninterrupted supply of experimental insects. This work was supported by a grant from the UK Biotechnology and Biological Sciences Research Council's Invertebrate Neuroscience Initiative.

References

Brown, B. E. & Starratt, A. N. (1975). Isolation of proctolin, a myotropic peptide from *Periplaneta americana. Journal of Insect Physiology* **21**, 1879–1881.

Cheung, C. C., Loi, P. K., Sylwester, A. W., Lee, T. D. & Tublitz, N. J. (1992). Primary structure of a cardioactive neuropeptide from the tobacco hawkmoth, *Manduca sexta. FEBS Lett*ers **313**, 165–168.

Cook, B. J. (1981). The action of proctolin and 5-hydroxytryptamine on the oviduct of the horsefly *Tabanus proximus. International Journal of Invertebrate Reproduction and Development* **3**, 209–212.

Cook, B. J., Holman, G. M. & Meola, S. (1984). The oviduct musculature of the cockroach *Leucophaea maderae* and its response to various neuro-transmitters and hormones. *Archives of Insect Biochemistry and Physiology* **1**, 167–178.

Cook, B. J. & Wagner, R. M. (1991). Some pharmacological properties of the oviduct muscularis of the stable fly *Stomoxys calcitrans. Comparative Biochemistry and Physiology* **102C**, 273–280.

Davey, K. G. (1965). Copulation and egg production in *Rhodnius prolixus*: The role of the spermathecae. *Journal of Experimental Biology* **42**, 373–378.

Davey, K. G. & Kriger, F. L. (1985). Variations during the gonotrophic cycle in the titre of the myotropic ovulation hormone and the response of the ovarian muscles in *Rhodnius prolixus. General and Comparative Endocrinology* **58**, 452–457.

Davies, N., Miller, T., Lehman, H. & Hildebrand, J. (1993). Crustacean car-dioactive peptide: immunological neurons and physiological actions in *Manduca sexta. Society for Neuroscience Abstracts* **16**, 856.

Dircksen, H., Müller, A. & Keller, R. (1991). Crustacean cardioactive peptide in the nervous system of the locust *Locusta migratoria*: an immunocytochemical study on the ventral nerve cord and peripheral innervation. *Cell and Tissue Research* **263**, 439–457.

Furuya, K., Liao, S., Reynolds, S. E., Ota, R. B., Hackett, M. & Schooley, D. A. (1993). Isolation and identification of a cardioactive peptide from *Tenebrio molitor* and *Spodoptera eridania. Biologische Chemie Hoppe-Seyler* **374**, 1065–1074.

Giebultowicz, J. M. & Truman, J. W. (1984). Sexual differentiation in the terminal ganglion of the moth *Manduca sexta*: role of sex-specific neuro-nal death. *Journal of Comparative Neurology* **226**, 87–95.

Girardie, A. & Lafon-Cazal, M. (1972). Contrôle endocrine des contractions de l'oviducte isolé de *Locusta migratoria* migratorioides (R. et F.). *Comptes Rendus Hebdomadaires des Séanes de l'Académie des Sciences, Paris* d **274**, 2208–2210.

Hildebrand, J. Lehman, H., Miller, T. Davies, N. (1990). A peptide related to crustacean cardioactive peptide in *Manduca sexta*: characterisation and distribution. *Society for Neuroscience Abstracts* **16**, 856.

Holman, G. M. & Cook, B. J. (1985). Proctolin. Its presence and action on the oviduct of an insect. *Comparative Biochemistry and Physiology* **84C**, 205–211.

Huesman, G. R., Cheung, C. C., Loi, P. K., Lee, T. D., Swidereck, K. M. & Tublitz, N. J. (1995). Amino acid sequence of CAP_{2b}, an insect cardioacceleratory peptide from the tobacco hawkmoth, *Manduca sexta*. *FEBS Letters* **371**, 311–314.

Johnen, C., von Glisynski, U. & Dircksen, H. (1995). Changes in haemolymph ecdysteroid levels and CNS contents of crustacean cardioactive peptide-immunoreactivity during the moult cycle of the isopod. *Oniscus asellus*. *Netherlands Journal of Zoology* **45**, 38–40.

Kalogianni, E. & Pflüger, H.-J. (1992). The identification of motor and unpaired neurones innervating the locust oviduct. *Journal of Experimental Biology* **168**, 177–198.

Kalogianni, E. & Theophilidis, G. (1995). The motor innervation of the oviducts and central generation of the oviductal contractions in 2 orthopteran species (*Calliptamus* sp. and *Decticus albifrons*). *Journal of Experimental Biology* **198**, 507–520.

Kataoka, H., Toschi, A., Li, J., Carney, R., Schooley, D. & Kramer, S. (1989). Identification of an allatotropin from adult *Manduca sexta*. *Science* **243**, 1481–1483.

Kingan, T. & Titmus, M. (1983). Radioimmunological detection of proctolin in arthropods. *Comparative Biochemistry and Physiology* **74C**, 75–78.

Lange, A. B. & Orchard, I. (1986a). Identified octopaminergic neurones modulate contractions of locust visceral muscle *via* adenosine-3′,5′-monophosphate (cyclic AMP). *Brain Research* **363**, 340–349.

Lange, A. B. & Orchard, I. (1986b). Ventral neurons in an abdominal ganglion of the locust, *Locusta migratoria*, with properties similar to dorsal unpaired median neurons. *Canadian Journal of Zoology* **64**, 264–267.

Lange, A. B., Orchard, I. & Adams, M. E. (1986). Peptidergic innervation of insect reproductive tissue: the association of proctolin with oviduct visceral musculature. *Journal of Comparative Neurology* **254**, 279–286.

Lange, A. B., Orchard, I. & Loughton, B. (1984). Neural inhibition of egg laying in the locust *Locusta migratoria*. *Journal of Insect Physiology* **30**, 271–278.

Lehman, H., Murguic, C., Miller, T., Lee, T. & Hildebrand, J. (1993). Crustacean cardioactive peptide in the sphinx moth, *Manduca sexta*. *Peptides*, **14**, 735–741.

Mesnier, M. (1972). Recherches sur le déterminisme de la ponte chez *Galleria mellonella* (Lepdoptère). *Comptes Rendus Hebdomadaires des Séances de l'Académie des Sciences, Paris* d **274**, 708–711.

Mesnier, M. (1984). Patterns of laying behaviour and control of oviposition in insects: further experiments on *Sphodromantis lineola* (Dictyoptera). *International Journal of Invertebrate Reproduction and Development* **7**, 23–32.

Mesnier, M. (1985). Origin and release sites of a hormone stimulating oviposition in the stick insect *Clitumnus extradentatus*. *Journal of Insect Physiology* **31**, 299–306.

Mesnier, M. & Provansal, A. (1975). Rôle des organes périsympathiques dans l'induction de la ponte chez *Galleria mellonella* (Lepidoptère). *Comptes Rendus Hebdomadaires des Séances de l'Académie des Sciences, Paris* d **281**, 905–907.

Mulye, H. & Davey, K. G. (1995). The feeding stimulus in *Rhodnius prolixus* is transmitted to the brain by a humoral factor. *Journal of Experimental Biology* **198**, 1087–1092.

Orchard, I. & Lange, A. B. (1985). Evidence of octopaminergic modulation of an insect visceral muscle. *Journal of Neurobiology* **16**, 171–181.

Orchard, I. & Lange, A. B. (1987). Cockroach oviducts: the presence and release of octopamine and proctolin. *Journal of Insect Physiology* **33**, 265–268.

Paeman, L., Tips, A., Schoofs, L., Proost, P., Van Damme, J. & De Loof, A. (1991a). Lom-AG-myotropin, a novel myotropic peptide from the male accessory glands of *Locusta migratoria*. *Peptides* **12**, 7–10.

Paeman, L., Schoofs, L., Proost, P., Decock, B. & De Loof, A. (1991b). Isolation, identification and synthesis of Lom-AG-myotropin II, a novel myotropic peptide in the male accessory glands of *Locusta migratoria*. *Insect Biochemistry* **21**, 243–348.

Platt, N. & Reynolds, S. E. (1986). The pharmacology of the heart of a caterpillar, the tobacco hornworm, *Manduca sexta*. *Journal of Insect Physiology* **32**, 221–230.

Schoofs, L., Vanden Broeck, J. & De Loof, A. (1993). The myotropic peptides of *Locusta migratoria*: structures, distribution, functions and receptors. *Insect Biochemistry and Molecular Biology* **23**, 859–881.

Sefiani, M. (1987). Regulation of egg laying and in vitro oviducal contraction in *Gryllus bimaculatus*. *Journal of Insect Physiology* **33**, 215–222.

Snodgrass, R. E. (1935). *Principles of Insect Morphology*. New York: McGraw-Hill.

Spittaels, K., Schoofs, L., Grauwels, L., Smet, H., Van Damme, J., Proost, P. & De Loof, A. (1991). Isolation, identification and synthesis of novel oviductal motility stimulating head peptide in the Colorado potato beetle, *Leptinotarsa decemlineata*. *Peptides* **12**, 31–36.

Stangier, J., Hilbich, C., Beyreuther, K. & Keller, R. (1987). Unusual cardioactive peptide (CCAP) from pericardial organs of the shore crab, *Carcinus maenas*. *Proceedings of the National Academy of Sciences USA* **84**, 575–579.

Stangier, J., Hilbich, C., Dircksen, H. & Keller, R. (1988). Distribution of a novel cardioactive neuropeptide (CCAP) in the nervous system of the shore crab *Carcinus maenas. Peptides* **9**, 795–800.

Taghert, P. H. & Truman, J. W. (1982). Identification of the bursicon-containing neurones in abdominal ganglia of the tobacco hornworm, *Manduca sexta. Journal of Experimental Biology* **98**, 385–401.

Thomas, A. (1979). Nervous control of egg progression into the common oviduct and genital chamber of the stick-insect *Carausius morosus. Journal of Insect Physiology* **25**, 811–823.

Thomas, A. & Mesnier, M. (1973). Le rôle du système nerveux central sur les mecanismes de l'oviposition chez *Carausius morosus* et *Clitumnus extradentatus. Journal of Insect Physiology* **19**, 383–396.

Thorn, R. S. & Truman, J. W. (1989). Sex-specific neuronal respectification during the metamorphosis of the genital segments of the tobacco hornworm moth *Manduca sexta. Journal of Comparative Neurology* **284**, 489–503.

Tublitz, N. (1989). Insect cardioactive peptides: neurhormonal regulation of cardiac activity by two cardioacceleratory peptides during flight in the tobacco hawkmoth, *Manduca sexta. Journal of Experimental Biology* **142**, 31–48.

Tublitz, N. J., Cheung, C. C., Edwards, K. K., Sylwester, A. W. & Reynolds, S. E. (1992). Insect cardioactive peptides in *Manduca sexta*: a comparison of the biochemical and molecular characteristics of cardioactive peptides in larvae and adults. *Journal of Experimental Biology* **165**, 265–272.

Tublitz, N. J. & Evans, P. D. (1986). Insect cardioactive peptides: cardioacceleratory peptide (CAP) activity is blocked *in vivo* and *in vitro* with a monoclonal antibody. *Journal of Neuroscience* **6**, 2451–2456.

Tublitz, N. J. & Truman, J. W. (1985a). Insect cardioactive peptides. I. Distribution and molecular characteristics of two cardioacceleratory peptides in the tobacco hawkmoth, *Manduca sexta. Journal of Experimental Biology* **114**, 381–395.

Tublitz, N. J. & Truman, J. W. (1985b). Identification of neurones containing cardioacceleratory peptides (CAPs) in the ventral nervous cord of the tobacco hawkmoth, *Manduca sexta. Journal of Experimental Biology* **116**, 395–410.

Tublitz, N. J. & Truman, J. W. (1985c). Intracellular stimulation of an identified neuron evokes peptide release in an insect. *Science* **228**, 1013–1015.

Veenstra, J. A., Lehman, H. K. & Davies, N. T. (1994). Allatotropin is a cardioacceleratory peptide in *Manduca sexta. Journal of Experimental Biology* **188**, 347–354.

Part IV

Peptidases, peptide and pseudopeptide mimetics: towards new strategies of insect pest control

R. ELWYN ISAAC, DAVID COATES,
TRACY A. WILLIAMS and LILIANE SCHOOFS

Insect angiotensin-converting enzyme: comparative biochemistry and evolution

Introduction

Proteases are essential components of many endocrine signalling pathways, being involved in the processing of pro-proteins to peptide/protein ligands, activation of cell surface receptors and providing mechanisms for terminating the biological activity of peptide/protein hormones and neurotransmitters (Hollenberg, 1996). Much of this knowledge has been gleaned from studies of mammalian systems, but it is now known that similar proteolytic mechanisms are also important for successful functioning of signalling pathways in insects. For example, homologues of mammalian proteolytic enzymes (prohormone convertase, angiotensin-converting enzyme and carboxypeptidase) involved in the biosynthesis of mammalian peptide hormones, have also been found in insects (Roebroek *et al.*, 1991; Lamango & Isaac, 1993; Lamango & Isaac, 1994; Settle, Green & Burtis, 1995) and, in the fruitfly *Drosophila melanogaster*, mutations in the carboxypeptidase genes (silver, *svr*) and angiotensin-converting enzyme gene (*Ance* or *Race*) result in lethal phenotypes. Insect homologues of mammalian peptidases responsible for terminating the signalling activity of peptide hormones and transmitters have also been described in the nervous system and in peripheral tissues of insects, indicating that similar mechanisms operate in insects for the metabolic inactivation of neuropeptides (Isaac, 1987; Isaac, 1988; Fox & Reynolds, 1991; Rayne & O'Shea, 1992; Yagi, Yu & Tobe, 1992; Lamango & Isaac, 1993; Strey, Hayes & Isaac, 1993; Masler, Wagner & Kovaleva, 1996). In general, these peptidases/proteases can hydrolyse a broad spectrum of peptide structures, and are widely distributed in different tissues and cell types. It is therefore not a simple task to elucidate all the biological roles of these enzymes and to understand their importance for the correct functioning of the neuroendocrine system.

Angiotensin-converting enzyme (ACE, dipeptidyl carboxypeptidase I, kininase II, EC 3.4.15.1) has an important role in the regulation of blood pressure, fluid and electrolyte homeostasis, and consequently it is one of the most studied mammalian peptidases. As part of the renin–angiotensin system, vascular endothelial ACE catalyses the last step in the biosynthesis of the potent vasopressor angiotensin II, by cleaving His-Leu from the C-terminus of angiotensin I. ACE also cleaves the C-terminal dipeptide from circulating bradykinin, resulting in the inactivation of this vasodilatory peptide (Erdös, 1990). The major form of mammalian ACE is classed as a type II ectoenzyme, with the bulk of the protein, including the active site, positioned on the extracellular surface of cells and connected to a short intracellular C-terminal tail by a hydrophobic sequence that crosses the plasma membrane (Hooper, 1991; Soubrier *et al.*, 1993a). The enzyme exists as two main isoforms, a somatic form (M_r 140 000–180 000, depending on species) which has a wide tissue distribution, and a smaller form (M_r 90 000–100 000, depending on species) found exclusively in the testes (Soubrier *et al.*, 1993a). Mammalian somatic ACE (sACE) is composed of two very similar domains (N- and C-domains), both of which are catalytically active (Wei *et al.*, 1991. Testicular ACE (tACE) or germinal ACE (gACE) is identical to the C-domain of sACE, apart from a short N-terminal sequence (64–72 amino acid residues, depending on species), and includes the transmembrane region and the intracellular C-terminal domain of sACE (Ehlers *et al.*, 1989; Kumar *et al.*, 1989; Lattion *et al.*, 1989). Both sACE and tACE are transcribed from a single gene with the testis-specific transcript under the control of an alternative intragenic promoter (Corvol, Williams & Soubrier, 1995). The physiological function of tACE with its single catalytic domain is not known.

ACE has a broad *in vitro* substrate specificity, hydrolysing the penultimate bonds of [Leu^5]enkephalin, [Met^5]enkephalin, enkephalin-Arg-Phe, neurotensin and the haemoregulatory peptide N-acetyl-Ser-Asp-Lys-Pro, in addition to the hydrolysis of angiotensin I and bradykinin (Erdös & Skidgel, 1985; Rousseau *et al.*, 1995). Although primarily a dipeptidyl carboxypeptidase, ACE can also be an endopeptidase, cleaving dipeptide amides and sometimes tripeptide amides from the C-terminus of α-amidated peptides such as substance P and cholecystokinin 8 (CCK8) (Hooper, 1991). An even more unusual ACE activity is the hydrolysis of luteinising hormone-releasing hormone (LH-RH) at two peptide bonds to release an amino terminal tripeptide as well as a carboxyl-terminal tripeptide (Skidgel & Erdös, 1985). However, it is only angiotensin I, bradykinin and N-acetyl-Ser-Asp-Lys-Pro that have been confirmed as *in vivo* substrates for mammalian ACE. Substance P is a candidate *in vivo* substrate for brain and lung ACE, although this role for ACE has not been substantiated. Selective inhibitors of mammalian ACE have been developed to block the formation of angiotensin II and inhibit the inactivation of bradykinin (Ondetti, 1994). These compounds are used in the treatment of hypertension and myocardial remodelling after myocardial infarction, and their success as therapeutic agents has focused a great deal of attention on the structure and function of mammalian ACE. Mammalian ACE is also found in non-endothelial cells, such as the absorptive epithelia of the small intestine and kidney proximal convoluted tubule, brain and male genital tract (Soubrier *et al.*, 1993b). Minor soluble forms of ACE are present in body fluids (plasma, seminal fluid, cerebrospinal flud and ileal fluid), and these are considered to be dervied from proteolytic cleavage of the membrane-bound enzyme (Corvol *et al.*, 1995).

The first report of an invertebrate ACE came from work on the metabolism of neuropeptides by membranes prepared from the heads of the housefly, *Musca domestica* (Lamango & Isaac, 1993, 1994). [D-Ala^2, Leu^5]enkephalin is an *in vitro* substrate for insect endopeptidase 24.11, which cleaves the peptide at the Gly-Phe bond and is inhibited by phosphoramidon. However, when [D-Ala^2, Leu^5]enkephalin was incubated with plasma membranes from *M. domestica* heads, phosphoramidon did not inhibit all of the hydrolysis of the Gly-Phe bond, indicating the presence of a second peptidase. This second activity was inhibited by captopril, a potent inhibitor of mammalian ACE, demonstrating for the first time the existence of an ACE-like dipeptidyl carboxypeptidase in insects.

Detailed characterisation of purified *M. domestica* and *D. melanogaster* ACE has established that the insect enzyme is structurally and enzymatically similar to mammalian ACE (Cornell *et al.*, 1995; Lamango, Sajid & Isaac, 1996a; Williams *et al.*, 1996). Comparisons with the mammalian enzymes do not immediately provide clues to the function of ACE in insects; insects do not have a closed circulatory system, and peptides that are structurally related to the established *in vivo* substrates of mammalian ACE (angiotensin I, bradykinin and *N*-acetyl-Ser-Asp-Lys-Pro) have not been found in insect tissues. However, there is strong evidence from the effects of ACE inhibitors and from the lethal phenotype of two *D. melanogaster* ACE mutants, that ACE has a vital role in insect growth and development (Lamango, 1994; Tatei *et al.*, 1995). In the pursuit of the biological role of insect ACE, work has focused on the substrate specificity, and the tissue and cellular localisation of the enzyme. These data, together with ongoing studies of *D. melanogaster* ACE mutants and the effects of ACE inhibitors on insect physiology, will lead to the elucidation of the role of insect ACE. This information is likely to provide an insight into the evolution of peptide hormone processing pathways, and might eventually lead to the discovery of hitherto unknown functions of vertebrate ACE that have been conserved in the course of evolution.

Biochemical properties

Substrate specificity

Musca domestica, housefly, ACE displays a broad *in vitro* substrate specificity, hydrolysing a range of mammalian insect peptide hormones/transmitters

Table 1 *The substrate specificity of housefly ACE. Initial sites of hydrolysis are indicated by bold arrows (↓) and subsequent cleavage positions are indicated by normal type arrows (↓). One unit of activity is 1 nmol of peptide hydrolysed per min*

Peptide	Structure and sites of hydrolysis by ACE	K_m μmol l^{-1}	V_{max} (units/mg)	V_{max}/K_m
Angiotensin I	Asp-Arg-Val-Tyr-Ile-His-Pro-Phe ↓-His-Leu	235	88	0.4
Bradykinin	Arg-Pro-Pro-Gly-Phe-Ser-Pro ↓-Phe-Arg	9	95	10.6
Bradykinin[1-7]	Arg-Pro-Pro-Gly-Phe-↓-Ser-Pro	33	176	5.3
[Leu]5]enkephalin	Tyr-Gly-Gly ↓-Phe-Leu	98	255	2.6
[Leu5]enkephalinamide	Tyr-Gly-Gly ↓-Phe-Leu-NH$_2$	557	888	1.6
[Met5]-enkephalin	Tyr-Gly-Gly ↓-Phe-Met	53	241	4.6
[Met5]-enkephalinamide	Tyr-Gly-Gly ↓-Phe-Met-NH$_2$	1056	510	0.5
Substance P	Arg-Pro-Lys-Pro-Gln-Gln ↓-Phe ↓-Phe ↓-Met-Leu-NH$_2$	222	171	0.8
LH-RH	pGlu-His-Trp ↓-↓ Ser-Tyr ↓-Gly-Leu ↓-Arg-Pro-Gly-NH$_2$	253	44	0.2
Cus-DP II	Asn-Asn-Ala-Asn-Val-Phe-Tyr-Pro ↓-Trp-Gly-NH$_2$	94	210	2.2
Leucokinin I	Asp-Pro-Ala-Phe-Asn-Ser ↓-Trp-Gly-NH$_2$	634	2352	3.7
Locustatachykinin I	Gly-Pro-Ser-Gly-Phe ↓-Tyr-Gly ↓-Val-Arg-NH$_2$	296	546	1.8

Note:

LH-RH, luteinising hormone-releasing hormone; Cus-DP, *Culex* depolarising peptide.

with different C-terminal amino acid sequences (Lamango *et al.*, 1996, 1997). The housefly enzyme cleaves the C-terminal dipeptides His-Leu from angiotensin I, Phe-Arg from bradykinin, Ser-Pro from bradykinin$^{1-7}$, Phe-Leu from [Leu^5]enkephalin and Phe-Met from [Met^5]enkephalin (Table 1). Of these peptides bradykinin has the highest affinity for the insect enzyme with a K_m of 9 μmol l^{-1}. The other peptides gave K_m values ranging from 33 to 235 μmol l^{-1}. The similarity between the insect and mammalian ACEs extends to the hydrolysis of α-amidated peptides like substance P, LH-RH, [Leu^5]enkephalinamide and [Met^5]enkephalinamide. Dipeptideamides are hydrolysed by *M. domestica* ACE from [Leu^5]enkephalinamide, [Met^5]enkephalinamide, and both a dipeptideamide and tripeptideamide are cleaved from substance P. It is clear from comparing the K_m data for the enkephalin peptides that α-amidation results in a much weaker binding of the peptide at the active site.

Insect peptide hormones often belong to families of closely related structures with many common amino acid sequences, especially at the C-terminus. It is therefore possible in many instances to predict whether whole families of peptides will serve as ACE substrates by studying the susceptibility of a few representatives of the peptide families to hydrolysis. The majority of characterised insect peptide hormones have an α-amidated amino acid at the C-terminus, which often forms part of a core C-terminal sequence containing the essential structural features for molecular recognition by its receptor. In addition, the C-terminal α-amide group can protect the peptide from attack by degradative carboxypeptidases. For such peptides, cleavage close to the C-terminus will inevitably result in the inactivation of the peptide hormone. Using *M. domestica* enzyme, it has been shown that a variety of insect peptide hormones are susceptible to hydrolysis by insect ACE (Lamango *et al.*, 1997). These include members of the leucokinin, locustatachykinin, adipokinetic hormone/red pigment-concentrating hormone (AKH/RPCH) families of peptides. Primary cleavage sites have been identified for leucokinin I (LK I), leucokinin II (LK II), locustatachykinin I (Lom-TK I) and *Culex* depolarising peptide II (Cus-DP II) (Table 1). Of the insect peptides tested Cus-DP II has the highest affinity for the enzyme with a K_m of 94 μmol l^{-1}. The preferred cleavage site for the endoproteolytic activity of *M. domestica* ACE towards Cus-DP II, LK I, LK II and Lom-TK I is at the penultimate peptide bond, releasing the dipeptideamide. Proctolin and crustacean cardioactive peptide (CCAP) are not readily hydrolysed by *M. domestica* ACE. Peptides with a penultimate C-terminal proline residue are quite resistant to hydrolysis by mammalian ACE, and it is also likely that the presence of a proline at the penultimate position in proctolin is a major factor contributing towards the resistance of this pentapeptide to attack by insect ACE. CCAP has a single cysteine bond that folds away the C-terminus, which is likely to sterically hinder binding to the active site of ACE.

Soluble and membrane-bound ACE

Although insect ACE activity was first found in crude plasma membranes prepared from heads of *M. domestica*, it now appears that soluble ACE is the major form of the enzyme in tissues of *M. domestica, D. melanogaster*, the hornfly, *Haematobia irritans*, and the locust *Locusta migratoria* (Lamango & Isaac, 1994; Cornell *et al.*, 1995; Wijffels *et al.*, 1996; I. Gonzales-Chamon, unpublished data). The phase separation properties of Triton X-114 at temperatures above 30°C, and the partitioning of amphipathic proteins into the detergent-rich layer, provide a convenient method for isolating integral membrane proteins. ACE associated with plasma membrane preparations from *M. domestica* does not behave like an integral membrane protein when the membranes are solubilised with Triton X-114; most of the enzyme activity partitions into the detergent-poor phase during detergent phase separation at 30°C (Lamango & Isaac, 1994). *Musca domestica* ACE has been purified from both a membrane and a soluble source, and there is no apparent difference in the size and enzymatic properties of the two enzymes (Lamango, 1994). Therefore there is still a question mark as to whether there is an integral membrane isoform of insect ACE in addition to the major soluble form of the enzyme.

Tissue distribution and cellular localisation of insect ACE

Tissue distribution

ACE activity is concentrated in certain tissues of the insect, but the tissue distribution is not identical in all species. For example, in the third-instar larvae of *D. melanogaster* the best source of ACE is the midgut section of the intestine, with the haemolymph and central nervous system showing significant but lower activity. Much weaker ACE activity is found in the Malpighian tubules, fat body and salivary glands (I. Gonzales-Chamon, unpublished data). In adult female *Manduca sexta*, the highest specific activity for ACE was seen in the haemolymph and the ovaries (Lamango, 1994). The blood and gonads of male adult *Manduca sexta* also contain high levels of ACE. In locusts, the adult testis contains very high ACE activity, and this tissue is the richest source of insect ACE found in the course of our studies on a number of insect species (R. E. Isaac, unpublished data). In adult *H. irritans*, ACE has been found in the fused ganglion, midgut and in the maturing testes of male insects (Wijffels *et al.*, 1996). The source of the ACE in the haemolymph of *D. melanogaster, M. sexta* and *M. domestica* is not known, and synthesis and secretion of the enzyme by haemocytes should be considered (Isaac & Lamango, 1994).

Subcellular localisation of ACE in the locust central nervous system

Determining the cellular localisation of ACE and identifying potential substrates of the enzyme through co-localisation studies is important evidence in any case that proposes a physiological role for insect ACE. *Schistocerca gregaria* and *L. migratoria* were chosen for immunocytochemical localisation studies, because of the large number of bioactive peptides (over 50) that have been isolated from these two locust species and the knowledge of the localisation of many of these peptides (Schoofs *et al.*, 1997). Antibodies were raised against ACE from *M. domestica*, and specificity for insect ACE was confirmed by preabsorption of the antiserum with pure recombinant *D. melanogaster* ACE (the *Ance* gene product). This antiserum is known to cross-react with recombinant *D. melanogaster* ACE obtained from COS-7 cells and the yeast *Pichia pastoris*, transfected with the *D. melanogaster Ance* cDNA (Cornell *et al.*, 1995; Williams *et al.*, 1996).

There was strong ACE staining in the brain, corpora cardiaca, corpora allata and suboesophageal ganglion of *L. migratoria* and *S. gregaria*, with very similar staining patterns seen for both species. Five groups of immunopositive cells were detected (Fig. 1). A first group of 40–50 immunoreactive cell bodies were observed in the pars intercerebralis. The axons of the median neurosecretary cells of the pars intercerebralis enter the nervi corporis cardiaci I (NCC I)-immunoreactive collaterals from the NCC 1. They extend deeply into the dorsal protocerebral neuropil and leave the brain, passing into the storage part of the corpora cardiaca. Immunoreaactive fibres leave the corpora cardiaca and continue via the nervi corporis allati I (NCA I) into the corpora allata. A second group of three to seven immunoreactive cell bodies form a linear pattern in the lateral edge of each posterior tritocerebral hemisphere close to the circumoesophageal connectives (Fig. 2a,b). The dendrites of these cells form a tract that projects ventrally and then ascends dorsally in the NCC I, to project into the inferior median protocerebrum. Alternate sections of these cells were stained with ACE antiserum and antiserum that recognises locustamyotropin I (Gly-Ala-Val-Pro-Ala-Ala-Gln-Phe-Ser-Pro-Arg-Leu-NH$_2$) and locustamyotropin II (Glu-Gly-Asp-Phe-Thr-Pro-Arg-Leu-NH$_2$), showing co-localisation of ACE with members of the locustamyotropin/PBAN (pyrokinin/pheromone biosynthesis-activating neuropeptide) family of peptides (Fig. 2a,b). This family includes locustamyotropins I–IV and locustapyrokinins I and II, all

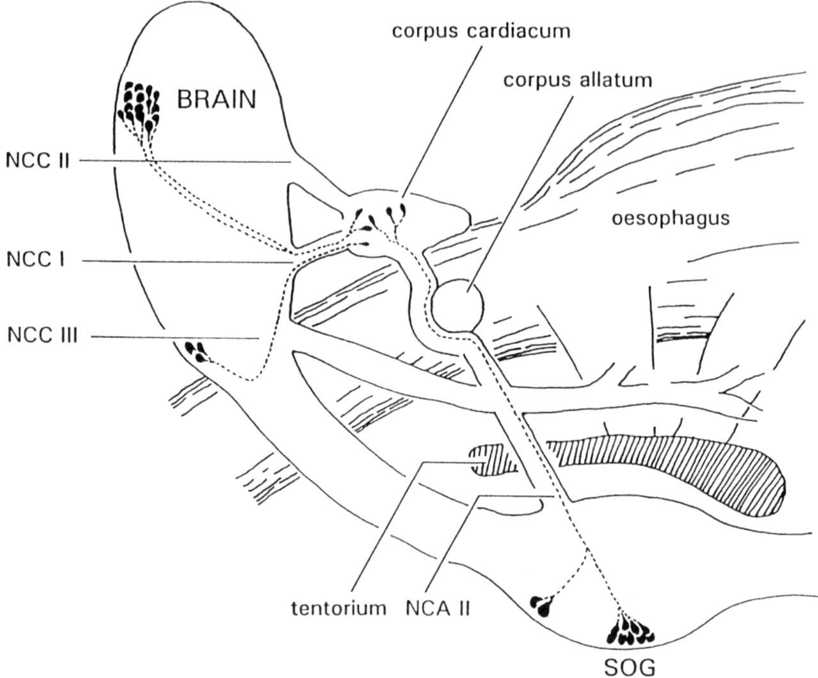

Fig. 1. Map of ACE immunoreactive cells in the brain–corpus cardiacum–corpora allata-subesophogeal ganglion complex of *L. migratoria*. NCA II, nervi corpori allati II; NCC I, II, III, nervi corpori cardiaci I, II and III; SOG, subesophogeal ganglion. Cells of the optic lobe are not indicated.

of which have a common -Phe-Xaa-Pro-Arg-Leu-NH_2 sequence (Schoofs *et al.*, 1997). A third group of about six immunoreactive cells was found in each optic lobe, on the dorsal anterior margin, near the medulla. Immunoreactive fibres project to the midbrain joining the tract passing ventrally around the lobula. Of the glomerular neuropils, the upper portion of the mushroom body calyces, and the fan-shaped body of the central body complex and its ventral ellipsoid region are invaded by immunoreactive fibres.

The fourth and fifth group of immunoreative cells were detected in the ventral midline of the suboesophageal ganglion: an anterior group of four larger cells just posterior to the hypopharyngeal roots and a posterior group of up to 15 cells, situated under the maxillary ventral commissure I. The axons of both cell clusters come together as they pass dorsally, then send fibres to the neuropilar region of the ganglion. In the anterolateral edge of the ganglion, the axon bundle bifurcates to exit in the nervi corporis allati II (NCA II). The immunoreactive axons in the NCA II cells bypass the corpora allata and project into the corpora cardiaca. Using double-labelling techniques, both anterior and posterior cells are shown to be identical with the suboesophageal ganglion cells detected in a previous study using the antiserum directed against locustamyotropin I and locustamyotropin II (Fig. 2c,d; Schoofs *et al.*, 1992). Interestingly, ACE immunoreactivity was also found in endocrine cells of the midgut, but the hormones synthesised by these cells have not been identified.

Molecular cloning of the *D. melanogaster* ACE gene (*Ance*) and the predicted protein structure

Partial and full length *D. melanogaster* cDNAs with homology to mammalian ACE were identified using the human ACE cDNA to probe a *D. melanogaster* 4–8 h embryo cDNA library (Cornell, Coates & Isaac, 1993; Cornell *et al.*, 1995) (Fig. 3). The *D. melanogaster* ACE gene is called *Ance*, because the acronym *Ace* had been previously awarded to the *D. melanogaster* acetylcholinesterase gene. COS-7 cells transfected with the *D. melanogaster Ance* cDNA expressed functional angiotensin-converting enzyme with properties similar to ACE purified from *D. melanogaster* embryos, confirming the identity of the cloned cDNA.

A number of structural features can be predicted from the *D. melanogaster* cDNA:

(i)   *Ance* codes for a single domain, 70 000 M_r protein;

(ii)   the proprotein has a hydrophobic N-terminal sequence serving as a signal for secretion;

(iii)   there are three consensus sequences for *N*-glycosylation, the position of one is conserved in mammalian ACE;

(iv)   there is a region of high homology to mammalian ACEs around the active site domain;

(v)   a C-terminal hydrophobic transmembrane domain is absent.

There is a good consensus site for cleavage of the signal peptide between residues 17 and 18 of the prohormone, which would result in a mature protein with an M_r of 68 921. This value is very close to the M_r of the ACE protein isolated from *D. melanogaster* embryos and from adult *M. domestica*. When *Ance* cDNA is expressed in COS-7 cells, the translated protein is around 5 kDa larger than the 67 kDa protein isolated from *D. melanogaster* and *M. domestica* (Cornell *et al.*, 1995). This indicates that some or all of the three potential *N*-glycosylation sites of the ACE synthesised by the COS-7 cells are occupied with oligosaccharides. A similar situation arises when *Ance* is expressed in the yeast *P. pastoris*, to give a recombinant protein with an M_r of 74 000 (Williams *et al.*, 1996). Complete digestion

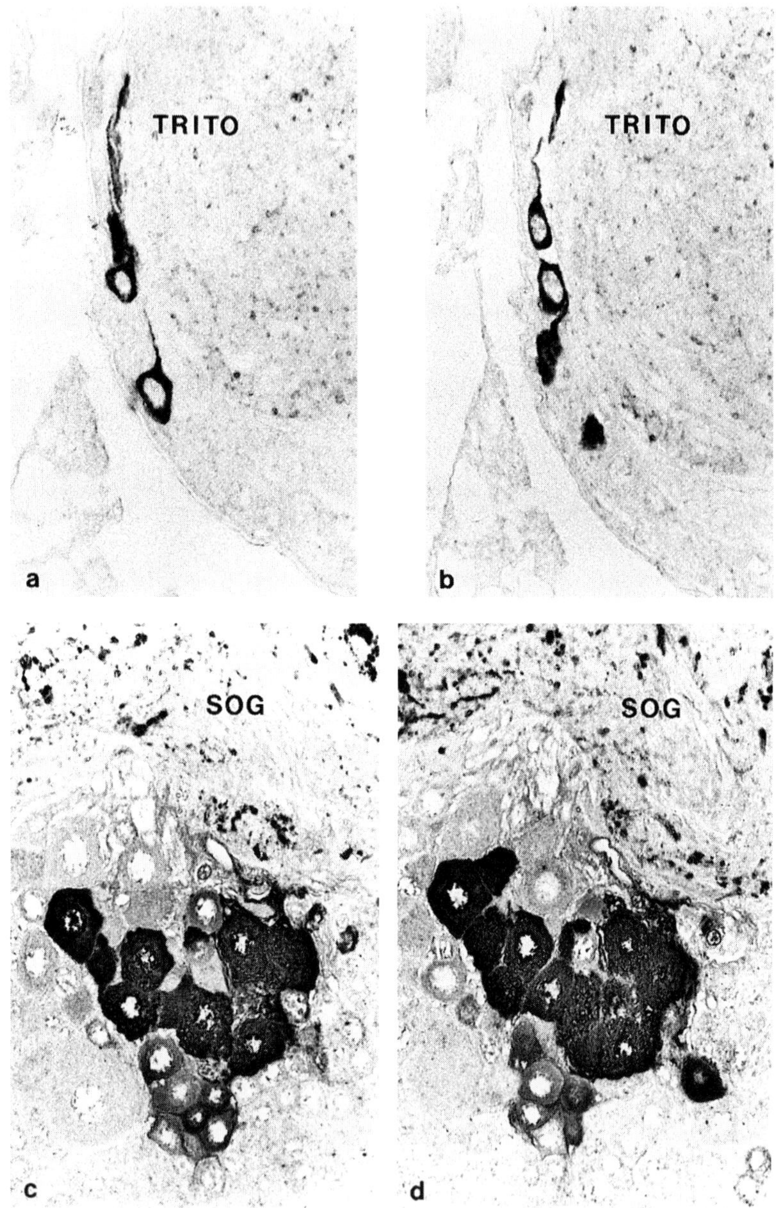
TRITO
TRITO
SOG
SOG
a
b
c
d

with endoglycosidase F yields a single 69 kDa protein, whereas partial digestion with endoglycosidase F results in four proteins in the size range 69–74 kDa. Thus, all three N-glycosylation sites of the yeast recombinant protein are linked to oligosaccharides. Although the M_r of the native insect protein is around the size expected for a non-glycosylated *D. melanaogaster* ACE, the slight fuzziness of the protein band on SDS-PAGE gels suggests that there is some microheterogeneity, which might result from the presence of carbohydrate chains at one or more of these glycosylation sites (Cornell *et al.*, 1995). The removal of the three N-glycosylation sites from AnCE by site-directed mutagenesis reduces the thermal stability of the enzyme, indicating that glycosylation influences protein structure (Williams *et al.*, 1996).

The amino acid sequence between residues 311 and 395 of *D. melanogaster* ACE displays 72% and 76% identity with the mammalian N- and C-domains, respectively. Conserved amino acids in this region include the zinc co-ordinating residues (His[367], His[371] and Glu[395]), an aspartic acid residue Asp[399] believed to have a role in the positioning of the first zinc binding histidine, and Glu[368] which is involved in catalysis (Corvol *et al.*, 1995).

An hydropathy plot of *D. melanogaster* ACE reveals the 17 amino acid residue signal sequence at the N-terminus and indicates the absence of a hydrophobic transmembrane domain in the C-terminal part of the protein, which in mammalian ACE provides the anchorage to the plasma membrane (Fig. 4). Thus, it was predicted that *Ance* codes for a soluble secreted protein (Cornell *et al.*, 1995). This prediction was confirmed when *Ance* was expressed in COS-7 cells and over 90% of the expressed enzyme activity was found in the culture medium. In mammals, soluble ACE is a minor form of the enzyme and is derived by proteolytic cleavage of the plasma membrane enzyme close to the C-terminal anchor. In contrast, the major form of ACE in insects is a soluble enzyme which is synthesised directly without post-translational processing of the protein.

ACE and development

Recent studies to identify target genes that are regulated by the homeobox gene *zerknullt* (*zen*) and the secreted growth factor gene, decapentaplegic

---

Fig. 2. Co-localization of ACE and locustamyotropin peptides in the brain and subesophogeal ganglion of *L. migratoria*. Consecutive sections of the tritocerebrum (TRITO) of the brain (a,b) and subesophogeal ganglion (SOG (c,d) showing co-localisation of ACE (a,c) and locustamyo-tropin (b,d).

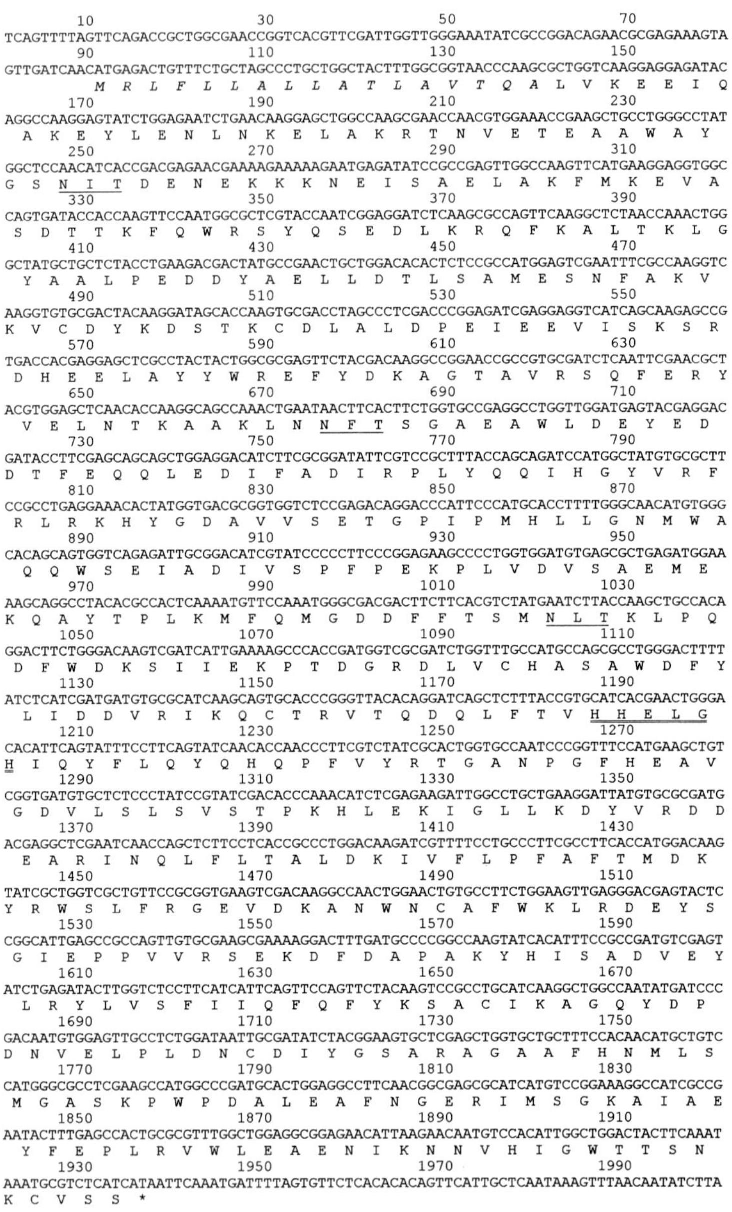

```
              10                  30                  50                  70
TCAGTTTTTAGTTCAGACCGCTGGCGAACCGGTCACGTTCGATTGGTTGGGAAATATCGCCGGACAGAACGCGAGAAAGTA
              90                 110                 130                 150
GTTGATCAACATGAGACTGTTTCTGCTAGCCCTGCTGGCTACTTTGGCGGTAACCCAAGCGCTGGTCAAGGAGGAGATAC
         M   R   L   F   L   L   A   L   L   A   T   L   A   V   T   Q   A   L   V   K   E   E   I   Q
             170                 190                 210                 230
AGGCCAAGGAGTATCTGGAGAATCTGAACAAGGAGCTGGCCAAGCGAACCAACGTGGAAACCGAAGCTGCCTGGGCCTAT
  A   K   E   Y   L   E   N   L   N   K   E   L   A   K   R   T   N   V   E   T   E   A   A   W   A   Y
             250                 270                 290                 310
GGCTCCAACATCACCGACGAGAACGAAAAGAAAAAGAATGAGATATCCGCCGAGTGGCCAAGTTCATGAAGGAGGTGGC
  G   S   N   I   T   D   E   N   E   K   K   K   N   E   I   S   A   E   L   A   K   F   M   K   E   V   A
             330                 350                 370                 390
CAGTGATACCACCAAGTTCCAATGGCGCTCGTACCAATCGGAGGATCTCAAGCGCCAGTTCAAGGCTCTAACCAAACTGG
  S   D   T   T   K   F   Q   W   R   S   Y   Q   S   E   D   L   K   R   Q   F   K   A   L   T   K   L   G
             410                 430                 450                 470
GCTATGCTGCTCTACCTGAAGACGACTATGCCGAACTGCTGGACACACTCTCCGCCATGGAGTCGAATTTCGCCAAGGTC
  Y   A   A   L   P   E   D   D   Y   A   E   L   L   D   T   L   S   A   M   E   S   N   F   A   K   V
             490                 510                 530                 550
AAGGTGTGCGACTACAAGGATAGCACCAAGTGCGACCTAGCCCTCGACCCGGAGATCGAGGAGGTCATCAGCAAGAGCCG
  K   V   C   D   Y   K   D   S   T   K   C   D   L   A   L   D   P   E   I   E   E   V   I   S   K   S   R
             570                 590                 610                 630
TGACCACGAGGAGCTCGCCTACTACTGGCGCGAGTTCTACGACAAGGCCGGAACCGCCGTGCGATCTCAATTCGAACGCT
  D   H   E   E   L   A   Y   Y   W   R   E   F   Y   D   K   A   G   T   A   V   R   S   Q   F   E   R   Y
             650                 670                 690                 710
ACGTGGAGCTCAACACCAAGGCAGCCAAACTGAATAACTTCACTTCTGGTGCCGAGGCCTGGTTGGATGAGTACGAGGAC
  V   E   L   N   T   K   A   A   K   L   N   N   F   T   S   G   A   E   A   W   L   D   E   Y   E   D
             730                 750                 770                 790
GATACCTTCGAGCAGCAGCTGGAGGACATCTTCGCGGATATTCGTCCGCTTTACCAGCAGATCCATGGCTATGTGCGCTT
  D   T   F   E   Q   Q   L   E   D   I   F   A   D   I   R   P   L   Y   Q   Q   I   H   G   Y   V   R   F
             810                 830                 850                 870
CCGCCTGAGGAAACACTATGGTGACGCGGTGGTCTCCGAGACAGGACCCATTCCCATGCACCTTTTGGGCAACATGTGGG
  R   L   R   K   H   Y   G   D   A   V   V   S   E   T   G   P   I   P   M   H   L   L   G   N   M   W   A
             890                 910                 930                 950
CACAGCAGTGGTCAGAGATTGCGGACATCGTATCCCCCTTCCCGGAGAAGCCCCTGGTGGATGTGAGCGCTGAGATGGAA
  Q   Q   W   S   E   I   A   D   I   V   S   P   F   P   E   K   P   L   V   D   V   S   A   E   M   E
             970                 990                1010                1030
AAGCAGGCCTACACGCCACTCAAAATGTTCCAAATGGGCGACGACTTCTTCACGTCTATGAATCTTACCAAGCTGCCACA
  K   Q   A   Y   T   P   L   K   M   F   Q   M   G   D   D   F   F   T   S   M   N   L   T   K   L   P   Q
            1050                1070                1090                1110
GGACTTCTGGGACAAGTCGATCATTGAAAAGCCCACCGATGGTCGCGATCTGGTTTGCCATGCCAGCGCCTGGGACTTTT
  D   F   W   D   K   S   I   I   E   K   P   T   D   G   R   D   L   V   C   H   A   S   A   W   D   F   Y
            1130                1150                1170                1190
ATCTCTCGATGATGTGCGCATCAAGCAGTGCACCCGGGTTACACAGGATCAGCTCTTTACCGTGCATCACGAACTGGGA
  L   I   D   D   V   R   I   K   Q   C   T   R   V   T   Q   D   Q   L   F   T   V   H   H   E   L   G
            1210                1230                1250                1270
CACATTCAGTATTTCCTTCAGTATCAACACCAACCCTTCGTCTATCGCACTGGTGCCAATCCCGGTTTCCATGAAGCTGT
  H   I   Q   Y   F   L   Q   Y   Q   H   Q   P   F   V   Y   R   T   G   A   N   P   G   F   H   E   A   V
            1290                1310                1330                1350
CGGTGATGTGCTCTCCCTATCCGTATCGACACCCAAACATCTCGAGAAGATTGGCCTGCTGAAGGATTATGTGCGCGATG
  G   D   V   L   S   L   S   V   S   T   P   K   H   L   E   K   I   G   L   L   K   D   Y   V   R   D   D
            1370                1390                1410                1430
ACGAGGCTCGAATCAACCAGCTCTTCCTCACCGCCCTGGACAAGATCGTTTTCCTGCCCTTCGCCTTCACCATGGACAAG
  E   A   R   I   N   Q   L   F   L   T   A   L   D   K   I   V   F   L   P   F   A   F   T   M   D   K
            1450                1470                1490                1510
TATCGCTGGTCGCTGTTCCGCGGTGAAGTCGACAAGGCCAACTGGAACTGTGCCTTCTGGAAGTTGAGGGACGAGTACTC
  Y   R   W   S   L   F   R   G   E   V   D   K   A   N   W   N   C   A   F   W   K   L   R   D   E   Y   S
            1530                1550                1570                1590
CGGCATTGAGCCGCCAGTTGTGCGAAGCGAAAAGGACTTTGATGCCCCGGCCAAGTATCACATTTCCGCCGATGTCGAGT
  G   I   E   P   P   V   V   R   S   E   K   D   F   D   A   P   A   K   Y   H   I   S   A   D   V   E   Y
            1610                1630                1650                1670
ATCTGAGATACTTGGTCTCCTTCATCATTCAGTTCCAGTTCTACAAGTCCGCCTGCATCAAGGCTGGCCAATATGATCCC
  L   R   Y   L   V   S   F   I   I   Q   F   Q   F   Y   K   S   A   C   I   K   A   G   Q   Y   D   P
            1690                1710                1730                1750
GACAATGTGGAGTTGCCTCTGGATAATTGCGATATCTACGGAAGTGCTCGAGCTGGTGCTGCTTTCCACAACATGCTGTC
  D   N   V   E   L   P   L   D   N   C   D   I   Y   G   S   A   R   A   G   A   A   F   H   N   M   L   S
            1770                1790                1810                1830
CATGGGCGCCCTCGAAGCCATGGCCCGATGCACTGGAGGCCTTCAACGGCGAGCGCATCATGTCCGGAAAGGCCATCGCCG
  M   G   A   S   K   P   W   P   D   A   L   E   A   F   N   G   E   R   I   M   S   G   K   A   I   A   E
            1850                1870                1890                1910
AATACTTTGAGCCACTGCGCGTTTGGCTGGAGGCGGAGAACATTAAGAACAATGTCCACATTGGCTGGACTACTTCAAAT
  Y   F   E   P   L   R   V   W   L   E   A   E   N   I   K   N   N   V   H   I   G   W   T   T   S   N
            1930                1950                1970                1990
AAATGCGTCTCATCATAATTCAAATGATTTTAGTGTTCTCACACACAGTTCATTGCTCAATAAAGTTTAACAATATCTTA
  K   C   V   S   S   *
```

Fig. 3. Nucleic acid sequence of *Ance* with the predicted translation product. The putative signal peptide is marked in *italics*, the three glycosylation sites are <u>underlined</u>, and the conserved HExxH (His-Glu-Xaa-Xaa-His) motif is <u>double underlined</u>.

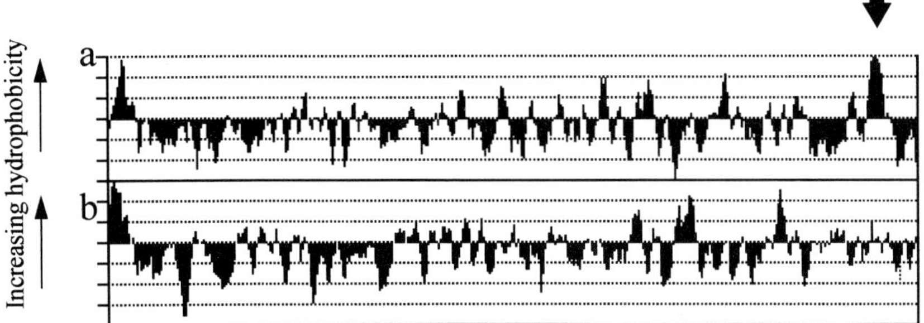

Fig. 4. Hydrophobicity plots of (a) human germinal ACE and (b) *D. melanogaster* ACE predicted from the cDNA sequence of *Ance*. The C-terminal transmembrane domain of human germinal ACE is arrowed.

(*dpp*), have identified *Race* (*R*elated to *a*ngiotensin-*c*onverting *e*nzyme) as a candidate target gene which is identical to *Ance* (Tatei *et al.*, 1995). *Race (Ance)* expression is first detected during cellularisation in the presumptive and differentiating amnioserosa. Expression is also seen in early developmental stages of the anterior and posterior midgut. In late embryos, *Race (Ance)* expression is expressed in epithelial cells of the midgut and the pericardial cells of the dorsal heart. Expression in the midgut correlates with the tissue localisation of ACE activity in *D. melanogaster* larvae (I. Gonzales-Chamon, unpublished data). Both *Ance* mRNA and ACE enzyme activity increase during development from embryos to adult insects, with peak levels occurring during pupal stages (Cornell, 1994; Cornell *et al.*, 1995; Tatei *et al.*, 1995; R. E. Isaac, unpublished data). In *zen⁻* and *dpp⁻* mutants, *Race* (*Ance*) expression in the presumptive amnioserosa, differentiated amnioserosa and pericardial cells is completely abolished, whereas normal expression is seen in the midgut after germ band elongation. Thus, *Race* (*Ance*) expression in the amnioserosa and pericardial cells is under separate control from *Race* (*Ance*) expression in the midgut.

Evolution

Although the presence of angiotensin-converting enzyme activity has been known about in vertebrates for some time, it was only when the human gene was characterised at the molecular level that it became clear that the vertebrate somatic isoform was transcribed from a tandem duplication. The ancestral copy was predicted to be the C-domain, equivalent to the germinal isoform (Lattion *et al.*, 1989). It was also clear that the two isoforms were not differentially spliced from the same promoter, but that there were separate promoters, with the somatic ACE transcribed from an upstream promoter, and the germinal form from a specific promoter within intron 12 (Hubert *et al.*, 1991; Langford *et al.*, 1991; Zhou *et al.*, 1995). Before the discovery of ACE activity in a dipteran, *M. domestica* (Lamango & Isaac, 1994), ACE had been identified in a variety of vertebrate species. In all the vertebrates examined, down to and including *Torpedo* spp. (Altstein, Dudai & Vogel, 1987; Turner *et al.*, 1987), the size of the enzyme was consistent with there being two-domain isoforms, whereas both *M. domestica* and *D. melanogaster* had smaller enzymes at all stages of development, consistent with their being only one, single-domain isoform of the enzyme. The lack of a duplication was confirmed when the *D. melanogaster* cDNA coding for fly ACE (*Ance*) was cloned and sequenced (Cornell *et al.*, 1993; Cornell *et al.*, 1995). Further confirmation that the tandem duplication was limited to the vertebrates came with the identification of ACE in another dipteran, *H. irritans* (Wijffels *et al.*, 1996), and in the cattle tick *Boophilus microplus* (Jarmey *et al.*, 1995), both of which contained only single-domain iso-forms. More recently, an ACE-like enzyme has been purified and partially sequenced from the leech *Theromyzon tessulatum* (Laurent & Salzet, 1995), and identified in the mollusc *Mytilus edulis* (V. Laurent, G. Stefano & M. Salzet, personal communication), and these also appear to produce only the small isoform.

All this suggested that the duplication which gave rise to the vertebrate two-domain isoform had occurred relatively recently in chordate evolution, and sequence comparisons of the core domains of the N- and C-domains of vertebrate ACE, and the single-domain forms of invertebrate ACEs are con-sistent with this view (Cornell *et al.*, 1995). On biochemical grounds as well as sequence analysis, it seemed likely that the C-domain of the vertebrate somatic isoform was indeed more similar to the ancestral single-domain isoform, as has been suggested previously (Lattion *et al.*, 1989). Although the initial comparative analysis implied a duplication time within the period of the evolution of the fish, the estimates were confounded both by the possibility that selection constrained the variation seen between the two domains, and the effect of concerted evolution, and it seems much more likely that the two-domain isoform arose during protochordate evolution. This would tie in with the observation that gene duplications seem to have occurred often between the two major protochordate lineages, the cephalochordates (*Branchiostoma* spp.) and the tunicates (Pendleton *et al.*, 1993; Holland *et al.*, 1994; Ruddle *et al.*, 1994; Holland & Garcia-Fernández, 1996).

However, the obvious conclusion, that a tandem duplication occurred between the cephalochordate and tunicate lineage, may no longer be so obvious, since the identification of a second ACE-like gene, called *Acer*

(Taylor, Coates & Shirras, 1996) in *D. melanogaster*. *Acer* was identified in a screen for early developmentally regulated genes, and the predicted protein shows 65% similarity to *Ance*. Indeed, *Ance* was serendipitously cloned in a very similar fashion independently (Tatei *et al.*, 1995), and both genes are developmentally expressed in the *D. melanaogaster* embryo. Although both on chromosome 2, *Ance* and *Acer* are not tandem repeats – *Ance* maps to 34E3–5, whilst *Acer* maps to 29D/E. The question arises as to whether this duplication is the same as the duplication seen in the chordates, and only looks like two independent duplications because the sequences have evolved in concert. As yet, there can be no definitive answer, but a couple of points are worth bearing in mind. First, there is the evidence that duplications which lead to isoforms of the gene product being expressed in different tissues appear not to evolve in concert, so dating on the basis of sequence differences may be valid (Iwabe, Kuma & Kiyata, 1996), at least for the vertebrate ACE, where there is complete separation of expression between somatic and germ line cells. Secondly, the appearance of the two-domain form, whether by duplication or rearrangement, coincides with other gene duplications during protochordate evolution. Assuming therefore that the two duplications are independent, it remains to be seen when duplication occurred in the invertebrate lineage. On the basis of comparative analysis, and assuming *Acer* and *Ance* are evolving independently, the prediction would be that all dipterans have two isoforms of the enzyme, whereas ticks only have one (see Fig. 5). If there is concerted evolution, then we may have to go much further back, and the possibility still exists that the duplication is very ancient.

So what was the ancestral gene? Currently there is no evidence for ACE-like activity inhibitable with captopril in nematodes, but one clue may come from the *Caenorhabditis elegans* sequencing project (Waterston & Sulston, 1995). At the moment, with some 40% of the genome sequenced, there is one potential candidate for an M2 zinc metalloendoprotease, a predicted coding sequence identified on cosmid C42D8 which shows greatest homology to the ACE-like endopeptidases, and for which several expressed sequence tags (ESTs) have been identified (Wilson *et al.*, 1994). Our laboratory is currently investigating this gene to find out exactly what it is.

Detailed analysis of the various ACE gene products has concentrated on the so-called core regions defined for the human somatic ACE, which are most easily shown, on the basis of exon/intron boundary conservation, to include the duplicated region (Hubert *et al.*, 1991). Within this region, there are significant stretches of amino acid residue identity between all sequenced ACEs. These include the zinc-binding region, which has the structure His-His - Glu - (Met/Leu) - Gly - His - Xaa - Xaa - Tyr - Xaa - Leu - Gly - Tyr - (11 Xaa) - Asn-Xaa-Gly-Phe-His-Glu-Ala-(Val/Ile)-Gly-Asp, a short region

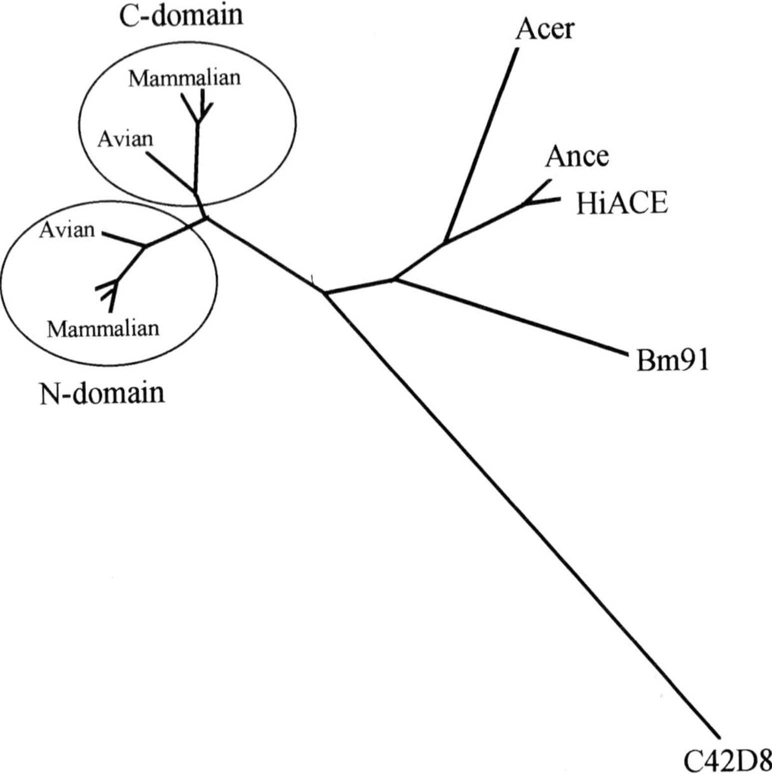

Fig. 5. Gene tree of ACE and ACE-like protein core sequences. Sequences were abstracted from the OWL non-redundant protein database (Bleasby, Akrigg & Attwood, 1994), aligned using the program ClustalW (Thompson, Higgins & Gibson, 1994), output in Phylip format (Felsenstein, 1989). The aligned sequences were compared for similarity using PROTDIST, and a tree derived using NEIGHBOR. The tree was drawn using DRAWTREE. The robustness of the branches was tested by bootstrapping 100 iterations and looking for the best tree using CONSENSE: the weakest node (separating the mouse and rabbit N-domains) scored 79%, all other nodes scored greater than 88%. Sequences were from mouse (OWL:ACE_MOUSE, residues 167–574, and OWL:ACET_MOUSE, residues 185–592), human (OWL: ACE_HUMAN, residues 162–569, and OWL:ACET_HUMAN, residues 186–593), rabbit (OWL:ACE_RABIT, residues 166–572, and OWL:ACET_RAB, residues 191–598), chicken (**Avian**: OWL:JC2489, residues 60–467 and 658–1065), *D. melanogaster* (***Ance*** is OWL:DMU25344, residues 139–552, *Acer* is from (Taylor *et al.*, 1996), residues 148–560), *H. irritans* (**HiACE**: OWL:HMBHIPR, residues 139–552), *B. microplus* (**Bm91**: P. Whitfeld (personal communication), residues 163–567) and *C. elegans* (**C42D8**, predicted gene C42D8.5, gi 1293847, residues 287–721).

upstream of this (Ser-Ala-Trp-Xaa-Xaa-Tyr-Xaa-Xaa-Xaa-Asp-Xaa-Arg-Ile-Lys-Xaa-Cys-Thr), and another block even more towards the N-terminus (Gly-Xaa-Xaa-Pro-Xaa-Xaa-Leu-Leu-Gly-Xaa-Met-Trp-Xaa-Glyn-Xaa-Trp), as well as a distributed set of conserved mono-, di- and tripeptide residues. What relevance these may or may not have is unknown, although the strict conservation of four cysteine residues throughout the length of the core region may be structurally important.

Physiological role

There is little doubt that ACE is vital for normal growth and development in insects, because mutations in the *Ance* gene result in death during larval/pupal stages (Tatei *et al.*, 1995). Furthermore, the injection of ACE inhibitors into larvae of *M. sexta* results in a reduction in growth rates and is sometimes lethal (Lamango, 1994). *Ance* expression peaks during papal development in *D. melanogaster*, suggesting a special role for the enzyme during metamorphosis.

Understanding the physiological role of insect ACE requires knowledge of the substrate specificity of the enzyme as well as the spatial and temporal expression of ACE in insect tissues. The existence of ACE in the CNS, haemolymph, midgut and reproductive tissues from a number of insect species suggests that ACE is of physiological importance in these tissues. The localisation of ACE to neurosecretory cells and neuropile regions is consistent with a dual role for the enzyme in the insect endocrine and nervous system. In neurosecretory cells, ACE is likely to be involved in the conversion of prohormones to active peptide hormones secreted into the haemolymph from the corpora cardiaca or corpora allata, whereas the neuropile enzyme might have a role in the metabolic inactivation of peptide neurotransmitters. The fact that all the locustamyotropin-containing cells of the locust brain and suboesophageal ganglion also contain ACE is evidence for a role for ACE in the biosynthesis of locustamyotropin-like hormones. Haemolymph ACE co-exists with a large number of circulating peptide hormones, many of which are susceptible to hydrolysis by ACE, at least *in vitro*. Whether these peptides are *in vivo* substrates for ACE will depend upon the kinetic parameters of the enzyme for each potential substrate. Other potential substrates for haemolymph ACE include the antibacterial peptides, which can reach high levels in the circulation of infected insects. These antibiotics are synthesised from proproteins and must undergo proteolytic processing to acquire antibacterial activity. In the locust midgut, ACE is found in endocrine cells suggesting a role for the enzyme in the processing of gut hormones. However, the high levels of ACE activity detected in midgut tissue of Diptera would also be consistent with a digestive role for ACE in these insects. ACE has an essential, but as yet undefined, role in mammalian reproduction, because male mice that do not express germinal ACE are infertile despite producing normal looking sperm (Krege *et al.*, 1995). The high levels of ACE seen in insect testes suggests that there may also be an important role for ACE in insect reproduction. Indeed, male *D. melanogaster* transheterozygotes of two mutant *Ance* alleles are known to be sterile (Tatei *et al.*, 1995).

Acknowledgements

We acknowledge the support of the Science and Engineering Research Council, UK, Fondation pour la Recherche Médicale, France, and the European Union for the support of work conducted in our laboratories.

References

Altstein, M., Dudai, Y. & Vogel, Z. (1987). Angiotensin-converting enzyme associated with *Torpedo californica* electric organ membranes. *Journal of Neuroscience Research* **18**, 333–340.

Bleasby, A. J., Akrigg, D. & Attwood, T. K. (1994). Owl – a nonredundant composite protein-sequence database. *Nucleic Acids Research* **22**, 3574–3577.

Cornell, M. J. (1994). Molecular characterisation of a peptidase gene from *Drosophila melanogaster*. Ph.D. thesis, University of Leeds.

Cornell, M. J., Coates, D. & Isaac, R. E. (1993). Characterisation of putative *Drosophila* angiotensin converting enzyme. *Biochemical Society Transactions* **21**, 243S.

Cornell, M. J., Williams, T. A., Lamango, N. S., Coates, D., Corvol, P., Soubrier, F., Hoheisel, J., Lehrach, H. & Isaac, R. E. (1995). Cloning and expression of an evolutionary conserved single-domain angiotensin-converting enzyme from *Drosophila melanogaster*. *Journal of Biological Chemistry* **270**, 13613–13619.

Corvol, P., Williams, T. A. & Soubrier, F. (1995). Peptidyl dipeptidase A – angiotensin-I-converting-enzyme. *Methods in Enzymology* **248**, 283–305.

Ehlers, M. R. W., Fox, E. A., Strydom, D. J. & Riordan, J. F. (1989). Molecular cloning of human testicular angiotensin-converting enzyme – the testis isozyme is identical to the C-terminal half of endothelial angiotensin-converting enzyme. *Proceedings of the National Academy of Sciences USA* **86**, 7741–7745.

Erdös, E. G. (1990). Angiotensin-I converting enzyme and the changes in our concepts through the years – Dahl, Lewis, K. Memorial Lecture. *Hypertension* **16**, 363–370.

Erdös, E. G. & Skidgel, R. A. (1985). Structure and functions of human angiotensin-I converting enzyme (kininase-II). *Biochemical Society Transactions* **13**, 42–44.

Felsenstein, J. (1989). PHYLIP – phylogeny inference package (version 3.2). *Cladistics* **5**, 164–166.

Fox, A. M. & Reynolds, S. E. (1991). Degradation of adipokinetic hormone family peptides by a circulating endopeptidase in the insect *Manduca sexta*. *Peptides* **12**, 937–944.

Holland, P. W. H. & Garcia-Fernández, J. (1996). Hox genes and chordate evolution. *Developmental Biology* **173**, 382–395.

Holland, P. W. H., Garcia-Fernandez, J., Williams, N. A. & Sidow, A. (1994). Gene duplications and the origins of vertebrate development. *Development*, 125–133.

Hollenberg, M. D. (1996). Protease-mediated signaling – new paradigms for cell regulation and drug development. *Trends in Pharmacological Sciences* **17**, 3–6.

Hooper, N. M. (1991). Angiotensin converting enzyme – implications from molecular-biology for its physiological functions. *International Journal of Biochemistry* **23**, 641–647.

Hubert, C., Houot, A. M., Corvol, P. & Soubrier, F. (1991). Structure of the angiotensin I-converting enzyme gene – 2 alternate promoters correspond to evolutionary steps of a duplicated gene. *Journal of Biological Chemistry* **266**, 15377–15383.

Isaac, R. E. (1987). Proctolin degradation by membrane peptidases from nervous tissues of the desert locust (*Schistocerca gregaria*). *Biochemical Journal* **245**, 365–370.

Isaac, R. E. (1988). Neuropeptide-degrading endopeptidase activity of locust (*Schistocerca gregaria*) synaptic-membranes. *Biochemical Journal* **255**, 843–847.

Isaac, R. E. & Lamango, N. S. (1994). Pepltidyl dipeptidase activity in the hemolymph of insects. *Biochemical Society Transactions* **22**, S292.

Iwabe, N., Kuma, K. & Miyata, T. (1996). Evolution of gene families and relationship with organismal evolution – rapid divergence of tissue-specific gene in the early evolution of chordates. *Molecular Biology and Evolution* **13**, 483–493.

Jarmey, J. M., Riding, G. A., Pearson, R. D., McKenna, R. V. & Willadsen, P. (1995). Carboxydipeptidase from *Boophilus microplus* – a concealed antigen with similarity to angiotensin-converting enzyme. *Insect Biochemistry and Molecular Biology* **25**, 969–974.

Krege, J. H., John, S. W. M., Langenbach, L. L., Hodgin, J. B., Hagaman, J. R., Bachman, E. S., Jennette, J. C., Obrien, D. A. & Smithies, O. (1995). Male–female differences in fertility and blood-pressure in ACE-deficient mice. *Nature* **375**, 146–148.

Kumar, R. S., Kusari, J., Roy, S. N., Soffer, R. L. & Sen, G. C. (1989). Structure of testicular angiotensin-converting enzyme – a segmental mosaic isozyme. *Journal of Biological Chemistry* **264**, 16754–16758.

Lamango, N. S. (1994). Characterization of neuropeptidases in the house fly, *Musca domestica*. Ph.D. thesis, University of Leeds.

Lamango, N. S. & Isaac, R. E. (1993). Metabolism of insect neuropeptides – properties of a membrane-bound endopeptidase from heads of *Musca domestica*. *Insect Biochemistry and Molecular Biology* **23**, 801–808.

Lamango, N. S. & Isaac, R. E. (1994). Identification and properties of a peptidyl dipeptidase in the housefly, *Musca domestica*, that resembles mammalian angiotensin-converting enzyme. *Biochemical Journal* **299**, 651–657.

Lamango, N. S., Sajid, M. & Isaac, R. E. (1996). The endopeptidase activity and the activation by Cl⁻ of angiotensin-converting enzyme is evolutionarily conserved – purification and properties of an angiotensin-converting enzyme from the housefly, *Musca domestica*. *Biochemical Journal* **314**, 639–646.

Lamango, N. S., Nachman, R. J., Hayes, T. K., Strey, A. & Isaac, R. E. (1997). Hydrolysis of insect neuropeptides by an angiotensin-converting enzyme from the house fly, *Musca domestica*. *Peptides* **18**, 47–52.

Langford, K. G., Shai, S. Y., Howard, T. E., Kovac, M. J., Overbeek, P. A. & Bernstein, K. E. (1991). Transgenic mice demonstrate a testis-specific promoter for angiotensin-converting enzyme. *Journal of Biological Chemistry* **266**, 15 559–15 562.

Lattion, A. L., Soubrier, F., Allegrini, J., Hubert, C., Corvol, P. & Alhencgelas, F. (1989). The testicular transcript of the angiotensin-I converting enzyme encodes for the ancestral, non-duplicated form of the enzymes. *FEBS Letters* **252**, 99–104.

Laurent, V. & Salzet, M. (1995). A comparison of the N-terminal sequence of the leech *Theromyzon tessulatum* angiotensin converting-like enzyme with forms of vertebrate angiotensin-converting enzymes. *Neuroscience Letters* **198**, 60–64.

Masler, E. P., Wagner, R. M. & Kovaleva, E. S. (1996). In-vitro metabolism of an insect neuropeptide by neural membrane preparations from *Lymantria dispar*. *Peptides* **17**, 321–326.

Ondetti, M. A. (1994). From peptides to peptidases – a chronicle of drug discovery. *Annual Review of Pharmacology and Toxicology* **34**, 1–16.

Pendleton, J. W., Nagai, B. K., Murtha, M. T. & Ruddle, F. H. (1993). Expansion of the Hox gene family and the evolution of chordates. *Proceedings of the National Academy of Sciences USA* **90**, 6300–6304.

Rayne, R. C. & O'Shea, M. (1992). Inactivation of neuropeptide hormones (AKH-I and AKH-II) studied in vivo and in vitro. *Insect Biochemistry and Molecular Biology* **22**, 25–34.

Roebroek, A. J. M., Pauli, I. G. L., Zhang, Y. & Vandeven, W. J. M. (1991). cDNA sequence of a *Drosophila melanogaster* gene, Dfur 1, encoding a protein structurally related to the subtilisin-like proprotein processing enzyme furin. *FEBS Letters* **289**, 133–137.

Rousseau, A., Michaud, A., Chauvet, M. T., Lenfant, M. & Corvol, P. (1995). The hemoregulatory peptide N-acetyl-Ser-Asp-Lys-Pro is a natural and specific substrate of the N-terminal active-site of human angiotensin-converting enzyme. *Journal of Biological Chemistry* **270**, 3656–3661.

Ruddle, F. H., Bentley, K. L., Murtha, M. T. & Risch, N. (1994). Gene loss and gain in the evolution of the vertebrates. *Development* **SS**, 155–161.

Schoofs, L., Tips, A., Holman, G. M., Nachman, R. J. & De Loof, A. (1992). Distribution of locustamyotropin-like immunoreactivity in the nervous-system of *Locusta migratoria*. *Regulatory Peptides* **37**, 237–254.

Schoofs, L., Veelaert, D., Vanden Broeck, J. & De Loof, A. (1997). Peptides in the locusts, *Locusta migratoria* and *Schistocerca gregaria*: a review. *Peptides* **18**, 147–156.

Settle, S. H., Green, M. M. & Burtis, K. C. (1995). The silver gene of *Drosophila melanogaster* encodes multiple carboxypeptidases similar to mammalian prohormone-processing enzymes. *Proceedings of the National Academy of Sciences USA* **92**, 9470–9474.

Skidgel, R. A. & Erdös, E. G. (1985). Novel activity of human angiotensin-I converting enzyme – release of the NH_2-terminal and COOH-terminal tripeptides from the luteinizing-hormone-releasing hormone. *Proceedings of the National Academy of Sciences USA* **82**, 1025–1929.

Soubrier, F., Hubert, C., Testut, P., Nadaud, S., Alhencgelas, F. & Corvol, P. (1993a). Molecular-biology of the angiotensin-I converting-enzyme. 1. Biochemistry and structure of the gene. *Journal of Hypertension* **11**, 471–476.

Soubrier, F., Wei, L., Hubert, C., Clauser, E., Alhencgelas, F. & Corvol, P. (1993b). Molecular-biology of the angiotensin-I converting-enzyme. 2. Structure–function – gene polymorphism and clinical implications. *Journal of Hypertension* **11**, 599–604.

Strey, A., Hayes, T. K. & Isaac, R. E. (1993). Metabolism of insect hyper-trehalosemic hormone in *Blaberus discoidalis* cockroaches. *Biochemical Society Transactions* **21**, 244S.

Tatei, K., Cai, H., Ip, Y. T. & Levine, M. (1995). *Race* – a *Drosophila* homolog of the angiotensin-converting enzyme. *Mechanisms of Development* **51**, 157–168.

Taylor, C. A. M., Coates, D. & Shirras, A. D. (1996). The *Acer* gene of *Drosophila* codes for an angiotensin-converting enzyme homologue. *Gene* **181**, 191–197.

Thompson, J. D., Higgins, D. G. & Gibson, T. J. (1994). Clustal-W – improving the sensitivity of progressive multiple sequence alignment through sequence weighting, position-specific gap penalties and weight matrix choice. *Nucleic Acids Research* **22**, 4673–4680.

Turner, A. J., Hryszko, J., Hooper, N. M. & Dowdall, M. J. (1987). Purification and characterization of a peptidyl dipeptidase resembling angiotensin converting enzyme from the electric organ of *Torpedo marmorata*. *Journal of Neurochemistry* **48**, 910–916.

Waterston, R. & Sulston, J. (1995). The genome of *Caenorhabditis elegans*. *Proceedings of the National Academy of Sciences USA* **92**, 10836–10840.

Wei, L., Alhencgelas, F., Corvol, P. & Clauser, E. (1991). The 2 homologous domains of human angiotensin-I-converting enzyme are both catalytically active. *Journal of Biological Chemistry* **266**, 9002–9008.

Wijffels, G., Fitzgerald, C., Gough, J., Riding, G., Elvin, C., Kemp, D. & Willadsen, P. (1996). Cloning and characterization of angiotensin-converting enzyme from the dipteran species, *Haematobia irritans exigua*, and its expression in the maturing male reproductive-system. *European Journal of Biochemistry* **237**, 414–423.

Williams, T. A., Michaud, A., Houard, X., Chauvet, M.-T., Soubrier, F. & Corvol, P. (1996). Characterisation of *D. melanogaster* wild-type and unglycosylated angiotensin I-converting enzyme expressed in *P. pastoris*: similarity with the C domain of the mammalian enzyme. *Biochemical Journal* **318**, 125–131.

Wilson, R., Ainscough, R., Anderson, K., Baynes, C., Berks, M., Burton, J., Connell, M., Bonfield, J., Copsey, T., Cooper, J., Coulson, A., Craxton, M., Dear, S., Du, Z., Durbin, R., Favello, A., Fraser, A., Fulton, L., Gardner, A., Green, P., Hawkins, T., Hillier, L., Jier, M., Johnston, L., Jones, M., Kershaw, J., Kirsten, J., Laisster, N., Latreille, P., Lloyd, C., Mortimore, B., O'Callaghan, M., Parsons, J., Percy, C., Rifken, L., Roopra, A., Saunders, D., Shownkeen, R., Sims, M., Smaldon, N., Smith, A., Smith, M., Sonnhammer, E., Staden, R., Sulston, J., Thierrymieg, J., Thomas, K., Vaudin, M., Vaughan, K., Waterston, R., Watson, A., Weinstock, L., Wilkinsonsproat, J. & Wohldman, P. (1994). 2.2 Mb of contiguous nucleotide-sequence from chromosome-III of *C. elegans*. *Nature* **368**, 32–38.

Yagi, K. J., Yu, C. G. & Tobe, S. S. (1992). Phosphoramidon enhances allatostatin-mediated inhibition of juvenile-hormone biosynthesis in the corpora allata of the cockroach, *Diploptera punctata*. *Experientia* **48**, 758–761.

Zhou, Y. D., Delafontaine, P., Martin, B. M. & Bernstein, K. E. (1995). Identification of 2 positive transcriptional elements within the 91-base pair promoter for mouse testis angiotensin-converting enzyme (testis ACE). *Developmental Genetics* **16**, 201–209.

RONALD J. NACHMAN, G. MARK HOLMAN,
and GEOFFREY M. COAST

Mimetic analogues of the myotropic/diuretic insect kinin neuropeptide family

Introduction

The insect kinin neuropeptide family shares the common C-terminal pentapeptide Phe-Xaa^1-Xaa^2-Trp-Gly-NH$_2$ (Xaa^1=His, Asn, Phe, Ser, or Tyr; Xaa^2=Ala, Ser or Pro) and has been isolated from such diverse sources as the cockroach *Leucophaea maderae* (Holman, Cook & Nachman, 1987), cricket *Acheta domesticus* (Holman *et al.*, 1991), locust *Locusta migratoria* (Holman, Nachman & Wright, 1990), corn earworm *Helicoverpa zea* (Blackburn *et al.*, 1995), and mosquitoes *Culex salinarius* (Hayes *et al.*, 1994) and *Aedes aegypti* (Veenstra, 1994). Although the first members of this peptide family were isolated on the basis of their ability to stimulate contractions of the isolated cockroach hindgut (Nachman & Holman, 1991), they have also been associated with diuretic activity in the cricket, mosquito and corn earworm moth. The *Leucophaea* hindgut preparation is extremely sensitive to these myotropic peptides, with thresholds in the range 10 to 100 pmol l^{-1} (Nachman & Holman, 1991). Hayes *et al.* (1989) have reported that leucokinins influence transepithelial membrane potential and rate of fluid secretion in isolated Malpighian tubules from the mosquito *Aedes aegypti*. Coast, Holman & Nachman (1990) have shown that the achetakinins at 1 nmol l^{-1} double the rate of fluid secretion by isolated Malpighian tubules of the cricket *Acheta domesticus*, and demonstrate EC$_{50}$ values between 20 and 320 pmol l^{-1}. While the insect kinins do not stimulate secretion maximally, as do the corticotropin-releasing factor (CRF)-related diuretic peptides, the two classes of diuretic peptides can act synergistically (Coast, 1995; see Coast, this volume). At threshold concentrations, one peptide class markedly potentiates the response of the other in locust Malpighian tubules. This could have significance on a physiological level because acting alone, maximal *in vivo* tubule secretion requires the release of 50% of the total store of *Locusta*-DH from the corpora cardiaca, whereas in the presence of 50 pmol locustakinin l^{-1} the same effect can be achieved with the release of only 2% of the stored peptide (Coast, 1995; see Coast, this volume). Therefore the insect kinins, in concert with insect CRF-related diuretic peptides, may regulate water and ion balance in addition to hindgut motility in insects. Recent work has established that members of a series of insect kinins isolated from the cockroach *Leucophaea maderae*, the leucokinins, also inhibit release of the carbohydrases invertase and α-amylase in the midgut of the red palm weevil, *Rhynchophorus ferrugineus*, at concentrations between 0.1 and 1 μmol 1^{-1} (R. J. Nachman & S. Sreekumar, unpublished data).

While the insect kinins modulate several physiological systems important for insect survival, they hold little promise as pest control agents due to their susceptibility to enzymatic degradation in the target insect, and an inability readily to penetrate the insect cuticle or gut wall. In this chapter, the active conformations adopted by the peptides at the receptor site are reviewed, and the development of active pseudopeptide and nonpeptide mimetic analogues with enhanced resistance to peptidase attack are described. In addition, new data are presented on a hydrophobic carboranyl analogue with the potential for topical activity. Disruption of peptide-regulated processes or behaviour by mimetic agonist and/or antagonist analogues could form the basis for future pest insect management strategies.

Structure–activity studies and mimetic analogues

The C-terminal pentapeptide sequence common to the insect kinins is all that is required to elicit a physiological response in myotropic and diuretic assays. In particular, the active core sequence Phe-Tyr-Pro-Trp-Gly-NH$_2$ is equipotent with the parent nonapeptide in cockroach hindgut myotropic (Nachman & Holman, 1991) and cricket Malpighian tubule secretion (Coast *et al.*, 1990) assays. As with the cockroach hindgut myotropic assay (Nachman & Holman, 1991), replacement by alanine of each of the residues of the leucokinin C-terminal pentapeptide Phe-Tyr-Ser-Trp-Gly-NH$_2$ did not result in complete loss of diuretic activity in the cricket, with the exception of the substitution in the Phe and Trp positions (Roberts *et al.*, 1997). Indeed, the leucokinin analogue <u>Ala</u>-Tyr-Ser-Trp-Gly-NH$_2$, lacking the critical Phe amino acid, antagonises the activity of the insect kinins in the cricket Malpighian tubule fluid secretion assay at concentrations between 0.1 and 1 μmol 1^{-1} (Nachman *et al.*, 1995b).

While the C-terminal acid analogue Ala-Phe-Phe-Pro-Trp-Gly-OH (AFFPWG-OH) proved inactive in the myotropic assay (Nachman *et al.*, 1994), it demonstrated weak activity in the cricket Malpighian tubule secretion assay with an EC$_{50}$ of 1 μmol 1^{-1} (Nachman *et al.*, 1995a). The negative charge of the C-terminal acid group was apparently detrimental to interaction of the insect kinins with the receptor sites. When the negative charge of the carboxylic acid was neutralised at the C-terminal group, improved activity in the cricket diuretic assay was observed. For example, the C-terminal ester analogue AFFPWG-OCH$_3$ was found to have an EC$_{50}$ value of 20 nmol l^{-1}. Notably, the ester analogue AFFPWG-SCH$_3$ (EC$_{50}$=5 nmol l^{-1}), containing a larger sulphur atom, proved to be four-fold more potent than the methyl ester analogue. The monosubstituted amide analogue AFFPWG-NHCH$_3$ stimulated Malpighian tubule secretion at an EC$_{50}$ of 28 nmol l^{-1}, while the hindered, branched analogue AFFPWG-N(CH$_3$)$_2$ proved inactive. Movement of the branched position away from the C-terminal carbonyl moiety, as in the ester analogue AFFPWG-OCH$_2$CH$_2$CH(CH$_3$)$_2$ did not result in a return of diuretic activity. However, by incorporating branched carbons into a planar, aromatic ring, as in the analogue AFFPWG-OCH$_2$(C$_6$H$_5$), diuretic activity (EC$_{50}$=50 nmol l^{-1}) could be restored. Thus, the C-terminal analogues containing -SCH$_3$ and -OCH$_2$(C$_6$H$_5$) moieties offered the greatest retention of diuretic activity, coupled with increased hydrophobicity and steric bulk (Nachman *et al.*, 1995a). These qualities are important for cuticle and/or gut penetration and peptidase resistance.

Evaluation of a series of insect kinin analogues containing varying numbers of methylene group spacers between the α-carbon and the amide group of the C-terminal Gly residue demonstrated that the cricket diuretic receptor can accommodate a shift in the position of the amide. The analogues AFFPWG—CH$_2$C(O)NH$_2$, AFFPWG—(CH$_2$)$_2$C(O)NH$_2$ and AFFPWG—(CH$_2$)$_4$C(O)NH$_2$ demonstrated diuretic activity at a range of EC$_{50}$ values between 0.5 and 2.5 nmol l^{-1}. The notation AFFPWG— is used to denote the molecular entity following the α-carbon of the C-terminal Gly position. However, the analogue AFFPWG—(CH$_2$)$_5$C(O)NH$_2$ proved inactive, suggesting that an α-carbon–amide distance spanning five methylene group spacers is incompatible with the receptor interaction required to elicit a diuretic response (Nachman *et al.*, 1995a).

Evaluation of a series of unnatural hexapeptide analogues Xaa-Phe-Phe-Pro-Trp-Gly-NH$_2$ in the cricket Malpighian tubule fluid secretion assay demonstrated the following ascending order of potencies: Xaa-empty (EC$_{50}$=100 pmpl l^{-1}), Xaa=Trp (EC$_{50}$=42 pmol l^{-1}), Xaa=Phe (EC$_{50}$=29 pmol l^{-1}), Xaa=Arg (EC$_{50}$=20 pmol l^{-1}), Xaa=Ala (EC$_{50}$=5 pmol l^{-1}), Xaa=Lys (EC$_{50}$=5 pmol l^{-1}), Xaa=Aib (EC$_{50}$=5 pmol l^{-1}), Xaa=Asp (EC$_{50}$=2 pmol l^{-1}), Xaa=Leu (EC$_{50}$=2 pmol l^{-1}), Xaa=Asn (EC$_{50}$=1 pmol l^{-1}) and Xaa=Val (EC$_{50}$=1 pmol l^{-1}) (R. J. Nachman & G. M. Coast, unpublished data). All of the analogues demonstrated maximal insect kinin stimulation of fluid secretion. The native achetakinins (AK) AK-I to AK-V stimulate fluid secretion with EC$_{50}$ values in the range of 18 to 350 pmol l^{-1}. Thus, hexapeptide analogues in which the Xaa position is occupied by Val, Asn, Leu, or Asp, containing neutral, aliphatic or acidic side chains, demonstrate super-agonist activity at a potency level about an order of magnitude greater than the

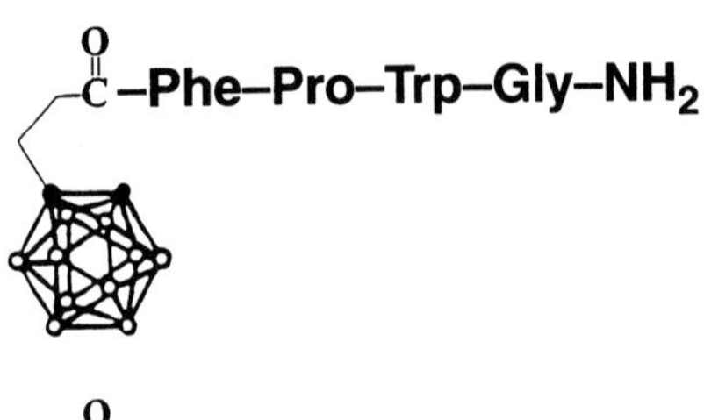

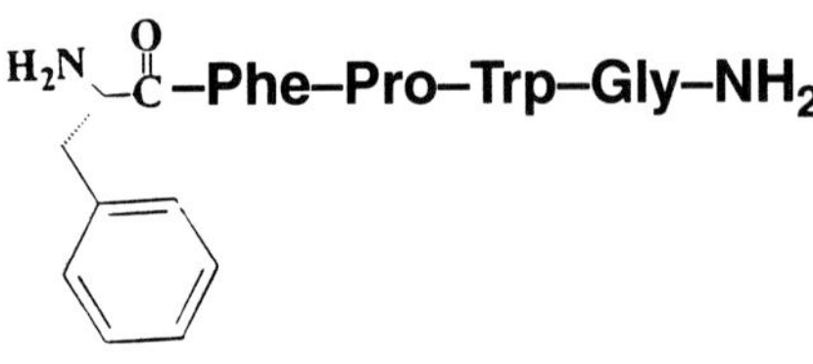

Fig. 1. An active carboranyl analogue, **Cbe**-Phe-Pro-Trp-Gly-NH$_2$ (**Cbe**=Cb-CH$_2$-CH$_2$-C(O)-(Cb=o-carborane)) (top) (R. J. Nachman & G. M. Coast, unpublished data) of the insect kinin C-terminal pentapeptide core fragment Phe-Phe-Pro-Trp-Gly-NH$_2$ (bottom). The hydrophobic, ball-shaped o-carborane moiety, containing 10 boron and 2 carbon atoms, replaces the phenyl ring of the Phe residue.

most active of the native achetakinins on the cricket Malpighian tubule fluid secretion assay. Analogues in which the Xaa position is occupied by Trp, Phe or Arg, containing side chains with aromatic or electron-delocalised character, demonstrate the least potency in this hexapeptide series.

The analogue Ala-Phe-Phe-βAla-NTrp-NGly-NH$_2$ is a peptide-peptoid hybrid structure in which 'peptoid' character is incorporated into the sensitive C-terminal region. The hybrid analogue nevertheless retains diuretic activity in the cricket Malpighian tubule fluid secretion assay at concentrations between 1 and 10 μmol l^{-1} (Nachman *et al.*, 1995b). In peptoids, the side chains are attached to the amide nitrogens as opposed to the α-carbon and their presence markedly increases the stability of the amide bonds to peptidase attack (Simon *et al.*, 1992).

The N-terminal Phe residue of the pentapeptide core fragment was replaced with a carboranyl, acyl moiety. The dodecahedral volume created by rotation of the planar phenyl ring is mimicked by the ball-shaped carborane structure. It also mimics the aromatic character of the phenyl ring, but is a considerably more hydrophobic structure. Specifically, the Phe residue was replaced by the carborane-ethanoylacyl group (**Cbe**=Cb-CH$_2$-CH$_2$-C(O)-(Cb=o-carborane) to yield the pseudotetrapeptide **Cbe**-Phe-Pro-Trp-Gly-NH$_2$ (Fig. 1). The retention time observed for the carboranyl analogue on a C$_{18}$ reversed-phase high-performance liquid chromatography column (Nachman *et al.*, 1993) was 42 min, compared with 16 min for the parent insect kinin pentapeptide fragment Phe-Phe-Pro-Trp-Gly-NH$_2$. As the two peptide analogues are of roughly the same amino acid content and length, a comparison of retention times provides a measure of the relative difference in the hydrophobic character. Despite this major structural modification and increase in hydrophobic character, the carboranyl pseudotetrapeptide analogue retained considerable activity in the cricket Malpighian tubule fluid secretion assay (EC$_{50}$=0.2 nmol l^{-1}; 60% maximal response) (R. J. Nachman & G. M. Coast, unpublished data). A similar carboranyl analogue of the core pentapeptide of the pyrokinin/pheromone biosynthesis-activating neuropeptide (PBAN) insect neuropeptide family, **Cbe**-Thr-Pro-Arg-Leu-NH$_2$, demonstrated strong activity in a *Heliothis virescens* (armyworm) pheromonotropic assay, with a potency 10-fold greater than the much larger, natural 33-membered PBAN molecule (HezPBAN). Aqueous solutions of the amphiphilic pyrokinin analogue solution readily 'wet' the lateral abdominal surface of female *H. virescens*. These analogue solutions also elicit pheromone production about 15 min after topical application with an EC$_{50}$ of 25 pmol female^{-1} and a maximal response at about 60 pmol female^{-1}. In contrast, 100 pmol doses of HezPBAN in aqueous solution at 15 min, 1 h, 2 h, and even 3 h after topical application had no statistical effect on pheromone production (Nachman *et al.*, 1997). The ability of the analogue to penetrate the cuticular surface of the moth means that female moths can be induced to produce pheromone without using invasive procedures, such as injection. The development of a reliable *in vivo* Malpighian fluid secretion assay could help to determine whether the hydrophobic, carboranyl (Fig. 1) and/or related analogues of the insect kinin family can also penetrate the cuticle and induce a physiological response following topical application to a target insect.

A bifunctional, heterodimeric insect kinin analogue that features a covalent linkage between the N-termini of the active cores of the insect kinin and pyrokinin/PBAN neuropeptide families has also been reported (Nachman *et al.*, 1993). Members of the pyrokinin (Phe-Xaa-Pro-Arg-Leu-NH$_2$) insect neuropeptide family have been associated with oviduct contractile activity in the cockroach, locust and cricket (Nachman & Holman, 1991; Nachman *et al.*, 1993; Schoofs *et al.*, 1994); pheromonotropic activity in the corn earworm, *Helicoverpa zea* (Raina *et al.*, 1989), and silkworm, *Bombyx mori* (Kitamura *et al.*, 1989), diapause induction activity in the silkworm (Imai *et al.*, 1991), and pupariation acceleration in the fleshfly *Sarcophaga bullata* (Zdarek, Nachman & Hayes, 1997). As previously stated, the insect kinins elicit stimulation of Malpighian tubule fluid secretion in several insects (Hayes *et al.*, 1989; Coast *et al.*, 1990; Blackburn *et al.*, 1995). Other than the cockroach hindgut myotropic assay, there is no known cross-activity between the insect kinin and the pyrokinin/PBAN peptide families. The

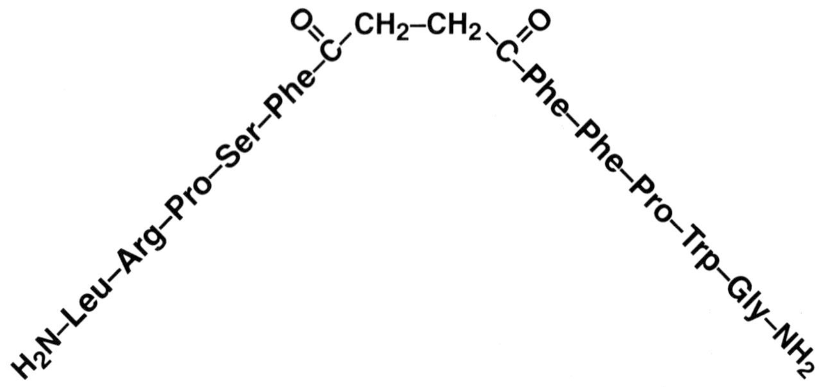

Fig. 2. A bifunctional heterodimeric analogue incorporates the C-terminal pentapeptide active core fragments of the insect kinin (Phe-Phe-Pro-Trp-Gly-NH$_2$) and pyrokinin/PBAN (Phe-Ser-Pro-Arg-Leu-NH$_2$) peptide families into a single molecular entity. They are linked together via their N-terminal amino groups with a succinyl group. The analogue elicits biological activity associated with both classes of insect peptides (Nachman *et al.*, 1993).

heterodimer   [H$_2$N-Gly-Trp-Pro-Phe-Phe-succinyl-Phe-Ser-Pro-Arg-Lys-NH$_2$] (Fig. 2), however, has been shown to elicit two physiological responses in one insect species that are normally the exclusive province of either the insect kinin or the pyrokinin/PBAN peptides. The heterodimer stimulated fluid secretion in cricket Malpighian tubules with a notable EC$_{50}$ of 0.6 nmol 1^{-1}, a typical response for the insect kinin family. On the isolated cricket oviduct preparation, the heterodimer demonstrated myostimulatory activity at a threshold concentration of 7.7 nmol 1^{-1}, six-fold more potent than the pyrokinin core fragment Phe-Ser-Pro-Arg-Leu-NH$_2$ (Nachman *et al.*, 1993). The heterodimer can potentially elicit any combination of the physiological responses of diuresis, pheromone biosynthesis, oviduct contraction, diapause induction, and pupariation in suitable insect species. Non-peptide and pseudopeptide mimetics fashioned after this and other insect neuropeptide-heterodimers would have the potential to disrupt more than one physiological process in a target insect.

Non-peptide modification of the insect kinin active core was first achieved by replacing the amide bond (-C(O)NH-) between Phe[1] and Phe[2] of the sequences Phe-Phe-Ser-Trp-Gly-NH$_2$ and Phe-Phe-Pro-Trp-Gly-NH$_2$ with a reduced amide bond linkage -CH$_2$NH-. The resulting modified pseudotripeptides Phe ψ[CH$_2$-NH]Phe-Ser-Trp-Gly-NH$_2$ and Phe ψ[CH$_2$-NH]Phe-Pro-Trp-Gly-NH$_2$ demonstrated threshold concentration values in the cockroach hindgut myotropic assay in the order of 3 nmol 1^{-1}, about 1% of the activity of the parent peptides (Nachman *et al.*, 1991). Subsequently, a series of modified pseudotripeptide insect kinin analogues were also shown to demonstrate activity in the cricket Malpighian tubule secretion assay (Cameron *et al.*, 1995).

Further modification of the N-terminal region was accomplished by synthesising a series of pseudopeptides in which up to three of the N-terminal pentapeptide core amino acids were replaced with aryl-acyl groups, such that the total length of the analogues could remain essentially constant. While the pseudotetrapeptide **Hca**-Tyr-Pro-Trp-Gly-NH$_2$ (**Hca**=hydrocinnamic acid) demonstrated a potent myotropic threshold concentration of 58 pmol l^{-1} and an EC$_{50}$ of 10 nmol l^{-1}, the pseudotripeptide **6Pha**-Pro-Trp-Gly-NH$_2$ (**6Pha**=6-phenylhexanoic acid) elicited myostimulation at a reduced threshold concentration of 8 nmol l^{-1} and an EC$_{50}$ of 2 µmol l^{-1}. The pseudodipeptide **9Pna**-Trp-Gly-NH$_2$ (**9Pna**=9-phenylnonanoic acid) elicits weak activity in the myotropic assay at a threshold concentration value of 5–7 µmol l^{-1} (Nachman *et al.*, 1993). The pseudodipeptide analogue lacks the Pro residue and several of the peptide bonds of the pentapeptide core fragment, leading to a loss of some of the conformational properties of the natural peptides. To return some of the conformational properties of the insect kinin core fragment, a pseudodipeptide analogue incorporating a carbocyclic proline mimetic component was designed, synthesised, and evaluated in the two bioassay systems. The pseudodipeptide analogue 4Pbm-**cChd**-Trp-Gly-NH$_2$ (4Pbm=4-phenylbutylamine; **cChd**=*cis*-cyclohexyldiacyl) evoked a stimulatory response in the cockroach myotropic assay at a threshold concentration of 790 nmol l^{-1}, an order of magnitude more potent than the straight chain pseudodipeptide 9Pna-Trp-Gly-NH$_2$. In the diuretic assay, the pseudodipeptide stimulated fluid secretion in cricket Malpighian tubules at an EC$_{50}$ of 750 nmol l^{-1} (Nachman *et al.*, 1995b). The biological evaluation of pseudopeptide analogues of the insect kinins demonstrate that successful interaction with the insect kinin receptors requires the C-terminal dipeptide Trp-Gly-NH$_2$ with an N-terminal extension that incorporates the equivalent of a phenyl ring to represent the Phe residue. Furthermore, the conformational relationship between the dipeptide and phenyl ring is of great importance for potent activity.

Insect kinin active conformation

Experimental spectroscopic and molecular dynamics studies on a conformationally restricted analogue of the insect kinins, active in both myotropic and diuretic assays, shed light on the conformational preferences of the insect kinins at the receptor site. Molecular dynamics simulations on the cyclic hexapeptide insect kinin analogue *cyclo*(Ala-Phe-Phe-Pro-Trp-Gly) suggested that it could adopt two turn conformations within the pentapeptide core region. The more rigid conformation features a *cis* Pro in the third position of a type VI β-turn encompassing residues Phe-Phe-Pro-Trp, whereas the other less rigid conformation contained a *trans* Pro in the second position of a type I-like turn over residues Phe-Pro-Trp-Gly. Nuclear magnetic resonance (NMR) spectroscopy established that the turn conformations exist in a 60:40 ratio (Nachman *et al.*, 1990; Nachman *et al.*, 1995b; Roberts *et al.*, 1997). In solution, the *cis* Pro conformation is apparently the more energetically favourable of the two. The cyclic analogue is about 20-fold more potent than the putative degradation product Phe-Phe-Pro-Trp-Gly-Ala-OH in both bioassays, demonstrating that the observed activity arises from the cyclic analogue rather than linear by-products of *in situ* peptidase degradation during the course of the assays. In addition, the activity of the cyclic analogue is equivalent to the linear-appended analogue Phe-Phe-Pro-Trp-Gly-Ala-NH$_2$ (Roberts *et al.*, 1997). Therefore, studies on the cyclic hexapeptide insect kinin analogue provide strong evidence that the peptides adopt a turn conformation about the active-core Pro residue during receptor interaction.

ACE endopeptidase resistant analogues

Recent experiments demonstrate that several members of the insect kinin family are hydrolysed, and therefore inactivated, by angiotensin-converting enzyme (ACE) from the housefly via removal of the C-terminal dipeptide amide fragment (Fig. 3) (Lamango *et al.*, 1997). Inactivation results because the hydrolysis site is located within the insect kinin C-terminal pentapeptide active core. In mammals, the Zn^{2+} metallopeptidase ACE is responsible for the conversion of angiotensin I to the active form angiotensin II, involved in the control of blood pressure (see Isaac *et al.*, this volume). In addition, ACE inactivates a wide range of mammalian peptide hormones such as bradykinin, cholecystokinin, [Leu^5]- and [Met^5]enkephalinamides, substance P and luteinising hormone-releasing hormone (LH-RH). The broad substrate specificity and widespread distribution of ACE in mammalian tissues suggests that it plays multiple roles in addition to controlling blood pressure, although these roles have yet to be defined. Similarly, the precise role of ACE has not been delineated in insects. However, the fact that a number of different C-terminally amidated insect neuropeptides are substrates of ACE, and the presence of the endopeptidase in haemolymph and other tissues, suggested that the endopeptidase may play a role in the degradation of regulatory peptides in insects (Lamango *et al.*, 1997; see Isaac *et al.*, this volume).

The analogues Phe-Phe-**Aib**-Trp-Gly-NH$_2$ and pGlu-Lys-Phe-Phe-**Aib**-Trp-Gly-NH$_2$, in which the Xaa^2 residue (Ser or Pro) is replaced with the sterically hindered aminoisobutyric acid (**Aib**) residue (Fig. 3), demonstrated complete

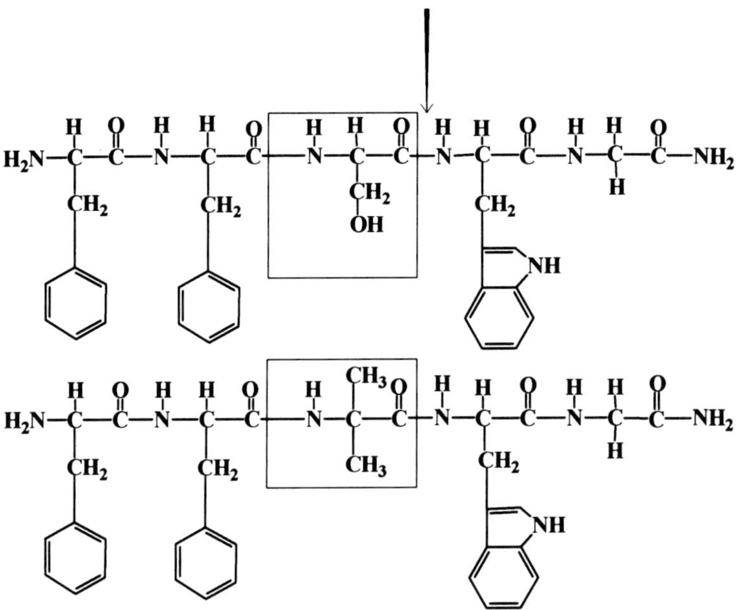

Fig. 3. The insect kinin C-terminal pentapeptide analogue Phe-Phe-Ser-Trp-Gly-NH$_2$ is pictured at the top. The arrow indicates the point at which the insect kinins are hydrolysed, and therefore inactivated, by the endo-peptidase angiotensin converting enzyme (ACE) from the housefly, *Musca domestica*. The insect kinin analogue Phe-Phe-**Aib**-Trp-Gly-NH$_2$ (below) features a sterically hindered aminoisobutyrl (**Aib**) residue (boxed), compatible with a turn in this region, and demonstrates both complete resistance to ACE and potent Malpighian tubule fluid secretion activity (Reprinted with permission from *Peptides* **18**, Nachman, R. J., Isaac, R. E., Coast, E. M. & Hotman, G. M. Aib-containing analogues of the insect kinin family demonstrate resistance to an angiotensin-converting enzyme and potent diuretic activity, pp. 53–7, 1997, Elsevier Science Inc).

resistance to hydrolysis by ACE from the housefly *Musca domestica* over an incubation period sufficient to hydrolyse most of the natural insect kinin leucokinin I (LK-I). Notably, the latter Aib analogue is blocked at the N-terminus with a pGlu residue, which confers resistance to another class of endo-peptidases, the aminopeptidases. The sequence of LK-1 is Asp-Pro-Ala-Phe-Asn-Ser-Trp-Gly-NH$_2$, which shares with AK-IV an Asn residue in the variable Xaa[1] position of the C-terminal pentapeptide core region. The two analogues demonstrated potent stimulation of fluid secretion on the isolated Malpighian tubules of the cricket *A. domesticus*, with EC$_{50}$ values of 5.6 pmol 1^{-1} and 2.8 pmol 1^{-1}, respectively. The two analogues are therefore about four to eight times more potent in the Malpighian tubule assay than the most potent naturally occurring AK peptide (Coast *et al.*, 1990; see above). Moreover, both produce a maximal diuretic response which was not significantly different from that obtained with the endogenous AK peptides. The Aib residue is compatible with the formation of a turn at this position in the active core which is important for the diuretic activity of the insect kinins. Unfavourable steric interactions between the branched chain of the Aib α-carbon and side chains of surrounding residues promote the formation of a turn or kink to alleviate strain (Toniolo *et al.*, 1983). This characteristic explains, at least in part, the potent biological activities observed for these kinin analogues.

In aggregate, the results suggest that incorporation of Aib residues in appropriate turn regions of these and other insect peptides can confer a measure of resistance to ACE and/or other endopeptidases that degrade them, potentially without loss of biological activity.

Summary and future directions

This chapter describes conformational aspects of the interaction of the insect kinin neuropeptide family with their receptor sites, and the development of mimetic analogues with enhanced resistance to peptidase degradation with either partial or, for several examples, complete retention of potency. Remarkably, incorporation of an Aib residue, compatible with a turn conformation in the middle of the active core fragment, can confer resistance to the endopeptidase ACE, which hydrolyses and inactivates the natural insect kinins, and retention of full biological activity. The presence of the sterically hindred Aib residue within the active core fragment may also confer resistance to other endopeptidases. Replacement of the phenyl ring with the ball-shaped carboranyl moiety markedly increases the hydrophobicity of the active core fragment, while the resulting analogue retains a significant portion of the Malpighian tubule fluid secretion activity of the natural peptides. A similar carboranyl analogue of the pyrokinin/PBAN insect neuropeptide family demonstrates the ability to elicit pheromone biosynthesis following topical application to a living insect. The development of a reliable *in vivo* Malpighian tubule fluid secretion assay can provide evidence as to whether this or related hydrophobic analogues can penetrate the insect cuticle. A bifunctional, heterodimeric analogue incorporating active core regions of the insect kinin and pyrokinin families, simultaneously elicits biological activity associated with both classes of peptides. Heterodimeric analogues open the possibility of simultaneous disruption of more than one insect physiological system. Enzyme-resistant mimetic analogues of the insect kinin and pyrokinin peptides can provide useful tools to insect physiologists studying the neuroendocrine control of such processes as diuresis, digestive enzyme release, pheromone production, oviduct contraction, and the developmental processes of diapause and pupariation, as well as the behavioural consequences of challenging an insect with a neurochemical signal it cannot degrade. If such analogues, whether in isolation or in combination with other factors, can disrupt the internal physiological balance of insects, they hold potential utility for the control of pest insect populations and the reduction of environmental residues from harmful pesticides.

Acknowledgements

The authors wish to acknowledge the financial support of NATO grant no. 90248 (R.J.N., G.M.C. and G.M.H.).

References

Blackburn, M. B., Wagner, R. M., Shabanowitz, J., Kochansky, J. P., Hunt, D. F. & Raina, A. K. (1995). The isolation and identification of three diuretic kinins from the abdominal ventral nerve cord of adult *Helicoverpa zea. Journal of Insect Physiology* **41**, 723–730.

Cameron, S., Coast, G. M., Ford, M. G., Goldsworthy, G. J., Khambay, B. P. S., Stone, J. & Watson, P. (1995). The influence of structure on stability, penetration and activity of analogues of the insect neuropeptide achetakinin-I. In *Insects: Chemical, Physiological and Environmental Aspects*, ed. D. Konopinska, pp. 248–252. Wroclaw: Wroclaw University Press.

Coast, G. M. (1995). Synergism between diuretic peptides controlling ion and fluid transport in insect Malpighian tubules. *Regulatory Peptides* **57**, 283–296.

Coast, G. M., Holman, G. M. & Nachman, R. J. (1990). The diuretic activity of a series of cephalomyotropic neuropeptides, the achetakinins, on isolated Malpighian tubules of the house cricket, *Acheta domesticus. Journal of Insect Physiology* **36**, 481–488.

Hayes, T. K., Holman, G. M., Pannabecker, T. L., Wright, M. S., Strey, A. A., Nachman, R. J., Hoel, D. F., Olsen, J. K. & Beyenbach, K. W. (1994). Culekinin depolarizing peptide: a mosquito leucokinin-like peptide that influences insect Malpighian tubule ion transport. *Regulatory Peptides* **52**, 235–248.

Hayes, T. K., Pannabecker, T. L., Hinckley, D. J., Holman, G. M., Nachman, R. J., Petzel, D. H. & Beyenbach, K. W. (1989). Leucokinins, a new family of ion transport stimulators and inhibitors in insect Malpighian tubules. *Life Sciences* **44**, 1259–1266.

Holman, G. M., Cook, B. J. & Nachman, R. J. (1987). Isolation, primary structure and synthesis of leucokinins VII and VIII: the final members of this new family of cephalomyotropic peptides isolated from head extracts of *Leucophaea maderae. Comparative Biochemistry and Physiology* **88C**, 31–34.

Holman, G. M., Nachman, R. J., Schoofs, L., Hayes, T. K., Wright, M. S. & De Loof, A. (1991). The *Leucophaea maderae* hindgut preparation: a rapid and sensitive bioassay tool for the isolation of insect myotropins of other insect species. *Insect Biochemistry* **21**, 107–112.

Holman, G. M., Nachman, R. J. & Wright, M. A. (1990). A strategy for the isolation and structural characterization of certain insect myotropic peptides that modify spontaneous contractions of the isolated cockroach hindgut. In *Chromatography and Isolation of Insect Hormones and Pheromones*, ed. A. R. McCaffery & I. D. Wilson, pp. 195–204. New York: Plenum Press.

Imai, K., Konno, T., Nakagawa, Y., Komiya, T., Isobe, M., Koga, K., Goto, T., Yaginuma, T., Sakakibara, K., Hasegawa, K. & Yamashita, O. (1991). Isolation and structure of diapause hormone of the silkworm, *Bombyx mori*. *Proceedings of the Japanese Academy* **67B**, 98–101.

Kitamura, A., Nagasawa, H., Kataoka, H., Isogai, A. & Suzuki, A. (1989). Amino acid sequence of pheromone-biosynthesis-activating neuropeptide (PBAN) of the silkworm, *Bombyx mori*. *Biochemical and Biophysical Research Communications* **163**, 520–526.

Lamango, N. S., Nachman, R. J., Hayes, T. K., Strey, A. A. & Isaac, R. E. (1997). Hydrolysis of insect neuropeptides by an angiotensin-converting enzyme from the housefly, *Musca domestica*. *Peptides* **18**, 47–52.

Nachman, R. J., Coast, G. M., Holman, G. M. & Beier, R. C. (1995a). Diuretic activity of C-terminal group analogs of the insect kinins in *Acheta domesticus*. *Peptides* **16**, 809–813.

Nachman, R. J., Coast, G. M., Holman, G. M. & Haddon, W. F. (1993). A bifunctional heterodimeric insect neuropeptide analog. *International Journal of Peptide and Protein Research* **40**, 423–428.

Nachman, R. J., Coast, G. M., Roberts, V. A. & Holman, G. M. (1995b). Incorporation of chemical/conformational components into mimetic analogs of insect neuropeptides. In *Insects: Chemical, Physiological and Environmental Aspects*, ed. D. Konopinska, pp. 51–60. Wroclaw: University of Wroclaw Press.

Nachman, R. J. & Holman, G. M. (1991). Myotropic insect neuropeptide families from the cockroach *Leucophaea maderae*: Structure–activity relationships. In *Insect Neuropeptides: Chemistry, Biology and Action*, ed. J. J. Menn & E. P. Masler, pp. 194–214. Washington, DC: American Chemical Society.

Nachman, R. J., Holman, G. M., Haddon, W. F. & Vensel, W. H. (1991). An active pseudopeptide analog of the leucokinin insect neuropeptide family. *International Journal of Peptide and Protein Research* **37**, 220–223.

Nachman, R. J., Holman, G. M., Hayes, T. K. & Beier, R. C. (1993). Acyl, pseudotetra-, tri-, and dipeptide active-core analogs of insect neuropeptides. *International Journal of Peptide and Protein Research* **42**, 372–377.

Nachman, R. J., Isaac, R. E., Coast, G. M. & Holman, G. M. (1997). Aib-containing analogues of the insect kinin neuropeptide family demonstrate resistance to an insect angiotensin-converting enzyme and potent antidiuretic activity. *Peptides* **18**, 53–57.

Nachman, R. J., Roberts, V. A., Holman, G. M. & Tainer, J. A. (1990). Consensus chemistry and conformation of an insect neuropeptide family. In *Progress in Comparative Endocrinology*, ed. A. Epple, C. G. Scanes & M. H. Stetson, pp. 60–66. New York: Wiley-Liss, Inc.

Nachman, R. J., Tilley, J. W., Hayes, T. K., Holman, G. M. & Beier, R. C. (1994). Pseudopeptide mimetic analogs of insect neuropeptides. In *Natural and Derived Pest Management Agents*, ed. P. Hedin, J. Menn & R. Hollingsworth, pp. 210–229. Washington, DC: American Chemical Society.

Roberts, V. A., Nachman, R. J., Coast, G. M., Hariharan, M., Chung, J. S., Holman, G. M. & Tainer, J. (1997). Consensus chemistry and β-turn conformation of the active core of the insect kinin neuropeptide family. *Chemistry and Biology* **4**, 105–117.

Schoofs, L., Holman, G. M., Nachman, R. J., Proost, P., Hayes, T. K. & De Loof, A. (1994). Structure, function and distribution of insect myotropic peptides. In *Perspectives in Comparative Endocrinology*, ed. K. G. Davey, R. E. Peter & S. S. Tobe, pp. 155–165. Ottawa: National Research Council of Canada.

Simon, R. J., Kania, R. S., Zuckerman, R. N., Huebner, V. D., Jewell, D. A., Banville, S., Ng, S., Wang, L., Roseberg, S., Marlowe, C. K., Spellmeyer, D. C., Tan, R., Frankel, A. D., Santi, D. V., Cohen, F. E. & Bartlett, P. A. (1992). Peptoids: a modular approach to drug discovery. *Proceedings of the National Academy of Sciences USA* **89**, 9367–9371.

Toniolo, C., Bonora, G. M., Bavoso, A., Benedetti, E., Di Blasio, V. & Pedone, C. (1983). Preferred conformations of peptide analogs containing α,α-disubstituted α-amino acids. *Biopolymers* **22**, 205–214.

Veenstra, J. A. (1994). Isolation and identification of 3 leucokinins from the mosquito *Aedes aegypti*. *Biochemical and Biophysical Research Communications* **202**, 715–719.

Zdarek, J., Nachman, R. J. & Hayes, T. K. (1997). The insect neuropeptides of the pyrokinin/PBAN family accelerate pupariation in the fleshfly (*Sarcophaga bullata*) larvae. In *Neuropeptides in Development and Aging*, ed. F. Strand, W. Beckwith & C. Sandman *Annals of the New York Academy of Science* **814**, 67–72.

Index

Note: page numbers in *italics* refer to figures and tables.

gonad-inhibiting hormone (GIH) (*cont.*)
  localisation at medulla terminalis X-
    organ 61
  mode of action 64
  oocyte growth 63
  perikarya 61
  physiological roles 63–4, *65*
  polymorphism 55
  preprohormone 55, *56*
    expression 59–60
  species specificity 64
  vitellin synthesis 63, 64
  X-organ sinus gland complex 59
gonad-stimulating hormone (GSH) 42, 62
gonadotrophic cycle, ixodid ticks 103
gonadotropin, ticks 108
Gryllus bimaculatus 22, 23
  peptides 23
gut hormone synthesis, ACE role 373

haemolymph 4
  ACE activity 362, 363
  allatostatin content 17
  amastatin-sensitive aminopeptidase
    18–19
  endopeptidase 19
  flow redirection with CCAP 304
  FMRFamide-like immunoreactivity 284
harvestman, CCAPir neurones 319, *324*
heart, inotropic effects of crustacean
    cardioactive peptides 311
heart muscle, SchistoFLRFamide 284–5
Helix pomatia 319
Heteroptera vitellogenesis 77
hindgut
  CavTKs 268
  CCAP effects 307, 311
  contraction inhibition by
    leucomyosuppressin 283
  innervation by allatostatin-
    immunoreactive neurones 16
  kinin neuropeptide actions 379
  LemTRPs 263, *264*
  Leu-callatostatin neurones *240*, 241
  motility regulation by Leu-callatostatin
    242, *243*, 244
HMG-CoA reductase cyclicity 81–2
HMG-CoA synthase cyclicity 81–2
5-HT *see* 5-hydroxytryptamine
HVFLRFamide *290*, 291, 294
  His residue 293, 294
  inhibitory biological activity 295
7-*S*-hydroprene 17–18
hydroxyecdysone 100–1
20-hydroxyecdysone 74
  insects 93

metabolism in ticks 97
sex pheromone role 102
structure *94*
tick moulting 95, 96
vitellogenin production
  inhibition 83
  stimulation 78, 79–80
yolk protein gene expression modulation
  80
3-hydroxykynurenine 37, 38, *39*
20-hydroxylation 107
hydroxyphenyl propionate (HPP), AKH-I
    modification 159
5-hydroxytryptamine (5-HT) 61–2
  diuretic function 191
  feeding cessation 201
  see also serotonin
hyperglycemia, CHH release 62
hypertrehalosaemic hormone 83, 156, 157,
    182
  message transduction 183
  signal transduction cascade 181
  see also AKH-I; AKH-II; AKH-III
hypertrehalosaemic peptides 176
hypocerebral ganglion 257

ileum myoinhibition by Leu-callatostatin
    242, *243*, 244
inositol lipid cycle 181
inositol phosphates
  calcium mobilisation 181–2
  concentration effects of AKH 183
inositol trisphosphate 131, *134*, 156
insects
  angiotensin-converting enzyme (ACE)
    357–9
  crustacean cardioactive peptide 311–13,
    314, 315, *316*, 317–18, *321–2*
    oviduct regulation 349
  developmental hormone systems 93–5
  flight activity 172
  kinin neuropeptide family 379–80
  LomTK-like immunoreactivity in
    cerebral structures *269*
  moulting 326
  signalling pathways 357
  urine production regulation 189
insulin, locust 162
insulin-like compounds 161–2
inulin 191
ion transport peptide (ITP) 59, *213*
  actions 217
  amino acid sequence 217–19
  biological actions 220–1
  cDNA expression 219–20
  distribution among insects 219